T0275643

Minding the Weather

Minding the Weather

How Expert Forecasters Think

Robert R. Hoffman, Daphne S. LaDue, H. Michael Mogil, Paul J. Roebber,
and J. Gregory Trafton

The MIT Press
Cambridge, Massachusetts
London, England

This book was set in ITC Stone Sans Std and ITC Stone Serif Std by Toppan Best-set Premedia Limited.

Library of Congress Cataloging-in-Publication Data is available.

ISBN: 978-0-262-03606-1 (hardcover)
ISBN: 978-0-262-54881-6 (paperback)

I dedicate this book to the rainbows in my life—my loving wife Robin and my special children Rachel and Eric. The Hoffman Clan was always on my mind and in my heart as I wrote for this book.

Robert R. Hoffman
Pensacola, FL
April 2016

My efforts are dedicated to the National Weather Service forecasters who work daily through struggles of dealing with imperfect guidance while they pour their hearts and souls into their work.

Daphne LaDue
Norman, OK
April 2016

Much as for any work of this magnitude, time and patience are must-haves. Thanks to my wife, Barbara, for providing both.

H. Michael Mogil
Naples, FL
April 2016

Thanks to my family—Kathleen Roebber and our two sons, Kevin and Chris—and also to my co-authors and to the many scientists whose work we cite, for their contributions.

Paul Roebber
Milwaukee, WI
April 2016

Thanks for all the support from my wife, Paula Raymond-Trafton, and our three children, Elena, Maia, and Jack.

Greg Trafton
Washington, DC
March 2016

Contents

About the Authors

Robert R. Hoffman, Ph.D. is a Senior Research Scientist at the Institute for Human and Machine Cognition in Pensacola, Florida. His primary specialization is in the field of Expertise Studies, a field that he pioneered with his first book, *The Psychology of Expertise* (1992). He is Lead Editor for the Department on Human-Centered Computing in *IEEE: Intelligent Systems* magazine. Among his books are the *Cambridge Handbook of Applied Perception Research* (2015), *Accelerated Expertise* (2014), *Perspectives on Cognitive Task Analysis* (2008), and *Working Minds: A Practitioner's Guide to Cognitive Task Analysis* (2006). Contact him at rhoffman@ihmc.us

Daphne S. LaDue, Ph.D. is a Research Scientist at the University of Oklahoma's Center for Analysis and Prediction of Storms and Director of the highly successful Research Experiences for Undergraduates Program at the National Weather Center in Norman, Oklahoma. Her primary specialization is in meteorological education and forecaster training. Her current research focus is on developing ways of utilizing advances in radar meteorology for prediction of and response to severe storms. Contact her at Daphne.S.LaDue-1@ou.edu

H. Michael Mogil is currently a Consulting Meteorologist focusing on meteorological education in the primary and secondary school systems with his not-for-profit venture, HOWTHEWEATHERWORKS [http://www.weatherworks.com]. He also posts a blog with weather forecasts, explanations of forecasts, and explanations of current and salient weather events and phenomena [http://www.weatherworks.com/lifelong-learning-blog/]. He has worked with the National Weather Service in various field office and Headquarters positions and served at NOAA's Storm Prediction Center, National Centers for Environmental Prediction and Satellite Applications Laboratory. His experience includes program management, forecasting, scientific research, and emergency readiness. He is renowned as one of the most proficient interpreters of weather satellite imagery; he trained entire generations of National Weather Service forecasters in the use of satellite data. He currently directs a nationwide weather camp program (now with more than a dozen camp sites in operation) and posts a forecasting blog. He was awarded the National Weather Association's (NWA) Public Education Award (2013) and the NWA Digital Media Seal (2014). He is the author of *Extreme Weather* (2011, Simon & Schuster). Contact him at ccm@weatherworks.com

Paul J. Roebber, Ph.D. is a Distinguished Professor of Atmospheric Sciences at the University of Wisconsin–Milwaukee. He served as Editor of *Weather and Forecasting* from 2004 to 2007 and as Associate Editor of *Monthly Weather Review* from 1999 to 2003. He has served with distinction on a number of standing committees of the American Meteorological Society and has received numerous awards for his research and teaching. He has published dozens of technical papers in meteorology journals and is widely respected as one of America's outstanding weather forecasters. Contact him at roebber@uwm.edu

J. Gregory Trafton, Ph.D. is a cognitive scientist at the Naval Research Laboratory in Washington, DC. His primary specialization is in human factors, human–machine interaction in particular. He has conducted research on a wide variety of topics, including human–robot interaction, error prediction and prevention, predictive user interfaces, and the cognition of complex visualization. Contact him at greg.trafton@nrl.navy.mil

Preface

The focus of this book is on the psychology of expertise at weather forecasting.

- What does it mean for someone to be an "expert"?
- What does it mean for forecasters to be experts at forecasting?
- How do people become forecasters?
- How do forecasters become expert forecasters?
- How do forecasters reason as they try to understand the weather?
- How does forecasting depend on situating the forecaster and forecasting technology in an interdependent relationship?

For each of these core questions, we consider the pertinent empirical and research. Although we do go into some technical details, we try to make the material broadly accessible.

It would take a second book, and more, to go beyond this focus, for example, to discuss the psychological aspects of broadcast meteorology (e.g., how can broadcast meteorologists express the details of a forecast without instilling bias or uncertainty on the part of the viewer; Demuth, Morrow, and Lazo, 2009). With the advent of the web, providing many sources of weather data, forecast information, and atmospheric visualizations, there has been a burgeoning of research on how people (laypersons, college students) interpret and misinterpret weather forecasts and information visualizations. There is a large research and policy literature on public understanding of weather forecasts and weather risks, how weather forecasts impact or influence human decision making and activity, and how forecasters can provide the public with forecasts that are clearly understandable, interpretable, and actionable (e.g., Daipha, 2012; Hoekstra et al., 2011; Joslyn, Nadav-Greenberg, and Nichols, 2009; Lazo, Waldman, Morrow, and Thacher, 2010; Martin et al., 2008; Murphy and Brown, 1983; Ripenberger et al., 2015; Savelli and Joslyn, 2013; Schröder, 1993; Stewart, 2009; Stewart, Pielke, and Nath, 2004;

Wynne, 1991; Zabini et al., 2015). A journal of the American Meteorological Society, *Weather, Climate and Society*, focuses on just these issues, as well as climate change.

One of the best ways to understand the concepts and methods of a science is to learn the history of the science, how the concepts and methods originated, and why. This book does not recount the history of meteorology, except selectively in chapter 2. There are a number of excellent and fascinating books on this subject, including Monmonier's *Air Apparent* (1999), Hamblyn's *The Invention of Clouds* (2001), and Moore's *The Weather Experiment* (2015). Also, weather forecasting science developed significantly in applications for the military. Winters' *Battling the Elements* (1998) recounts the impact of weather-infamous battles. Two recent sociological-ethnographic studies provide narratives dissecting the culture of the National Weather Service: one by Daipha (2007, 2012, 2016) and one by Fine (2007).

This book summarizes, reviews, and integrates current empirical knowledge about the reasoning processes and capabilities of professional weather forecasters. We have composed this book for a broad readership, including the general public and policymakers as well as people who work in areas of meteorology and its kindred disciplines and professions. Although we do not shy away from scientific details, we try to provide succinct explanations for the material that is more technical. Readers are referred to other books that are excellent primers on meteorology and forecasting. *An Observer's Guide to Clouds and Weather* (2014) by Carlson et al. and *The Cloudspotter's Guide* (2006) by Pretor-Pinney are both informative and entertaining reads.

Acknowledgments

Portions of chapters 1, 3, and 5 are adapted from LaDue (2011). Portions of chapter 7 are adapted from Hoffman et al. (2014) with permission from Taylor and Francis/CRC Press. A number of forecasters and meteorologists contributed information reported in this book and participated as experts in the experiments and interviews. This includes forecasters with the U.S. Navy Meteorology and Oceanography Command; meteorologists at the U.S. Air Force Geophysics Laboratory, Hanscom Air Force Base in Bedford, Massachusetts; instructors at the U.S. Navy Aerographer School at Keesler Air Force Base in Biloxi, Mississippi; Weather Forecast Offices of the National Weather Service in Mobile, Alabama, Tallahassee, Florida, Upton, New York, and Albany, New York; and forecasters at the National Weather Service Eastern Region Headquarters in Bohemia, New York. The authors would like to acknowledge the enthusiastic assistance provided by the administrative staff and the dedicated participation of the operational personnel of the Naval Training Oceanography and Meteorology Facility, Naval Air Station Pensacola. The researchers would especially like to thank Capt. Daniel J. Soper, AGCS Jerome J. McNulty, and AGC Jeffery S. Fulson. The authors would also like to thank the individuals who consulted on the methodology for cognitive task analysis and knowledge modeling: Joseph Novak (Florida Institute for Human and Machine Cognition), Kim Vicente (University of Toronto), Jim Richmond (USN, Ret.), and William Clancey (Florida Institute for Human and Machine Cognition). William Clancey also provided insightful comments and edits on drafts of chapters 3, 11, 13, and 14. The authors thank James Correia, Jr., of the University of Oklahoma's Cooperative Institute for Mesoscale Meteorological Studies and Jim LaDue of the NWS Warning Decision Training Division for their guidance concerning computer models and forecasting technologies. The authors thank Kevin Lipton of the Albany New York NWS Weather Forecast Office for his guidance concerning current forecasting procedures and GOES image interpretation. The authors thank Pam Heinselman of the National Severe Storms Laboratory for input on chapter 10. The authors thank Anthony Lyza and Ryan

Wade of the Department of Atmospheric Sciences, University of Alabama–Huntsville for a summary of the severe weather outbreak of 31 March 2016. For their assistance in researching some topics covered in this book, the authors thank Matthew J. Bolton (an undergraduate student at Saint Leo University and intern at HOW THE WEATHER WORKS) and William G. Blumberg (a graduate student in the School of Meteorology at the University of Oklahoma). The authors thank Marc Rautenhaus of The Computer Graphics and Visualization Group, Technische Universität München, Garching, Germany, for providing the references that appear in appendix D.

Personal Acknowledgment

I would like to express my abiding appreciation to John W. Coffey of the University of West Florida for his many contributions to the weather forecasting research reported in chapters 7 and 8. Our collaboration and friendship mean a great deal to me.

Pen first met paper for this book quite some years ago. I want to express my appreciation to my co-authors. This has been a magnificent learning experience. I am honored to have benefitted from their involvement and our collaboration. And their patience. Likewise, I want to express my deep appreciation to the MIT Press for its support and patience.

R. R. Hoffman
March 2017

1 Introduction

The human information processing system is the least understood, yet probably the most important, component of forecasting accuracy. (Stewart et al., 1989, p. 24)

Psychological research of the past few decades has demonstrated limitations in people's ability to engage in critical thinking, and especially reasoning about probabilities. This is important because many weather forecasts express probabilities (e.g., of rain). Dozens of "cognitive biases" have been demonstrated, and new ones appear in the scientific journals on a regular basis. People get anchored by previous evidence or experience and do not try to disconfirm hypotheses (*I simply do not believe in climate change*). People ignore base rates or frequency of occurrence when predicting events (*Weather guessers are no good*); people miscalibrate their own understanding and tend to be overconfident (*I'm really sure it will not rain today*), and so on. In sum, the literature on human cognitive psychology suggests that the broad swath of humanity is not capable of sound critical thinking, and this extends to the claim that human reasoning is inherently and necessarily limited (see, e.g., Evans, 1989; Fischoff and Beyth, 1975; Gilovich, Griffin, and Kahneman, 2002; Kahneman, Slovic, and Tversky, 1982; Kahneman and Tversky, 2000; Slovic, 1969).

Our aim in this book is to understand what human reasoning can achieve when performing at its highest level of achievement and expertise. The primary motivation for this book stems from our interest in understanding the concept and phenomena of expertise.

We are interested in how people acquire massive and highly organized knowledge and develop the reasoning skills and strategies that enable them to achieve the highest levels of performance. We are not likely to ever perfectly sample the atmosphere (whether for lack of instrumentation, will, or funding) and not likely to ever "perfectly" predict weather even with the most advanced computer systems. Hence, forecasting is an interdependence of humans and technologies. Human expertise will always be necessary. This theme is developed across the chapters of this book.

Our focus is on the knowledge, perception, and reasoning of forecasters who work in both the public and private sectors—anyone whose job it is to make a projection about the weather. We refer to studies of the reasoning and perception of forecasters who work for the National Weather Service (NWS), the National Oceanic and Atmospheric Administration (NOAA), and the military.

Just what are the distinctions among meteorologists, forecasters, and broadcast meteorologists? Meteorologists [or meteorology researchers] conduct scientific research. Some forecasters conduct research while primarily working in a forecasting role. Most broadcast meteorologists call themselves meteorologists because they really are (degree to prove it). Some, however, are primarily announcers who label themselves as meteorologists. Some forecasters work as consultants. Certified Consulting Meteorologists (CCMs; [https://wcdirectory.ametsoc.org/specialties]) provide the meteorological input into weather-based litigations and insurance claims. They conduct "hindcasting," that is, they reconstruct events and may reassess warning actions, public response, data availability, and mode performance following a significant weather event in a given forecast area. Reconstructions can involve air and vehicle accidents, slip and falls, snow-loading, high winds, flooding, and more. They prepare reports, can be deposed, and can be required to appear at trial.

Box 1.1
What Makes for Expertise?

As we will elaborate in chapter 7, expertise is a level of proficiency above that of the journeyman. A journeyman is an individual who can perform reliably and competently without supervision. Hence, the name "journeyman." The expert is a distinguished or brilliant journeyman, highly regarded by peers, whose judgments are uncommonly accurate and reliable, whose performance shows consummate skill and economy of effort, and who can deal effectively with certain types of rare or "tough" cases. Also, an expert is one who has special skills or knowledge derived from extensive experience with subdomains.

We discuss several different literatures (meteorology, cognitive science, computer science), and we propose models of the reasoning undertaken by expert forecasters. Forecasters do not just issue forecasts. They also issue warnings. The two activities involve more than different time scales; they involve different mindsets and the reliance on different products and types of data (Klinger, Hahn and Rall, 2007). There are also many different kinds of forecasts, encompassing a range spanning seasonal climate (long-range forecasting) to warnings just minutes in advance (nowcasting). In addition, forecasting procedures are a moving target because of continual advancements in technology and our understanding of the weather. In other words, there is no single model that captures how forecasters reason: There are many.

That integration points to potentially useful avenues for cognitive research that might help train the next generations of expert forecasters and might suggest ideas for new technological aids and display systems. The main goal of this book is to explain, to both the general public and a technical readership, how forecasters understand and reason about the weather in an interdependence relationship with their computers and observational/display tools. The following excerpt from an interview with a U.S. Navy aviation forecaster illustrates this.

Interviewer: Can you remember a situation where you did not feel you could trust or believe certain data, such as a computer model or some other product—a situation where the guidance gave a different answer than the one you came up with?
Forecaster: All the time! For example, Hurricane Georges. The National Hurricane Center (NHC) said the eye would go one way, but it hit Biloxi. It was a Sunday that it made landfall. I was Forecaster on midwatch duty Saturday evening. The Airstation was going to Condition of Readiness-2. Planes had sortied out and ships had left the port. The forecast office was boarded up. We had to provide information to local people (e.g., Disaster Preparedness). We had blankets, books, food, flashlights, and camped out all weekend. The NHC had Georges tracking west-northwest. (See National Hurricane Center Advisory #48 in box 1.2.)

Forecaster: We could see the eye coming up on the radar. We had to go with the official forecast, but I did my own track. Georges was off the southeast shore of Louisiana. The National Hurricane Center had the wrong track. They said the eye would go one way, but it ended up hitting Biloxi. We looked at buoy data every few hours and did our own charts. See the chart I did at midnight (00Z). [figure 1.1]

The NHC had it shifting northwest to Louisiana, more of a westward track. Based on buoy data, we could tell that it was heading north. We could see it heading due north toward Biloxi. We saw the eye coming up. At 2:00 AM the NHC shifted the track a little to the east out to Gulfport but we were leery about that track… It picked up speed right after the NHC conference call, so there was not much they could do. The hurricane sped up and head[ed] straight north. The NHC had it shifting northwest to Louisiana, more of a westward track. But we could see it heading due north toward Biloxi. The models had it going every which way after landfall. The NHC was off by about four hours on predicted landfall. Still, it hit within the area of their forecast. You can't blame the NHC. They already had their forecast out and they had to follow it. So did we. The NHC could always update every three hours, changing where they put the storm surge watches… It made landfall Monday morning between Biloxi and Ocean Springs.

This example shows how forecasters depend on technology but also rely on their own perception, reasoning, and judgment. This is a main theme of this book. A primary

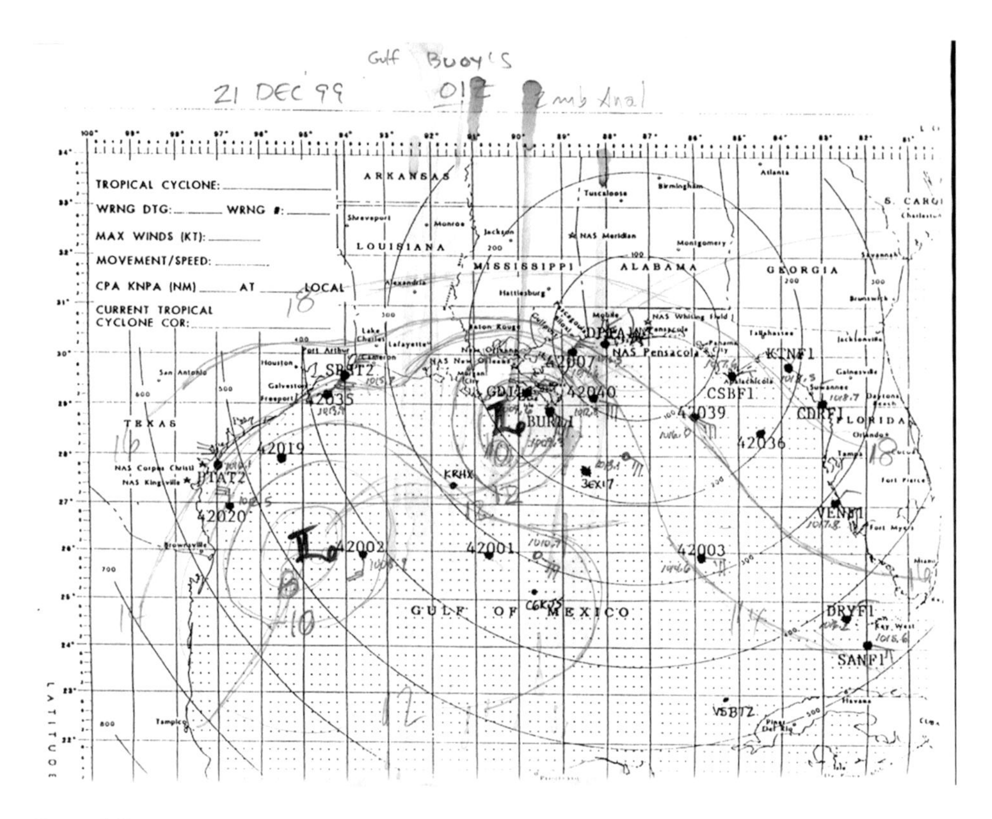

Figure 1.1
A chart showing a forecaster's analysis of a hurricane.

concept on which we rely is the concept of "expertise," and we argue that it is possible for forecasters to achieve genuine expertise. We show how it is achieved, and we show how it is applied in weather forecasting.

Motivation for the Study of Forecaster Reasoning

The question of what it means for a forecast to be "accurate" or "reliable" is discussed in detail in chapter 5. Setting this question aside for the moment, there is no doubt that severe weather has severe costs, illustrated in the United States by the disasters caused by Hurricanes Katrina (New Orleans, 2005) and Sandy (New Jersey, 2012). The following summary statistics (from reports by the National Research Council, 2001, 2006, 2010; and Risk Management Solutions, Inc., 2008) describe the impact of weather and climate:

Box 1.2

HURRICANE GEORGES ADVISORY NUMBER 48

ZCZC MIATCPAT2 ALL

TTAA00 KNHC DDHHMM

BULLETIN

NATIONAL WEATHER SERVICE MIAMI FL

4 AM CDT SUN SEP 27 1998

...DANGEROUS HURRICANE GEORGES APPROACHING THE WARNING AREA...BE PREPARED...

A HURRICANE WARNING IS IN EFFECT FROM MORGAN CITY LOUISIANA TO

PANAMA CITY FLORIDA. A HURRICANE WARNING MEANS THAT HURRICANE CONDITIONS ARE EXPECTED IN THE WARNED AREA WITHIN 24 HOURS. PREPARATIONS TO PROTECT LIFE AND PROPERTY SHOULD BE RUSHED TO COMPLETION...

AT 4 AM CDT...0900Z...THE CENTER OF HURRICANE GEORGES WAS LOCATED NEAR LATITUDE 28.1 NORTH, LONGITUDE 87.6 WEST...

GEORGES IS MOVING TOWARD THE NORTHWEST NEAR 10 MPH AND THIS MOTION IS EXPECTED TO CONTINUE TODAY WITH SOME DECREASE IN FORWARD SPEED. OUTER BANDS SHOULD GRADUALLY BEGIN TO SPREAD ACROSS THE COASTAL

SECTIONS WITHIN THE WARNING AREA SOON AND HURRICANE FORCE WINDS SHOULD BEGIN TO AFFECT THE AREA LATER TODAY...

HURRICANE FORCE WINDS EXTEND OUTWARD UP TO 115 MILES FROM THE

CENTER...AND TROPICAL STORM FORCE WINDS EXTEND OUTWARD UP TO 175 MILES MAINLY TO THE EAST...

- Industries sensitive to weather and climate account for approximately 25% of the U.S. gross domestic product. Industries with direct sensitivity account for almost 10%.
- Estimated losses due to drought are $6 to $8 billion annually.
- Estimated losses due to hurricanes average $1.3 billion per year for the years 1949–1989, $10.1 billion for the years 1990–1995, and $35 billion per year for the years 2001 to 2006. Hurricane Katrina pushed damage for 2005 over the $100 billion mark.
- Tornados, hurricanes, and floods account for $11.4 billion in losses each year. According to the Property Claim Services unit of the Insurance Services Office, in October 1993, close to 900,000 claims were filed in 24 states for a total insured loss of $1.75 billion due to wind, hail, tornado, flooding, snow, ice, and freezing perils. Losses included damage to the residential, commercial, auto, and inland marine lines of business.

Additionally, the National Flood Insurance Program had more than 11,000 claims for $186 million in flood-related damage (see Risk Management Solutions, 2008).

- The strong El Niño of 1997–1998 resulted in $2.6 billion in losses, $2 billion of which were from crop losses.
- 70% of air traffic delays are caused by weather, resulting in $4.2 billion in lost economic efficiency.
- Between 1980 and 2009, 96 weather disasters in the United States caused at least $1 billion in damage, with total losses exceeding $700 billion.
- Adverse effects of weather on roads and highways can be associated with more than 7,000 deaths per year and $24 billion in economic losses.
- Between 1999 and 2008, an average of 629 weather-related fatalities occurred per year.
- More than 60,000 deaths per year can be attributed to poor air quality due to pollution.

As the accuracy and timeliness of forecasts improves, so does the economic value of the forecasts. "The estimated annualized benefit [of investment in public weather forecasts] is about $31.5 billion, compared to the $5 billion cost of generating the information" (National Research Council, 2010 p.v1; see also Lazo et al., 2009). In the simplest of cases, with good forecasts, a newspaper can save money by knowing it will not need to use extra plastic sleeves for delivery on a potentially rainy day. An electric utility can anticipate peak need and can hedge against the risk of having to purchase extra electricity on short notice at higher cost. Corporations such as Wal-Mart operate in-house Emergency Operations Centers to continually determine how to best mitigate losses and prepare for weather impacts (Jackson, 2006; LaDue, 2011). People can plan appropriately for going to an evening baseball game (see Dutton, 2002; Roebber and Bosart, 1996a; Sheets, 1990). An Article in *The New York Times* in 2015 said:

> Two consecutive years of volatile weather... have proved disastrous for companies that rely on predictable temperatures to sell cold-weather clothing like sweaters and coats. So the $200 billion American apparel industry, which is filled with esoteric job titles like visual merchandiser and fabric assistant, is adding a more familiar one: weather forecaster. Liz Claiborne, the apparel company, has hired a climatologist from Columbia University to predict weather for its designers to better time the shipments of seasonal garments to retailers. The discount retailer Target has established a "climate team" to provide advice on what kind of apparel to sell throughout the year. More and more, the answer is lighter weight, "seasonless" fabrics. And the manufacturer Weatherproof, which supplies coats to major department stores, has bought what amounts to a $10 million insurance policy against unusually warm weather, apparently a first in the clothing business. [Downloaded 16 January 2015 from http://www.nytimes.com]

The advancement of science and technology for better observing, understanding, and predicting of the weather has been a U.S. national priority for some years (National Research Council, 2006, 2010). Advances in capabilities are nowhere more salient than the advent of NEXRAD—the NEXt generation weather RADar that most of us see presented on televised weather reports. It has dramatically improved the ability of forecasters to observe and understand what is going on in the atmosphere in four dimensions and with new data arrays. Similarly, Geostationary Operational Environmental Satellites (GOES) imagery has allowed forecasters to better assess upper levels and wind patterns (satellite-derived winds), better determine the position and track of hurricanes, and better interpret the evolution of mesoscale storm systems. The public has benefited by being better able to visualize what TV meteorologists are discussing and for TV meteorologists to provide better information to their viewers.

The field of meteorology is obviously of great importance to society—the prediction of the tracks of hurricanes (and associated evacuation impacts), the prediction of tornado outbreaks, and warnings for widespread winter storms, to name a few. The importance of weather to the economy, human activities, and human well-being cannot be underestimated, although one can cite the statistics of lives lost (thousands per year) and property damaged (tens of billions per year) because of a large mix of different types of severe weather. As forecasting technology and capabilities have advanced in recent years, there has been a corresponding ramp-up in consideration of the importance of the socioeconomic aspects of weather, as reflected in a number of interdisciplinary research projects (Morss et al., 2008). This includes the new NOAA program titled "Weather-Ready Nation," which has the goal of increasing society's responsiveness and resilience to extreme weather events (Lindell and Brooks, 2012)

Likewise, there has been more research on how weather forecast information plays into human decision making (e.g., Lazo, Morss, and Demuth, 2009). Based on a survey, researchers at the National Center for Atmospheric Research estimated that the U.S. households obtain several hundred billion forecasts each year, linked to many billions of dollars in benefits even though the majority of forecasts (more than 70%) are obtained simply because people want to know what the weather will be. The remaining 30% of forecasts are obtained because of a need to plan various activities or respond to dangerous weather. For the National Weather Service (NWS), one of the most important beneficiaries of weather forecast information are Emergency Managers (see Baumgart et al., 2008).

Like many modern domains of work, significant workforce issues have arisen in recent years (Bordogna, 1999; Florida, 2002, 2005; Lachance, 2000; National Science Board, 2004; Stokes, 1997). One challenge is the need to support and expand minority

Box 1.3
Acronyms Galore

In mentioning NEXRAD and the National Weather Service, we find ourselves immediately apologizing for the acronyms, knowing that many more are yet to come in the pages of this book. A list is presented in appendix A. Meteorology is rife with acronyms, making their use unavoidable. We introduce acronyms only when the terms are relied on subsequently. We must also rely on the technical jargon of meteorology. Digression would be too easy. Even the widely known term "weather front" represents a complex and widely misunderstood concept. We try to use most terms in a way that lets the context imply meaning when full technical definitions are not really necessary. In some places, we provide information in sidebars such as this one.

Readers can also refer to the online Meteorology Glossary of the American Meteorological Society
[http://glossary.ametsoc.org/wiki/Main_Page]

involvement in the sciences, given the changing national demographics (Armstrong and Thompson, 2003; National Science Board, 2004). Another challenge has to do with the fact that our nation's pool of experts is aging (Hoffman and Hanes, 2003). In some sectors, such as the utilities, more than 50% of the senior personnel are at or near retirement age (Fisher, 2005; Moon, Hoffman, and Ziebell, 2009; National Public Radio, 2005). This has triggered an interest in the capture, preservation, and reuse of expertise via a process called "knowledge management" (Becerra-Fernandez, Gonzalez, and Sabherwal, 2004; Brooking, 1999; Choo, 1998; Coffey and Hoffman, 2003; Crandall, Klein, and Hoffman, 2006; Davenport and Prusak, 1998; Klein, 1999; Nonaka and Takeuchi, 1995; O'Dell and Grayson, 1998).

One thing we hope to show in this book is that the field of meteorology, and forecasting in particular, is ripe for applications of psychology, human factors engineering, and cognitive ergonomics to help address these national workforce issues. The motivation for applied psychological research on topics in meteorology is quite multifaceted, but we introduce this book by pointing to just one outstanding theme. After briefly setting the stage in this way, we will describe the organization of the book.

The Data Overload Problem

In many domains of work, new technology has resulted in more multisourced data than decision makers can effectively interpret and use. By the late 1980s, there were already enough environmental satellite data in archives to keep the available image analysts busy for decades (Hoffman, 1990). One design for the Earth Observation Satellite system involved the collection of more than a terabyte of multispectral data per day. This is an order of magnitude greater than the amount of data that can fit through the pipelines to the forecasting offices. Computer models today are generating petabytes of computations.

Perhaps nowhere has the data overload problem been more salient than in the domain of meteorology (Bosart, 1989; Errico, 1999; Hoffman, 1991; Monmonier, 1999). The data overload problem began to emerge in the National Weather Service (NWS) in the mid-1970s. Since then, observing and measuring systems have been improved and expanded continuously, resulting in floods (pardon the pun) of atmospheric, hydrologic, and oceanic data. Beginning in 2006, an attempt was made to establish the Constellation Observing System for Meteorology, Ionosphere, and Climate (COSMIC). It consisted of six small satellites that measured microwave and infrared energies by using signals from Global Positioning Systems satellites viewed edge-on and passing through the atmosphere, providing soundings (measures of pressures, temperatures, winds). COSMIC provided thousands of soundings per day, covering the entire globe. This filled in crucial data for regions where it is not possible to launch instrument-carrying balloons (radiosondes) (Serafin, MacDonald, and Gall, 2002). Although the COSMIC satellites have been decommissioned, the program is suggestive of how atmosphere observing systems have been expanding in scope in recent decades. Signals from the Global Positioning System continue to be used to probe the dynamics of the atmosphere (e.g., phase delays in the signals can be converted to estimates of water vapor content) (Ware et al., 2000).

GOES Satellite Products

Images captured by the Geostationary Operational Environmental Satellites (GOES) operated by NOAA revolutionized forecasting (and televised weather reporting as well). The first-generation GOES satellites provided relatively low-resolution images in selected visible and infrared bandwidths, once every half-hour. The current generation of GOES satellites includes high-resolution radiometers and spectrometers, yielding visible and infrared image data (in multiple bandwidths) every few mintues for significant events such as tornado outbreaks and hurricane landfalls (Menzel and Purdom, 1994). The next generation of GOES satellites (GOES-R Series, designations GOES-14 and GOES-15; planned for launch in 2016) will collect data at faster scan rates, high accuracies, and high resolutions, resulting in scores of products, such as aerosol particle size, aircraft icing threat, cloud top temperatures, low cloud and fog, magnetospheric protons, rainfall potential, reflected solar radiation, sea and lake ice, turbulence, and many others (Schmit et al., 2005). GOES-R is close enough to becoming operational that the NWS has been simulating the new products and is already talking about training on the new products (Goodman, 2012; Schmit et al., 2013, 2015). (During the final editing of this book the training was nearly complete.)

GOES data are processed by a number of enhancement algorithms relating radiated energy to pixel brightness value. For instance, clouds with heavy precipitation potential typically involve stronger updrafts and have higher cloud tops. Higher means colder, and so cold infrared sources (partialing out the contribution of reflected solar illumination) can be enhanced to suggest areas where heavy precipitation is most likely. For many years, the forecasters printed the GOES images by fax. As a 1980s generation of workstations (AFOS-PROFS; described in detail in chapter 2) was phased out and a 1990s generation was phased in (AWIPS; also see chapter 2), images were viewed on cathode ray tube displays, still using gray scale (which actually had certain advantages in terms of how the forecaster could perceive the heights of clouds; see Hoffman et al., 1993). But it was not long before display technology made possible the use of color. An example colorized (i.e., enhanced) GOES infrared image appears in figure 1.2 (plate 1). Note that for figures 1.6, 1.7 (plate 5), and 1.8 (plate 6), we have used images taken at about the same time on January 3, 2015. This was done to support the comparison of different data types and displays.

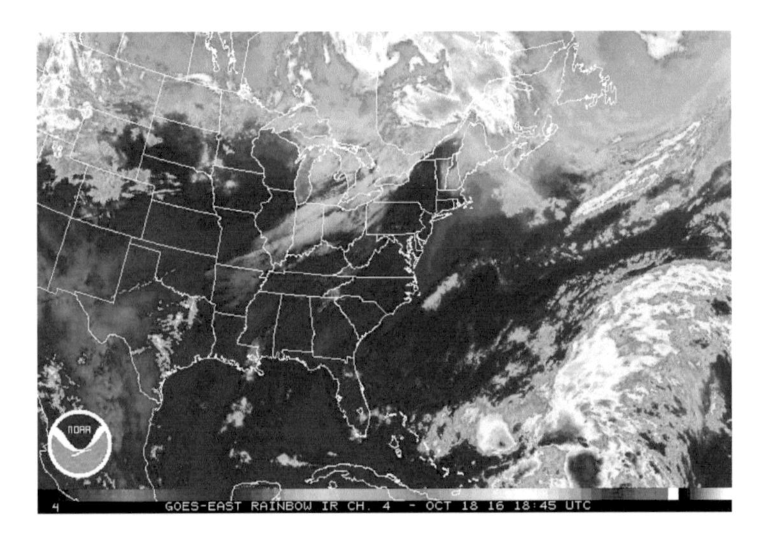

Figure 1.2
(plate 1) A colorized infrared GOES satellite image [downloaded 18 October 2016 from http://www.goes.noaa.gov/goes-e.html].

The infrared data depicted in figure 1.2 (plate 1) are sometimes shown in televised weather forecasts using a locally determined, nonstandard coloration (usually green) to show areas of likely precipitation, although what the infrared actually shows are areas in which the coldest, highest thunderstorm cloud tops are located. Sometimes the televised images are composites of satellite and radar data, which also use green coloration to show areas of precipitation. Such images show clouds as detected by the satellite and precipitation as detected by radar.

An important GOES product is the water vapor image, which deliberately measures in the bandwidth of infrared energy in which there is greatest energy absorption by water vapor. This enhances the contrast among moist, dry, and cloudy regions and offers an easy-to-see view of atmospheric circulations on small to large scales. A sample water vapor image appears in figure 1.3 (plate 2). This too typically utilizes a standardized color-coding scheme. Dryer air appears orange (depicted using saturation shades), and air containing water vapor is depicted using a palette primarily of white, gray, blue-gray, and blue-green hues (with yellow red, violet, and blue used to represent the extremes).

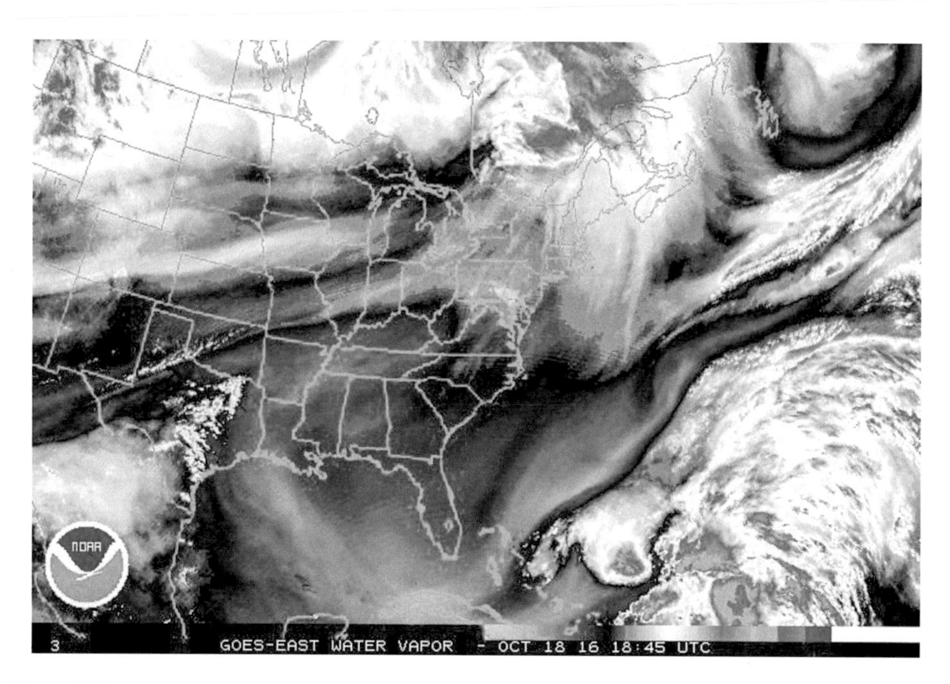

Figure 1.3
(plate 2) A colorized GOES water vapor satellite image [downloaded 18 October 2016 from http://www.goes.noaa.gov/goes-e.html].

NEXRAD Radar Products

NEXRAD, the NEXt generation RADar, is a remarkable system. One forecaster referred to it as the "Cadillac" of radars, meaning it has powerful functionalities. The capabilities of NEXRAD were significant when the system was approved by Congress in 1988. Through operational use and research studies, its capabilities have continually expanded. As a result, NEXRAD is powered by a wide array of special algorithms that can be tailored to satisfy a variety of specific purposes, (e.g., not only whether it will hail, but the size of the hailstones) and local constraints (e.g., terrain effects).

As in all weather radars, the energy in NEXRAD radar's microwave pulses are reflected back to the radar, providing information about, *What's there?* But in NEXRAD, the reflectivity is more sensitive than previous radars, so the "Base Reflectivity" products of NEXRAD represented a major advance. NEXRAD also capitalizes on the Doppler effect, the shifting of the frequencies of the return microwave energy as a function of the relative movement, toward or away from the emitter, of the objects that reflect the pulse back to the receiver. Thus, the NEXRAD "Relative Velocity" products provide answers to questions such as, *"How fast are the winds in the storm blowing and from what radial direction?"* An even more advanced capability is provided by using a dual polarized set of pulses, that is, the microwaves in one pulse are perpendicular to that of the other. This permits better characterization of hail, improved thunderstorm warnings, improved rainfall estimation, improved service in mountainous regions, and overall improved data quality (i.e., decluttering from birds, insects, etc.), all of which mean that better data will be input to the computer models [see http://www.roc.noaa.gov/wsr88d/dualpol/DualPolOverview.aspx]. The new products are complex. Some experienced forecasters have not yet fully incorporated them into their warning decision-making process, but many have.

A sample NEXRAD Base Reflectivity display is shown in figure 1.4 (plate 3), which shows regions of precipitation. Variants on this product (sometimes based on radars operated by TV stations themselves) are usually the ones utilized in televised weather reports.

The importance of NEXRAD to forecasting activities is matched by the degree of its contribution to data overload. The "Boston Area NEXRAD Demonstration Project," conducted by the U.S. Air Force in 1985, involved the generation of 450 hours of display products over 59 days, for about 8.3 hours of products per day (Forsyth et al., 1985). Each NEXRAD system is capable of producing dozens of different kinds of analyses—precipitation, winds, storm cell characteristics, hail, and others. Even in its early builds across the 1980s, NEXRAD could create more than 100 different products

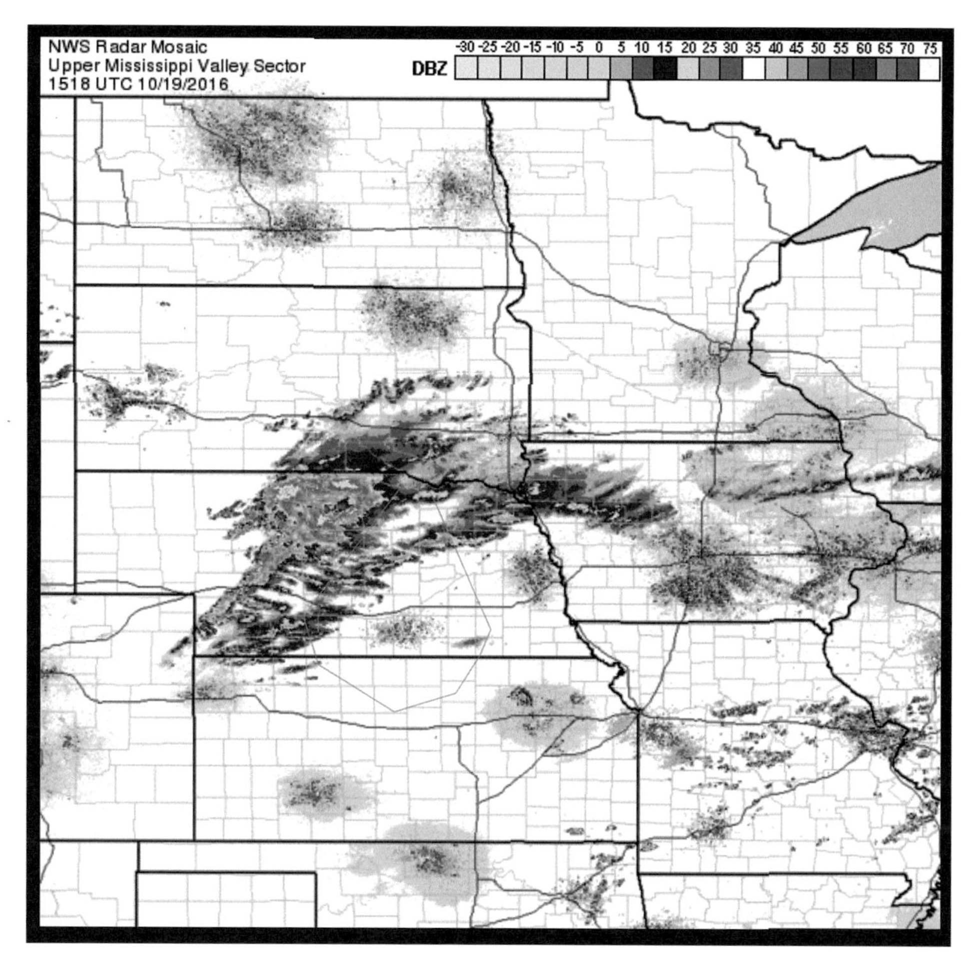

Figure 1.4

(plate 3) An example of a NEXRAD Base Reflectivity product [downloaded 3 January 2015 from http://radar.weather.gov/ridge/Conus/full.php/].

in a given five-minute period (Hoffman 1987a; Steadham, 1998), and new products and algorithms, to this day, are constantly being created at Weather Forecast Offices (WFOs) as well as at the NOAA/NWS Radar Operations Center in Norman, Oklahoma.

In addition to new sensor systems, a host of new information processing workstations permit the creation of nearly endless combinations of the new data types. For instance, the "Sat-Rad" display (now being included in many televised weather reports) overlays data from the national NEXRAD network on an infrared satellite image. The Regional and Mesoscale Meteorology Advanced Meteorological Satellite Demonstration

Box 1.4
The Scales of Atmospheric Dynamics

Atmospheric dynamics must be considered on a number of scales based on areal extent and time. Global or *Planetary scale* is 2,500 kilometers or more in horizontal extent and involves weather events that span weeks—the extent and movement of large air masses and major frontal systems across continents and oceans. *Synoptic scale* is continental or smaller—the scale of many low-pressure systems. This involves wavelengths of atmospheric troughs (relatively low pressure) and ridges (relatively higher pressure) ranging from 1,000 to 2,500 kilometers and spanning days to weeks—the size range and lifetime of most cyclonic systems, their attendant fronts, the movements of smaller air masses, and the like. *Mesoscale* involves regional or local weather (a few to several hundred kilometers) over a period ranging from a day or two. Finally, *Microscale* is roughly the size of a neighborhood (2 km or less); events play out in a matter of minutes to hours and are addressed by what are called "nowcasts." At each scale, the dynamics must be appropriately described but also linked to forcing events (called "boundary conditions") at larger scales.

and Interpretation System (RAMSDIS) workstation supports the analysis of radar and satellite images and image loops, overlaid with observation charts.

In addition to new data types, a number of computationally intensive computer models run daily on supercomputers at the National Centers for Environmental Prediction and other meteorology laboratories and forecast centers worldwide, and they are used to make predictions (see Barnston et al., 1999). There are many different models and a diverse array of products that can be derived from them. New models, using more and more advanced physics packages, are constantly being developed, and these produce products at many resolutions. An example output from a computer model is presented in figure 1.5 (plate 4).

With the advent of the new data types and display systems, the computer model outputs can be shown on displays in such a way as to compare the computer predictions with actual observations—resulting in yet another layer of combinatorics to the data display and overload problem. As will be explained in more detail in chapter 12, various graphs, charts, and tables showing such things as wind fields, temperatures, and air pressure at different heights in the atmosphere can be generated from these computer models. Computer models can be used to produce literally hundreds of products that a forecaster can request. The model runs involve millions of regression calculations, resulting in many thousands of forecasted data points per day, and thousands of forecast products made available per day for hundreds of locations, including

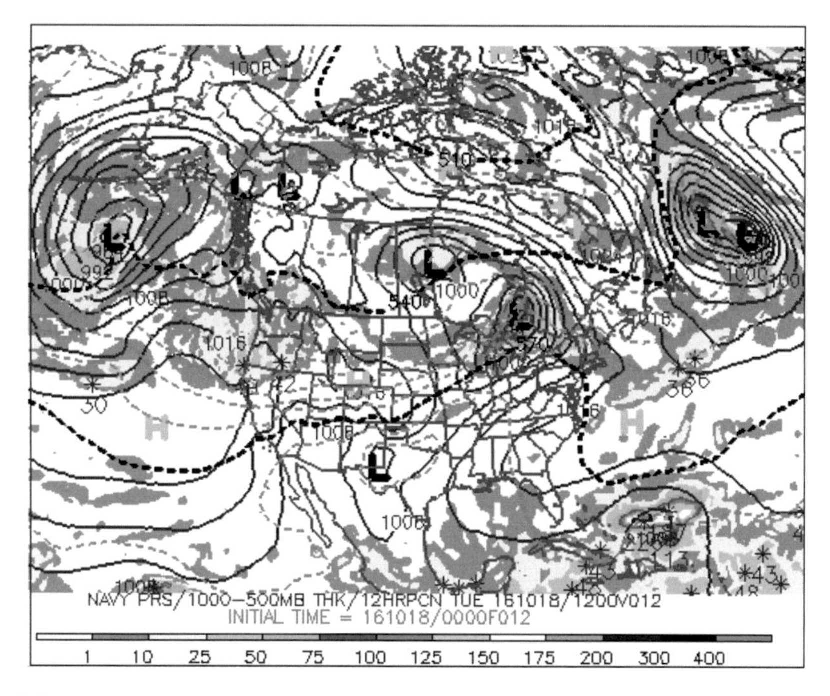

Figure 1.5
(plate 4) This display shows the 12-hour forecast for upper level winds generated on 18 October 2016 by the U.S. Navy's Navy Operational Global Atmospheric Prediction System (NOGAPS). [http://mp1.met.psu.edu/~fxg1/NOGAPS_0z/nogapsloopw.html]

all of the NWS WFOs in the United States (including Puerto Rico and the Virgin Islands).

The model results are subject to postprocessing, in which statistical regression analyses take local trends, regional climate, and local terrain into account and correct for certain kinds of biases and errors in the computer model. The result is a data table that presents the "short and sweet" of a model's predictions, called Model Output Statistics (MOS; see Klein and Glahn 1974). An example MOS guidance is presented in figure 1.6. In this example, the rows refer to the specific weather variables that are predicted: At the top, DT and HR = hours across a span of three days; then TMP = temperature, DPT = dewpoint, CLD = cloud cover, WDR = wind direction, WSP = wind speed, P06, P12 = percent chance of precipitation over a specified time interval, Q6, Q12 = quantitative precipitation (rainfall and/or transformation snow to its liquid equivalent), CIG = ceiling, VIS = visibility, and OBV = obstruction to vision.

```
PHHI   NAM MOS GUIDANCE    1/25/2016  0000 UTC
DT /JAN  25            /JAN  26                /JAN  27           /
HR   06 09 12 15 18 21 00 03 06 09 12 15 18 21 00 03 06 09 12 18 00
X/N                    79          58          79          60    79
TMP  71 68 67 65 65 76 78 75 68 63 62 60 62 76 78 75 70 65 64 63 77
DPT  67 66 65 63 62 60 61 62 63 62 62 60 62 62 63 65 66 64 63 62 64
CLD  SC OV FW SC FW SC BK SC SC CL CL FW FW SC BK BK SC CL CL FW BK
WDR  33 33 34 33 28 07 33 36 30 30 34 31 26 24 20 25 23 18 23 24 23
WSP  04 04 06 09 02 06 06 07 02 02 04 03 01 05 07 07 04 04 05 01 07
P06        12    12     2     3     0     7    12     6    11 13  9
P12              21           3           7          12       16
Q06         0     0     0     0     0     0     0     0     0  0  0
Q12               0           0           0           0        0
CIG   8  7  8  7  8  6  6  8  8  8  8  8  8  8  6  6  8  8  8  8  6
VIS   7  7  7  7  7  7  7  7  7  7  7  7  7  7  7  7  7  7  7  7  7
OBV   N  N  N  N  N  N  N  N  N  N  N  N  N  N  N  N  N  N  N  N  N
```

Figure 1.6
An example of Model Output Statistics (MOS) guidance. These statistics were derived from the outputs of the North American Mesoscale Forecast System (NAM) [downloaded from ftp://tgftp.nws.noaa.gov/SL.us008001/DF.anf/DC.mos/DS.met].

Although MOS guidance can be considered one more contributor to data overload, it should be noted that MOS guidance is important in forecasting. MOS predictions are especially used for weather at the surface and are thus handy for forecasters. The MOS guidance can include quantifications of the uncertainty of each particular prediction. Overall, MOS guidance is considered to be more reliable than the "raw" outputs of the computer models. The MOS essentially summarizes what a computational model is saying. Indeed, some "forecasts" are basically recapitulations of the MOS (see chapter 12).

The data overload problem increased more with the advent of the Interactive Forecast Preparation System (IFPS) in NWS operations, which requires that forecasters construct seven-day graphical representations of weather forecast variables on grids of about 5 kilometers (Mass, 2002; Ruth, 2002). Complicating things even further, a technique called ensemble forecasting combines the outputs of several computer models into a single forecast that compensates for some of the tendencies of biases of the individual computer models (see Tracton and Kalnay, 1993). For example, one model might tend to overpredict the depth of low-pressure centers that form off the Mid-Atlantic coast after the low "jumps" over the Appalachian Mountains and reforms over Gulf Stream waters. Another model might perform poorly at predicting hurricane tracks because it does not take into account sea surface temperatures, and so on. The ensemble concept relies on the principle that combined information from multiple sources, given that the errors from individual sources are not too highly correlated, leads to improved forecasts (Leutbecher and Palmer, 2008).

An example ensemble forecast product is shown in figure 1.7 (plate 5). Model outputs can represent precipitation and/or winds at various heights in the atmosphere. Outputs can depict results for any of 20 or so different computer models. The display in figure 1.7 is from the Global Forecast System (GFS) of the National Centers for Environmental Prediction. This data field shows "500mb heights and vorticity." Vorticity (in the Northern Hemisphere) is the counter-clockwise spin or curvature of air parcels or wind flow. One can see areas of curvature fairly clearly in figure 1.7. The map essentially shows a surface at which the air pressure is 500 millibars and how the air at that layer is moving. Weather dynamics at 500 millibars are crucial in forecasting because they provide a picture of the main weather dynamics at a continental scale: 500 millibars is approximately the height in the atmosphere that divides half of the mass of the troposphere above and below (roughly about 18,000 feet or about 5 kilometers). (Sometimes atmospheric data at various heights are expressed relative to "geopotential heights," which is an adjustment based on the variation of gravity as a function of latitude.)

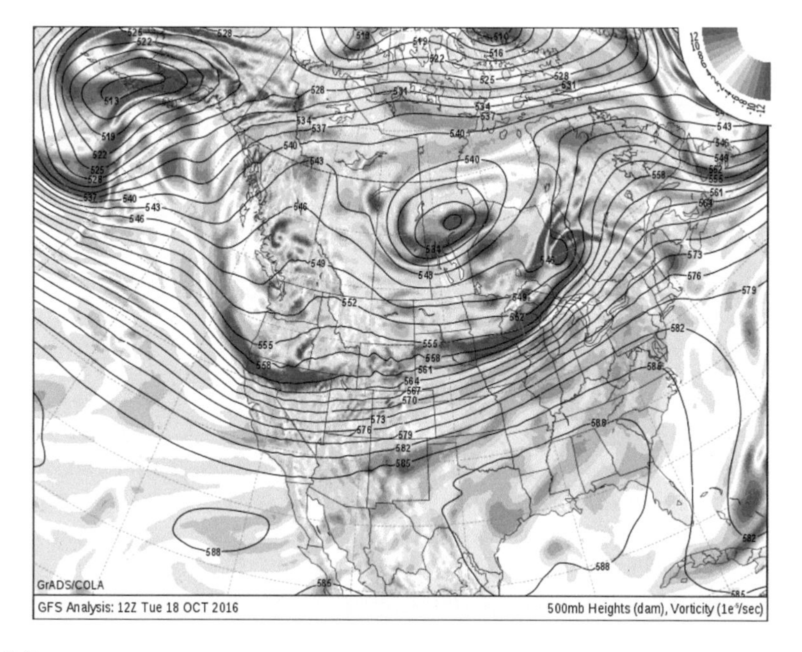

Figure 1.7
(plate 5) An example 500-millibar product [downloaded 18 October 2016 from http://radar.weather.gov/Conus/full_lite.php].

Box 1.5
What Is a Millibar?

In the metric system, the bar is a unit of pressure. The unit it is based on is the pascal, named after Blaise Pascal, Renaissance genius who pioneered hydrodynamics and hydraulics. He conducted experiments which proved that barometers work because a vacuum is created above the mercury column inside the barometer's glass tube. Pascal studied air pressure by making barometric measurements at ground level and also in a church bell tower. He also experimented on how the air pressure changes during the day. The pascal unit expresses force per unit area. A bar is 100,000 pascals, and a millibar is 100 pascals. Air pressure relies on a millibar (thousandths of a bar) scale because air pressure decreases considerably as elevation increases. The average air pressure at the Earth's sea surface is about 1,000 mb (varying between about 970 mb and 1,050 mb); 500 mb is the height in the atmosphere at which pressure and atmospheric mass is half that at sea level. At that height in the atmosphere, one can easily see the "troughs" and "ridges" that characterize air masses and their interactions at the continental scale, as in figure 1.6. At the 500 mb height, the air temperature is rarely above freezing.

The variety and vibrancy of weather products can be seen at [http://spaghettimodels.com/]. The new information-processing and display systems have been motivated by a decades-long plea from the meteorology community for richer and more timely data and forecasts. A result, however, has been that forecasters can now get overwhelmed by the flood of information. This became salient more than 25 years ago, even while it was clear that in actual forecasting contexts particular data types and data sets are pertinent depending on the forecasting problem at hand:

> Despite the flood of data from satellites and radar, forecasts [have] barely improved. This suggests shortcomings in our understanding of the atmosphere as expressed in our conceptual and numerical models ... Meteorology texts and case studies neatly and often mathematically link divergence, vorticity, [etc.]; scientific certainty pervades the pages ... academically-trained forecasters thought they understood the weather; they usually believed that the lack of sufficient data ... caused wrong predictions ... they generally viewed statistical aids as no more than temporary, inadequate stop gaps, and thought that more data would automatically reduce forecast errors ... since the 1960s, the unexpectedly small improvement in forecasts stemming from the vast supply of satellite data threw doubt on the most important assumptions. (Ramage, 1993, pp. 1863–1865)

This situation has not been unique to academic meteorology and civilian forecasting. Ramage was echoed by Dyer (1987) in her comments on forecasting by the U.S. Air Force:

> Increasing the amount and quality of the data available to the meteorologist by the introduction of weather radars and satellites and a denser network of surface observation stations has not improved the average [forecast accuracy] as much as might be hoped. Indeed, the modern operational forecaster often feels overwhelmed by the amount of information that must be assimilated in a short period. (p. 20)

Nor is this situation confined to North America. In a study of aviation forecasting procedures in Sweden, Perby (1989) found that, "When talking to meteorologists it is striking that they say, on the one hand, that they cannot assimilate all the available information and, on the other hand, they want more information" (p. 50). This apparent contradiction—of wanting more data but being overwhelmed by the data available—can be understood as a consequence of the forecaster's need to make sense of the weather. The forecasting process can be likened to exploration; forecasters are always interested in finding new data or new data types that might help them make sense of things. Given that new data products (new sensors, algorithms, visualizations, etc.) appear all the time, the exploration never stops. As one forecaster put it, "Fishing, fishing, always fishing for something better, something that'll let us know what's going to happen!"

Related to this search for meaning on the part of experienced forecasters, data overload is acute for less experienced forecasters, ones who are less facile at search, selection, and sensemaking of the data. In the military, many forecasters have limited opportunity to develop expertise due to both the limited tours of duty and assignment rotation (Dyer, 1989; Fett et al., 1997; Peak and Tag, 1989; Pliske et al., 2004: Pliske et al., 1997). A similar situation arises in the NWS when forecasters are transferred to new assignments—it takes months to regain proficiency when having to forecast in a new region (Dyer, 1987). Data overload is also acute when forecasting must be done under time pressure, which occurs often in both military and civilian forecasting contexts. In time-pressured situations, the forecaster cannot always afford to take the time to develop a thorough understanding of the weather situation (Uccellini et al., 1992). In addition to impact on forecasting quality, another effect of the flood of new data, new data types, and new technologies is an increase in the mental workload of forecasters (see Lee, 1977).

One of the main goals of the forecaster is not just to look at data but to understand it, integrate it into a relatively coherent whole, and figure out what will happen next. Simply having a ton of data only helps so much. Without the tools to help this integration process, weather forecasters are overwhelmed. One way of thinking about this is that experts should know which data source to look at under different circumstances, and forecasters do know this. However, there is simply so much data and so many different ways of looking at the data that most forecasters often do not have the time to do a full integration. Some idea of the number and types of forecasting products that are available can be gained from surfing a number of websites. We suggest the website of the U.S. Navy's Coupled Ocean/Atmosphere Mesoscale Prediction System (COAMPS) [http://www.nrlmry.navy.mil/coamps-web/web/home]. We also recommend the website of NOAA's Aviation Digital Data Service [http://www.aviationweather.gov/]. This site posts satellite and radar imagery, but it also posts pilot reports, airfield reports, and forecasts of convection, turbulence, icing, winds, temperatures, and other data types. In addition, products can be viewed in different modes: as individual images or time series loops or animations.

Figure 1.8 (plate 6) is a composite screenshot from the COAMPS that gives some idea of the sheer number and diversity of data types available to forecasters. What printed images of this type cannot show is all the image loops and graphic animations. As the

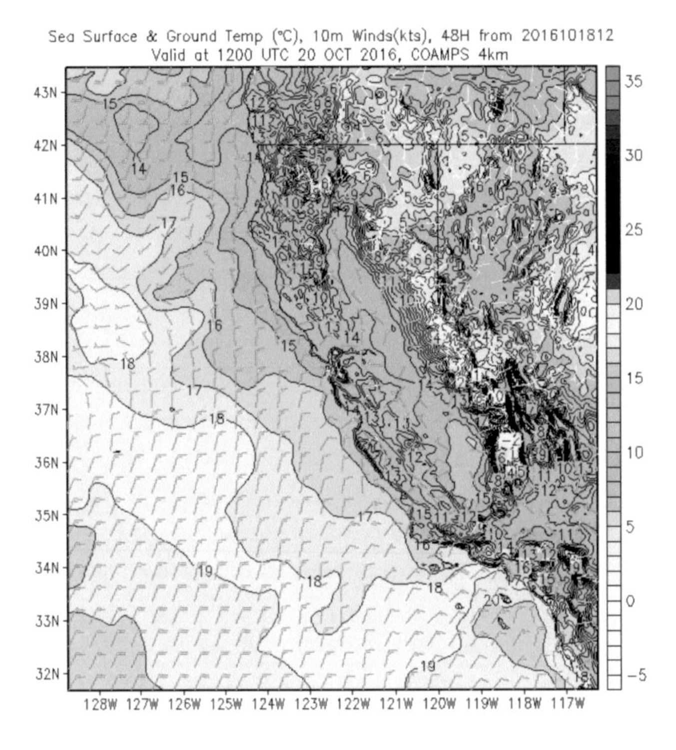

Figure 1.8

(plate 6) An image from the COAMPS web site showing the 2-day forecast for surface temperatures and winds. [Downloaded 18 October 2016 from http://www.nrlmry.navy.mil/coamps-web/web/home].

reader can surmise, there are many kinds, including time series satellite loops, dynamic model outputs, and so on.

Another great example is Penn State's Weather e-Wall, in which every hyperlink takes one to map images or animations [http://mp1.met.psu.edu/~fxg1/ewall.html].

So, How Much Data Are There?

Exactly how much data come into a Weather Forecast Office on a daily basis? An NWS report from the Alaska Region (Curtis, 1992) showed that in a given 12-hour time period, forecasters could access about 400 graphic products, about two dozen satellite images, about a thousand radar products, and a few hundred sets of observations. The report went on to predict, on the basis of systems then under development, that within ten years, 1,400 graphics products would be available per 24-hour period, over a thousand satellite images, over 1,200 radar products, and over 4,000 sets of observations.

If one were to count individual displays of individual data fields (e.g., surface observations, winds at various heights, etc.), but count each animation (satellite loop or successive radar scans) as a single data field (as opposed to a set of images), then the number of data fields that *could* potentially be grabbed and used at a WFO is certainly in the many hundreds. Some of the data products that come into WFOs are standard and some arrive on a regular schedule (e.g., forecast discussions). Some products are alerts (e.g., for severe weather). The number of different data displays/fields that are grabbed depends on the forecaster, the weather problem of the day and shift, the climate, the region, and other factors as well, so it is hard to put a single number on it. Instead, what we can ask is: What data did a particular forecaster look at during a shift, perhaps as a representative example? Table 1.1 lists the data that were examined by a forecaster in the northeast United States on a November 2015 night shift. This was chosen because it was a relatively quiet shift. (Table 1.1 references a number of different computer models. These will be explained in more detail in chapter 12.)

The data overload problem is not just a problem as it is, but a problem that is growing. Just in regard to radar, for example, new experimental radars and radar networks are being developed to allow forecasting of severe weather at an even finer scale (Brotzge et al., 2010; Heinselman et al., 2012). Efforts are underway to assimilate radar data into the mechanisms of computer models (e.g., Kain et al., 2010). New display systems are being developed to support weather radar information processing (Maese et al., 2007). New software tools for data analysis are being developed using the open source development model (Hesterman et al., 2015).

Table 1.1
Data examined on a night watch by a WFO forecaster in the northeast United States

Data Type	Data	~ Number of data type products per shift
Observations	Surface observations once every hour for 9-hour shift plus once every hour for the 3 hours prior to beginning of shift to see trends that occurred.	12
	Upper Air Rawinsonde data two per shift from each of three sites.	6
	Satellite Data (continuous loops of water vapor and infrared loops, every 15 minutes over 12 hours).	120
	Radar (continuous loops from our office and four surrounding offices every 6 minutes, just base reflectivity viewed over five hours on days of benign weather).	50
Global Computer Models	Global Forecast System Model out to at least 168 hours (6-hour increments).	2
	European Centre for Medium Range Weather Forecasting Model out to at least 168 hours (6-hour increments).	2
	Canadian Global Environment Multiscale Model out to at least 168 hours (6-hour increments).	2
Regional and Mesososcale Computer Models	Rapid Update Model (RAP 13) out to 18 hours (1-hour increments out to 18 hours).	8
	High Resolution Rapid Refresh Model out to 12 hours (1-hour increments out to 12 hours).	8
	North American Model (NAM12) (out to 84 hours).	8
Ensemble Model Outputs	Global Ensemble Forecast System out to 168 hours (6-hour increments).	2
	Short-Range Ensemble Forecasts out to 87 hours (3-hour increments).	2–3
Text Products	Alphanumeric data (called "statistical guidance") from the Global Forecast System and the North American Mesoscale Model out to 72 hours (3-hour increments).	2–3
	TOTAL	60–62

Our discussion of weather data overload is intended not only to introduce some ideas and technologies that we will refer to in this book, but also to lead us to some of the real psychological questions that are the focus of this book: How do forecasters cope with and integrate all this information? How might new technologies help in their integration?

Coping with Data Overload

Because humans can have difficulty processing large volumes of data, there has been an ongoing push for increased reliance on technology in the forecast process. Some data processing is a trivial task for even simple computers. This has led to an ongoing discussion concerning the future role of humans in weather forecasting (e.g., Brooks, et al., 1992; Brooks and Doswell, 1993; Doswell, 1986a, 1986b, 1986c; Doswell et al., 1981; Glahn, 2003; Hoffman, 1991; Hoffman et al., 2006; Mass, 2003a, 2003b; Tennekes, 1988, 1992) and the effect of such technological overload on forecaster mental workload and hence on morale. Many forecasters actively appraise their conceptual understanding of forecast problems and, through the achievement of expertise, are able to add value to the forecasts from the computer models, even as the computer models improve (e.g., Bosart 2003; McIntyre 1999). This persistent advantage stems from the human ability to deal with information at the level of meaning, something that (still) goes beyond the capabilities of even the most sophisticated computer systems. The persistent advantage also stems from the decision support services that the NWS provides to its stakeholders and everyone who uses its information. There are web pages for predicting climate trends [for instance, https://www.climate.gov/decision-support] and websites to support forecaster decision making for particular events or circumstances, such as the Superbowl [http://www.nws.noaa.gov/com/weatherreadynation/news/140220_super_bowl.html#.VNfZgkLG0q8]. The persistent advantage also stems from the adeptness of skilled practitioners at the *interpretation* and *evaluation* of information that the computer models provide, another thing that remains outside the capabilities of automated systems. Many studies have shown that forecasters can add value to automated systems by adjusting their reliance on particular pieces of information according to the meteorological situation and their local experience (Roebber, 1998; Roebber and Bosart, 1996a, 1996b; Roebber et al., 1996). In recognition of this, the NWS has been emphasizing the importance of providing decision support services for forecasters, government agencies, and the general public. The website climate.gov lists more than 30 tools for mapping, monitoring, and forecasting of weather-related events such as crop moisture stress, energy demand, air quality, and so on [https://www.climate.gov/decision-support]

These natural advantages of human expertise can be blunted when forecasters are constrained by data overload and poorly conceived and poorly designed automation. Writing in 1992, J. C. Curtis argued that the increase in data by an order of magnitude clearly mandated a multidisciplinary approach to the development of new workstation systems, to support data integration and the design of new ways to facilitate communication, and to generate new ideas concerning the duties and tasks of meteorologists. Curtis pointed out that there is an important role to be played here by applied experimental psychologists, cognitive scientists, and human factors engineers.

Some Key Terms

When forecasters inspect data, they build a model of what is going on in the atmosphere (Morss et al., 2015). For decades, forecasters have called this a *conceptual model* to distinguish it from formal or mathematical equations or computer models of the atmosphere (see chapter 4). The term "mental model" comes from cognitive psychology (e.g., Anderson, 2005; Gentner and Gentner, 1983; Gentner and Stevens, 1983; Schumacher and Czerwinski, 1992; Stevens and Collins, 1978). Forecasters are quite comfortable with the notion, acknowledging that they form mental images of such things as fronts and air masses and storms, and imagine the forces playing out according to known principles. Conceptual models in forecasting take the form of diagrams (see figure 4.1, plate 3), but forecasters also regard equations of atmospheric dynamics as being conceptual models. Ironically, it is some psychologists who have not been comfortable with the notion of a mental model precisely because it is mentalistic and subjective. This stance is a reflection of psychology's lingering hangover from behaviorism. In this book, we are unabashedly mentalistic (especially see chapter 10).

The formation of a mental or conceptual model is just one element within a larger process of forecaster reasoning. Thus, the reasoning models include "mental model formation" as one element or subprocess. But there are additional kinds of "models" that have to be considered. We have *reasoning models* that describe forecaster reasoning (chapters 4, 10). These describe sequences of data examination, hypothesis formation, hypothesis testing, and so forth. Some such models have been discussed by forecasters, whereas others come from psychological research.

Finally, we have *computational models* that meteorologists develop to mathematically express the dynamics of the atmosphere and thereby generate predictions of the weather.

This may seem like a lot of subtly different but substantively interrelated models, but they are all necessary considerations for our topic. Across the chapters of this book, we are explicit about which sort of model is under discussion at any given point.

Sources of Information Concerning Forecaster Cognition

Our empirical understanding of the cognition of forecasters comes primarily from sources that form the organization of the chapters in this book:

- Analysis of the forecasting workplace (chapter 2),
- Studies on the question of how people come to be forecasters (chapter 3),
- The literature on meteorology in which atmospheric scientists describe their own reasoning and strategies (chapter 4),
- Research in which the quality of forecasts is evaluated (chapter 5),
- The literature on the nature of expertise (chapter 6),
- Studies of how forecasters come to be expert forecasters (chapter 7),
- Cognitive research on forecaster knowledge (chapter 8),
- Cognitive research on forecaster perceptual skills (chapter 9),
- Cognitive research on forecaster reasoning (chapter 10),
- "Expert Systems" designed to imitate forecaster reasoning (chapter 11),
- Computer models of the atmosphere that are used in forecasting (chapter 12).

Culminating this book are two chapters that present the concept of *human–machine interdependence* (in contrast to "man-versus-machine" viewpoints), and prospects for further research and progress regarding the development of forecasting expertise and the forecasting workforce.

2 What Is the Forecasting Workspace Like?

The forecaster sits at the center of a web of information gathering equipment, absorbing and integrating the often conflicting information as it arrives, and from this distills the essential ingredients used to produce a forecast (Targett, 1994).

Of particular relevance to the question of expertise at weather forecasting is the nature of the forecasting workstation technologies and computer-driven visualizations of weather data. We hinted at this in chapter 1 in the discussion of the data overload problem, and we present more details here because we will have need to refer to these technologies and visualizations in the subsequent chapters of this book.

Evolution of the Modern Workstation-based Workplace

The forecasting workspace layout has influences not just on forecasting procedures but on the process by which forecasters learn how to make forecasts (LaDue, 2011). The workspace at both governmental and commercial forecasting services is largely open, usually with a central desk, intended to facilitate collaborative sensemaking and forecasting. Traditionally, the workspace included a chart wall arrayed with maps and many clipboards. Each clipboard would have printouts of one or another data type, which forecasters could page through to see changes over time. Examples are shown in figure 2.1.

The Rise of the Workstation: AFOS, McIDAS, and PROFS

The NWS relied for many years on a workstation system called Automation of Field Operations and Services (AFOS; Giraytys, 1975; Wilkins and Johnson, 1975). It was a multiconsole workstation system that could store and display weather observations and the outputs of computer forecasting models; it also supported communication

Figure 2.1
Traditional chart walls in weather forecasting workspaces (photos by R. Hoffman).

among facilities of the National Oceanic and Atmospheric Administration (NOAA), of which the NWS is a part.

Figure 2.2 is a photo of the forecasting workstation at Rockefeller Center in New York City, taken in 1992. This picture shows one of the first AFOS systems (the rightmost workstation), still in use at that time, and to the left shows the radar "scope" and related displays of the previous generation of technology, essentially Korean war-era technology.

A photo of an expanded AFOS workstation is shown in figure 2.3. AFOS had one terminal for the display of alphanumeric data, a display for communications, and a display for graphics. The dominant theme in AFOS was multiple, large, and (very) heavy steel cabinets.

AFOS was taken a step further by technologies created by the Program for Regional Observing and Forecasting Services (PROFS). Partly related to the limited graphical

Figure 2.2
The weather forecasting workstation at Rockefeller Center, New York City, 1992.

display and manipulation capabilities of AFOS, new forecasting workstation technology developed in the PROFS program was operationalized in the early 1980s with the purpose of providing additional services and forecast products, especially satellite image products (Brundage, 1986; Reynolds, 1983). The dominant theme for PROFS was the computer-based workstation, but still relying on cathode ray tube display technology. Figure 2.4 shows a PROFS workstation. PROFS enabled the storage, display, and dissemination of many hundreds of data products (see Wilkins and Johnson, 1975).

Schlatter (1985) described the forecasting process from the perspective of decision making and saw the need for forecasters to refine their ability to perceive meaningful patterns in data. He regarded the (then-new) workstation technology as providing a significant opportunity for forecasters to anticipate developments at the synoptic scale (continental scale of weather spanning days to weeks). Before long, however, it became clear that the workstation approach:

> as it has evolved, makes it difficult if not impossible for the forecaster to assimilate and weigh all the available guidance information properly … and discourages him from using his training and experience to depart from this guidance. (Golden et al., 1978, p. 1336)

Following the introduction of PROFS, even greater advances in the information sources and systems became available to weather forecasters.

Figure 2.3
The AFOS workstation, the NWS workhorse for nearly 30 years (photo by R. Hoffman).

Beginning in the 1970s, researchers at the University of Wisconsin developed a computer-based weather analysis and forecasting workstation called the Man Computer Interactive Data Access System (McIDAS). At the beginning of the project (Chatters and Suomi, 1975; Smith, 1975), it was recognized that forecasters and meteorologists needed a new tool to support satellite image processing—WFOs were still getting satellite imagery via facsimile. It was also recognized that forecasters needed to spend their time conducting analyses and not learning how to operate complex computational equipment. Hence, there was an eye toward what we would call user-friendliness. The innovations associated with McIDAS included the use of color displays (using color to enhance monochrome displays), the ability to rapidly display sequences of images (satellite image loops), and capabilities to support interactive analysis (e.g., measuring the motions and heights of clouds), with the computer automatically accessing requested data, conducting the analysis operations, and displaying the results.

Initially, McIDAS did not live up to its initial dream; it was regarded as clumsy and difficult to use primarily because of its reliance on a command-line interface using cryptic command codes (see Doswell, 1992). Hence, there were waves of revamping

Figure 2.4
A NOAA-NWS operations facility circa 1980, based on the PROFS technology (photo by R. Hoffman).

(see Suomi, Fox, Limaye, and Smith, 1983). McIDAS in its fifth version release [https://www.ssec.wisc.edu/mcidas/] was installed at dozens of locations worldwide, including a number of NOAA and U.S. Air Force installations. The main innovations introduced by McIDAS were carried over into the workstation developed for NOAA's Program for Regional Observing and Forecasting Services, known as PROFS (see Schlatter, Schultz, and Brown, 1985)

Another workstation system, called Advanced Weather Interactive Processing System (AWIPS; Brundage, 1986; Bullock et al., 1988; Lee, 1997; Lusk, 1993; Lusk et al., 1999), was designed to integrate and organize various data types (radar, observation charts, etc.) and soon began replacing the AFOS-PROFS system. A photo of the AWIPS prototype workstation is shown in figure 2.5. The innovation for AWIPS was geolocation: showing all the various data fields (winds, temperatures, etc.) as "layers" projected on a common map.

Figure 2.5
A prototype AWIPS workstation (photo by R. Hoffman).

In a test of the AWIPS workstation conducted over a 4-month period, forecasters at the Denver NWS office requested and examined more than100,000 displays involving literally thousands of different types of products (i.e., different types of weather observations, various sorts of radar images, maps, etc.) (Roberts et al., 1997). AWIPS provided capabilities to generate forecasts in text form and supported communication among NOAA facilities. These were traditional functions carried out using the AFOS system. A main new capability was to permit the overlay of diverse data types—literally hundreds of different types of data sets and products (including satellite images, observational charts, radar and MOS, to name just a few)—and to easily animate and zoom these. This suite of capabilities was frequently used and was generally highly rated by operational forecasters (see Roberts et al., 1997).

Compared to previous systems, AWIPS was more user-friendly because it was iterated over a long period (about 15 years), originating in the design of PROFS (Brundage,

1986) and then evolving into a prototype that was refined based on testing in the WFO field setting (Bullock et al., 1988; Kucera and Lusk, 1996; Lusk, 1993). "[C]are was taken to make the system user-friendly" (Bullock et al., 1988, p. 70). The main method was re-prototyping based on end-user feedback (see Maximuk, 1997), which resulted in some salient developments. For instance, the ability to combine various data types in a single display (e.g., surface pressure, wind, and vorticity) and animate the display proved to be "popular" among NWS forecasters (Bullock et al., 1988). Problems were also discovered in field-testing, especially the loss of legibility of alphanumerics during zooming.

One reason that the AWIPS prototype was positively evaluated in the field-testing phase was because its capabilities for displaying NEXRAD data represented an improvement relative to the AFOS system. Up to that point, NEXRAD had its own workstation called the Principal User Processor (PUP). AWIPS made it easier for forecasters to actually use the radar data in their forecasting operations (see Maximuk, 1997). The PUP, shown in figure 2.6, suffered from a number of poor human factors design aspects. Nevertheless, once forecasters could view NEXRAD images from their main workstation, the PUP came to be used less often and then primarily for fine-tuning of the algorithms and other special functions.

Related to AWIPS was another highly advanced system called WFO-A because it had been created for use at NWS WFOs (Grote and Bullock, 1997) and because it was "advanced," that is, based on lessons learned from AWIPS and other workstation projects (Bullock et al., 1988)—even though AWIPS had become operational. The WFO-A interface design philosophy included an emphasis on functional organization and simplicity for use. For example, unlike previous workstation systems that necessitated the use of multiple cryptic command-line entries to perform operations such as animating a satellite image display, in WFO-A, such operations could be performed with single mouse clicks. Demonstrations and evaluations of AWIPS served to show which operations and functions were important to forecasters, and so WFO-A supported fast access to frequently used information. Additional new capabilities of WFO-A included a browser to support the selection of models, data types, and data sets, and a product maker that permitted the creation of special graphic products that combine any of a number of numerical models and data types. Additionally, the technology had come a long way from command-line interfaces and, with WFO-A, involved command windows and icons.

With each generation of technology, the human factoring was more explicit and thorough, insofar as human factors considerations were taken into account from the onset, a new design philosophy. Furthermore, the human factoring was more thorough

Figure 2.6
The NEXRAD Principal User Processor of PUP.

insofar as systems were created through a process of consensus design, in which teams composed of forecasters and end-users as well as computer scientists collaborated in prototyping, evaluation, and then re-prototyping (e.g., Sanger et al., 1995).

The human factoring was also more thorough in that the evaluations did not rely on a simple satisficing criterion (i.e., it is "good enough"). In much software and system development work, once a new system was built, it was presented to end-users who would work with it for some time and then provide a subjective evaluation in what is called "usability analysis" (see Bias and Hoffman, 2013). Because newer systems were invariably more clever, more capable, and fancier than older systems, subjective ratings typically showed that new systems were better overall. Cognitive dissonance and demand characteristics certainly played a role in such evaluations: A person who is evaluating a new technology is hardly likely to assume they are being shown a badly designed system. A developer or user who is involved in the system development effort that led to a new workstation is not likely to respond to a satisficing evaluation by saying strongly negative things. A more thorough evaluation includes subjective evaluations but also a formal evaluation of performance (e.g., failure in an attempt to display a product). It included an evaluation of performance during training evaluations (e.g., how quickly can forecasters learn to use each system function) as well as performance in an operational context (i.e., weather briefings and/or forecasting operations) (e.g., Bias and Hoffman, 2013; Grote and Bullock, 1997; Sanger et al., 1995; Lusk, Kucera, Roberts, and Johnson, 1999). Evaluation of new workstation systems during and immediately after training was critical (as opposed to attempting to examine performance after longer periods of practice and use) because users could learn to work with any poorly designed interface with sufficient practice.

Prototyping for meteorological information processing systems never ceases, it just sort of trickles off (Ballas, 2007; Doswell, 1990). Even after commissioning and

Box 2.1
Cognitive Systems Engineering Implications for Procurement

The discipline of cognitive systems engineering emerged from traditional human factors engineering in response to changes in the modern workplace: work became more cognition-intense and computer-dependent (see Hollnagel and Woods, 1983). Studies of diverse cognitive work systems such as NASA Mission Control, air traffic control, and emergency response demonstrated the importance of involving the intended end-users in the design process. Most weather forecasting technologies are designed and developed by teams that include meteorologists and forecasters along with the technologists and engineers. One example is McIDAS, whose initial design concept was developed at the University of Wisconsin (Chatters and Suomi, 1975). A subsequent improved version was adopted by the National Severe Storms Forecast Center. Another example is the NEXRAD network. Meteorologists were involved in the research and development activities conducted at the National Severe Storms Laboratory and by the U.S. Air Force (Forsyth et al., 1985) and subsequently commercialized by private sector. Two new technologies currently in development are phased array radar, a potential replacement for the NEXRAD network (Zrnić et al., 2007), and Probabilistic Hazards Information tool (Karstens et al., 2014). Both of these technologies are undergoing testing by forecasters to ensure that their designs best serve the weather community (e.g., Bowden et al., 2016; Heinselman et al., 2015). This deep and continuous involvement of meteorologists and forecasters in the procurement process (designing, protoyping, testing, re-prototyping, and operationalizing) of information processing and workstation technologies sets meteorology apart from some domains for which the government supports large-scale procurements. Failure to engage "end-users" and, more broadly, failure to fully integrate cognitive systems engineering methodologies accounts for a number of procurement failures in which extremely expensive information processing systems were procured and only later found to be lacking in usability, usefulness, and understandability (see Cooke and Durso, 2008; Hoffman, Cullen, and Hawley, 2016; Hoffman and Elm, 2006; Hoffman, Neville, and Fowlkes, 2009; Neville et al., 2008).

operationalization, systems are continuously refined, operators create kludges at the local level (i.e., WFOs) through means as simple as the use of Post-its© to means as complex as rebuilds, adding new algorithms and adding local programs. The trickling off of the prototyping process is typical of most modern complex sociotechnical domains (Koopman and Hoffman, 2003). Furthermore, new forecasting workstation and software support systems are always being developed for various particular applications (e.g., Ballas et al., 2004).

Figure 2.7 shows one of the current NWS forecasting workspaces. Among the newest technologies is a nationwide lightning detection network that has further aided in forecasting and analysis of severe storms. WFOs are now using wall-mounted, large-format, high-resolution color LED displays, allowing anyone to view satellite, radar, and other products simultaneously from just about any place on the operations floor.

Figure 2.8 shows the watch floor at the U.S. Navy's Fleet Weather Center in San Diego, CA. Fleet Weather Centers provide full-spectrum weather services to facilitate risk management, resource protection, and mission success for fleet, regional, and unit

Figure 2.7
A photo of a current WFO forecasting facility (photo by R. Hoffman)

Figure 2.8
The watchfloor at the U.S. Navy's Fleet Weather Center, San Diego, CA (U.S. Navy photo by Mass Communication Specialist Seaman Bradley J. Gee/Released).

commanders. Comparison with the WFO (figure 2.5) reveals a high degree of similarity in terms of layouts and the data types shown on the main displays.

How Many Displays?

AWIPS initially had only one primary display, and in its next iteration had three primary displays. In the first AWIPS, WFO-A, and other systems, the solution to the issue of "how many displays" was finessed through screen sectoring. A large main sector presented one data set or product while as many as four smaller sectors off to the left side of the screen presented any of a number of other data types or products (see Grote and Bullock, 1997). With a point and click operation, any of the data types could be swapped from one of the smaller sectors to the main larger sector. Was this a good solution to the problem of multiple displays?

In a U.S. Air Force (USAF) Air Weather Service project, Hoffman (1991) conducted nonintrusive observations of 14 weekly synoptic/mesoscale forecasting deliberations conducted by a group of forecasters and research meteorologists at the Air Force Geophysics Laboratory (Hanscom Air Force Base). In these deliberations, the forecaster would present a summary of the weather situation, an analysis of the pertinent dynamics, and a forecast that often focused on mesoscale (regional) weather, but could include salient phenomena occurring anywhere. From the standpoint of psychology, in these sorts of deliberations (called "forecast briefings"), the forecaster thinks out loud, a natural parallel to the think-aloud problem solving method (Ericsson and Simon, 1993) that has been widely used in the experimental study of expertise (see, e.g., Chi, Feltovich, and Glaser, 1981). Hoffman recorded which displays/data types were examined, for how long, and for what reasons.

A main result was that the forecaster who was in a diagnosis or prognosis mode needed to refer to two or three data types/displays per minute. Using a similar methodology, Trafton et al. (2000) found that forecasters who used a computer for forecasting looked at an average of six visualizations per minute, whereas forecasters who used a chart wall looked at an average of three visualizations per minute. Hegarty et al. (2010) reported that forecasters examined an average of eight different displays/data types to generate a forecast. Hoffman recommended that a next-generation workspace should retain the traditional chart wall, on which a variety (dozens) of data type/fields are posted and could be inspected at a glance (see Wilkins and Johnson, 1975).

In fact, there seems to be a huge advantage in using multiple displays *including* chart walls over single-display computer visualizations. Trafton et al. found that forecasters who used a chart wall took 35% less time than forecasters who used a single-display computer system. This rather large time difference was not due to a speed/accuracy tradeoff: forecasters were equally accurate regardless of whether they used either a computer or a chart wall. Jang, Trickett, Schunn, and Trafton (2012) suggested that different interfaces have different access costs, and these access costs have a large impact when forecasters need to integrate information across different data sources.

Note in figure 2.7 that the workstation in the foreground has four displays. Forecasting workstations needed to include at least five displays for general forecasting situations—one display for alphanumerics (i.e., surface observations), one for data fields (e.g., computer models), one for satellite imagery, one for radar, and one on which diverse data types could be overlaid and integrated. For radar, forecasters still rely, and rely heavily, on sectoring the display: A storm or frontal system would be examined from more than one radar, at more than one height, and using different Doppler capabilities (e.g., rainfall, relative motion, etc.). This enabled forecasters to explore the data and develop their conceptual model of storm structure. But it came at a cost in resolution: The smaller sectored fields were of small size and lower resolution (see Maximuk, 1997). The forecaster needed to manage the sectors by point-click-zoom operations because the data types that needed to be inspected were a function of the weather scenario. When specific data values needed to be inspected, one of the previously sectored images could be dragged back over to the large sector, but this meant yet more pointing and clicking. In AWIPS and WFO-A, data types could be downloaded only through the main screen sector, so actual use of the workstations entailed even more pointing and clicking than would be implied by the claim that system functions involve minimized click operations (see Grote and Bullock, 1997). In the evaluation of WFO-A, conducted over a four-month period, forecasters at the WFO engaged in about 10,000 product swaps for just the top ten most frequently examined displays (i.e., forecast maps, satellite images, local observations, radar, and the like) (Roberts, Kucera, Lusk, Johnson, and Walker, 1997). These findings made a strong argument in favor of keeping something like the traditional chart wall, which one can still see in the forecasting workspaces at most WFOs, at most commercial forecasting services companies, and at some TV stations.

As the previous discussion suggests, it is effectively impossible to separate the discussion of workspace and workstation design from the topic of display and visualization design.

Visualization Design

It has become widely recognized that complex data visualization is critical in many areas of science (e.g., Davies et al., 1990; Durrett, 1987; Friedhoff, 1991; Klein and Hoffman, 1992) and in the practice of countless professions, such as radiology (see Lesgold, Rubinson, Feltovich, Glaser, Klopfer, and Wang, 1988) and, of course, meteorology (Harned, Businger, and Stephenson, 1997). Psychological factors became an important consideration in the display of multidimensional data (see Hoffman, 1990; Hoffman and Conway, 1990; Ware and Beatty, 1988), including cartographic data (Bertin, 1983; Curran, 1987; MacEachren and Ganter, 1990; Olson, 1987) and topographic data (Eley, 1988). Psychological research has yielded abundant confirmation of the instructional value of illustrations that preserve dynamical information and link that information to explicit causal explanations (Mayer, 1989; Mayer and Anderson, 1991; Mayer and Gallini, 1990).

The same decades that have seen increased emphasis on data visualization, display technology, and interface design have also seen the creation of literally dozens of software tools for the display and analysis of weather information, especially involving new ideas about the three- and four-dimensional display of weather data (Lavin and Cerveny, 1986; Love and Mundy, 1997; Pearce and Hoffert, 1997). Looking back to the 1980s, as computer and information display technology improved, a number of research teams developed visualization systems that would allow forecasters to choose and integrate whatever variables the forecaster may want to inspect. An example is the "MERCURY" proof-of-concept system, one of the first to use three-dimensional perspectival maps (Fields et al., 1992).

The basic vision of the McIDAS and PROFS (Schlatter Schultz, and Brown, 1985) programs—that a computer system could support the forecaster in a process of data exploration and integration—was preserved in subsequent projects. Lessons learned from PROFS (see Brundage, 1986) and McIDAS (see Bullock et al., 1988; Lazzara et al., 1999) helped inform the creation of a newer generation of workstation systems, largely through a more thorough and explicit consideration of interface design and the human factors of data visualization (see Corbett, Mueller, Burghart, Gould, and Granger, 1994; Grote and Bullock, 1997; Lazzara et al., 1999; Sanger, Steadham, Jarboe, Schlegel, and Sellakannu, 1995).

For example, a system called Zeb was developed at the National Center for Atmospheric Research (NCAR) to support meteorological research, largely through the use of superpositioning of data types (Corbett et al., 1994). For instance, a field of graphic elements called "wind barbs" (which show winds speeds and directions; see Figure 9.1) could be overlaid on a surface map that also showed the locations of radars, with displays from the individual radars shown in a sector below the map; a satellite image could be overlaid with a map of temperature contours, and so on. Looking across all the various systems, capabilities ranged from mesoscale forecasting to specialized synoptic-scale analysis and forecasting (e.g., of lightning maps, hurricane tracks, and marine conditions). The capabilities of these systems typically included data acquisition (from NOAA and other sources, including satellites), data display and analysis, and product generation. As far as user-friendliness is concerned, all of the new systems had online help (not necessarily helpful help) and relied on pop-up menus (not necessarily friendly) and direct-manipulation icon systems (also not necessarily friendly), although interaction could also be at a command code level for individuals who were more familiar with system operations and functions.

A great many specialized weather information display systems have been developed at universities, by commercial firms, and for civilian and military forecasting in a host of countries, including Australia, the United Kingdom, Canada, France, Germany, as well as the United States. Some systems, such as the Cloud Scene Simulation Model

(Raffensberger, Cianciolo, Schmidt, and Stearns, 1997), were designed to support training rather than operational forecasting. For some systems, the displays and products were electronic versions of traditional meteorological charts. Technical reports on new visualization systems emphasized the utility of the display and data integration capabilities and the usability of their interfaces (see, e.g., Bullock et al., 1988; Jesuroga, Drake, Cowie, and Himes, 1997; Kelly and Gigliotti, 1997; Steadham, Swartz, Schlegel, Roberts, and Hoffman, 1997). Commercial products include the Advanced Meteorological Image and Graphics Analysis System (AMIGAS) of Control Data, The Automated Weather Distribution system of GTE, the RADAC 2100 radar display and analysis system, the SURECAST forecasting support system, and the TRIMETS display system all of Kavouras, Inc., the WEATHER systems of WSI Corporation, and various systems and displays created by The Weather Channel. The proliferation of websites that provide weather information has stimulated research on the design of visualizations specifically for the web, web page usability, and the understandability of atmospheric data by the general public (e.g., Oakley and Daudert, 2016).

Many private sector weather companies have developed 3-D and 4-D display systems that take government- or research-focused systems to new levels because their focus also includes the public and other end-users. We exemplify some of the most recent developments in weather data visualization by describing the Perceptual Rule-based Architecture for Visualizing Data Accurately (PRAVDA) system created by a collaboration of scientists from the NOAA Forecast Systems Laboratory and IBM (Rhyne et al., 1992; Rogowitz and Treinish, 1993, 1996; Treinish, 1997, 2000, 2002; Treinish and Rogowitz, 1997) and by describing Met.3D, an open-source tool for the interactive three-dimensional visualization of the predictions generated by computer models (Rautenhaus et al., 2013, 2015a, 2015b).

PRAVDA and Met.3D

One of the striking capabilities of PRAVDA was to go from radar data to generate a four-dimensional picture of cloud structures, in which translucence and desaturated (pastel) hues are used to depict, for instance, isosurfaces of radar reflectivities (i.e., cloud water density), overlaid on a depiction of the terrain using a computer-generated map. Laid over the map are vectors showing wind speeds and directions, as well as colored regions showing total precipitation. In displays of this sort, a number of data types can be depicted in a single display, in a way that supports pattern recognition and also the precise analysis of numerical data. Examples appear in figures 2.9 (plate 7) and 2.10 (plate 8).

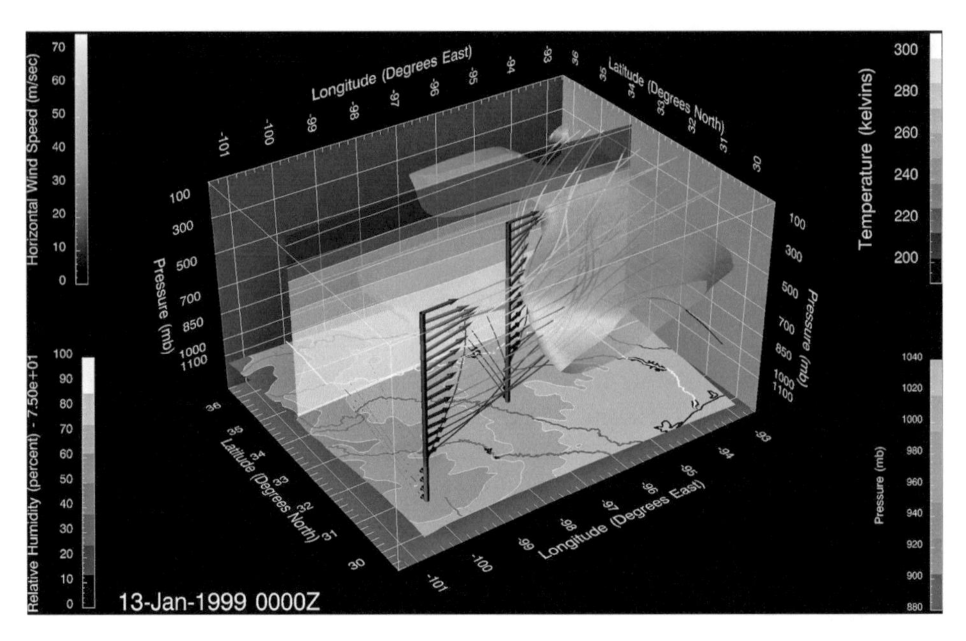

Figure 2.9
(plate 7) An example PRAVDA display (courtesy of Lloyd Treinish, IBM) [http://www.research.ibm.com/dx/bonuspak/html/bonuspak295.html].

The display techniques used in PRAVDA were based on perceptual principles (Rhyne et al., 1992) and relied on a human-centered strategy for visualization based on the need to preserve the fidelity of the original data and the need to take into account known facts about human perception and cognition. The PRAVDA project was initially conceived to help forecasters cope with the data overload problem. The approach to solving the problem was based on the belief that properly designed displays could support the interpretation and integration of weather data via perceptual capabilities rather than through the sorts of analytical thinking that is mandated in traditional displays (i.e., the need to inspect and analyze multiple, static, two-dimensional contour maps). At the same time, the system had to preserve the fidelity of the data, especially observational data, and it had to define coordinate systems onto which different data types could be registered across time in a topologically invariant space—all to permit precise analysis.

PRAVDA also served as an example of the role that human-centering considerations could play in display design. They created an advisory tool for the specification of appropriate color-to-data mappings depending on whether the goal of visualization is exploration or presentation. PRAVDA included a rule base and a library of color maps

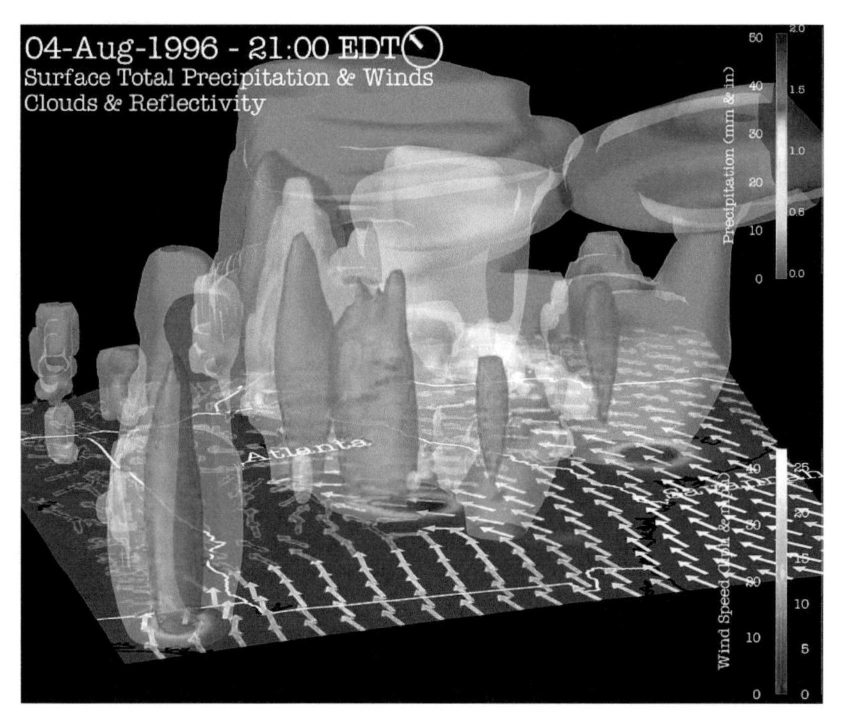

Figure 2.10
(plate 8) An example PRAVDA display (courtesy of Lloyd Treinish, IBM) [http://www.research.ibm.com/dx/bonuspak/html/bonuspak295.html].

that together permitted users to make decisions about the visualization of data without requiring them to become experts in human vision, data structures, visualization algorithms, or color theory. In other words, PRAVDA placed the visualization design process in the hands of the meteorologist. The rule-base ensured that data content was reflected in the image displays and that perceptual artifacts were not erroneously interpreted as data features (e.g., the artifacts that often occur in the use of the standard cartographic palette of highly saturated primary hues—the "rainbow" code; see Hoffman et al., 1993). This was accomplished by including in the color map library a set of specifications for mapping as a function of the scalar nature of the data (i.e., ratio, interval, monotonic) and specifications based on psychophysical scaling data on color discriminability (e.g., S. S. Stevens, 1966). A result of the rule-based mappings was that luminance and saturation were determined by the spatial frequency of the data that were to be depicted (e.g., mesoscale humidity is a low spatial frequency data type, whereas radar reflectivity is a high spatial frequency data type). For example, in a map for a monochrome display, a monotonic increase in pixel luminance goes hand in hand with a monotonic increase in perceived magnitude. Thus, the resulting display makes details apparent, details that can be masked by traditional uses of color (see figures 1–12 in Treinish and Rogowitz, 1997). (Although not a focus here, it is important to recognize that color vision weakness affects about 10% of the U.S. population, mostly male. It can affect both forecasting and dissemination systems.)

The PRAVDA system also provides advice on representation depending on whether the goal of visualization is exploration, analysis, or presentation. Specifically, the developers of PRAVADA provided guidelines for how to collapse multiple variables and data types into individual displays and guidelines to support the user in defining coordinate systems onto which data may be registered in space and time. One of PRAVDA's perspectival displays portrays horizontal winds (using a color palette of saturation shades of violet), relative humidity (using saturation shades of brown), surface temperature overlaid on the base map (using a two-tone palette of saturation shades of blue and green-blue), and air pressure (indicated in a semi-transparent vertical plane using saturation shades of blue-violet and green and a palette of saturation shades of green-blue and blue-green). Also depicted are three-dimensional cloud structures. For all of their graphic products, the use of perspective, depth pseudoplanes, and animation permits the perceptual discrimination of the multiple variables (images can be viewed at http://www.research.ibm.com/people/l/lloydt/).

Met.3D

This open source software system was created for the purpose of allowing meteorologists to explore the outputs of computer models developed at the European Center for Medium-Range Weather Forecasts (ECMRF). The primary application has been to aviation forecasting. The system enables users to explore the contributions of individual computer models to a merged or ensemble forecast, especially with respect to uncertainties that are inherent to computer modeling. Figure 2.11 (plate 9) illustrates how MET.3D coordinates multiple perspectives. Figure 2.11 (plate 9) also shows wind speeds at various heights, with a vertical cross-section that employs colors to show potential temperature. Ordinarily, winds speeds are depicted in a two-dimensional chart winds as a particular height in the atmosphere, drawn over a surface map. In MET.3D, the cross-section can be interactively moved by the user to explore changes in wind speed at heights relative to the surface map.

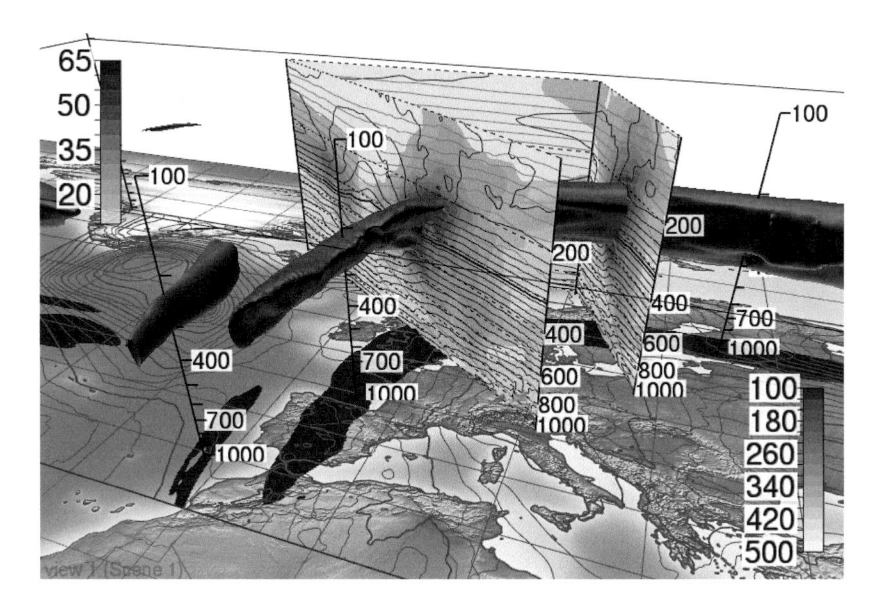

Figure 2.11
(plate 9) An example display in MET.3D (courtesy of Marc Rautenhaus, Technische Universität München).

Conclusions

Over the decades in which forecasting became dependent on computational systems, the human factors of workstations and visualizations became well understood (Eggleston, Roth, and Scott, 2003.;Hoffman, 1987a, 1991, 1997; Hoffman et al., 1993; Scott et al., 2005). Despite the need for standardization, there remained a critical need for flexibility. Onsite (or regional) expertise should be utilized to the fullest extent to craft "locally tailored" variations on any standardized workstation system, its display default features, menus, and so on (Doswell, 1992; Lazzara et al., 1999). Human–computer interaction should be guided by the use of human-centered interface and menu schemes (Steadham, Swartz, Schlegel, Roberts, and Hoffman, 1997). The visualization software needs to permit the concatenation of diverse data types: System capabilities that aid synthesis include animation and looping, re-mapping so that various data sets (e.g., satellite and radar) can be superimposed on one another, and three-dimensional perspectives of selected data sets (Bullock, 1985). Features from diverse data types need to be combined (e.g., an area of strong convection shown on a satellite image and an area of high winds taken off a radar image) through such operations as point and drag (e.g., clicking on a feature such as a frontal boundary or a low-pressure center symbol and dragging it onto a conceptual modeling screen).

Above all, the technology needs to support the forecaster's process of forming a mental conceptual model of what is going on in the atmosphere:

the use of conceptual models during the hypothesis step can greatly assist the meteorologist in understanding what is currently happening. ... Since there is rarely time to look up schematic representations of conceptual models ... the workstation should support the hypothesis step by storing schematic representations of conceptual models so that they can be displayed on the workstation and compared with current conditions. (Bullock, 1985, p. 4; see also Schlatter, 1986)

A Day in the Life: A Cautionary Tale about Work System Design

It is important to not get stuck in the notion that the unit of analysis is the "one person-one machine" dyad. The cognitive work has to be considered with reference to the larger workplace and the collaborative teamwork activities engaged therein. Forecasting has a strong social component within the forecasting office, as frequent conversations take place while forecasts are formulated (Daipha, 2007; Hahn et al., 2002; Morss and Ralph, 2007). Forecasters incorporate others' knowledge, experience, and interpretations into their forecasts (Daipha, 2007; Fine, 2007; Hahn et al., 2002; Morss and Ralph, 2007). Forecasters get direct, immediate feedback from others, and they conduct "postmortems" to relate data and information available to the actual weather outcomes (Daipha, 2007; Hahn et al., 2002). The true cognitive work is not limited to the mental efforts of the individual or individuals who "work at" a workstation. This was highlighted in a study of the watchfloor operations at the weather forecasting facility at Pensacola Naval Air Station (Hoffman, Coffey, and Ford, 2000). Hoffman and his colleagues observed watch floor operations over a period of about a year. The actual work of the forecasters—filling out weather information on a form used to give preflight information to pilots and pilot trainers—was often disconnected from the "true work" that occurred when pilots came into the weather facility to talk to the forecasters.

Pilots (and pilot trainers) would often come to the forecasting facility, even though they could get the preflight weather information over the local Internet). As revealed in the in-depth interviews Hoffman et al. (2000) conducted, pilots sought guidance *in person* from a forecaster in whom they had developed trust. Figure 2.12 shows a pilot in discussion with a forecaster. Note that they can engage (more or less directly) but cannot co-refer to the data displays. They are not referencing the completed preflight

Figure 2.12
A pilot getting preflight weather information from a Forecast Duty Officer (photograph by R. R. Hoffman).

weather briefing form (which is lying on the counter). The pilot is asking the forecaster particular questions (e.g., How bad will the turbulence really be as I make it over the Rockies?).

When workstation and display systems were introduced at this facility, they created a "Wall of Thunder," shown in figure 2.13. The doorway into the facility is just to the left in this image. The advantage is that pilots and pilot trainers could enter the weather forecasting office, turn to their left, and immediately see the weather data displays. But there was a crucial disadvantage: They could not directly engage with the forecaster, who was mostly hidden behind a partition.

The workspace layout made it impossible for forecasters and pilots to co-reference the displays. Thus, the recommendation was to change the layout, and this involved re-creating a traditional chart wall (figure 2.14), which had been removed when the Wall of Thunder was emplaced. This rearrangement also allowed the forecasters to work adjacent to the pilots and pilot trainers.

Figure 2.13
The "Wall of Thunder."

Figure 2.14
The watchfloor layout was rearranged to enable forecasters and pilots to reference charts and displays during the in-person briefings.

3 How Do People Come to Be Forecasters?

> A good forecast is a necessary precursor to the myriad of decisions that individuals, corporations, organizations, and governments make, yet despite its importance, forecasting is rarely explicitly taught and there is an absence of literature describing how one learns to forecast. (LaDue, 2011, p. 2)

Many meteorologists and forecasters explain that they developed an interest in weather in childhood, often by experiencing striking events such as tornadoes or severe storms. In one way or another they get hooked, and weather events become salient to them (Stewart, 2009). They routinely watched TV weather reports, watched the skies above, and read book after book about weather. Some built their own weather stations and/or started keeping records of wind, temperature, precipitation, and other weather variables. Some even started to chase storms. We know of a number of meteorologists and forecasters who were making reasonably good local area forecasts even while they were in high school. Some of these became their school's morning weather person or even wrote weather forecast columns for local newspapers.

The National Weather Camp Program (Morris, Mogil, and Tsann-Wang, 2012) brings together students with interests in weather and related sciences into commuting or residential settings for one- or two-week-long programs. Whether middle school or high school level, students are almost always amazed that so many others share their weather interest; almost all students have either experienced a life-altering weather event or have become been a TV weather junkie. Many weather campers go on to college and enter into weather or related programs.

It is common for farmers, pilots, fishermen, mountain climbers, and others to develop forecasting skills by virtue of long periods of exposure to the elements while working outside or pursuing their main interests (Hontarrede, 1998). An observant person might notice, for example, increasing numbers of puffy low-level cumulus clouds that begin to extend taller than they are wide—a good signal that the atmosphere is becoming conducive to thunderstorm development. The air might become still as the cloud base above becomes dark—a good sign you are under an updraft of a storm, and rain or lightning might be imminent. Forecasting skills may be developed through perceptive observation, correlating events together in time and remembering previous evolutions or outcomes of observations. The atmosphere provides many clues.

Countering such early developing and widespread intrinsic motivation for understanding and predicting the weather has been the way in which forecasting training has been approached.

Historical Background

Historically, there has been a disconnect between academic education in the science of meteorology and on-the-job training in the profession of forecasting.

Just when meteorology was becoming a science in the mid-eighteenth century, the British government nearly banned forecasting (Hontarrede, 1998). British scientists had been pressuring their government to stop what they saw as an activity similar to that of astrologers and charlatans. Yet even while being disdained by some scientists into the mid-1850s, forecasting was becoming an important application of meteorology for maritime activities. Meteorology emerged as an academic discipline in the 1940s, when a handful of graduate programs were established (Allen, 2001). Historical writings on the founding of meteorology departments at colleges and universities (e.g., Koelsch, 1996) described a struggle for an identity for the discipline—was it a natural science or a physical science? Was academic meteorology the place for forecasting?

Forecasting became more sophisticated after development of the telegraph and other long-distance means of real-time communication meant that weather maps could be constructed. World wars benefited from training and utilization of forecasting skills, after which military forecasters sought civilian applications for their skills (Spiegler, 1996). Over this period of approximately 100 years, forecasting became an increasingly legitimate activity and useful for a variety of purposes. Schools temporarily shifted their focus to prepare forecasters for operations during World War II (Allen, 2001), but how they did so was not well documented. Yet there was still opposition from the academics. In an address to the World Meteorological Organization, Baum (1975) asserted that forecasting was an application of the science and therefore outside the purview of the university. "Many research meteorologists pursue an entire career without ever having to analyze a real weather map" (Doswell, Lemon, and Maddox, 1981, p. 985). The assumption made by colleges and universities was that "...the student seeking a career in forecasting generally is regarded as one who should terminate his or her education at the bachelor's level" (p. 985).

Knox and Ackerman (2005) conducted a survey of 750 students taking introductory meteorology at two major U.S. universities. The sample was demographically representative of the student populations of the universities, although many students were taking the course to satisfy a science requirement. The questionnaire asked, "What specific question about the weather and climate would you most like to have answered in this class?" Interestingly, the most frequent response was how to do weather forecasting. Various forms of severe weather (e.g., tornados, floods, etc.) combined accounted for only about 25% of the responses, but after forecasting tornadoes was the second most frequently cited interest. "The strong interest in forecasting rivals students' better-known fascination with severe weather" (p. 1434).

Forecasting contests are a common and favorite activity, and many universities support them.

Forecasting Training within Meteorological Education

Although local collegiate forecasting contests had been held prior to 1970, an intercollegiate forecasting contest was started by the mid-1970s (Meyer, 1986). Studies of the results from the contests have generally shown that enthused students can achieve reasonable levels of forecasting skill quickly. In an experiment that investigated learning during a forecasting contest, students showed a significant rise in skill for precipitation forecasting on days when they had to write a forecast discussion (Market, 2006). Apparently, students can achieve a surprising level of skill and do so more rapidly than faculty expect (Sanders, 1973).

Roebber and Bosart (1996b) examined data from a forecasting contest conducted at the State University of New York–Albany over the years 1988–1992, in which groups of 10 to 20 students and meteorology faculty created forecasts of daily high and low temperatures and Probability of Precipitation (POP) for about 66 days in each of the two academic semesters each year. Each student was placed into a high- or low-experienced group according to background (i.e., interest in weather, familiarity with weather data types, etc.). There was a statistically significant difference in the forecast skill scores comparing participants when grouped according to high versus low experience but not when grouped by education level (i.e., students versus faculty). In other words, there was not a significant difference between university faculty and their students, despite faculty having much deeper knowledge of meteorological science, but this needs to be understood in context. The skill scores of forecasts by experienced students and meteorologist faculty were all uniformly high (in the range of 0.98 to 0.91) and did not suffer from conditional bias, that is, any tendency for forecasts to take on extreme values when one or more individual data values (e.g., surface dewpoint, windspeed, etc.) took on an extreme value. The results also confirmed the finding that skill scores of consensus forecasts made in the academic setting can approximate those of the NWS forecasts (see also Bosart, 1983; Sanders, 1986), and that the development of proficiency at forecasting precipitation involves a longer time frame than the development of proficiency at forecasting temperatures.

This study also showed that forecasting skill is determined by experience that happens beyond the baseline provided by meteorological training, that is, experience that occurs in the few years after the apprentice stage, just before the journeyman stage is reached (see chapter 7). The first ten or so forecasts showed high errors relative to the consensus forecast, a break-in period when the forecaster learns the basic forecasting process. In another study, Gedzelman (1978) found that students gain appreciable forecasting skill by the 30th forecast. The skill advantage that develops subsequently seems related to developing a consistent procedure, but also on the ability to recognize when to adjust the computer model outputs based on the weather pattern and know when deviations from a standard forecasting procedure are called for (see also Roebber, 1998).

Roebber, Bosart, and Forbes (1996) examined data from the 1992–1993 National Collegiate Weather Forecasting Contest. In this contest, teams of student forecasters from a number of North American colleges generated forecasts for a range of sites. Roebber et al. (1996) examined errors in daily temperature forecasts as a function of distance from the team home site. Although high experience-level participants (faculty and graduate students) suffered less from moving their forecasts to a distant site, for both high- and low-experience (undergraduate) groups, distance from the familiar site significantly impacted forecast accuracy—the differences in errors comparing distant to familiar sites were half again as much as the differences in errors comparing the high- and low-experience groups. Roebber et al. (1996) concluded that greater experience is reflected in:

- a greater ability to take weather conditions into account in understanding the causation of precipitation (e.g., precipitation due to fronts, troughs, weak warm fronts, upslope winds, intensifying cyclonic activity, etc.), and
- a greater ability to adjust computer model guidance in light of the weather situation, that is, to take computer model biases or limitations into account (these and other abilities of experts are discussed in more detail in chapters 8, 9, and 10).

The results of the research of Roebber et al. (1996) also speak to the importance of understanding the weather down to the level of particular local effects—prevailing and/or seasonal winds and the effects of snow cover in nearby mountainous regions. Using local knowledge, the more experienced forecasters can make better use of the data cues in forming mental models that link observations to the forecasted events via causal explanation. An example would be to explain why showers and thunderstorms are anticipated for a region by saying that a low-pressure system was bringing moisture into the region or by saying that diurnal solar heating will make storms most likely just after sunset [see, e.g., http://www.wpc.ncep.noaa.gov/html/discuss.shtml].

Whereas the Roebber et al. (1996) studies looked at skill development in the context of forecasting competitions, another study examined the development of forecasting skill in the academic context. Bond and Mass (2009) studied the development of skill at daily forecasting on the part of seniors in a course on atmospheric dynamics and thermodynamics that had a forecasting laboratory associated with it. The researchers were able to tap a large data set: courses for the years 1997 to 2007. Over the academic quarter, students made next-day forecasts of ceiling/visibility, winds, temperatures, precipitation probability, and severe storm probability. Available to the students were the data on which forecasters ordinarily rely: surface and upper level observations, satellite images, radar, and outputs from certain computer models (i.e., the Global Forecast System and the North American Mesoscale Forecast System model [NAM]; see table 12.1). The students were prohibited from relying on what are called Model Output Statistics (MOS; see chapter 1): "based on the idea that inexperienced forecasters will use model output statistics as a crutch … using model output statistics likely delays the understanding of how various elements of the weather relate to larger-scale aspects of the atmosphere" (Bond and Mass, 2009, p. 1142). Student forecasts were scored using thresholds and ranges for the various parameters and a correction based on a "persistence forecast" of what would be expected in the current weather dynamics persisted into the next day (e.g., tomorrow's high temperature is likely to be the same as today's if no weather event such as the passage of a front changes things).

In a given academic quarter, skill scores showed an increase, but most of the improvement happened early in the quarter, and there was considerable variability and a marked difference comparing the best and worst performing students. The better forecasters tended to be the students who did better on the tests given in the lecture portion of the course, but again there was considerable variability. The top students had an immediate edge, continued to improve, and by the end of the term had skill approaching that of their instructor:

the typical student requires about 6 weeks or about 25 forecasts to gain basic proficiency in next-day forecasts of clouds, winds and temperatures ... it appears that this proficiency arises from both practice in the drill of forecasting and from the development of local knowledge, that is, of the nature of the weather in particular locations. While the best student forecasters have comparable skill to the instructor during the latter portion of the class, his prior experience gives him a sizable advantage early in the class ... the flat learning curves for [certain regional forecasts] reflect presumably their preexisting knowledge of the [local] weather ... typical students have almost immediate skill at [precipitation forecasts, which] may be attributable to all forecasters relying on basically the same [computer model output]. (Bond and Mass, 2009, p. 1147)

Confirming the results of competitions studies, proficiency was gained by students over the academic term. Forecasting contests are certainly a way to assess student learning at forecasting (Harrington, Cerveny, and Hobgood, 1991), but are they, or should they be an element of meteorological education?

College-Level Education

Meteorology programs are found today in a variety of departments ranging from geography to math, physics, and even engineering. There are about 100 undergraduate and graduate programs (American Meteorological Society, 2003). (By comparison, there are about 300 collegiate programs in mechanical engineering.) As for any academic specialization, the evaluation and development of courses and curricula in meteorology has been a focus of colleges and universities (e.g., Ulanski, 1993). The benefits of exposure to meteorology for general education and skill development are often noted (e.g., Spaid, 1994). The AMS has both education committees and regular conferences on just meteorological education. Suggestive of the perceived importance, the *Bulletin of the American Meteorological Society* regularly publishes articles on educational innovation and has a regular department on Educational Affairs. A search through just the years 1999 through and 2001 revealed a host of publications on meteorology education and innovation (e.g., Brown et al., 1999; Croft, 1999; Gallus et al., 2000; Ibarra et al., 1999; Morss, 2000; Mullendore and Tilley, 2014; Phoebus et al., 2001; Smith, 2000; Takle, 2000). Educators often report on new approaches to instruction specifically in forecasting (e.g., Yarger et al., 2000). Training in computer science and its applications in meteorology has become a salient and important component in curricula (e.g., Koval and Young, 2001).

What about college-level training in forecasting? In a 2000 presentation at the New York Academy of Science, Joe Bastardi worried about the state of forecasting education from his perspective as a senior forecaster at AccuWeather. He had seen generations of forecasters just entering the workforce and had noted their training shortfalls:

the way forecasting is being taught today is a problem ... [there is a need to] intensify the forecasting emphasis in the curriculum and make it hard. I am not suggesting that we get rid of the math and physics [but] the idea that one can send someone out with 90% of their major course curriculum having nothing whatsoever to do with what they are working on, to me is nuts. (Bastardi, 2000, handout memo)

We conducted a web-based informal survey of the course offerings of undergraduate meteorology programs in existence today and confirmed, to our surprise, that few schools list courses that are explicitly about instruction in forecasting. Some of the few undergraduate institutions that have courses on forecasting have their students take those courses even before much of the science of meteorology has been learned. For example, St. Cloud State requires only an introductory meteorology course as prerequisite to its forecasting course. Iowa State University went so far as to incorporate a forecasting activity into an introductory meteorology course taken by nonmajors (Yarger et al., 2000). The University of Oklahoma encourages meteorology students to start forecasting as freshmen through the student-run Oklahoma Weather Lab [http:// http://owl.ou.edu].Although the latter school uses forecasting as a way to maintain students' interest in meteorology during the time students must take several necessary prerequisite mathematics and physics courses, Yarger and his colleagues used forecasting in a different way: to encourage problem solving, collaboration, and communication among students in the course. All these collegiate cases suggest that an ability to anticipate weather changes does not require an extensive background in the science of meteorology, and forecasting experience and practice contribute to the acquisition of knowledge in meteorology.

Thompson (1987) pointed out in an address to the 67th annual meeting of the American Meteorological Society that much of what is taught in university courses comes from research, particularly that of the professors. Especially on the graduate level, Thompson said, today's students become tomorrow's researchers not tomorrow's forecasters. Many of those researchers go on to work in universities, and the result is a continued focus in university courses on the latest scientific endeavors. Most atmospheric science programs do well on the science side of things, but the forecasting side (and especially connecting the two to make scientific forecasting possible) seems ad hoc and idiosyncratic. Meteorology programs are not placing much emphasis on forecasting—that is, when universities attempt to teach it at all.

We searched the web pages of major universities known for their meteorology programs and found course listings for eleven of them. (Finding course listings is actually a nontrivial endeavor.) A summary is presented in table 3.1. There was considerable consistency in the listings of core requirements, introductory courses, and

Table 3.1
Examples of courses listed in college- and university-level programs in atmospheric sciences or meteorology

Prerequisite Courses	Reading & Writing Calculus and Analytic Geometry Linear Algebra and Differential Equations Physics Statistics Computer Science Programming General Chemistry Remote Sensing Science Ethics
Meteorology Introductory Courses	Atmospheric Science (Introduction) Weather Analysis and Forecasting (Introduction) Introduction to Meteorology
Meteorology Courses	Climatology/Climate Change Microclimatology Instrumentation Oceanography Air Pollution Meteorology Atmospheric Thermodynamics Dynamic Meteorology Synoptic Meteorology Mesoscale Meteorology Physical Meteorology/Atmospheric/Cloud Physics Atmospheric Radiation and Remote Sensing Computational Meteorology
Forecasting-Related Courses	Current Weather Discussion Weather Forecasting Operational Meteorology/Operational Forecasting

specialist courses, likely because the federal government has specific course requirements for the GS-1340 Meteorology Series [https://www.opm.gov/policy-data-oversight/classification-qualifications/general-schedule-qualification-standards/1300/meteorology-series-1340/]. For some of the programs, the meteorology courses are part of the departments of Earth Sciences or Environmental Sciences. In such departments, it is perhaps not surprising to see a requirement that students take a course on Dinosaurs, Environmental Policy, or Geology, but one wonders why such broad Environmental/Earth Sciences programs would be listed by professional meteorology societies as being programs in meteorology.

Three of the six listed one or two courses specifically in forecasting. Two of these were universities having a close association with an operational Weather Forecast Office of the NWS. Typical of the listings for programs that are more clearly focused on meteorology was a school listing 28 courses of the kind listed in the top three rows in table 3.1 and not a single course in forecasting at all or a program listing 32 courses and only one specifically on forecasting. Only one program listed a course on the history of meteorology. A number of programs list seminars or internships, which are described in such a way as to suggest that they engage students in the forecasting process, but this is left open as when, for example, a course is described as "Supervised practical experience in a professional meteorological agency. Experiences may include providing weather information for radio, TV, utilities, government agencies, construction, or agribusiness." In contrast, one of the universities listed a course on forecasting and also one about weather specifically for sailors and one specifically for pilots. It is fairly clear that those would involve activities of the sort in which forecasters engage. A final note worth mentioning is that courses on forecasting often state that physics and math are prerequisites, but it is arguable as to the extent to which forecasters rely on those areas of knowledge and skill to enable them to generate good forecasts.

Given this disconnect between meteorology and forecasting, it is ironic that forecasting contributes to meteorology and vice versa.

The Interdependence of Meteorology Education and Forecaster Training

Historically, professional forecasters had to develop both knowledge and skill in the absence of theory. Doswell et al. argued that forecasting needed more interaction with meteorology to advance both the science and art of forecasting (see also Ramage, 1978). Although it is arguably true that forecasters can develop and have developed skills without the benefit of deep knowledge of meteorology, it is also true that ideas and hypotheses coming from the experience of forecasters can inform the science of meteorology. For example, military forecasters identified many factors for tornado formation, some of which were subsequently verified by meteorological research. Other examples include civilian forecasters identifying how northwest flow can result in tornado outbreaks, revival of the term "derecho" to describe particularly long-lived damaging wind events, and the conditions under which bow echoes can form from convective storm complexes (Johns, 1993). Some of the factors in tornado formation that the forecasters identified were also shown by meteorological research not to hold true (e.g., dry intrusions of air in the middle levels of the atmosphere; Gilmore and Wicker, 1998).

Not only do meteorology and forecasting inform one another, but forecasting training contributes to meteorology education. Navarra, Levin, and Navarra (1993) taught introductory meteorology to college freshmen by abandoning the traditional lecture format and using a problem-based learning approach in which students worked as teams to pursue questions specifically about weather forecasting. The method was based on what educational psychologists call the "constructivist" theory of learning, which states that knowledge and reasoning skills are actively constructed by the learner and cannot be passively absorbed by rote learning. As a result of problem-based learning, students may do less well than traditionally instructed students on such assessments as multiple-choice exams. However, they can outperform traditionally instructed students in terms of realistic problem solving and data interpretation (see Watkins, 1989).

Navarra et al. (1993) began by asking the college students, "How accurate are NWS forecasts?" Across the semester, each team of students went on to redefine and refocus this initial problem as a way of structuring their investigation. The Navarra, et al. discussion of this project included many details about what students did (e.g., access data from AccuWeather©, generate maps, compare forecasts to actual weather, learn how to use spreadsheets, etc.). The researchers' impression was that learning proceeded in three steps:

1. expansion of the knowledge base (i.e., learning about concepts such as the standard error),
2. crude implementation of the knowledge base, and
3. advancement to a more sophisticated level of understanding.

Our impression is that sharing and critiquing ideas conveys a more realistic insight into how knowledge is developed and acquired In this kind of marketplace setting, the students learn to appreciate criticism and learn from the errors they make (Navarra et al., 1993).

On the one hand, most jobs available to meteorologists with a bachelor's degree are in forecasting or broadcasting. On the other hand, the role of the university is to educate meteorologists generally, not weather forecasters specifically (Baum, 1975). There are many applications of meteorology, and forecasting is just one of them. As one might expect, standardized meteorology exams focus on meteorological concepts, not forecasting methods or methodology (see Davenport, Wohlwend, and Koehler, 2015).

Forecaster Training

Concern about the need for a workforce of highly proficient forecasters reached a flashpoint in 1981 when Doswell, Lemon, and Maddox published an article on training issues, and a number of other forecasters subsequently chimed in. Doswell et al. (1981) asserted that forecasters were undertrained, the available training was outdated (i.e., just making guidance documents available), the training was misdirected (cookbook exercises in the use of technology), and shift work made continuing learning nearly impossible. Doswell et al. (1981) expressed a concern that their article would not elicit any response, but it did. Commentators strongly agreed with the issues, problems, and proposed solutions that Doswell et al. articulated. The schism between research and operational forecasting was uniformly regarded as a key issue. Commentators argued for changes in such things as workload, incentives, and the overall approach that the NWS was taking with regard to training.

There were proposals for new structures within the NWS and new approaches to training (see *Bulletin of the American Meteorological Society*, 1982, pp. 781–786; Grice, 1983). This included more requirements for training in forecasting in the academic meteorology programs and upgrades to the National Weather Service Training Center:

> If we want talented and educated people to go into operational forecasting, and those already in forecasting to continue to advance their education, we must make forecasting expertise and experience a requirement for advancement. (Ellsaesser, 1982, p. 782)

The introduction of the PROFS workstation concept and the NEXRAD radar, the availability of new data types, and new information-processing technologies (see chapter 2) resulted in considerable concern about the forecasting workforce. Significant changes were needed in meteorological education, including a concern for how to best train severe storm forecasters, how to use the new radars, and how to use the computational models (Fritsch, 1992). As new technologies and methods come along, there became a need for training to higher levels of technical proficiency (Hallett, Wetzel, and Rutledge, 1993; Rothfusz et al., 1992). As the WFO system was built up, there was a need for more and more highly trained forecasters. J. Michael Fritsch (1992) said that this "presents the threat of a shakeout in the academic community" (p. 1846).

There are now opportunities for "overall" forecasting training in case study workshops held at conferences of the American Meteorological Society (AMS) and the National Weather Association (NWA). New developments have also mandated a lengthening of the period of education and the creation of new degrees in applied meteorology.

The NWS designates certain individuals at its WFOs and national centers as science officers. These individuals are responsible for training forecasters at local forecast offices, regional centers, and national centers. They spend at least one third of their time developing, delivering, and facilitating staff training that is not already offered through the training branches of the NWS or through the COMET program (Cooperative Program for Operational Meteorological Education and Training). Much of the training concerns the use of individual tools and products, such as how to use the NEXRAD radar.

The NWS's Instruction 20–103 (National Weather Service, 2002) required new interns to complete a Forecaster Development Course [see http://www.meted.ucar.edu/nwp/course/]. The training covers: (1) the NWS organization, its structure, personnel and administration policies and various communication tools; (2) operational instrumentation, data collection, and management; (3) "Numerical Weather Prediction" Software for issuing forecasts, troubleshooting, and so on; and (4) customer service and outreach, computer security, use of email, and so on. The units appear to neglect how to apply this important knowledge about technology to the actual creation of a forecast.

The Forecaster Development Program (National Weather Service Training Center, 2006) focuses on the computational elements required for the task of forecasting, especially the computer models and particular software tools available, rather than the forecasting task itself. In the material on forecasting, there is only one subsection on the forecasting process. The remainder covers atmospheric dynamics, with a heavy emphasis on numerical weather prediction rules and how to use a software tool that defaults to the computer model's forecast. The forecasting process is conceived superficially, in terms of a "funnel" in which the forecaster's understanding moves from the "big picture" of longer term and hemispherical spacetime scales down to shorter-term local scales (more of this description of the forecasting process appears in chapter 4). More positively, the National Weather Service's Radar Operations and the Advance Warning Operations Courses do integrate knowledge with hands-on simulations and problem-solving activities for forecasting on very short time scales.

At each of the 2004 and 2005 annual meetings of the AMS, a forum of approximately 200 members representing the international meteorological community concluded that the changing role of humans in the forecast process made ongoing education and training imperative (Stuart et al., 2006; see also Stuart et al., 2007). Entry-level forecasters should have some familiarity with the forecast process and mechanics of producing forecasts for various sectors of the field. Career-long education should include two elements: the science—including diagnosis and prognosis—and operations, or mechanical production of forecasts elements of the job. As duties shift, forecasters will also need strong communication skills as they become key in the dissemination of forecasts to anyone who might benefit from forecast information.

These training issues and concerns are not limited to professional venues. In their observations of forecasting operations in the U.S. Air Force (as well as the NWS WFOs),

Pliske et al. (1997, 2004) found that training specifically for the forecasting task was deficient. The near absence of formal training on the forecasting task leaves professionals to self-direct the majority of their learning and to learn by doing their jobs. Pliske et al. (1997, 2004) recommended that there should be more "embedded training." This notion emerged clearly in a recent ethnographic study of how people come to be forecasters.

An Ethnographic Study of How People Get to Be Forecasters

LaDue (2011) interviewed a number of forecasters at various stages in their careers. She asked how they developed their forecasting skills and the reasoning strategies they use. She asked about their experience at being mentored and how they self-directed their learning as they navigated their careers. The fundamental question, "How do meteorologists learn to forecast?", was broken down into the following:

- What initiates efforts involved in learning to forecast?
- Why do forecasters make the efforts they do to learn to forecast?
- How do forecasters go about choosing resources and forming strategies to learn how to forecast?
- What is the role of social interaction in learning to forecast?
- What is the role of context in learning to forecast?

Eleven professional forecasters were identified through personal networks. Seven of the forecasters were with the NWS. They had between 1 and 18 years of experience. Four of the participants worked in the private sector. They had between 1 and 8 years of experience. One public sector forecaster did specialty forecasting for a small and specific geographic area in the central plains; another did specialty forecasting at a national center. The others forecasted all types of weather for forecast offices in the western, central, southwestern, and eastern United States. One participant had first worked in the private sector for three years before moving to the public sector, and another had worked in a different profession before returning to a childhood interest to become a forecaster. The private sector forecasters did seasonal or specialized forecasting for specific clients.

Three of the participants were women, and they had only four or fewer years of experience. As is typical for a physical science, women and racial/ethnic minorities remain underrepresented in meteorology as compared to the general population. Where they are engaged as forecasters, they may advance quickly out of forecast positions. Forecasting usually involves shift work, resulting in a tendency for women to move into other positions when they start families (this is one of many manifestations of the need for reconstruction of our nation's science, technology, engineering, and math fields; see http://www.ed.gov/stem).

A total of 101 stories were provided that had sufficient detail to identify how and why learning occurs. The stories the participants told included learning events prompted by curiosity, changes in technology, changes in organizational structure, a request to give a talk to a particular audience, and personal observations of atmospheric anomalies. Other stories were events that would arise during a critical incident, such as investigating some apparent inconsistencies of information during weather events, unexpected damage from a storm that did not appear to be severe, or large errors in forecasted temperature.

The interview transcripts were analyzed using methods described in the literature on qualitative data analysis (i.e., Charmaz, 2006; Corbin and Strauss, 2008; Lincoln and Guba, 1985; Ryan and Bernard, 2000). This included the construction of conceptual models of learning and cognitive processes. Quality of the data and analysis was triangulated through discussion with three Science and Operations Officers (the position responsible for onsite training in an NWS forecast office), forecasters who were not participants, and an individual involved in professional education of forecasters in a private sector company. (In the following discussion, we use pseudonyms.)

The Drive to Be a Weather Forecaster Often Kick-Starts in Childhood

Five of the eleven interviewees had a strong interest in weather before majoring in meteorology in college; four others had an interest in weather and science more generally. One interviewee said he had been interested in storms since he was a "baby" and related stories of pursuing weather topics throughout his school years. Childhood experiences included cloud watching, looking at weather online, participating in weather forecasting discussion boards, getting a weather-related job at a local science center, reading books about clouds and weather, and actively attempting to forecast the weather.

Six interviewees told a number of stories of exciting weather experiences, at observing the same types of clouds and storms they had read about. One forecaster said he had visited his grandfather's farm daily to see what the barometer was doing. One participant retained strong memories of weather from his childhood. His family had moved often, giving him wide-ranging experiences from blizzards to tropical storms and frequent summer thunderstorms.

Four participants described a childhood interest to a lesser extent. However, they also engaged in activities based on their interest in weather: weather projects with 4-H

Box 3.1
Case Studies in Childhood Experiences

Cassie said: "Well, you know, me being a huge weather dork, when we'd have pretty bad weather I'd go outside and I'd look at the clouds." Tyler's kindergarten teacher told his parents she thought he would be a meteorologist after he routinely cut the forecast from the paper and took it to class. He said, "I'm being honest here, for as long as I can remember, I've always loved the weather. So me getting into meteorology was just a natural thing." Mike said he had been interested in storms as long as he could remember. "It probably gave somebody a clue that I was always doing my science fair projects on tornadoes every year." Lisa thought she was around 9 or 10 when she saw a rare tornado for her area: "I was always looking at clouds and telling myself that those types of clouds brought rain, this type of cloud formation was rain. And I knew, of course, what tornadoes and stuff looked like. I was always reading stuff about the weather, about tornadoes. ... There was actually a tornado that touched down [nearby] back then. And I saw this. I was sitting there in my front yard and I climbed up the tree and saw the great big supercell out there, and you could actually see this thing rotating. You could see the rotation and everything in it, and I said, there's got to be a tornado close by! ... And sure enough, they had put out a tornado [warning] ..." Cassie felt a thrill when watching weather. "I'm one of those," she laughed. Mike and Tyler had intense "passion" for weather. Both said they had been interested "for as long as [they] could remember."

clubs, reading books about weather, watching the movie *Twister*, and noticing what the weather was like on days with good surf. Only one forecaster stated that he did not have a childhood interest in weather.

Early experiences with weather allow forecasters to do local, near-term forecasts. Stories told by the younger forecasters comprised most of the instances of this, in part, because interview questions focused on recent learning.

Forecasters Learn How to Forecast on the Job

Human factors psychologists have noted that significant learning occurs on-the-job in many professional domains (e.g., Derouin, Parrish, and Salas, 2005). This is true for the domain of weather forecasting. Writing in 1994, Australian meteorologist Phil Targett said, "While the technical aspects and theory of meteorology are regimented and taught in a formal way, the actual skill of forecasting is acquired by on-the-job training while working with experienced forecasters who pass on their practical knowledge and experience" (p. 48). LaDue's interviews revealed a number of major conditions for learning and means by which learning is conducted. Perhaps the most surprising thing is that forecasters actually have to teach themselves how to forecast. Young forecasters who were interviewed said they avidly read a routine product issued by the NWS called an Area Forecast Discussion, as well as Internet weather discussion boards. Forecasters do get training, but it is not in forecasting. Some training modules that NWS forecasters have to take emphasize a specific level of detail and do not apply generally; other modules reiterate only the simplest, most basic concepts.

As we mentioned earlier, many forecasters grow up doing their own forecasting and participating in forecasting contests. Although forecasters have professors who deliberately integrate forecasting for the local area into courses so that meteorology students can experience first-hand the weather that they attempt to forecast, many forecasters never have the opportunity to take a forecasting course during formal schooling.

Graduate-level courses in synoptic meteorology can build connections between theory and weather, resulting in a deep knowledge base that forecasters can use to forecast, but forecasters have to find those connections on their own. An example of this apparent disconnect between meteorological knowledge and forecasting skill was provided by one interviewee forecaster. During his childhood, he once recorded barometer readings that seemed anomalous and that he could not explain based on his awareness of the major forces that were determining his regional weather at the time. Years later, a professor in one of his meteorology courses mentioned that a distant tropical cyclone can generate pressure wells that move well ahead of its location and generate pressure waves higher than otherwise would be expected. This disconnect means double trouble for budding forecasters: Not only do they have to learn how to create good forecasts, but they have to discover connections between meteorological science and the weather phenomena that are the focus of forecasts. All of the interviewee forecasters told stories of either stumbling onto such connections or having to actively search for them in order to explain what happened to "bust" a forecast, or in other words, to get the forecast wrong.

Individuals who are new to the profession often find themselves in offices that are short staffed, leaving little time to work through training resources for forecasters.

Box 3.2
Case Study in Learning on the Job

Lisa felt that her schooling had left her ill prepared to forecast. She "met all the qualifications [to be a forecaster]," but, she explained, her school did not "have people instruct you and show you the different features and have people instruct and show you the different things that are happening that you've learned [about]."

Some but not all private sector employers have training programs and focused mentoring. Of the eleven interviewees, only one reported feeling somewhat prepared for the job, explaining that his college was oriented toward forecasting and had integrated forecasting exercises into several courses. The participants all spoke of the steep, initial learning curve that forecasters encounter. Younger forecasters sometimes felt they had to create strategies to learn the job and understand the science, at times without help. One of the private sector forecasters worked at a company that had no training program. After a harsh six-month review, he formed a strategy of working through training modules and reading other materials on his own time.

To generalize, the first ten years are a struggle. Given that forecasters have to learn how to forecast even while in the role of "newbie" forecaster, it is not until around three years into the job where they get past the "learning hump"—acquiring the majority of the basic knowledge and skills needed to reliably issue good forecasts.

The 2003 report of the National Academy of Sciences Panel on the Road Map for the Future of the National Weather Service (Gordon et al., 2003) was a major review of an NWS continuing modernization and restructuring effort. In discussing training, the Panel acknowledged that on-the-job training was crucial, taking the new forecaster beyond the training received at the National Training Center, the NEXRAD Operational Support Facility, and/or the COMET program:

> On-the-job training, which takes place informally all the time, will continue to be an important part of the continuing education and training of National Weather Service staff ... a more formal journeyman/apprentice model would have advantages for learning and would also recognize the contributions of the trainer as well as the accomplishments of the trainee. (p. 60)

One of the Most Important Conditions for Learning Is Exposure to Weather

Science officers at Weather Forecast Offices are responsible for training forecasters at local forecast offices. One of the science officers interviewed by LaDue pointed out that forecasters encounter difficulty when moving from one location to another. They have to change their mindset—expecting one type of weather based on their previous regional/climatological experience. Over time, they discover different conditions and seasonal trends. Forecasters are best able to become good at forecasting types of weather they see frequently. Thus, experience with weather sometimes reveals learning needs, just as lack of experience with weather may hide learning needs. Eight of the participants had vivid stories where personal experiences with weather (such as the forecasting contests) allowed them to learn faster or more deeply. Their professors focused contests on local weather so students would both forecast and experience resulting weather first-hand. Two of the interviewees said that they forecasted outside their schooling for their own storm chases, leading to a daylong engagement and immersion in the weather they were attempting to forecast.

Forecasters Learn How to Forecast by Mentoring and Collaborative Learning

Mentoring is also significant for learning. Mentoring activity has been observed to take place in every study in which cognitive task analyses have been conducted in weather forecasting offices (see, e.g., Klinger, Hahn, and Rall, 2007). Learning happens faster when experienced forecasters share their knowledge. A few of the forecasters that LaDue interviewed reported that experienced forecasters had sought them out and initiated a mentoring effort. Forecasters routinely help each other catch up after they had a few days off, and actively share resources they have discovered helped them forecast. All public sector forecasters were involved in teaching each other, either from their job specialty or assigned focal point duties.

The science officer we mentioned previously had a long and distinguished career engaged in mentoring with a considerable number of junior forecasters, and this gave him a special perspective. He knew that a shift in mindset was particularly important, and he would choose cases for training that forced forecasters to work through such situations. He saw the differing approaches to forecasting of forecasters as an asset during training. He valued that they learned from each other during training simulations. He also mentioned that regardless of how a forecaster considered data, there were times when they had to return to—or discover—an empirical basis for the weather in a particular region or at a particular time of day. He had led some of these studies himself, motivated by a deep desire to improve the state of art.

Forecasters are often mentored because they have been hired to replace someone who was retiring, and that person focuses attention during his last few months on helping the replacement learn the job. Forecasters are also routinely mentored by listening to the daily weather briefings given by the lead or senior forecasters in their office.

A number of the interviewees mentioned having benefited from multiple mentors. Mentoring resulted in a deeper conceptual understanding of weather processes and an ability to more quickly focus on the most important data and processes for the particular situation. Mentoring experiences are especially crucial for forecasters who are in the early career stage:

> I can expect—every time I'm on shift with this person—that I'm gonna learn a whole bunch of new things. And it's awesome! The older forecasters know things, they have seen things, they recognize things a lot quicker than you do.

One of the interviewee science officers asked better forecasters to mentor new interns and encouraged all forecasters to fulfill a mentoring role. The science officer facilitated each forecaster's growth into a mentoring role. He said that expertise at forecasting *and* skill at mentoring were to be grown, managed, and seen as a resource. Some forecasters are intrinsically and highly motivated to train others. They go out of their way to develop training materials, document cases of challenging forecast situations, and share them with their juniors. "To me it's just important to pass it on ... of mentoring someone else. But it's something I learned the same way." But not all forecasters are good mentors. One interviewee forecaster remembered how he had once wanted to improve his ability to forecast snow and was frustrated that another forecaster could not articulate what he was thinking as he looked at several parameters and decided on a forecast snow amount. Three forecasters told stories of weather events they or someone in their office was unable to explain, leaving them with an implausible, weak explanation, likely misapplied, that they had heard somewhere before. Complementary to this is the fact that mentors (like exceptional teachers) learn from teaching. Forecasters began engaging in peer learning as they became increasingly competent. Two of the participants said that they found the shift to teaching younger forecasters to be a milestone in their own competence.

Learning Happens Faster When Forecasters Engage in Collaborative Analysis

Weather is complex, so being exposed to others' thinking was enormously helpful in learning to forecast. When forecasters are able to interact with peers and experts, they figure things out faster. For example, the forecasters spoke of learning through the daily NWS Area Forecast Discussion. The three forecasters who spoke of their learning as if it were a solo endeavor still had strong social aspects to their learning; they just tended to speak of their accomplishments in a personal rather than a collective way. The social aspects were revealed through follow-up questions that probed their stories.

As the case for experts in all professional domains, forecasters learn from their mistakes. Every participant mentioned some type of review to figure out what they had missed when their forecast turned out to be deficient. It is not always easy to identify why a forecast was deficient and then determine a plausible explanation for why the busted forecast made sense at the time. Even with the help of experienced colleagues, forecasters cannot always figure out what caused an event, and certain forecasting situations remain difficult because of poor data.

Forecasters told of instances where the review failed to resolve the cause of an event. Some of these stories were from younger forecasters, and in "busted forecast" situations they would often take a passive role in following the efforts of a more experienced forecaster who did the review. The more experienced they became, however, the more likely they were to successfully identify the cause of their missed forecast. They could reason through what they knew, key in on the most important data and concepts, and figure out the causal mechanism to explain an event. They would act on their new-found knowledge, for example, by displaying data in a new way after figuring out what had not been considered in a previous forecast.

Reviews could become quite extensive for middle career forecasters who were dissatisfied with the state of the art. Mysteries were a seed for major, longer-term learning efforts to advance the state of the science. For example, when thinking things through did not come to a resolution, a strategy was to take notes on what they did so that they might later determine the efficacy of various forecast strategies. Most participants' stories involved difficulties rather than ease in connecting pieces of knowledge. Connections were occasionally easy but more frequently needed facilitation and effort.

Not Being Exposed to the Connections between Meteorological Knowledge and Forecasting Reasoning Can be Devastating in Its Implications

One forecaster said she became desperate to learn when she started her career. She described asking every forecaster in her office how they forecasted various things. The other forecasters were shy and reluctant, but they would share, prefacing their help with, "Well, I do it this way, because it is the fastest" or "I do it this way because it seems to work the best." Apparently there was no accompanying explanation of the underlying reasoning. Seeing connections allows for the beginnings of true forecasting skill, when forecasters realize the meaningful relations between observations and future events.

Over Time, Forecasters Develop a Sense of Professional Identity, and This Is Very Important to Them

Although it seems easy to label someone as a weather geek, and for weather geeks to label themselves as such, this is actually the start of the development of a sense of professional identity, years before actual entry into the profession. One of the first and most important things is "affirmation"—some sort of signal from others that the student's interest in weather suggests a life path, a path that is a good one. Interviewees reported that classmates started asking them what the next day's weather would be. Most of the interviewees were affirmed clearly and readily. They said that how others reacted to them in their childhood and early adulthood helped shape their identity

Box 3.3
Case Studies in the Development of a Sense of Professional Identity

Mike and Forest had nearly identical statements, with Mike saying, "Well, if I'm going to answer questions [about what the weather will be], I better actually try to figure out how to forecast!" Tyler got a high school job with a local science center and began forecasting. His boss noted his interest and skill and let him update the center's website forecasts. With Cassie, interest was more than helping her friend. For Cassie, her role became a deep and meaningful part of a friendship: "My friend was so scared. That I just took it upon me to try to calm her fears. ... I felt a strong urge to comfort her in whatever way I could. ... I guess that kind of fueled my interest ... in something I wanted to learn more about [anyway]." Cassie continues in this role today: "Even into our young adult lives, she's still looking for answers from me. It's kind of fun." Janet's friends liked her career choice and thought she should "be on TV." She said to them it was the "holy grail" for a meteorologist and that she appreciated their support. Travis said his parents became excited about his career choice as they learned more about it. Early on the job, Forest, Tyler, Henry, and Shawn benefited from what they saw as personal, high-quality mentoring. Lisa and Cassie described older forecasters responding to their questions and sometimes taking initiative to share explanations and insights with them. Mike said his lack of success in storm chasing during college became a joke among friends, leading to his current conservative forecasting style: He had seen—first-hand—dozens of ways that weather could fail to come together to produce tornadoes. A few participants spoke of affirmation regarding their developing skill. Jordan reported he could often send a specialty forecast without it being checked. Janet had gotten a positive reaction from a local business college, who says her interests in combining meteorology with a business degree are in high demand at the moment.

and resolve. Their interest was persistent, and this was noticed by those around them. Teachers, parents, and friends began calling them the weatherman or weathergirl and asking what the weather was going to be like on a given day. That affirmation by others began a sense of identity, and they were pleased by this.

Two of the 11 interviewees maintained their resolve despite affirmation being misplaced or late in coming (e.g., persisting through a college curriculum in meteorology designed to "weed out" those with poor math skills). All 11 interviewees told stories of how interactions with people who were interested in forecasting helped them see forecasting as a profession they wanted to enter. Through interactions with others, forecasters learned what others value and need. As children, these forecasters wanted to help childhood friends who were interested in or afraid of weather.

Early in Their Career, Forecasters Become Mindful of What Is Most Important to Their Customers

Nine of the 11 interviewees spoke about interactions they or others in their office had had with emergency managers, pilots, departments of transportation, native peoples on tribal lands, and cooperative observers about their need to better understand the impact of weather. Interactions with customers or the beneficiaries of weather forecasts helped them understand the value of their forecasts, which was particularly motivating for learning. Forecasters thrive on it, often considering pursuing advanced degrees in business or economics to better understand customer's needs.

Overall, knowledge of customers' weather information needs is particularly motivating, although of course it can sometimes be a background consideration in the motivation of some forecasters. That being said, all of the interviewees spoke about their work as if it were part of who they were; they have a strong, encompassing sense of identity. Expressions of this sense of identify ranged from self-identifying as being "really" a researcher to self-identifying as a "huge weather dork." Those with the strongest interests refer back to their childhood, as discussed earlier. (All of the interviewees of course had outside interests, but the deliberate pursuit of balance, or the desire to achieve a balance, was only mentioned by one interviewee.)

Forecasters Progressively Deepen Their Understanding

Every interviewee provided at least one example of deepening their understanding. Deepening begins with the childhood experiences, realizing simple causal connections between observations and things that had been learned about weather concepts and dynamics. Once working in a professional setting, the deepening of understanding takes different forms, and not just deepening in the sense of enriching one's knowledge about weather. Forecasters tell stories of having to learn major constructs, new predictive parameters, new forecasting algorithms, phenomena that may be generally known but that had not been taught in school, and so forth. Forecasters always find themselves feeling like novices when they move to some new region and have to learn about new climate and weather tendencies. There is often a new surprise, a phenomenon they have never heard of and do not understand, or some new forecasting strategy or method comes along. Sometimes it is just the interactions between the atmosphere and the local geography that create new weather and climate scenarios.

After just one to a few years of experience in professional forecasting, once competence has been achieved for common processes and weather types, the forecaster's learning moments occur when something unexpected happens. The forecaster can recognize precursors to common phenomena that have major impacts on the weather in areas forecasted, and learning shifts to being caused by surprise when a forecast goes

Box 3.4
An Example of the Deepening of Understanding

Red sky at night, sailor's delight. Red sky in morning, sailor's warning. This simple association is based on the general tendency for northern hemisphere mid-latitude weather systems to move from west to east. High moisture content in the air, due to an incoming storm system (A), can cause a red sky in the morning (B). In contrast, the dust and clouds of departing weather systems can cause a red sunset. This analogy fails rather dramatically, even at mid-latitudes, if the storm system is a westward-moving hurricane.

The magnitude of the pressure gradient at 850 millibars is a good proxy for forecasting strong and gusty surface winds. The stronger the pressure gradient (A), the stronger and/or gustier the winds at the surface (B) would be. This association sometimes works, and sometimes it fails.

Box 3.5
Deepening of Understanding as a Result of Surprise

Forest initially had large forecast errors in high-temperature forecasts along a coastal area where marine fog events occurred. His errors became smaller as he learned some fundamental aspects of fog formation, but he still had to learn and understand the nuances of how fog dissipates. In one example, he mentioned that clouds did not clear completely, moderating the high temperature below what it would have been had the fog completely cleared. "Nine out of ten times it's gonna verify. There's that one out of ten times when it doesn't. ... That's when you have to go back and look at all the data and try to figure out what happened. That's when you learn and improve as a forecaster." Cassie described beginning to learn the nuances of strong winds. As she was beginning to understand how to forecast them, she came across an instance where the winds were not going to be strong, despite the pattern seeming similar to her. Another forecaster explained the nuance to her to help her understand. Raymond and Shawn both missed forecasting severe events because of subtle changes in instability. Travis learned that instability had now become a nuance, whereas in his first forecasting location essentially any instability led to severe weather.

wrong. All forecasters enter this phase of learning mainly from surprises. The learning of nuances can be rapid at this point, taking one or just a few experiences and rarely requires experience across more than one full seasonal cycle.

By middle career, forecasters are not often surprised at how the weather unfolds after making a forecast, and stories of forecasting challenges tend to focus on instances where they were dissatisfied with the state-of-art. Forecasters want to improve the data and tools, but not all forecasters become strivers, individuals who seek to advance the state-of-the-art. The extent to which they engaged in such activities might be dependent on the collusion between a strong sense of identity with social affirmation (more is said on this topic of "forecasting styles" in chapter 7). Whatever proficiency level they are at, the realization that they lack forecasting skill usually causes forecasters to review an event to try to figure out why the weather evolved differently than they had forecasted. The extent to which they engage in such investigations varies primarily by how much time they have and how difficult it is to figure out what went wrong. Time is a challenge to learning for all forecasters.

To summarize, the following core phenomena interact and feed into one another over the course of development, education, and training:

- Experiencing the weather,
- Seeing connections,
- Receiving affirmation by others,
- Becoming intrinsically motivated to learn how to forecast,
- Developing a sense of role or identity,
- Self-learning of knowledge and skill on the job,
- Creating and applying learning strategies on the job, and
- Benefiting from mentoring.

As child "weather geeks," budding forecasters show the features of novices that have been seen in studies of other professional domains: Their understandings are simple associations and causal connections, and their knowledge is limited. The culmination of the experiences can be genuine expertise: the ability to understand weather in terms of multiple and complex interactions, the ability to adapt to circumstances and develop new strategies and methods for forecasting. These are all defining features of expertise (see Hoffman, 1998; Simon, 1973).

Step 1: Recognition of an Inability to Forecast. Young forecasters begin with a general inability to forecast, as do more experienced forecasters who, after moving to a new region, are dealing with phenomena they had not forecasted or experienced before. The realization that they did not know something could trigger learning regardless of their time-in-service or experience with a similar phenomenon.

Step 2. Receiving Support. Following recognition of an inability to forecast, strong support from experienced forecasters is crucial to help them learn. This results in knowledge of weather and the appropriate processes to access and use that knowledge effectively — if the experienced forecasters have useful knowledge structures. This learning path is relatively fast. Even brief input from a veteran can trigger fast learning of some unfamiliar phenomenon; forecasters at this stage often commented that they became aware

Box 3.6
Case Studies in Affirmation

Fortunately, most forecasters find help readily available. The forecaster who needs to learn something new has to sometimes actively seek help and receive affirmation that reaching out for help is acceptable. One senior older forecaster reportedly told a junior forecaster, "You're seeing something, you're picking it up but you're not exactly clear on what it is. Let me elaborate a little bit and tell you what is causing this." This forecaster knew that no one coming into their office forecasted precipitation correctly in certain circumstances because it was a local effect of the geography. In saying this to the junior forecaster, the senior thereby affirmed her, making it okay that there was something she did not understand but needed to learn. She was relieved that someone finally understood how she had been feeling. Travis did not speak of having a mentor, and the science officer in his current office was initially unavailable. Travis described asking questions of older forecasters but none of them initiating explanations. None of his stories described particularly complex learning events or deep engagement in thinking about weather processes. It is perhaps not surprising that Travis did not seem to have a strong sense of professional identity.

that "learning got a lot faster." Stories were told of significant learning moments that took just minutes.

It is particularly challenging to young forecasters when others could not or did not help them learn. The knowledge needed to forecast the weather is extensive and complicated. It is difficult to learn without someone helping you learn how to think through complex processes. A few forecasters told of their feeling of desperation of not having a mentor and having to indiscriminately shadow experienced forecasters to absorb what they could because their reaching out did not seem welcomed. Sometimes support is not readily available. Interviewees described this as their most challenging yet significant kind of learning experience because the forecaster had to create a learning strategy; resourcefulness is an apparently solo effort. Younger forecasters create strategies to learn how to link the science to the forecast and to do the job, whereas experienced forecasters create strategies to build on their ability to resolve local or situation-specific forecasting challenges. Experienced forecasters also sought to extend the science by promoting research, reviewing scientific literature, publishing reports or articles, and/or attending scientific conferences to share their research.

Forecasters with the strongest senses of professional identity not only persisted through these challenges but also created effective learning strategies. Although their learning was primarily self-directed, they eventually needed to rely on others to some

Box 3.7
Case Studies in Self-Directed Learning

Raymond was interested severe weather, a particularly difficult forecast challenge for his region due in part to poor data. Forecasting severe weather in his area was also difficult because low population density may, as he was discovering, mask the occurrence of tornadoes. Although much of the effort was his own, he had spoken with forestry officials to learn about instances of tree damage. He was in the process of figuring out how to better forecast and identify tornado occurrences in his area based on tree data. Henry told the story of a forecast problem: "I started thinking something was going on in the boundary layer. I'm no expert ... but something was going on there. So we [put our observations out there] ... in hopes that someone would grab ahold of that and ... try to figure it out. ... That's an example of a case where I didn't do that project myself because I didn't have that in my area of expertise. ... And that happens fairly often."

degree. Sometimes they had to almost force someone else to help them figure it out, see connections, and learn. Sometimes episodes that would seem to trigger self-directed learning ended with no learning. All forecasters in LaDue's study, regardless of how strongly they identified as a forecaster, were bothered by these occurrences. Those with the strongest identities took the most extensive actions to learn.

LaDue's research sets the stage for further work on important questions. For example, the interviewees did not convey any stories where they later realized something someone taught them was incorrect or unproductive. A more systematic or controlled study would be needed to investigate this. An additional question has to do with mentoring. Although mentoring is seen as necessary, there is no scientific base or established methodology for determining how to identify individuals who might become good mentors. Experienced mentors provide insight into the mentoring process and its value, but here too there is a need for follow-on research that is more systematic or controlled.

New Learning Venues

The concern has been raised that any four-year education program could not possibly allot enough time for students to really learn about new technology. There has been a shift from traditional training (emphasis on forecasting based on maps and charts) to a new education model that adds course work on remote sensing, satellite image interpretation, the use of Doppler radar, and the interpretation of computer model outputs (see Fritsch, 1992). In addition, the process for involving meteorology students in field weather observation experiments was being formalized and expanded (see Mullendore, Tilley, and Carey, 2013.

Computer-assisted learning and distance learning have become a major methodology in meteorology education and forecaster training, from the college to the professional levels. For example, EUROMET is a project funded by 23 countries to provide network-based education and training (Gimeno and Garcia, 1998). In the United States, it became clear that training of professional forecasters was becoming more crucial and yet travel was becoming more expensive (Fritsch, 1992). Thus, the NWS established a training division and programs such as the Virtual Institute for Satellite Integration Training (VISIT) (Mostek et al., 2004). The Cooperative Program for Operational Meteorological Education and Training (COMET) began in 1989 as part of the modernization of the NWS.

COMET

COMET was implemented to improve meteorological education through the use of new workstation technology, the integration of new data types, and attempts to generate displays that graphically and symbolically depict atmospheric dynamics in such a way as to represent conceptual models of the atmosphere (e.g., three-dimensional models of the structure of clouds, based on radar data) (Johnson and Spayd, 1996; Spangler et al., 1994; Wash et al., 1992). Goals of COMET included:

- Training forecasters in the use of new systems (such as NEXRAD) and interpretation of images from new sensor systems (i.e., the GOES Advanced High Resolution Radiometer);
- Serving as a clearing house to provide new training data types/sets and training products to the educational community for use in courses on synoptic and mesoscale meteorology;
- Providing new training data types/sets and training products to the NWS and the Naval Meteorology and Oceanography Command for in-residence training; and
- Laying the foundations for a weather forecasting infrastructure, whereby there can be greater interaction among the research, academic, and forecasting communities and accelerated transfer of research results into operational forecasting.

Given these multiple objectives, the COMET program has funded dozens of forecasting research and development projects: Cyclogenesis in the Gulf of Mexico, measurement of rainfall rate using Doppler radar, prediction of the movement of volcanic ash clouds, development of a geographic information system that can be integrated with meteorological data, and so on. COMET has dozens of training modules available online. They are used in modules that are included in the National Weather Service's Forecaster Development Course (more on this below), and these are updated every few years.

Hints at the importance of cognitive factors in forecasting appeared in recommendations from the first wave of COMET efforts: The then-new workstations needed to be accompanied by better documentation, there was a need for a means to integrate satellite imagery with various data fields, and there was an outstanding need to integrate all the diverse software systems being developed (Wash et al., 1992, Table 1). (For a review of the first decade of the COMET efforts, see Johnson and Spayd, 1996.)

The COMET training system includes a great many case study data sets for multimedia distance learning (e.g., training modules on Doppler radar interpretation and the initiation of convection) (Serafin, MacDonald, and Gall, 2002). The emphasis is on new software for visualization, interactive computing, and networking. The data sets integrate surface, upper air, satellite, and radar data. New visualization tools include the graphical comparison of various computer model outputs, depiction of temporal changes in surface data fields, analysis of air parcel trajectories, navigation of historical data sets, time sequences of radar data, depiction of isentropic vorticity fields, and so on (see Spangler et al. [1994] for examples). The COMET project has also generated a number of striking animations, much of it now web-based in a virtual classroom [see http://www.comet.ucar.edu and https://www.youtube.com/user/cometmeted].

Discussions of various COMET projects and programs include testaments to their success, that one or another course was "rigorous" (Wash et al., 1992), or that a new display "improved student understanding of atmospheric dynamics" (Wash et al., 1992, p. 1446).

A result that seems to percolate up from the COMET programs is the value of using numerous case studies (numbering in the dozens) as the foundational exercise in courses. This finding fits with those of Navarra et al. (1993), described earlier. In addition, a benefit is seen in having the case studies presented in a format that mirrors that of operational forecasting. Finally, value is seen in formatting the learning modules in such a way as to promote the:

> interpretation of patterns associated with fronts, convergence and divergence, etc. ... Practice cases contain video-based discussions in which the content experts address important points about the case. Lesson exercises focus on ... quickly recognizing significant patterns that might be observed on Doppler radar (such as veering or backing winds with height). These exercises require the learner to interact with the system by responding to questions about concepts, match-

ing correct wind speeds with selected points on a velocity image, identifying events such as warm air advection, etc. ... Learners utilize a number of topic-specific conceptual models and tutorials that allow them to build their knowledge and skill levels. Content experts ask questions, give hints that guide the learners to the correct answer, provide expert answers, and explain techniques and concepts about various features and processes ... [it] challenges the learners' ability to correctly and rapidly identify convergence boundaries. (p. 1253.

These conclusions are reinforced by trainees' postinstruction evaluations and dovetail with ideas confirmed in other research on instructional design. Indeed, fundamental to our understanding of expertise, in general, is that the higher levels of proficiency are achieved only after a great deal of case-based practice (Ericsson et al., 2006).

INNOVATIVE WEATHER

The Atmospheric Sciences program at the University of Wisconsin at Milwaukee has a forecasting class, and its students are involved in the national forecasting contest. This continued for about 10 years until a need was perceived to connect strong student interest in forecasting, their need to earn income, and weather decision support forecast services in the community. The result (in 2007) was a service known as INNOVATIVE WEATHER, and it does all of the above [www.innovativeweather.com]. INNOVATIVE WEATHER engages high school students and Atmospheric Science clubs through talks, conferences, storm chasing, weekly forecast discussions, and forecasting competitions. It has a pre-internship program for high schoolers and an internship program for college students, through which students can develop skills in forecasting while still satisfying the requirements for a degree in atmospheric science. Interns have to do actual forecasting "shift work," thereby serving as a window on the day-to-day work of professional forecasters. This weather decision support service is provided to paying community clients across a wide variety of weather-sensitive sectors, including energy, transportation, and entertainment. In addition, there is another internship for students interested in broadcast forecasting. In all these internship programs, students work through the forecast process with their mentors and have specific products that they must generate and disseminate. Culminating the INNOVATIVE WEATHER program is the opportunity for students to work as a limited term staff meteorologist in a leadership position in the organization, thus providing peer mentoring to younger students and further professional development for all students. This university–private sector partnership allows the private sector partner to benefit by the service of its best trained and most experienced staff members, especially in the severe weather season, and it benefits the university, giving students more opportunities to gain forecast experience and build a strong resume by stepping into leadership roles within an operational program.

Since its inception in 2007, INNOVATIVE WEATHER has proved to be effective, giving more than 50 students real-world experience in weather forecasting and helping them connect what they learn in formal classes to actual forecasting practice. This requires knowledge of weather risks, understanding probabilities as well as possibilities, and helping clients factor that information intelligently into their decisions. This collaboration between student forecasters and decision makers requires students to have a clear understanding of the weather risks specific to a client and be able to translate the technicalities of the meteorology in a clear way to intelligent but not meteorologically sophisticated users of that information. Consequently, the value of the service is as much about understanding their needs and clear communication as generic forecast accuracy as measured by the Brier Skill Score or the root-mean-square-error. At the same time, it helps a set of students mature professionally in a manner similar to what happens for students who become involved in undergraduate research experiences.

Programs such as COMET and INNOVATIVE WEATHER reflect a wider recognition of the need to reconnect meteorology education with forecasting training, or at least to see these as two distinct yet interdependent educational needs.

Lamos and Page (2012; see https://www.meted.ucar.edu/training_detail.php), in a concept paper discussing how professional development of forecasters should be designed, promoted focused training that was directly tied to what forecasters did. Forecasters must first be taught how to apply an understanding of key atmospheric factors to the forecast and then to synthesize a large amount of information using tools provided. Lamos and Page asserted that scientific understanding was necessary to evaluate models and other tools; forecaster education needed to help forecasters build a complex understanding so they could visualize atmospheric processes.

Doswell (2003, 2004) also provided a vision for improving forecaster education. He proposed a creative method for learning, suggesting that forecasters would quickly gain a much deeper understanding of both atmospheric dynamics and model limitations if they could repeatedly change the input to locally run models and see the resulting outcomes. Doswell's vision is that all forecasters would be mentors to incoming forecasters, and meteorological education would include learning how to mentor effectively. Among the characteristics he promoted were high-level visualization and conceptualization skills, a passion for the subject, and continuous learning. This work is consistent with the findings of others.

But until such a time as the reconnection of meteorological science and forecasting is fully realized, we are still left with the question of how today's forecasters learned to become forecasters.

Some Not-so-Formal Learning Venues

Although formal learning has both positive and negative attributes, there are many other opportunities for forecasters (and others) to remain on a path of "continuing education." Mentioned earlier are various workshops and training sessions at AMS and NWA conferences. Indeed, the entire annual NWA conference and many specialized AMS conferences incorporate presentations and/or workshops that support learning (e.g., new ways to use GOES-R data, case study analyses, lessons learned from a summer field research program). Currently, these conferences also focus on social science and social media aspects of forecast dissemination and forecaster sharing/learning.

There are also a myriad of internships in forecasting, research and management (e.g., run by companies and/or coordinated by various AMS committees), webinars, archiving conference presentations online, and published NWS and private sector articles and storm summary reports available for forecasters to access (even on "slow" forecast days). In fact, both the NWA and the AMS require their TV Broadcasters, Certified Consulting Meteorologists, and Digital Media Sealholders to document continuing education activities (from a myriad of possible settings—conference attendance, publication of research, and publishing blogs) to retain their "seals." Links for seals and certifications are [http://www.nwas.org/seal/index.php] and [https://www.ametsoc.org/ams/index.cfm/education-careers/ams-professional-certification-programs/].

Finally, all WFOs and National Centers publish their "Forecast Discussions" online, making them readily available for forecasters, students, TV meteorologists, and others to use within self-learning frameworks.

Conclusions

Historically, there has been a disconnect between education in the science of meteorology and training in the profession of forecasting. Although new initiatives and programs are aimed at remedying this situation, the near absence of formal education and training on the forecasting task leaves professionals to self-direct the majority of their learning and learn by doing their jobs. Learning to forecast is affected by how well meteorologists can regulate their learning, their career stage, and contextual, sociocultural factors. Forecasters have to learn from other forecasters. Younger forecasters need to learn how to change their mindset when expecting one type of weather and another occurs. But the converse also holds true: Senior forecasters need to learn from younger ones (e.g., how to adapt to new software systems and tools). A repeating theme in the stories that forecasters tell about their careers is that a strong sense of identity with their professional role as a forecaster is important to how they engage in learning, particularly when they were poorly supported and had to create strategies to learn. Learning to forecast is faster, forecasters are happier, and their resulting knowledge is better connected and more thorough if they have social support: hearing how other forecasters think about the weather and how they use data in different situations. Forecasters are more likely to persist through adverse work conditions and poor social support if they had a strong sense of identity, going so far as to create their own strategies to learn.

Forecasting is an application of the science of meteorology, yet much of what meteorological education offers is unnecessary for acquiring short-range or even long-range forecasting skills. The separation of the meteorological science from the forecasting art has the actual effect of impeding the acquisition of knowledge about the weather. One can, and indeed should, say that forecasting is both a science and an art (Doswell, Lemon, and Maddox, 1981). To borrow a turn of this phrase (and fuzzy up the artificial basic vs. applied science distinction), it might be more appropriate to say that weather forecasting is a subtle science and an exact art (Bennett and Flach, 2011).

This chapter has described how people get to be forecasters and how they grow in that role. But how do they get to be *expert* forecasters? This is the topic of chapter 7. However, before addressing that topic, we need to say more about how forecasters describe their reasoning (chapter 4) and how well they perform (chapter 5). That information will put us in a position to ask whether forecasters, the really good ones, qualify as experts.

4 How Do Forecasters Describe How They Reason?

The literature of meteorology includes many discussions attesting to the importance of forecaster reasoning, skill, and knowledge in determining forecaster performance. For example, the forecasting of severe local storms by the National Storm Prediction Center has been said to depend on the forecaster's "interpretation and modification" of numerical model analyses, "subjective" surface analyses, and the "close examination" of satellite imagery (Johns and Doswell, 1992, p. 589). Even when forecasting is based on one or another predictive algorithm, there is a fundamental reliance on human experience and judgment. For instance, Gaffney and Racer's (1983) algorithm for forecasting the outbreak of severe storms relied on key parameters (e.g., 500 millibar vorticity advection) that were chosen because they "are thought by various meteorologists to be indicative" (p. 274).

To concretize this avowed importance of forecaster reasoning, a number of forecasters had attempted to describe the forecasting process or workflow, by referring to elements of team work, knowledge, perceptual skill, conceptual models, principles of forecasting, and reasoning processes.

Teamwork

Forecasting is a team effort, especially during outbreaks of severe weather or in impactful long-term, large-scale weather events (e.g., hurricanes, winter storms). In the Weather Forecast Offices (WFOs) of the National Weather Service (NWS), multiple forecasters share the workload. They will each have particular storm cells to track or regions on which to focus in the attempt to detect storms and track their evolution. Typically, the forecasting responsibility is distributed by customer (e.g., aviation, marine, public). Forecasters share views of data fields of various kinds, share their findings, and engage in discussions of their conceptual models of the weather (Andra et al., 2002), with the goal of generating a consistent and reliable forecast across the board. Hence, the forecasting workspace is designed around the fact that the activities are those of teams (chapter 2). Collaboration is also fundamental to on-the-job learning and the development of careers through mentoring (chapter 3). As we will describe in chapter 5, forecasts are the result of collaborative decision making and judgment. As we will show in chapter 7, collaboration is fundamental to the achievement of expertise at forecasting. Our focus now, however, is on how forecasters describe their individual reasoning, with the understanding that the cognition is "distributed" and "situated" in a collaborative context.

Knowledge

Charles Doswell, a senior meteorologist with the National Severe Storms Laboratory of NOAA (1986c; Brooks, Doswell, and Maddox, 1992), and Rosemary Dyer, a senior meteorologist with the U.S. Air Force (1978, 1990), made repeated reference to the fact that forecaster skill depends on knowledge of the principles of meteorology:

> As an example of how knowledge is applied ... [consider the forecast problem of whether a mesoscale convective system] will continue or dissipate. ... Having a knowledge of the system's history, the forecaster can employ simple extrapolation ... knowing whether it has been intensifying or dissipating is clearly helpful. However, there are more complex questions that must be answered by nonlinear [i.e., human] methods. Is some large-scale process sustaining the system? Are there any topographic features that might alter the system? How might diurnal cycles within the boundary layer modify the system's evolution? Does the present system fit any physical model? ... the forecaster uses knowledge of meteorology to answer such questions. (Doswell, 1986c, pp. 699–700)

The forecaster needs to know about local effects, patterns, and trends; and the effects of the regional geography. This has been known since the earliest days of forecasting. For instance, Moore (1922) presented a number of rules, including, "A low from the Northwest that reaches western Minnesota and western Iowa without precipitation or clouds will pass over Wisconsin as a dry low, unless the isobars are closer than five-eighths of an inch" (p. 154). Such local knowledge has extended in modern times to an awareness of the ways in which the computer models over- or underpredict certain things in certain regions:

> Operational forecasters quickly become aware of problems with [computer] models that affect their forecast area (Fawcett, 1969). For example, the weather forecast for Newport, Rhode Island, which sits on an island at the mouth of Narragansett Bay, uses the forecast for Providence, 25 miles away, but is often insufficiently adjusted for the sharp influences of the land-ocean interface. Forecasters deal with this by taking note of phenomena that are not handled properly by the

model, and their confidence in the model's prediction is a function of what they know about the model and the weather situation. (Brooks et al., 1992, p. 121)

The importance of sensemaking is also noted in discussions of military forecasting. For example, Colonel Beth McNulty (2005), commander of the U.S. Air Force Weather Agency, said that:

Systematic forecast development employs the ... available data to eliminate personal biases anticipate change, and explain the reasoning behind the forecast. A forecaster creates a mental concept of how the weather should develop over the next few hours or days. ... This mental image is the first stage in developing a forecast. (p. 5)

As we mentioned in chapter 2, a critical role is played by the forecaster's formation of what meteorologists refer to as a conceptual model. A conceptual model is said to support the perception of weather phenomena and the testing of hypotheses (see the classic text by Pettersen, 1940, ch. 11). In their discussion of information-processing systems for forecasting, Chisholm et al. (1983) concluded that forecasters' understanding is in terms of a "mental subjective integration" of data. Doswell and Maddox (1986; Doswell, 2004) regarded the formation of conceptual models as the critical step in forecasting, the "vital link" between objective or quantitative data and qualitative or intuitive information.

Giraytys (1975) described "mental integration" in a way that makes it clear that what these meteorologists are talking about is what cognitive scientists refer to as a mental model: "The forecaster develops ways of compensating for shortcomings in his information processing system. One [way] is to mentally integrate two dimensional 'pictures' into a four-dimensional forecast" (p. 112). Many meteorologists and forecasters are explicit on this point (e.g., Morss et al., 2015).

Roebber and Bosart (1998) examined NWS forecasts of precipitation during cyclogenesis events—when a trough of low-pressure deepens and begins a counterclockwise rotational circulation (in the Northern Hemisphere). Their analysis suggested that the patterns of precipitation cannot be derived from the overall weather pattern at the scale of continental low-pressure systems. Processes that are at a more local scale and that occur prior to and during cyclogenesis must be taken into account. Roebber and Bosart argued that proficient forecasters are those who are better able to do this—their conceptual models of forecast problems are extraordinarily rich. The forecaster's mental model integrates the large amounts of information (Dyer, 1987; Giraytys, 1975). In chapter 1, we discussed the data overload problem: the fact that forecasters literally have access to more information and data than they can possibly deal with. Forecasters know and understand this problem well, of course. Forecasters suggest that they are able to create a conceptual or mental model in order to integrate information from many sources.

We see that forecasters assert that they reason imagistically, in four dimensions (Doswell, 1986c; Godske et al., 1957).

Perception and Recognition

Dyer (1987) made a case for the importance of mental integration by presenting a challenge to the reader:

> Watch any weather forecast on television ... satellite photos, radar, and charts all dance across the screen. ... Visit any forecast office and note the rows of charts ... the computer displays with overlays and split screens. ...What is [the forecaster] doing?—what all forecasters do—image processing. (p. 23)

The importance of perception cannot be emphasized enough, and it shows in Ellrod's (1989) discussion of how to predict clear air turbulence, which emphasizes the visual inspection of satellite images and the visual appearance and shape of clouds (e.g., the shape of a comma cloud, the degree of curvature, the appearance of a dry slot, the presence of transverse cirrus cloud bands near the trailing edge of the comma, and the location of shear). The Dvorak technique for estimating tropical cyclone intensity is based on the visual inspection of satellite images (Dvorak, 1975; Velden, et al., 2006). Even the basic "hook echo" shown on radar as an indicator of a tornado (Glickman, 2000) is identified by visualization.

The perceptual foundations of meteorology and forecasting stand out in the history of the standard methods of charting and depicting weather features, and in discussions of chart-making skills and the "art" of perceiving weather features in collections of observations (see Gregg and Tannehill, 1937; Monmonier, 1999). The perceptual foundations are perhaps most salient in the literature on the interpretation of satellite images. In high-resolution imagery, the experienced forecaster can determine a great many things that escape the eye of the novice until they are pointed out (and sometimes even after they are pointed out—even the informed novice needs time to learn the cue configurations). Scores of examples could be listed (see American Meteorological Society, 1996; Bader and Waters, 1987). A dark notch in sunglint over the ocean reveals regions of calm seas (Fett, White, Peak, Brand, and Tag, 1997). The patterns in large-scale and fine-scale cloud structures reveal the development of cyclonic systems and the ways in which the dynamics of air masses result in the cloud features that one can observe (Conway, 1997). From the perceptible shape of cyclonic cloud systems, one can even estimate the surface pressure at the cyclone center (Smigielski and Mogil, 1991)—a useful tool for oceanic analysis and forecasting. Meteorologists have often commented on their ability to engage in rapid recognition of patterns and configurational cues (see Pliske et al., 1997, 2004).

Conceptual Models

In chapters 2 and 3 we referred to the forecaster's conceptual model of weather phenomena and the dynamics of the atmosphere (Hoffman et al., 2000; Lowe, 1994; Perby, 1989; Trafton et al., 2000). Some psychologists, especially behavioral psychologists, are uncomfortable with notions of a mental representation. This is at least ironic given that psychology is supposed to be about mental life. Meteorologists have no difficulty with the notion of a "conceptual model" at all. As forecasters assimilate information about what is going on in the atmosphere, they develop a subjective, imagistic representation of their understanding. Apart from the fact that meteorologists do not suffer from the historical baggage that has burdened some psychologists since the days of behaviorists John Watson and B. F. Skinner (see Rachlin, 1991), the main reason that meteorologists are comfortable with the notion of a mental model is that the concept has been in the literature of meteorology for decades.

Pioneers of meteorology of the late 1800s and early 1900s distinguished between their mathematical or formal models of the weather (based primarily on hydrodynamics) and their conceptual models. For example, in Norway, Vilhelm Bjerknes

> and the group of Bergen meteorologists set up a network of weather stations across Norway which recorded weather observations and reported the measurements back to Bergen. The measurements for each location were then plotted on a Norway map to give a picture of weather over a wide area. As Bjerknes and the others studied the picture maps, they noticed that different air masses—a warm and a cold—existed, and also that the most active weather conditions were found along narrow zones in-between these air masses. In military fashion, (World War I was of course happening during this period time) they called these boundary zones weather "fronts"—an analogy to the battlefronts of war. (Means, 2015, p. 1)

Bjerknes (1919) drew diagrams such as the one shown in figure 4.1, a depiction of his conceptual model of a cyclone, showing clouds, streamlines, precipitation, and a pair of vertical cross-sections, one to the north and one to the south of the cyclone center. This is not fundamentally different from today's weather maps and displays: "the shape of a frontal cyclone is indicative of its stage of development and can thus give information about its future behavior" (Eliassen, 1995, p. 9). This most basic weather model of all remains a mainstay of meteorology to this day.

The pioneers of meteorology were strong believers in the power of running hydrodynamic computations forward in time as a means of creating weather forecasts.

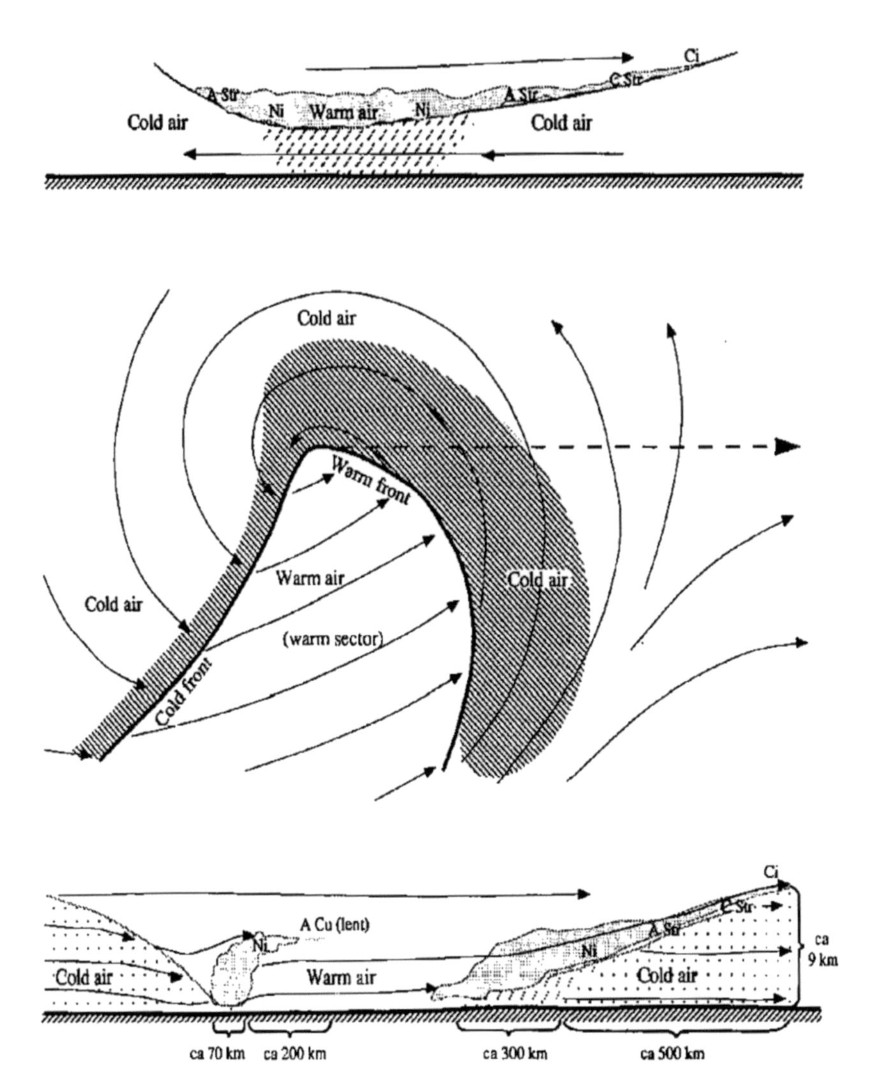

Figure 4.1
Vilhelm Bjerknes's (1919) "ideal cyclone" diagram. (Reproduced with permission from The Royal Swedish Academy of Sciences.)

But they were equally strong believers in the notion of conceptual models, and they applied these models to what they did in research and/or operations in the forecasting services they championed and academic departments they established. In their classic text on the science of meteorology, Godske, Bergeron, Bjerknes, and Bundgaard (1957) wrote:

> Paramount in importance are the special talents with which the analyst should be endowed: a faculty of combining a large number of observations into the most logical three-dimensional mental pictures. (p. 653)

In the 1960s, Vernon Dvorak developed a system for estimating the intensity of tropical cyclones based on the visual pattern of the cyclone as seen from above (in satellite images) (Dvorak, 1973; see also Smigielski and Mogil, 1991; see figure 4.2). The Dvorak Technique, still used today by NOAA, also incorporates a decision-tree process and a data-collection spreadsheet.

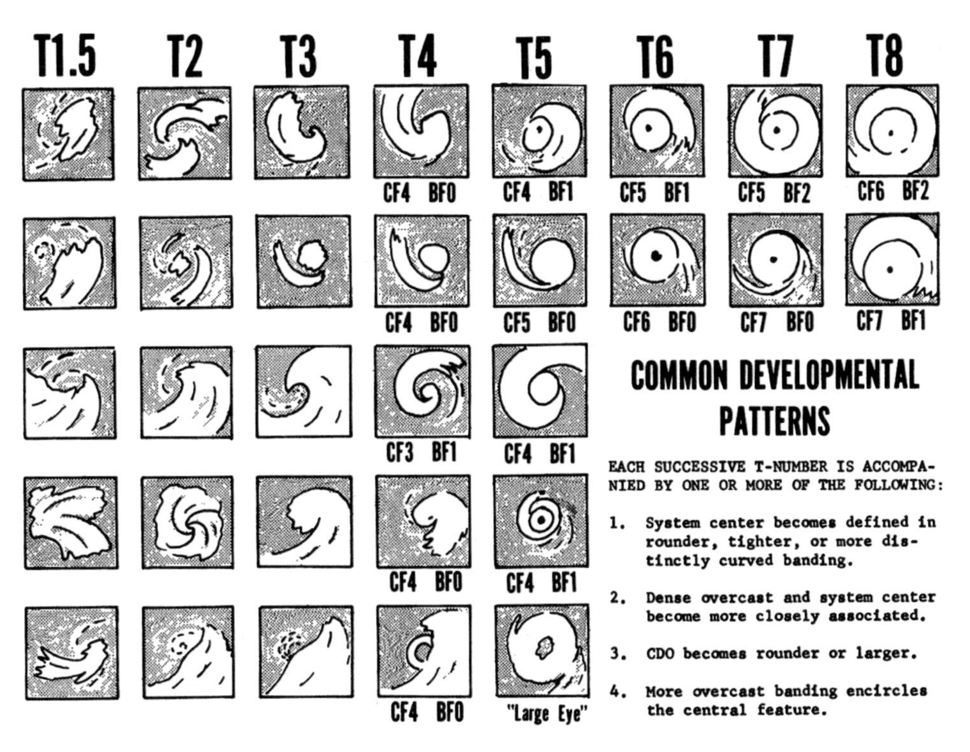

Figure 4.2
Vernon Dvorak's (1973) diagram of his scheme for estimating the intensity of tropical cyclones.

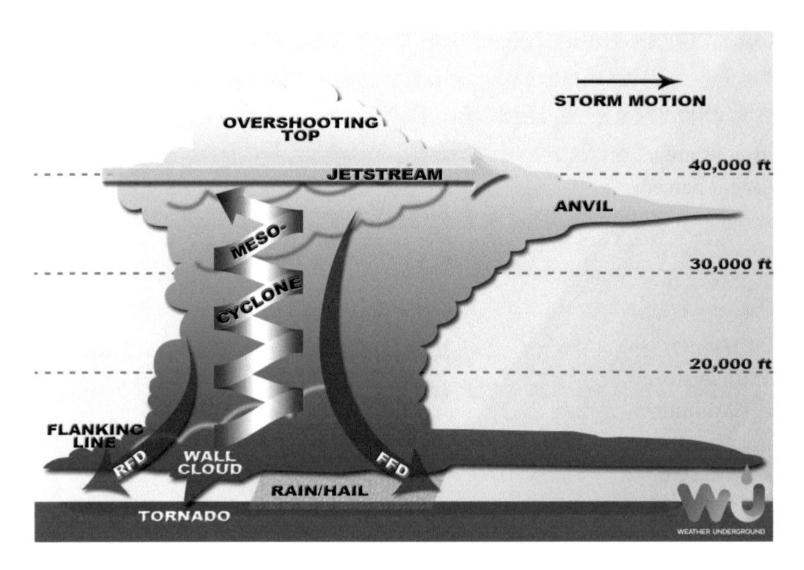

Figure 4.3
A conceptual model of the structure of supercell storms (reproduced with permission from the Weather Underground). RFD and FFD are rear and forward flank downdrafts, respectively.

Another example of a conceptual model is shown in figure 4.3. This is a model of the structure of "supercell" storms. Note that this attempts to convey dynamics as well as structure.

The development of a good conceptual model for the formation of supercells emphasizes the detection of convection as early as possible in the development of the storm system and the anticipation of the kinds of warnings that might have to be issued (e.g., flooding, damaging winds, tornadoes) (Doswell, 1992). The forecaster builds a mental model of the storm structure (Moller et al., 1994) (see figure 4.3). "Without conceptual models, the meteorologist does not have the means to anticipate intelligently storm evolution and threat and therefore the range of potential outcomes necessary to determine warning content" (Andra et al., 2002, p. 561). An example is the tendency for a line of storms along a cold front (a "squall line") to produce damaging winds rather than tornadoes.

Forecasting relies on a number of widely accepted conceptual models, and not all of them are models of things like storm structure (as in figure 4.3 and also figures 2.9 and 2.10 [plates 7 and 8]). For instance, there is a widely accepted model of the El Niño oscillation of warm and cold temperatures in the Pacific, which is used in the analysis of tropical cyclones as well as in the prediction of weather in the continental United States (see http://www.cpc.ncep.noaa.gov/products/precip/CWlink/MJO/enso.shtml).

Box 4.1
Thunderstorms, Supercells, Mesocyclones, and Tornadoes

Thunderstorms are a form of convection, which is the vertical movement of parcels of air. Two adjacent parcels of air can have different temperatures and moisture content, meaning that one of them is less dense and therefore more buoyant. These differences can arise from many factors, including solar heating during the day or a breeze coming off the ocean. As the parcel rises, it cools, but as long as it is rising into air cooler than it is, the parcel will remain buoyant. As the parcel cools, the moisture within it condenses, releasing a form of heat that tempers the rate at which it is cooling. If there is sufficient vertical instability, then a thunderstorm can develop.

Supercell thunderstorms, a special type of severe thunderstorm, contain steady, rotating updrafts called a *mesocyclone*. Supercells and the mesoscyclones within them are often associated with severe weather (Walsh, Charlevoix, and Rauber, 2014). Their rotation allows the storm to persist much longer than the time required for a single, buoyant air parcel to rise to the top of a thunderstorm. Supercells only form in environments characterized by vertical instability and vertical wind shear. The rotation in supercell updrafts begins from horizontal rotations within the parcel of rising air and is sustained by the vertical wind shear that the updraft experiences. The resulting mesocyclone can be relatively large: 2 to 10 km in diameter. The speed of rising air within these mesocyclonic updrafts can be quite high, peaking around 170 mph, enabling the storm to loft hailstones until they grow to considerable size. Sometimes the air feeding the updraft of a supercell occludes (much like a large-scale middle-latitude cyclone does), leading to the formation of a new updraft and mesocyclone adjacent to the original one. This is, in part, what leads to tornado families—a series of tornadoes from one supercell thunderstorm. Supercell thunderstorms produce all the violent (EF4–EF5) tornadoes, but perhaps fewer than 30% of supercells spawn any tornado at all.

Mesocyclone [http://glossary.ametsoc.org/wiki/Mesocyclone]
Supercell [http://glossary.ametsoc.org/wiki/Supercell]

Forecasters of the "old school," who grew up doing hand chart work, have told us that their mental images of the weather are rather like a traditional weather chart but are animated in the mind's eye. It might be easiest to think of the forecaster's mental model as a dynamic depiction of masses of air, some large, some small, some warm, some cold, some dry, some wet, all of them interacting, merging, and slipping over and under and/or mixing with one another. All of these dynamics as they are visualized are "governed" by (the forecaster's knowledge of) the principles of atmospheric dynamics (e.g., how a large dome of high-pressure behaves at the trough that forms at its interaction with another air mass).

Interestingly, the most recent technologies in weather data display are now approximating what we describe here as a conceptual model of the atmospheric dynamics. The visualizations of PRAVDA and Met.3D presented in chapter 2 are examples (figure 2.9, 2.10, and 2.11 [plates 7, 8, and 9]). Although these are computer-generated displays of computer model outputs, and as figures in a book we cannot present them as dynamic animations, they might be regarded as approximations of forecaster's conceptual models in that they are principled combinations of concepts, data, and dynamics and are imaginable. Interestingly, a display of the same type and character as those shown in these figures is used at a NASA website to illustrate Bjerknes's ideas of mathematical weather forecasting [http://earthobservatory.nasa.gov/Features/Bjerknes/].

All of the "conceptual models" that meteorologists have developed, to distinguish their understanding from their externalized, mathematical, and computational models, can be likened to the psychologist's notion of a memory "schema" and similar notion of a "mental model" (Bartlett, 1932; Gentner and Gentner, 1983; Gentner and Stevens, 1983; Johnson-Laird, 1983; Klein and Hoffman, 2008; Stevens and Collins, 1978). A schema is regarded as a framework or prototype for understanding something. The mental model notion places more emphasis on the visual imagery aspect to comprehension. The forecaster's notion of a conceptual model and the corresponding psychological notion of a mental model are important themes that weave across the chapters of this book.

The distinction between the forecaster's conceptual model and the psychologist's theory of mental models is subtle. For example, consider the Fujita scale for tornado intensity. In 1971, Tetsuya Fujita of the University of Chicago and Allen Pearson of the National Severe Storms Forecast Center (now NSSL) presented a conceptual model and a scale for rating tornado intensity. The scale is based on the damage to buildings, trees, and so on. Previous scales had been based only on estimates of maximum winds. The conceptual model relies on the ways in which different types of buildings respond to strong winds as a function of the building construction, foundation, and so on. This is regarded as a conceptual model, although it is rather unlike, say, the Dvorak model of cyclogenesis (figure 4.2) or the general model of supercell structure (figure 4.3). These two conceptual models clearly illustrate how the forecaster's conceptualization of weather dynamics relies on visualization and visual imagery and hence tie similarity to the psychologist's notion of a mental model. But not all conceptual models in meteorology have this aspect.

We therefore have four types of "models" to keep clear about:

- the forecasters' *conceptual models* or schema for understanding the atmosphere,
- the forecasters' *descriptions of forecaster reasoning*,

- the forecasters' formal *computer models*, and
- the psychologists' *reasoning models* that describe how forecasters think.

In the following chapters, we consistently use these phrases to refer to these different kinds of models.

General Forecasting Principles

Armstrong (2001a, b; see also Harvey, 2001) listed 139 "principles" of forecasting derived from entries in a forecasting handbook and subsequent open peer commentary, including commentary by 20 individuals who were said to be experts. This listing is a valuable compendium that spans all aspects of forecasting. Principles cover everything from policy (forecasts should be independent from politics) to process (i.e., the use of graphical displays).

The principles are clustered into categories, which include: Problem formulation, problem structuring, information acquisition, information analysis, integrating quantitative analysis with human judgment. Principles refer to the preference for developing causal models, the importance of assessing the validity of a forecast, the preference for relying on quantitative analysis, the advisability of being conservative in uncertain situations, rely on the concept of statistical significance, etc.

It is noteworthy that a quarter of the principles refer not just to the notion of forecasting expertise, but to the *value* of forecasting expertise.

- Experts should be tapped to brainstorm about risk or bias situations.
- Experts should be asked to do a decomposition analysis of their forecasting processes.
- Impartial experts should agree in their forecasts.
- Data analyses should conform to the experts' expectation .
- Experts can determine which forecasting methods are most appropriate for a given situation.
- Experts determine which data and which variables are the most relevant and important to the forecast situation.

Some of the principles that refer to experts seem to take the perspective of the person who is seeking a forecast and is consulting a forecaster. The decision maker should assess the validity of a forecast by seeing whether experts agree. The decision maker is advised that expert forecasts can be influenced by the way the question is framed. The decision maker should seek out experts with different backgrounds. The forecast should be famed in a way making it easy to understand.

In the phrasing of some of these principles, it seems to be assumed that all forecasters are expert. On the other hand, the principles are rather clear that experts are individuals who have "good domain knowledge and are not subject to bias" (p. 25). But the forecaster/expert is not always to be trusted, as in principles that recommend that the decision maker find experts who are unbiased and impartial. One principle asserts that experts can get confused by spurious data. Experts should be asked to justify their forecasts in writing. Do a reliability check by having the expert do their forecast twice, some days apart. The decision maker should be sure to consult more than one expert (the range of 5–20 is mentioned). The decision maker should rely on the forecaster's subjective judgments only when there is expertise that is brings to bear knowledge in addition to that provided by the formal (model) analyses.

Armstrong notes that some of the principles verge on common sense (e.g., graphs can be confusing, it is important to rely on the notion of statistical significance is important). The principles fall at a general or high level, and as such are quite valuable as advice, and reveal aspects of the overall job of the forecaster. However, in their form as a decontextualized list, they do not fit together to describe a coherent a workflow or start-to-finish reasoning process. That said, the principles are consistent with the reasoning or workflow models that have been proffered by meteorologists, and that we describe in this chapter.

Reasoning Process

In many discussions by forecasters of their forecasting procedures, one sees descriptions of their reasoning processes and strategies:

> the subjective forecasting of any weather phenomenon first requires an understanding of the relevant physical processes. ...The meteorologist observes, evaluates, and thinks ... thus, there are a number of activities which directly involve the cognitive processes of the meteorologist. (Lusk, Stewart, Hammond, and Potts, 1990, p. 627)

In their classic text on the science of meteorology, Godske, Bergeron, Bjerknes, and Bundgaard (1957) discussed at length such activities as the preparation of maps and reports, but when it came down to discussing what the meteorologist actually *does*, their presentation was in reference to reasoning:

> the data must be fitted into the kinematically and dynamically most probable system ... models are found by experience. ... Paramount in importance are the special talents with which the analyst should be endowed: a faculty of combining a large number of observations into the most

logical three-dimensional mental pictures, an intimate knowledge of the dynamics of the tropospheric models and of the local factors affecting them, expeditiousness in achieving the analysis in the shortest time possible without sacrificing quality. (pp. 651–653)

In his review of the short-range forecasting methods, Doswell (1986c) made similar statements about forecaster reasoning:

In contrast [to numerical modeling] the human-based forecast process is more complicated ... the human does not weight equally all the data on four dimensions. Instead, pattern recognition is used to assimilate complexity. Humans blend experience, theory, concepts, conjectures, and all the available data ... into a four-dimensional image of the atmosphere ... pattern recognition is crucial to the production of the trend ... knowing how the atmosphere will evolve may depend on knowing what processes are responsible for the observed distributions ... the human can make correct assessments and predictions with limited data—something that no purely objective approach can accomplish. This ability is counterbalanced by the capacity for disastrously incorrect assessments and predictions. (p. 690)

Additional instances of descriptions of forecaster reasoning are the decision trees for short-range forecasting (Belville and Johnson, 1982; Ellrod, 1989; Miller, 1972). Such trees represent specific sequences of decision points involved in each of a variety of forecasting situations (e.g., heavy snow). Each decision point in a tree involves a parametric question (e.g., about relative humidity, air temperature at various geopotential heights, turbulence, vorticity, and stability), each of which is to be answered in terms of weather data. Proceeding down a tree beginning with the most important or most diagnostic features, one ends up at terminal nodes that contain diagnostic conclusions (e.g., the occurrence of a convective event). Although the decision points are expressed in terms of weather parameters, it takes a forecaster to inspect the data and determine what parameter values to enter into the decision process.

Alan Murphy and Robert Winkler (1971) conducted a survey of forecasters at a commercial weather services company. The questionnaire asked about the information sources that were used and their importance, the relationships between judgment and forecasting, the meaning of probability forecasts, and the relation of the forecasts to the eventual weather. Salient findings were:

- the information deemed to be most important (e.g., hourly data, depending on the season) was also the information that the forecaster examined first,
- the information deemed least important in easy forecasting situations was regarded as more important in difficult forecasting situations,
- forecasters sometimes "hedged" when creating their forecast products, that is, the forecasts did not completely represent their judgments (see also Murphy, 1993),

Table 4.1
Proposed characteristics of a good forecaster

Technical Skills
Technical proficiency
Adaptability (to technology)
Ability to synthesize knowledge to useable information
Ability to learn from past events
Good diagnosis and prognosis skills
Ability to assimilate and integrate wide variety of data/information
Retain objectivity about forecast
Personality Components
Are aware of user needs, knowledge, expectations
Learn from peers
Have a strong interest and passion for meteorology
Have good management and people skills
Acknowledge others' perspectives
Are honest in communication with other forecasters
Can withstand criticism
Accept accountability for mistakes
Have stamina for shift work and long hours
Are dedicated to the profession
Provide feedback to developers/researchers

Source: Adapted from Stuart et al. (2006).

• forecasters differed about the meaning of POP forecasts—as subjective probability or "fair bet" as to whether precipitation would occur on the forecast area or as relative frequency of occurrence versus nonoccurrence across similar weather situations,
• forecasters agreed that the meaning of the POP forecast depends on the situation being forecast (shower vs. nonshower forecasts, point vs. areal forecasts), and
• forecasters were willing to alter their forecasting procedure if their forecasts were incorrect (i.e., spend more time inspecting data, reexamine past instances of good performance).

At the 2004 annual meeting of the American Meteorological Society, approximately 200 members representing the international meteorological community showed a "remarkable consensus" on the characteristics of a good forecaster (Stuart et al., 2006). These characteristics were placed into two broad categories: meteorological/technical skills and personality. Although we find this distinction to be psychologically naive, we retain it in table 4.1. The listing of technical skills basically seems to say two things: (1) good forecasters are good at forecasting, and (2) good forecasters are good at learning and reasoning. The personality components listing is somewhat more interesting, saying that good forecasters are motivated, collaborative, and perseverant. But it should be noted that most of the personality components perhaps say as much about forecasting organizations as they do about the personality of forecasters.

Forecasters' Descriptions of Forecaster Reasoning

By tradition, dating to the classic work of Godske et al. (1957), forecasting is often described using the medical metaphors of "diagnosis" and "prognosis." For instance, Smith, Zuckerberg, Schafer, and Rasch (1986) described the forecasting process as involving three sequential questions:

1. What is going on? (Diagnosis) →
2. Why is it happening? (Diagnosis) →
3. How is it going to change? (Prognosis).

Likewise, Doswell (1986a, 1986b, 2004) analyzed forecaster reasoning in terms of the achievement of a diagnosis (a conceptual model of the weather) leading to a prognosis (forecast). Lance Bosart (2003) characterized the weather analysis and forecasting task by the following six elements:

1. What happened?
2. Why did it happen?
3. What is happening now?
4. Why is it happening?
5. What will happen?
6. Why will it happen?

Bullock (1985) and Curtis (1992) described forecasting in terms of an iterative cycle of six steps:

1. Observation →
2. Analysis →
3. Diagnosis (synthesis) →
4. Hypothesis formation and testing →
5. Prognosis (forecasting) →
6. Back to 1.

Meteorologists agree that forecasting is not an activity that can be completely prescribed or proceduralized (LaDue, 2011). Only generalized guidance has been offered describing the forecasting process. For example, renowned forecaster Leonard Snellman (1982, 1991) presented what he called the "forecasting funnel." The basic idea is that a forecast process starts with an attempt to get the "big picture" of what is happening in the hemisphere in terms of major forces and dynamics and then inspect data to focus the understanding down to the continental scale (synoptic scale), at which point one asks, "What will be the forecasting problem of the day?" at regional scales (mesoscale). In parallel with the focusing of scale is an expansion of time. That is, the "big picture" can be determined easily and quickly, but more time has to be taken to make sense of what is happening at the smaller spacetime scales. In addition, Snellman listed some specific strategies, including: (1) the "persistence forecast," which assumes that the weather of the immediate future is likely to resemble the immediate past; (2) the "climatological forecast," which relies on climate data and averages for various weather parameters; and (3) the "consensus forecast," which relies most heavily on the search for agreement in the predictions of multiple computer models. Snellman noted that forecasters rely on a combination of these approaches. Snellman's reasoning model is described graphically in figure 4.4.

In reaching for "the perfect forecast," senior forecaster Joe Bastardi (2000) recommended a process reminiscent of Snellman's funnel, but Bastardi emphasized the first step: Getting the big picture or the "teleconnections" that span the entire globe, the major forces at hemispherical and continental scales, influences that the pattern seen in one season has on the pattern seen in the following season, the source regions of the larger air masses, and so forth.

> To assure the perfect forecast, the forecaster must: a) be married to the weather, b) have a hands-on knowledge of the actual hemispheric pattern ... c) first determine the preliminary forecast based on the weather data and pattern, but without reference to the model results, then d) use the modeling to fine tune the forecast. Is this contrary to the method most forecasters use today? It is people, not machines, that can do this in a consistent basis for extreme weather situations. (Bastardi, 2000, handout page)

Using a different sort of metaphor, Doswell et al. (1996) proposed an "ingredients" approach to the forecasting process. The forecaster is advised to look daily at the ingredients necessary for certain types of weather events and assess whether those ingredients are present in sufficient quantity or balance to cause those types of weather. This approach corresponds roughly to the activities Snellman grouped into the synoptic and mesoscale activities. It has been shown to be useful in certain types of forecasting tasks such as flash flooding. To use an ingredients-based approach, a forecaster must

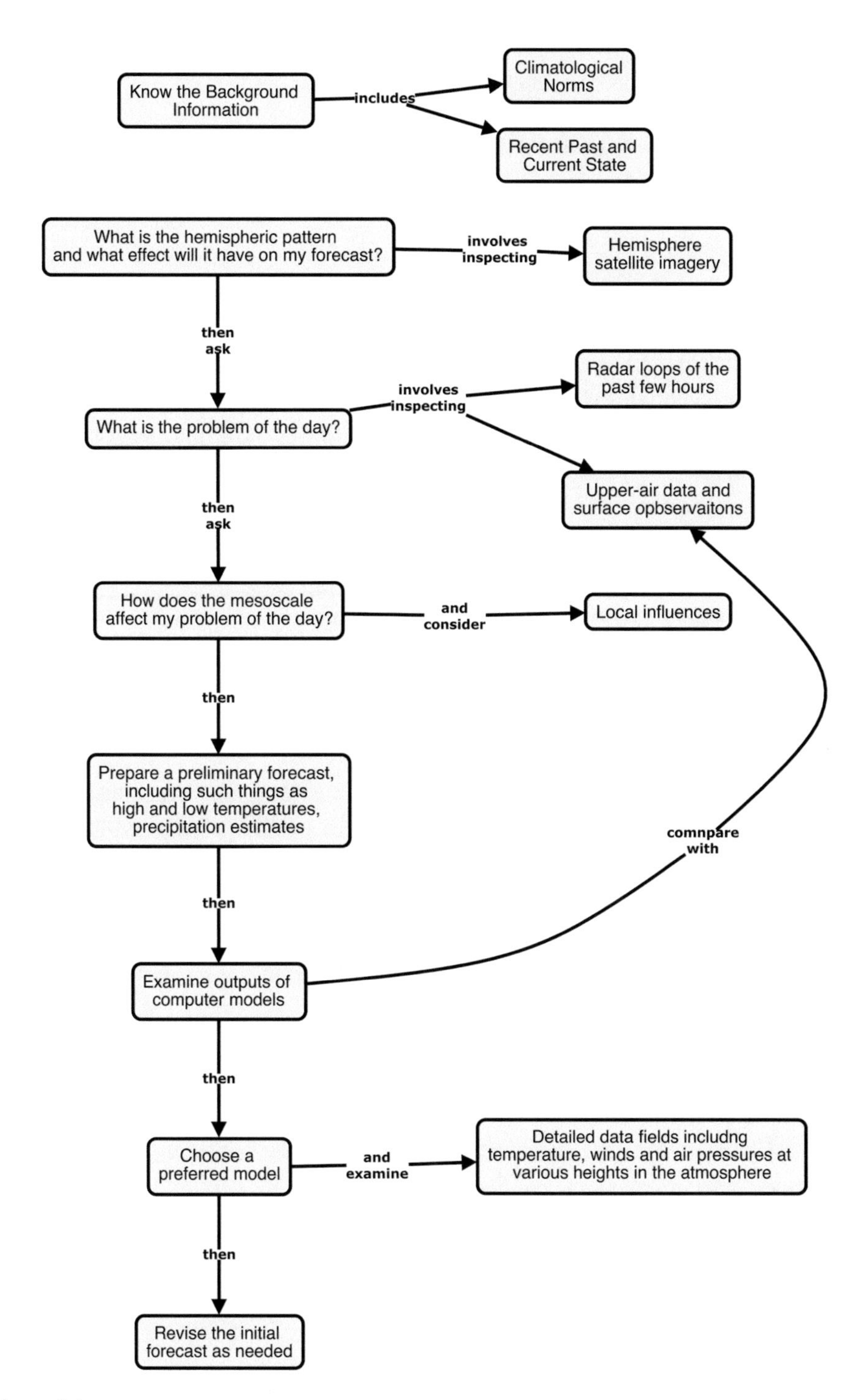

Figure 4.4

A workflow rendering of Leonard Snellman's "forecasting funnel" approach.

know which ingredients are critical in each particular weather event and region; many ingredients for something are nearly always present.

Conclusions

Meteorologists' discussions of the processes of forecasting have relied on cognitive factors. The foundations of forecasting are said to lie in mathematical analyses—this being the impression given by texts. But the foundations are also said to lie in "subjective" analysis—analysis that depends on knowledge, reasoning skill, and perceptual skill. These factors form the core of the modern conception of expertise in cognitive science (see Ericsson and Smith, 1991; Glaser, 1987; Hoffman, 1992).

Discussions of the forecasting process by meteorologists point to the crucial roles of reasoning skill, perceptual skill, knowledge, and experience wrapped up in a larger process that forecasters refer to as conceptual modeling. It is striking that meteorology (long considered a "hard" science) suggests that some of the most difficult aspects of the forecasting process concern the human component (Doswell, 1986c). Further, it is clear that most meteorologists believe that the forecasting process will never be at a level where the human component can be dismissed.

Meteorologists describe forecasting as a process by a "diagnosis-prognosis" analogy to scientific and medical hypothesis testing (cf. Bullock, 1985; Curtis, 1992; Doswell, 1986a, 1986b, 2004; Snellman, 1982). They use a great deal of information to make an informed decision, a form of abduction, that is, inference to the best possible explanation based on some evidence. In addition, meteorologists place considerable emphasis on perceptual skill, that is, the ability to perceive meaningful patterns in data and displays.

Meteorologists seem to agree that forecasting is a collection of many activities that morph over time as technology changes, and not a single activity that can be locked down as a set of prescribed or mandated steps. However, they have outlined general approaches to creating a forecast (e.g., the forecast funnel by Snellman, 1982; see chapter 8), which is to look daily at the ingredients necessary for certain types of weather events and assess whether those ingredients are available in sufficient quantity or balance to cause those types of weather. This ingredients-based approach has been shown to be useful in certain types of forecasting tasks such as flash flooding (Doswell, Brooks, and Maddox, 1996). To use an ingredients-based approach, a forecaster must know which ingredients are critical in each particular weather event and region because many necessary ingredients are present.

Murphy and Winkler (1971) drew the following conclusion from the answers that forecasters gave to their questionnaire about the forecasting process:

> [they] provided very little information regarding the nature of the assessment process itself. ... In view of the fact that weather forecasting is, and is likely to remain, a process in which forecasters assimilate information from a variety of sources and formulate judgments on the basis of this information, the nature and "efficiency" of the assessment process are "problems" of considerable importance. (p. 163)

Research on forecaster reasoning addresses this gap and is discussed in chapters 8 and 9.

5 How Well Do Forecasters (and Forecasts) Perform?

A forecast issued 26 January 2015 said that an imminent blizzard of "historic proportions" is predicted with seemingly total (100%) certainty to bury cities from Philadelphia to Portland. Rarely does one hear forecasts of snowstorms described with complete confidence being of historic, disastrous, life threatening, unprecedented, massive, etc. proportions even when only 24–36 hours in advance—not even with "historic" preceded by "likely," "probably," "potentially," etc. (Tracton, 2015)

"Weather forecasts today are at the point where their reliability and dependability are really good" (Lorditch, 2009, p. 25). However, it is common to hear complaints that weather forecasters are slouches, and how wrong their forecasts are. In the survey of undergraduates taking introductory meteorology course, Knox and Ackerman (2005) found that questions such as "Why is the meteorologist always so wrong?" were frequently mentioned as being of interest. Here's a story about a storm in 2000, about "how they got it wrong" (from Ladue, 2011):

Human forecasters ignored the signs in real-time weather data and continued to trust the models. About the time that snow began falling at surprising 1 to 2 inch per hour rates in North Carolina on the evening of the 24th, models finally began to correctly place the storm track over land. Human forecasters scrambled to change their forecasts during the evening hours, but Washington, D.C. was particularly affected by the timing: most people in that area go to bed before the late night news. Unaware the forecast had changed, they did not leave extra time for shoveling driveways or a longer commute. Officials responsible for activating sand and salt crews in the D.C. area were able to call in staff and mitigate some of the effect (Sipress, 2000), but the snow fell hard and fast during the overnight hours and into the morning, overwhelming snow-clearing efforts on the Metro rails and major roadways (Layton and Sipress, 2000a, 2000b). It was hardly a routine commute to work for D.C. area residents the next morning.

The population of the Washington, DC, area is more than 6 million. It was hit the hardest because of the timing. New York and Boston were also hit hard, but they had a little more time to get ready.

Heard far less often are proclamations of forecaster and forecast successes. Consider, for example, the blizzard of January 26–28, 2015, which struck the northeast United States. It moved up the eastern seaboard from North Carolina to the Washington, DC, New York, and Boston metro areas. The storm dropped a record 20.3 inches of snow near where it developed in the Raleigh-Durham, North Carolina, area. Even as real-time weather data began to foreshadow an ominous change in the expected evolution of the event, all three numerical weather prediction models in use at the time continued to develop the storm sufficiently far offshore to avoid impacting the dense East Coast population areas.

This storm event and the sheer magnitude of the expected snowfall (two to three feet of snow) appeared on the forecasting horizon three to five days in advance. The timing [of the predicted snowfall] was almost right on for most locales and the storm's central location was only about 100 miles or so from the forecast position. Further, the storm (which hadn't even formed yet when the first snow forecasts were issued) did undergo rapid cyclogenesis (deepening), verifying the "meteorological bomb" forecast. Snowfall reached the three-foot depth in some locations. Blizzard conditions (forecast days in advance) occurred. All in all, this was a superb forecast. Such a forecast, with this degree of overall accuracy, would not have been possible 30 to 40 years ago. (Mogil, 2015, p. 5)

The NWS forecasters issued strong warnings of potentially huge snowfall amounts. The amounts and the places where the snow was heavy were all forecast well. But there is a thing about blizzards, called the "steep snowfall gradient." Even just two or three miles distance can mean an inch or less snowfall. If the storm center had shifted 50 miles one way, Boston would have gotten less than a foot of snow, but instead it got the largest snowfall in recorded history (records dating back more than 130 years). Central Massachusetts was slammed with as much as 34 inches of snow. Had the storm shifted 50 miles another way, New York City would have been buried in snow, but it only received a few inches. Due to the steep gradient, the region at the western edge of the snowfall, northwest of New York City, did not receive the huge snowfall amounts that had been predicted. Same for New York City, generating much public complaint about what was the one miscue in the forecast. There were lots of people who could believe that the forecast was lousy. In fact, it was a huge success for severe weather forecasting.

Airlines, school systems, agencies responsible for snow removal and others need to make plans in advance. Forecasters need to and must convey their best assessment of a situation. A few words like "… the expected snowfall gradient on the western side of the storm will be very intense. Small shifts in storm movement can cause significant changes to the forecast in these areas," would be preferred to "there's a 20% chance of more than 18 inches of snow but an 80% chance we'll get at least six inches." (Mogil, 2015, p. 6)

Thus, one sees that forecasters are in a constant struggle to provide accurate, useful, and understandable forecasts while explaining the reasons for uncertainties *and* dispelling popular myth about how bad forecasts are.

While working in the Fort Worth TX forecast office, *The Dallas Morning News* called and asked why our forecasts were so bad for the month. ... I described the forecast process to him and explained that data was limited in parts of Texas. The reporter went back to his desk and reported, not about the forecast errors, but rather, about the problems involved in forecasting a region with high temperature and moisture variability. (Mogil, 2015, p. 10)

The evaluation of forecasts for heavy snowfall (or heavy rainfall) involves the calculation of "threat scores" (see Hamill, 1999). A threat score is the proportion of two areas: The area where a forecast of heavy precipitation was made compared to the area where heavy precipitation actually occurred. Lee Grenci of Pennsylvania State University examined threat score data from the years 1961 through 2000 (Grenci, 2001). "A group of experienced forecasters typically earns a threat score of less than 0.2 out of a possible 1.0 for predicting areas where snowfall will be 12 or more inches 36 to 60 hours in advance" (p. 51). Data from the Weather Prediction Center for 2015 show monthly average threat scores in the range of –0.39 to 0.43 [http://www.wpc.ncep.noaa.gov/html/hpcverif.shtml]. In short, it is difficult to forecast where heavy snowfalls (or heavy rainfalls) will occur. But there are successes. One spectacular success was for the March 1993 winter superstorm that hit the northeastern United States. A timeline for this event is presented in table 5.1.

Cyclogenesis along the East Coast was predicted up to five days in advance. The unusual intensity of the storm was highlighted three days in advance, with snowfall amounts exceeding 12 inches predicted over a large area with unprecedented lead times. Numerous blizzard watches and warnings were also issued with unprecedented lead times, allowing the media and government officials to prepare the public, aviation and marine interests to take necessary precautions. ... The forecasts for heavy snow and rate of snowfall were consistent across the entire event, although the snow in eastern Kentucky was underforecast. The winter storm watches issued by WFOs in the regions expected to receive the most snowfall were issued with 25 to 40 hours lead time. The winter storm warnings and special weather statements issued by the WFOs on 11 March were issued with 10–20 hours lead time, before a single snowflake had fallen. The long lead times allowed emergency response to coordinate with utilities, implement shelter plans, advise health centers to stock additional supplies, activate emergency broadcast systems, etc. ... the increasing confidence of forecasters to predict major storm events, although hard to quantify, was perhaps the key ingredient for the unprecedented lead times ... that led people to believe the forecasters and take appropriate action. (Uccellini, et al., 1995, pp. 197–199)

Forecasts have definitely improved over recent decades but not enough to satisfy forecasters. Writing in 2001, Lee Grenci said:

Table 5.1
A timeline for forecasting the March 1993 superstorm

Date	What the Technology Said	What the Forecasters Thought	What the Forecasters Did
7 March	Storm would develop along the East Coast of the United States. The new forecasts and forecasts based on statistical analysis predicted an 85% chance of precipitation (2–3 inches of snow) in West Virginia.	Forecasters at the National Meteorological Center (NMC) felt that cyclogenesis along the East Coast of the United States would be unlikely because the models had overforecasted previous events of this type. Also, the weak cyclones that had developed tracked further inland, and stronger ones that had developed tracked out over the Atlantic Ocean.	Continued analysis of model outputs and observational data.
7–11 March	Models consistently predicted a major cyclone along the East Coast.	NMC forecasters' skepticism of the model outputs diminished.	Continued analysis of model outputs and observational data.
10–11 March	One model predicted cyclogenesis in the Gulf of Mexico, whereas other models predicted cyclogenesis off the East Coast.	Consternation over the inconsistency of the outputs of the various computer models.	Local Weather Forecast Offices (WFOs) along the East Coast began issuing discussions of the potential for a severe storm, with blizzard conditions. They commenced frequent briefings with local emergency response managers. NWS Eastern Region Headquarters advised the Federal Emergency Management Agency of the possibility of blizzard conditions.

Table 5.1 (continued)

Date	What the Technology Said	What the Forecasters Thought	What the Forecasters Did
11 March	The storm event began to unfold. A jet stream pattern in the western United States that the models had predicted developed as predicted and would influence the East Coast cyclogenesis.		The National Meteorological Center (NMC) began to issue storm summary statements, predicting "unusually severe" and "perhaps record-breaking" snowfall of "historic proportions." Hurricane hotline was activated. WFOs along the East Coast began to exchange information and develop a consensus on which all of them could rely. WFOs issued the first winter weather watches.
11–12 March	Details of the developing storm differed from what computer models were predicting.	Difficulty in predicting the location, intensity, and track of the developing cyclone.	
12–13 March	The model that was predicting storm development along the East Coast underestimated the rapid cyclogenesis occurring in the Gulf of Mexico.		The NMC relied more on the outputs of the computer models that model the entire globe, rather than those that model just North America, because those had tended to overpredict the central low pressure of the storm.
13 March	Models began to converge on cyclogenesis in the Gulf, off the Louisiana coast.	The model differences were significant. Forecasters continued to compromise between their own analyses and the predictions of the computer models about cyclone position and the central pressure of the low.	NMC forecast a developing low pressure center in the southeastern United States. Adjusted the predicted position of the rain-snow line further south into central Alabama. With each successive model run, the forecasters predicted a lower and lower central pressure. Blizzard warnings were issued by all eastern region WFOs.

Table 5.1 (continued)

Date	What the Technology Said	What the Forecasters Thought	What the Forecasters Did
13 March	The Sterling, VA WFO had one of the first NEXRAD radar installations. Individual bands of snowfall could be tracked.	Although the Sterling, VA forecasters were busy because of the weather event and the need to work using a new technology, NEXRAD enabled forecasting that was previously impossible.	The Sterling, VA WFO issued frequent location-specific half-hourly nowcasts. They confirmed radar scan data with surface observations made by a network of cooperating observers.
14 March	Models began to converge in forecasting a major storm along the East Coast but differed in predicting its track. Earlier that winter, two of the models correctly predicted a track west of the Appalachian mountains when a third model had kept the storms along the East Coast.	Forecasters had to make judgments concerning snowfall amounts and the timing of when snow would change to ice or rain then back to snow. Snow amounts were of special concern because a record-breaking storm was expected. Forecasters felt that the models were showing underdevelopment of the cyclone. Different models were predicting different scenarios. Forecasters relied on different models to predict different things (low pressure, storm track, etc.) based on past experience with the models' successes.	NMC forecasters compromised among the models and their own judgments about the central pressure and location of the cyclone. Forecasters began to shift the predicted rain-snow line further to the north.
14 March	Models began to converge on the storm track, placing it along the New England coastline.		

Source: After Kocin et al. (1995) and Uccellini et al. (1995).

> Despite decades of progress in temperature forecasting, improved tornado warnings, and hurricane predictions, and so forth, the science has made little or no headway in the category of predicting areas of heavy precipitation over the past 40 years. ... If truth be told, we know a lot more than we used to know about a lot of things, but we still can't predict areas of heavy snow (or heavy rain) worth a hoot. (p. 50)

It is an overstatement to say "no headway." In the case of predicting where heavy snow will fall, it is important to consider the gradient effect we described earlier as this reminds us that some things can be of intrinsically low predictability (see Brooks, Doswell, and Maddox, 1992; Ehrendorfer, 1997; Mylne, 2006; Palmer and Hagedorn, 2006). (We will have more to say about intrinsic predictability later in this chapter.)

Stories that "they got it wrong" versus stories that "they got it right" can both be correct. After all, all forecasts have some degree of uncertainty associated with them, and the expression of uncertainty in useful ways is an active area of meteorological research (Novak, Bright, and Brennan, 2008). As the narratives we have presented show, some forecasters get some bits of it wrong, and those bits can impact many people. But some forecasters get most of it right most of the time. Forecasts are good and accurate far more often than not. Although they are sometimes off the mark on details, the ways in which they can be off the mark can be subtle and impactful.

Challenges in Measuring the Goodness of Forecasts

Consider again the March 1993 superstorm. The snowfall amounts in Eastern Kentucky and the intense snowbursts in northern Alabama were underforecast. "Nevertheless, these forecasts were consistent in alerting the public of the extreme nature of the event" (Uccellini et al., 1995, p. 194). In approaching the question of "How good are the forecasts?" it is important to keep in mind that some weather events are highly predictable and others are not. Low predictability can be incidental, that is, there may be limitations to the nature of the data available (i.e., data may be inaccurate or sparse). Bosart (2003) argued that the sometimes spectacular failures of expert forecasters result from a lack of real-time data of sufficient resolution and quality, making it difficult to exercise their skill in pattern recognition.

Low predictability can also be intrinsic to the weather phenomena. A given weather variable might be highly predictable or nearly unpredictable depending on the nature of the event (intrinsic predictability) The snowfall gradient effect is a case in point, as we discussed above: Predictions of which areas will get heavy snow get threat scores of only about 0.4.

The question of whether forecaster performance is any good must take predictability into consideration. There are situations where both the computer models and the human forecasters get it wrong, but not because of any misuse of the models or deficiencies on the part of the humans. Rather, it has to do with the limits of predictability and understandability.

"Goodness" and the Context of Use

Whether a forecast is good depends on the context of use. For a public forecast for daily high and low temperatures, the errors (usually plus or minus just a few degrees) are fairly minimal and not noticeable by most public consumers of the forecast. But for an application in forecasting energy demand for the electric utilities, say, small temperature errors have a big impact. This is also true for weather parameters that have critical thresholds (e.g., freezing mark for temperature).

To approximate an answer to the question of what makes for a good forecast, we need to distinguish technical accuracy and correctness from the public's conceptions and misconceptions about the meaning of forecasts. For example, there have been many studies of how people (laypersons, college students) interpret and misinterpret "chance of rain" or "probability of precipitation" (POP) forecasts (Adams, 1973; Gigerenzer et al., 2005; Josslyn et al., 2009; Maunder, 1969; Murphy et al., 1980; Namm, 1979; Rogell, 1972; Savelli and Josslyn, 2013; Stewart, 2009; Wagenaar and Visser, 1979). In converse to the issue of the interpretation of probability, there are issues in comprehension when uncertainty information is *not* provided.

> We live in a forecasting culture in which specificity routinely exceeds the skill of the science. Routinely, seven-day forecasts are represented on television as a single icon (cloud or sun), a single number for a high temperature, and a single number for a low temperature. (Grenci, 2001, p. 51)

Public forecasts of daily high and low temperatures are generally not accompanied by any expression of uncertainty, such as, "Today's high temperature will be 41 degrees but could be as high as 44 degrees or as low as 38 degrees." Savelli and Josslyn (2013; Josslyn and LeClerc, 2012) demonstrated that people's proper understanding of temperature forecasts is significantly aided by presenting such predictive intervals. Essentially, the single value deterministic forecasts leave people not really knowing what will happen.

The Challenges of Measurement

A number of studies have been conducted to gauge the goodness of weather forecasts issued by the NWS (and likewise the weather services in other nations). Conducting this research is not entirely straightforward. Suppose a forecast for a region is that the daily high temperature will be 80 degrees Fahrenheit, but the daily high as measured at an official weather station ends up 82 degrees. What if the forecast for a region is a 50% chance of rain, and it does rain, but only in roughly half of the region? Were the 50% of people who decided not to carry an umbrella upset? Were the farmers who saw no rain upset because they thought there was a chance they would see rain? Is a forecast of "partly sunny" the same as a forecast of "cloudy intervals"?

What these questions suggest is that "errors" in forecasting are as much related to the difficulty of measuring forecast accuracy as the inherent accuracy or correctness of the forecasts themselves. Nevertheless, one has to try and measure success somehow. Many evaluations look at one or another single weather parameter, such as rain or temperatures. Any evaluation must determine what range of values for things such as daily high temperature counts as a "hit." Plus or minus three degrees is typically used as a metric. Generally speaking for short-term forecasts (covering a period of one to a few days), plots of observed values against predicted values for temperatures, rain likelihood, and precipitation amounts form a nice 45-degree angle line (i.e., showing a strong correlation) for forecasts from the NWS and commercial forecasting services such as The Weather Channel (R. Olson, 2014). There are some indications that forecasts for rain tend to be overforecasts, that is, more rain is predicted than occurs, but there are also indications that computer models do the same (Williams, 2013). According to R. Olson (2014), broadcast weather forecasters are especially prone to overforecast rain amounts. A 2009 study by Intellovations LLC [http://www.forecastadvisor.com/blog] measured temperature forecast accuracy in terms of error and found that forecasts from the NWS were accurate (i.e., within the 3-degree metric) for forecasts up to two days in the future. For about 10 days in the future, the forecasts were no better than what one would predict on the basis of climatological data. Similar findings were reported for a study conducted by MINITAB, a statistical services company [http://www.minitab.com/en-us/].

Short-term forecasts are almost always more accurate than longer-term forecasts because of the inherent unpredictability of weather. Larger and more slowly evolving weather systems, such as the ones associated with heat waves and cold spells are more predictable at longer times than are the day-to-day variations that control rain showers. These systems last longer and trigger relatively gradual meteorological changes that can be more easily seen in the data ahead of time. (Postel, 2012, p. 1)

Another way of looking at this matter is to evaluate how forecast accuracy has improved over some period of years. Forecasts have been improving since the advent of computer models (see Charba and Klein, 1980; Monmonier, 1999, chapter 5; Silver, 2012). There are footnotes to this, of course. For example, forecasts reaching out over longer periods of time can be good for slow-moving weather systems and weather related to large air masses. Forecasts for tropical weather are also generally quite accurate, especially now that the computer models take into account the ocean surface temperature, which is a key factor. The 2015 Statement on Weather Analysis and Forecasting issued by the AMS said the following:

For lead times of approximately twelve hours to two days, short-range forecasts [for] tropical storms, hurricanes, and frontal systems and their accompanying weather elements (e.g., temperature, wind, and precipitation) are significantly improving: two-day National Hurricane Center hurricane track forecasts issued in 2012 had an average error of 79 miles as compared to 140 miles in 2002 and 192 miles in 1992. Likewise, two-day NOAA Weather Prediction Center forecasts of 24-hour accumulated precipitation issued in 2012 were as accurate as one-day forecasts in 2006. Medium-range forecasts with lead times of two to seven days are most successful for meteorological phenomena that stretch across areas of a thousand miles or more, or for larger-scale conditions that set the stage for development of smaller phenomena, such as severe thunderstorms. Over the past three decades, the skillful range of medium-range forecasts has been extended by roughly one day per decade. Specifically, five- and six-day surface temperature forecasts issued by the NWS had the same level of accuracy in 2012 as did three- and four- day surface temperature forecasts, respectively, in 1992. Extended-range forecasts are typically issued for meteorological phenomena that cover areas ranging from thousands of miles to the size of a continent and involve lead times of one to two weeks. Presently, forecasts of daily or specific weather conditions do not exhibit useful skill beyond eight days, meaning that their accuracy is low. However, probabilistic forecasts issued to highlight significant trends (e.g., warmer than normal, wetter than normal) can be skillful when compared to a baseline forecast. For example, the NOAA Climate Prediction Center operational 8–14 day temperature forecast skill in 2013 was approximately equal to that of operational 6–10 day temperature forecasts from the late 1990s, again demonstrating an increase in forecast success over time. Finally, monthly and seasonal forecasts are typically issued for meteorological phenomena that cover areas ranging from the size of a continent to the planet as a whole. Skill in monthly and seasonal forecasts is extremely variable from period to period, but the skill of NOAA Climate Prediction Center one- and three-month forecasts of temperature and precipitation increased by more than 25% between 2006 and 2013. Increases in forecast skill at these lead times can largely be attributed to improved understanding of and ability to forecast major modes of large-scale climate variability such as the El Niño-Southern Oscillation. (©American Meteorological Society, 2015, Used with permission)

Forecast Verification Research

A great number of formal experiments have been conducted to evaluate the quality and correctness of weather forecasts. These are referred to as "forecast verification" experiments (e.g., Colucci, Knappenberger, and Cepa, 1992; Jolliffe and Stephenson, 2003, 2005; Murphy and Daan, 1985; Murphy and Winkler, 1987; Ralph et al., 2005; Winkler, Murphy, and Katz, 1977; Tracton, 1993). In this research, the focus is on the accuracy of the forecasts (things such as temperature outlook and precipitation amounts). For each forecast, typically composed by a small group of forecasters rather than a single individual, a single "skill score" is derived.

Skill score does not mean skill in a human performance or psychological sense. It is a property of forecasts, not forecasters. A forecast has skill if it does better than what one would predict on the basis just of the climatological data (Sanders, 1958). Thus, for a locale with highly predictable weather—such as an island in the South Pacific—one can accurately predict such things as daily high temperature, rainfall, and winds just by looking up the mean values for that day of the year in the climate tables. A weather forecast (which could be from a computer model or a weather forecaster) would not do any better and would therefore have no "skill," although it might come from a good computer program or good forecaster.

The skill score represents the difference between the forecast values and the eventual (actual) weather, but the difference or error is relativized to the climatological norm (see Brier, 1950; Murphy, 1992; Stewart, 1990). For example, without even looking at a weather map, one could predict zero rainfall for most of the days of the summer for Los Angeles and receive a high hit rate, but a low skill score would be given because the forecast would be so close to the climatological average. As another example, suppose that a forecast for a given day is that there is a high likelihood of rain: up to 2 inches. Suppose further that climatological data show that on that day, on average, rainfall is likely and rainfall amounts are upward of 2 inches. Suppose further that it does indeed rain, and it rains that much. In this case, the forecast has no skill even though it was valid. However, if the forecast was for no rain, and it didn't rain, then the forecast would have had skill because the climatological forecast did not verify. As another example, skills scores can be computed for variables as forecast by radar (Keenan, Potts, and Stevenson, 1992). (For discussions of the mathematics of skill scores and different ways of calculating them, see Hamill, 1999; Heideman, Stewart, Moninger, and Reagan-Cirincione, 1993; Jolliffe and Stephenson, 2003; Manzato, 2005; Mielke et al., 1997; Murphy, 1988, 1992, 1993; Murphy and Daan, 1985; Murphy and Winkler, 1970, 1987; Stephenson, 2000; Stewart and Lusk, 1994; Wilks, 1995.)

This concept of a skill score highlights the fact that some things are simply more predictable than others (Mylne, 2006; Palmer and Hagedorn, 2006). This fact must be kept in mind when considering claims that weather forecasts are "no good." This is because the "skill" can be primarily a function of the things-in-themselves, such as the highly predictable weather in certain regions. Predictability can also be a consequence of an interaction between the limits of our capacity to predict and the nature of the weather events. An example would be tornados. They can develop fast, which by itself makes them hard to predict. Although the precursor factors are well understood (e.g., the dynamics within severe thunderstorms), our ability to detect tornados before they happen is still limited. However, detection is far superior than it was 30 years ago.

The routine analysis of NWS forecaster skill scores since the 1960s makes the domain of meteorology stand out from many other domains of professional expertise, in that there is a ready-made database that can be used to derive objective measures of proficiency (Brier, 1950; Charba and Klein, 1980; Jolliffe and Stephenson, 2003; Roebber, 2009; Wilks, 1995). Analyses of skill scores have repeatedly shown that the forecasts of highly experienced forecasters are greater than those produced by less-experienced forecasters (Dyer, 1987; Hoffman, Coffey, and Ford, 2000; see also chapter 7).

Furthermore, the skill score of individual (human) forecasts is generally close to (although a bit lower than) the "consensus skill score," which is based on an average of the forecasts of members of a team or an average of the forecasts of a select group of experts (see Clemen, 1989; Clemen and Murphy, 1986; Murphy, 1993; Sanders, 1973). The "bit lower than" needs to be understood in terms of factors that can cause a consensus forecast to not show greater skill than that of some individual forecasters. Specifically, for some weather events and under some circumstances, different forecasters will use different strategies and rely on different data or cues. Hence, the "crowd" is not always the wisest.

Forecast verification research has investigated a variety of topics, including the improvement of skill scores across the period during which college students receive their education in meteorology (see Roebber and Bosart, 1996b), and the effect of new technology and forecast products (e.g., Colucci, Knappenberger, and Cepa, 1992; Vislocky and Fritsch, 1997.

For example, Roebber and Bosart (1996b) approximated the effect on skill score of moving a forecaster to a new location—a drop in skill at forecasting temperatures by 5% to 10% relative to a consensus forecast and a drop in skill at forecasting precipitation by 10% or more. This result conforms to a view of meteorologists (e.g., Doswell, 1986c, 2004) that local knowledge is critical in forecasting. The same conclusion comes from first-generation expert systems work (see chapter 12; also see Elio, de Haan, and

Strong, 1987) and particular operational experiences such as in the application of the PRAVDA system (see chapter 2) to forecasting for the 1996 Summer Olympics (Treinish and Rothfusz, 1997).

The main goal of this avenue of research is to generate ideas for how to improve forecasting accuracy (or "resolution") and reliability, including changes in the methods used to calculate skill scores and generate consensus forecasts (Baars and Mass, 2005; Clemen, 1985, 1989; Clemen and Murphy, 1989; Murphy, 1992; Roebber, 2010; Stephenson, 2000). The ultimate goal is to develop methods for generating forecasts that are "good" in three senses (Murphy, 1993):

1. the forecasters are self-consistent, that is, the forecasts reflect the forecasters' best judgments, including expressions of uncertainty;
2. the conditions that are forecast match the actual conditions during the valid time of the forecast; and
3. the forecast takes into account the information requirements and decision-making problems of the end-users.

Additional goals are to encourage new approaches to training (e.g., the use of perceptual search tasks, feedback, and methods for reducing judgment bias) and suggest ideas for changes in information systems (i.e., improved displays) (see Stewart and Lusk, 1994).

We now review two major forecast verification projects that have the clearest implications regarding forecaster cognition.

How Good Are Those Probability Forecasts?

In 1905–1906, W. Ernest Cooke, an Australian astronomer, suggested that weather forecasts be accompanied by an expression of "the weight or degree of probability which the forecaster himself attaches to the prediction" (Cooke, 1906). Cooke's proposed scale was five "degrees of doubt." His main concern was that the inclusion of such judgments, for multiple weather parameters, would mean cramming more information into what were already crammed telegraph messages. In 1951, Philip Williams reported a study in which U.S. Weather Bureau staff in Salt Lake City accompanied their precipitation forecasts with expressions of their degree of confidence (expressed as percentages). The results were "favorable," as Williams put it: When the probability of rain was forecast at 100% confidence, the relative frequency of rain was 88%. When the probability was zero, the frequency was 3%. Each forecaster showed the greatest percentage of hits with the forecasts rated at the highest confidence. Williams also suggested that confidence factors would also be used for temperature forecasts.

Since Cooke put forth his proposal, there has been continuing debate about what probability forecasts mean; some forecasters even claim that some forecasters do not know what probability forecasts mean (see Stuart, Schultz, and Klein, 2007). Does probability of precipitation mean: (1) the proportion of the forecast area in which precipitation *will* occur, (2) the percent of time over the day in which there will be precipitation, (3) the likelihood that precipitation *might* occur anywhere in the area, (4) the chance that there will be precipitation but averaged over the forecast area, (5) the proportion of days just like today on which it rained somewhere in the region, or (6) the product of the proportion of the area where it will rain multiplied by the forecaster's confidence in that judgment? (For a discussion of what the National Weather Service means by POP, see http://www.srh.noaa.gov/ffc/?n=pop.)

Alan Murphy and his colleagues (Murphy, 1985; Murphy and Brown, 1983; Murphy and Daan, 1985; Murphy and Winkler, 1970, 1971, 1974a, 1974b, 1977, 1982, 1984; Winkler and Murphy, 1973) comprise the major group of researchers who have empirically examined aspects of forecaster reasoning and performance with regard to probability-based forecasts. Within the forecasting community, subjective probability in forecasting is how well people are able to predict the weather, whereas objective probability in forecasting is how computer models are able to predict the weather (for a history of subjective probability-based forecasting vs. objective probability forecasting, see Murphy and Winkler, 1984). Murphy et al. focused their investigations on the use of subjective probability in forecasts. In their initial research, Murphy et al. focused on the evaluation of regional probability of precipitation (POP) forecasts. Some issues in probability forecasting were discovered in an initial questionnaire survey of forecasters at a commercial weather services company (Murphy and Winkler, 1971). The questionnaire asked about the information sources that were used and their importance, the relationships between judgment and forecasting, the meaning of probability forecasts, and the relation of the forecasts to the eventual weather.

Forecasters disagreed about the meaning of forecasts—as subjective probability or "fair bet" as to whether precipitation would occur on the forecast area or as relative frequency of occurrence versus nonoccurrence across similar weather situations. Forecasters agreed that the meaning of the POP forecast depended on the situation being forecast (shower vs. nonshower forecasts, point vs. areal forecasts).

In their subsequent research, Murphy and his colleagues focused on the problem of the meaning and interpretation of probability forecasts, as well as the psychology of probabilistic reasoning (see Winkler and Murphy, 1973a, for a discussion of the mathematical representation of alternative strategies for aggregating conditionally dependent information into subjective probabilities). They solicited data nationwide from NWS forecasters (Murphy and Winkler, 1974a). The forecasters reported that POP forecasts were most often interpreted in terms of relative frequency, and they acknowledged that POP forecasts were a kind of "hedge." Most fundamentally, "Forecasters do not "think" in terms of probabilities. ... Forecasters prefer to obtain a general, overall picture of weather situations ... before making their POP forecasts" (Murphy and Winkler, 1974a, p. 1451).

Murphy and Winkler (1974a, 1974b; see also Murphy, 1985; Winkler and Murphy, 1973b) conducted an experiment to see whether NWS forecasters could create "credible temperature interval forecasts," reflecting the degree of belief that the actual daily temperature would fall in some predicted range (today's high, tomorrow's low). First, four experienced Denver Weather Forecast Offices (WFO) forecasters were trained on one or another method of generating credible intervals (e.g., making "indifference judgments" to determine the median and then assigning probabilities to intervals of either fixed or varying size). Then the forecasters used this procedure (in addition to conducting their usual forecasting tasks) over a four-month period, resulting in data for more than 120 forecasts. The results showed that there was a tendency for forecasters to underestimate minimum temperatures and overestimate maximum temperatures. However, the forecasts were valid and reliable overall, in the sense that the interval medians came close to the actual observed temperatures (an average difference of only –0.2 degrees Fahrenheit), and the sense that the new forecasting procedure resulted in forecasts that beat out the predictions that would be made on the basis of climatological data (i.e., the skill scores were positive).

Murphy and Winkler (1977) repeated this experiment at WFOs in Chicago and Milwaukee. The main result of this series of studies was a proof of concept—that forecasters could express their uncertainty in temperature forecasts in probabilistic terms comprising reliable and accurate forecasts. This finding is important from the perspective of the field of judgment and decision making because a great deal of research has attested to the fact that most people, most of the time, are woefully inadequate when it comes to probabilistic reasoning (for a review, see Fraser, Smith, and Smith, 1992).

In the series of studies by Murphy et al., forecasters were (essentially) forced to engage in probabilistic reasoning. The assumption was that the use of subjective probability forecasts should be expanded in terms of incorporation into NWS operations and extended to other types of forecasts that traditionally rely on verbal qualifiers such as "likely." Murphy and Winkler (1982) determined that NWS forecasters could learn to forecast tornadoes in terms of subjective probability. Although the subjective probability method seemed to have induced a slight tendency to overforecast for smaller areas and underforecast for larger areas, the subjective probability forecasts were reliable (i.e., the forecast probability increased as the number of actual tornadoes went up). This finding was obtained despite the fact that the forecasters received no feedback during each of the two five-month periods of the experiment.

Daan and Murphy (1982; Murphy and Daan, 1984, 1985) also conducted studies of probability forecasting in the Netherlands, including an eight-year study of precipitation probability forecasting. Again, forecasts proved to be reliable (forecast probabilities corresponded with observed relative frequencies) and had positive skill scores. The researchers also had forecasters use the subjective probability method in forecasting wind speed and visibility. Even in the case of this novel use of subjective probability forecasting, the probability forecasts were reliable and skilled. There was a tendency, or one might say a bias, to overforecast, in part a reflection of the desire to avoid underforecasting because severe weather events in the Netherlands often can have significant impact. At the end of the first year of conducting the novel forecasts, the forecasters were provided with feedback (one-on-one discussions concerning skill, reliability, and resolution). The forecasts over the second year showed an improvement in reliability and skill and a reduction in overforecasting.

Murphy and Brown (1984) sampled NWS forecasts nationwide and compared subjective probability forecasts (Probability of Precipitation, cloud amount, maximum and minimum temperatures) with those from the computer models, in an examination of trends (over the 1970s) in skill score. As was expected, skill scores tended to decrease as the forecast interval lengthened. Overall, both objective and subjective methods were reliable, and were about equally so for POP. The computer models' objective probabilities beat out the subjective probabilities for amount of cloud cover. With regard to temperature forecasts, skill score for the subjective probabilities depended on season (actual temperatures are more variable in the cool season). Over the 1970s, skill scores for both the subjective and objective methods increased, but the computer models showed a bigger increase in performance, representing improvements in computer modeling.

The research of Murphy et al. demonstrated that experienced forecasters can manifest superior performance (especially for shorter-range forecasts). The performance of individuals often approximates that of the consensus forecast, and they can engage in probabilistic reasoning appropriate to the domain. The increasing accuracy of computer-based forecasting techniques has entailed less value added by the adjustments of the human forecaster (Baars and Mass, 2005; Roebber and Bosart, 1996a, 1996b) for forecasting such things as precipitation and temperature. However, depending on the context, the value added by the human forecaster can still be substantial (e.g., Reynolds, 2003; Roebber, 2010). Situations where the human input to a forecast significantly increases the value of the forecast are those in which the weather deviates from the local norm—showing atypical patterns including severe weather—or where the decision-making context amplifies the importance of incremental forecast improvements. Thus, the increasing accuracy and value of computer-generated forecasts has entailed ideas about a new division of labor—distinguishing between routine-forecast specialists and severe weather or other forecast specialists (see Mass, 2003a, 2003b; Sanders, 1986). This finding fits with a claim made in the literature of meteorology and recapitulated in first-generation work on expert systems: that knowledge of local effects and trends is an important contributor to forecasting expertise.

Evidence has been adduced that weather forecasters are generally "well calibrated" and not highly overconfident in their use of probabilities, although skill scores for POP forecasts can be low if the climatological expectancy is uncertain (e.g., Sanders, 1958). Presumably the general reliability of POP forecasts has to do with the fact that estimating the probabilities of weather events and parameters is part of forecasters' familiar task, and they get daily feedback that allows for good calibration (see Lichtenstein, Fischhoff, and Phillips, 1982; Williams, 1951). Nevertheless, the interpretation of POP forecasts by professionals remains a matter of discussion (De Elia and Laprise, 2005; Gigerenzer et al., 2005; Josslyn et al., 2009; Roulston and Smith, 2004). A key point to that discussion is that many forecasters assert that they do not "think in terms of probabilities." How they reason while trying to make sense of the weather is separate from their judgments in the probability estimation task that is part of the process of generating a forecast. Forecasters can get good at issuing valid probability forecasts because that is a part of their regular job, and they get immediate or nearly immediate feedback from the weather. Another suggestion is that forecasters' creation of subjective probability forecasts has come to rely on examination of the computer models, which have probabilities as one of their direct results. Of course this complicates any comparison of the skill scores of the computer-generated forecasts to the skill of the human-generated subjective probability forecasts (Murphy and Winkler, 1974b): "the use of probabilities in highly recurrent events has produced innumerable benefits thanks to the availability of skillful forecasts whether from ... expert advice or from ensemble prediction" (De Elia and Laprise, 2005, p. 1225).

The interpretation of probability forecasts by laypersons also remains a matter of discussion. This is the current conundrum of probability forecasting. The probabilities can be useful and reliable, yet the public's interpretation of them challenges the forecasting community to communicate clearly to the public. Murphy et al. maintained that the language of probabilities—and an understanding of the theorems of probability—would benefit forecasting, that the forecasts of events other than precipitation could be profitably couched in probabilistic terms. Yet a great deal of psychological research shows that people, in general, are quite bad at reasoning in terms of probabilities; they commit all sorts of reasoning errors and are subject to all manner of biases (see, e.g., Tversky and Kahneman, 1982). Such findings need to be qualified by two facts. First, the "subjects" in the psychological experiments on bias in probabilistic reasoning are almost always college freshmen (taking introductory psychology classes), and, second, the tasks presented to them—various probability puzzles—are not their daily decision-making fare.

In his survey of the research, Murphy (1993) tried to answer the question of what it means for a forecast to be "good." He asserted that, "measures of accuracy or skill do not and cannot tell the whole story regarding forecasting performance" (p. 292). This conclusion sets the stage for a discussion of the research conducted by Alan Stewart and his colleagues.

Predictability versus Understandability

Some weather events can be hard to anticipate because they are of low intrinsic predictability, and some events can be hard to anticipate because they are hard to understand. Both of these factors, and their interaction, play into the question of forecast "goodness." Thomas Stewart and his colleagues have conducted a great many forecast verification experiments (Heideman, Stewart, Moninger, and Regan-Cirincione, 1993; Lusk and Hammond, 1991; Lusk, Stewart, Hammond, and Potts, 1990; Stewart and Lusk, 1994; Stewart, Roebber, and Bosart, 1997). They regard the accuracy of a forecast as a conjoint function of:

- the predictability of the environment (weather events can be more or less predictable),
- the reliability of the information that is presented,
- the accessibility or interpretability of the information, and
- the forecaster's cognition (e.g., knowledge, experience).

Heideman et al. (1993; see also Stewart, Heideman, Moninger, and Regan-Cirincione, 1992) engaged 12 meteorologists from NOAA and the U.S. Air Force Geophysics Laboratory (AFGL, Hanscom Air Force Base) in the nowcasting of severe storm events (convection/tornadoes) using test cases taken from archived data. The forecasters worked under one of four information conditions:

1. Radar data showing storm contours and storm centroid tracks;
2. Radar data plus data about each of the indicated storms (e.g., reflectivity, max shear, convergence line)

3. Radar data, storm data, upper air analysis, and computer model outputs, all presented as a set of 30 slides taken from workstation displays, or

4. Full data as in condition (3) but presented in an interactive computer workstation context.

The research was based on an intriguing suggestion from the literature on the psychology of expertise (e.g., Hoffman, 1992; Shanteau and Stewart, 1992)—that experts sometimes show an increase in confidence over time as more information is acquired, but this increase is not necessarily matched by an increase in judgment accuracy. One possibility, of course, is data overload. Another possibility is that additional information in the form of new data types and formats forces the forecaster to relearn how to perceive: There are new and important cue configurations to integrate into conceptual models. Heideman et al. (1993) tested the assumed benefit of new meteorological data types and graphic displays.

The NOAA (Boulder, Colorado) forecasters outperformed the AFGL forecasters, and this was to be expected due to differences in experience with tornadic storms. As a group, the Boulder forecasters performed better in the Workstation Condition. However, two of the Boulder forecasters and two of the AFGL forecasters performed better in Condition 3 (charts-with-profiles) than in the Workstation Condition. In addition, agreement among the forecasts was less in the Workstation Condition. A consensus forecast based on the forecasts in the Workstation Condition outperformed each of the individual forecasts made in that condition (i.e., the forecasts had not reached the skill limit). Although there was a uniform increase in skill scores across the information conditions, the increase was modest and not statistically significant. Despite difficulties in interpreting the results of this study (due to possible confounds in the experimental design), the clear implication of the results is that the improvement of forecasting may hinge on "devoting resources to improving the *use* of information over and above those needed to increase the *amount* of information" (Heideman et al., 1993, p. 35).

Findings such as these give one pause to wonder whether forecasting expertise can ever manifest high levels of accuracy and agreement. Experts in diverse domains of course can manifest high levels of accuracy (Shanteau, Johnson, and Smith, 2004; Stewart, Roebber, and Bosart, 1997). But experts in *certain* other domains can show poor performance (see Hoffman, 1992; Shanteau, 1992b). These tend to be domains where the primary task involves predicting human behavior (e.g., counseling, jurisprudence, economics, etc.), but some studies show good expert performance at clinical judgment (e.g., Christensen-Szalanski et al., 1982). It is certainly true that experienced forecasters sometimes show relatively poor performance. For instance, Uccellini, Corfiodi, Junker, Kocin, and Olson (1992) presented charts of surface observations to a group of experienced NWS forecasters, who then had to draw their estimates of the positions of fronts. The results showed considerable disagreement and engendered considerable debate on the meaning of the fundamental concept of the "front." The key to this may well be the season and location of the forecast analysis challenge. Locating fronts should be mostly an easy task, but whether a surface front actually represents a trough is not always easy to determine.

These findings affirm the ideas of Jim Shanteau (1992b; see also Hammond, 1980): If a task is highly predictable (in terms of the predictability of the events in question but also in terms of what the expert does—a specific sequences of activities, specific sequence of inspections of particular data types, etc.), then one can predict consensus along with high levels of accuracy and reliability. In the case of weather forecasting, the predictability of events can account for more of the variation in forecast accuracy than individual differences between forecasters (Stewart et al., 1997)—with the footnote that all of the forecasters being compared are experienced and proficient. In other words, individual differences in forecasting skill may show up in forecasting something that is less predictable (i.e., precipitation) and not in forecasting something that is more predictable (e.g., daily high temperature).

Many studies of human judgment and decision making have focused on the single question of whether judgments are correct (or accurate). Analysis often involves linear modeling.

Linear Models: Can the Human Forecaster Outperform a Simple Linear Model?

Multivariate linear models are mathematical formulae that add up a set of individual terms or factors (that might be individually weighted) and use that summation to predict some other variable. So, for example, we might try and predict whether a given person would go see a movie on a Friday night based on population statistics about personal interests, spending habits, and other factors. There is some evidence that such models can outperform experts on criterion tasks (e.g., Fischhoff, 1982). Combined with evidence showing that people are biased and limited in their reasoning ability, the argument has been made that linear models are preferred to expert judgment.

Cynthia Lusk and her colleagues (Lusk et al., 1990; see also Lusk and Hammond, 1991) had five forecasters at the National Center for Atmospheric Research predict microbursts given a set of (randomly generated) values of each of six microburst predictor variables (e.g., descending vs. nondescending core, low vs. high convergence values). Forecasters largely agreed on their predictions for the hypothetical cases (consistency coefficients in the range of 0.8 to 0.9). A regression analysis showed that the forecasts could be closely matched by a linear equation that combined the weighted values of the six predictor variables.

One must take into account the fact that some forecasting tasks require few inputs—inputs in the form of individual cues or data values—whereas other tasks require the inspection and integration of many inputs of diverse data types (both alphanumeric and graphic), and the perceptual foundation involves dynamic cue configurations across multiple data types, rather than isolated static cues. Stewart et al. (1997) examined data from forecasting exercises conducted at the Department of Atmospheric Science of the State University at Albany. POP and temperature forecasts were made by students and faculty in a realistic forecasting setting with full information available (i.e., NWS forecast products, radar and satellite images). Performance of the forecasters was gauged in terms of a full set of cues that were deemed important for the tasks (23 cues in all, including climatological temperature, surface dewpoint temperature, and 250 millibar vorticity advection), as represented in a linear model.

The Stewart et al. (1997) results "painted a positive picture of expert judgment" (p. 216). There were high rates of overall agreement among the forecasters. (The combined or group forecasts were about as good as those of the best forecaster.) The computer model outputs played a key role in allowing the forecasters to generate accurate

Box 5.1
What Is Vorticity?

Vorticity is the degree to which the flow of air is spinning. The presence of the horizontal transport of the spinning fluid (vorticity advection) implies that the air may be forced to move either upward or downward, which is a primary indicator of the formation of clouds and precipitation. The 250 millibar cue is a reference to height in the atmosphere ("upper levels") as depicted in a "skew-T, log p" diagram. This diagram uses a clever trick—measuring elevation or height in terms of pressure measured in millibars. This makes the y-axis rather like a rubber ruler. When a mass of air is relatively warm, the ruler is stretched (higher altitudes), whereas when a mass of air is relatively cold, the ruler is scrunched (lower altitudes). Temperature is the x-axis, and the interpretation of the diagram involves looking for patterns that appear as changes in temperature (skews) as a function of height in the atmosphere as measured in terms of pressure. "skew-T, log p" diagrams also show wind profiles and other atmosphere parameters. To those who are unfamiliar with the skew-T diagram, its appearance and interpretation are a mystery. To those who are familiar with it, the diagram provides immediately perceptible clues to atmospheric dynamics.

forecasts. However, the forecasters won out over the shorter-range computer-generated forecasts perhaps because the forecasters had "developed an understanding of when [the computer model] predicted well and when it did not, and ... had access to information that was not included in [the computer program]" (p. 214), an interpretation that was later confirmed in a different forecast context (Roebber, 1998). The forecasters showed good overall performance (i.e., accuracy) at the temperature forecasting task, which involved a reliance on *fewer* cues. In contrast, the POP task seemed to be more complex, and performance was not as good as in the temperature prediction task, nor was the rate of forecaster agreement as high. A linear model combining the weighted cues did not predict POP forecast performance as well as it had in the case of the temperature forecasts (i.e., there was a significant nonlinear component to the skill score), leading the researchers to conclude that the POP task involved the "configural use of cues."

Stewart et al. (1989) presented data for 75 radar scans to seven meteorologists who were asked to make probability forecasts of hail. Their performance was compared to that of a simple linear model that calculated as a weighted sum of a set of cues (e.g., reflectivity at levels in the atmosphere, rotation or convergence, storm tilt, Doppler hail signature, divergence in the upper atmosphere). Results from the linear model were compared to the human forecasts and with the predictions of the HAIL expert system (see chapter 11). The findings were that:

1. the forecasters generally agreed (correlations between 0.75 and 0.95);
2. the forecasters were able to distinguish storms that produced hail from those that did not, although skill scores showed only about an 8% improvement over climatology, and there was considerable bias (i.e., the tendency for the probability estimates to be either too large or too small for the events being forecast);
3. the simple linear model was able to reproduce forecaster performance accounting for 80% to 92% of the variance in the human forecasts; and
4. the correlations of the human forecasts with those of the expert system were in the range of 0.63 to 0.79. For the least accurate forecasts, the expert system won out, whereas for the most accurate human forecasts, the human won out.

It is interesting that the forecasters did not report making their decisions on the basis of anything like the summation of weighted, separate cues, referring instead to "synergistic aggregation." Yet the linear model approximated their performance. Accounting for some of the variation in the results, different forecasters associated different importance to the different cues, and some of the cues were intercorrelated, that is, they form meaningful patterns and are not considered separately. This fits with what is known about pattern perception in other domains of expertise (see chapter 9).

The Stewart et al. (1989) study involved presenting data about storms (i.e., decibels of radar reflectivity, meters per second of rotation, etc.), not actual radar displays. We know that the performance of experts depends on how information is presented and whether the task and the information format match those in their usual or familiar task (cf. Ettenson, Shanteau and Krogstad, 1987; Hammond et al., 1987; Shanteau, 1992a). The study mentioned above by Lusk et al. (1990), which showed that a linear model worked well, involved a second experiment in which microbursts were predicted in an evolving situation more like that of actual forecasting (using Doppler radar displays from archived data). Based on their analysis of time series of radar data, forecasters made judgments concerning the six precursor variables (e.g., descending core) and gave confidence judgments as well as a forecast of microbursts. In this more realistic situation, forecasters' judgments should theoretically increase in accuracy over time (as more information is accumulated), and the forecasters should show increased confidence in their judgments. However, the results showed that forecaster agreement was low and increased (only slightly) over time, even though they showed moderate to high confidence in their judgments. Most interesting was the finding that the forecasters disagreed about which precursor parameters were important and how the parameter values were to be combined. They even disagreed about the importance of a descending core, which the first experiment had shown was the most important precursor variable.

This suggested to the authors that "there are indeed practical advantages to be gained from a better understanding of the precursor identification and prediction phases of the forecasting process" (Lusk et al., 1990, p. 627). Indeed, the clear implication is that different forecasters, in the more realistic situation of Experiment 2, engaged in differing reasoning sequences and perceived different configurations (i.e., the judgment of descending core requires a comparison of reflectivity values across time and height). Lusk and Hammond (1991) hinted at this when they stated, "Some secondary cues may be more subjective" (p. 68). Lusk et al. (1990) checked to make sure that the forecasters were able to accurately determine the radar reflectivity data values. In their reports, Lusk and colleagues emphasize that forecaster accuracy and confidence do not hinge on the perception of specific features or data values but rather the inferences that are made on the basis of them.

Forecast Quality Is More Than Just "Hit Rate"

No analysis of the process of perception (bottom-up or data-driven inference) or the process of inference (top-down or knowledge-driven perceptual search) could be complete by looking *only* at the final accuracy or hit rates of forecasts and by attempting to predict hit rates using linear models. For a given level of skill, you can only increase hits by also increasing false alarms, and too many misses suggest some degree of bias (Roebber, 2009). Hit rate, false alarm rate, and bias are tied together. Doswell (2004) refers to this as the "duality of error":

> Uncertainty carries with it the inevitability of both false positives and false negatives, depending on where the thresholds fall. This relationship constitutes the duality of error: At a given level of forecast accuracy ... false negatives [i.e., misses] can only be reduced by increasing the false positives [i.e., false alarms], and vice versa. (p. 1118)

The general finding of forecasters often doing as well as, and sometimes better than, the computer models is interesting in the context of psychological decision-making research. A great deal of research conducted in the academic psychology laboratory has shown that people (i.e., college freshmen, mostly) are misled and make errors under the influence of extraneous or irrelevant information (e.g., Evans, 1989; Fraser, Smith, and Smith, 1992; Gilovich, Griffin, and Kahneman, 2002; Kahneman, Slovic, and Tversky, 1982). Mathematical models of decision making designed to predict hit rate can outperform the humans' decisions (Swets et al., 2000). Probably the best example of this is assessment by clinical psychologists: A linear model based on the analysis of factors that contribute to hit rate can reliably outperform a clinician's performance on certain tasks (Dawes and Corrigan, 1974; Dawes et al., 1989; Meehl, 1954). For example, a simple linear model might use data on persons' interests and habits—say, frequency of movie going—to predict whether they would go see a movie on a given Friday night. Mathematical models of this type result in predictions that are consistent, but when they fail, they fail spectacularly. In reference to the example of the movie goer, the model could not possibly take into account the fact that this given person had broken his or her leg that morning. Furthermore, the mathematical models are based on variables and parameters that were initially suggested by *human* experts as being the key variables to include in a predictive model. This fact is often overlooked in studies that proclaim the superiority of mathematical models to human judgment based solely on the analysis of hit rates.

The literature on forecast verification points to the lesson suggested by recent research in the new paradigm of "naturalistic decision making" (see Klein, Orasanu, Calderwood, and Zsambok, 1993; Zsambok and Klein, 1997)—that a full understanding of real-world expertise cannot stop at the analysis of the correctness of final judgments or decisions (whether in the form of hit rate or skill score). This is shown clearly in the research of Thomas Stewart and his colleagues, demonstrating important interactions of task variables in determining the flows of reasoning and whether performance is at a high level. In the view of some forecasters (e.g., Roebber, 1999b), cases where an averaged forecast is less expert than the forecasts of individual forecasters suggest that consensus skill scores should be based on the analysis of conceptual forecasts rather than such things as probability of precipitation. Although two individuals' forecasts might result in different skill scores (i.e., the forecasters disagreed in their prediction) for the probability of precipitation, they could nonetheless have agreed on their underlying mental models of the weather situation, that is, whether the evolving scenario was one of convection producing severe weather or heavy rain. The research program of Stewart and his colleagues straddles the paradigms of forecast verification in meteorology and decision making in experimental psychology and points to the importance of examining forecaster reasoning in the actual forecasting context.

A Cautionary Tale

We have referred to ideas and research coming from the field of judgment and decision making. It is important to keep a few things in mind regarding research on prediction, probability estimation, and cognitive biases.

First, the bulk of the cognitive research involves having college students conduct unfamiliar, unpracticed, and artificial reasoning tasks and probability puzzles in the academic laboratory. Yet the results are almost always generalized from the decision making of "subjects" to decision making by "people." Presumably all people. (We have more to say about whether experts show cognitive biases in chapter 6.)

Second, the psychological studies that claim to reveal facts about expertise typically fail to provide convincing evidence that the so-called experts actually qualify as experts. We are almost always only given hints about "years of experience."

Third, it is critical to distinguish *domains* of expertise as a function of predictability. For some domains, it is relatively easy to decide how to measure performance (and thereby identify the genuine experts) because the primary tasks involve the analysis of deterministic events (e.g., engineering, physics problem solving, computer programming, etc.). For other domains, it is less easy, and many of those are domains where the ostensive primary task involves anticipating individual or aggregate human activity, such as jurisprudence, counseling, clinical psychology, financial forecasting, and others.

Weather forecasting seems to straddle the two types of domains, with some variable aspects being fairly predictable and others less so. Thus, we see that there is evidence that weather forecasting experts are consistently able to outperform linear models, but evidence indicates that sometimes they do not (e.g., Lusk et al., 1990). Another manifestation of this straddling has to do with the disconnect between judgments of confidence and judgment accuracy. Research has shown that the two are often unrelated, that is, people tend to be overconfident, and this extends to expert judgment (at least in Type 2 domains) (e.g., Oskamp, 1965). Decision-making researchers have suggested that a partial remedy for overconfidence is for the reasoner to try to imagine ways in which their judgments might be wrong (Lichtenstein et al., 1982). The forecast verification studies show that experienced weather forecasters are able to do that very thing: to assess uncertainty (Doswell, 2004; see also Novak, Bright, and Brennan, 2008).

> There is some evidence that weather forecasters have some characteristics that make them quite different from the typical subjects chosen for judgment and decision making studies (Lichtenstein et al., 1982; Stewart et al., 1997). There is also evidence that weather forecasters may not be so different from the usual subjects of such research. (e.g., Hammond, 1996; Stewart et al., 1997) (Doswell, 2004, p. 1120)

Conclusions

Predictions involve multiple parameters: temperature (daily high and low), winds (strength and direction), and precipitation (type and amount). Each of these parameters can be associated with some range of plus-and-minus values, within which a prediction might count as being "correct" (e.g., plus or minus the degrees Fahrenheit). Each of the parameters can refer to smaller or larger regions, and each can have some degree of certainty or uncertainty associated with them. As one forecaster (over)stated it, one could derive "zillions" of functions on forecast accuracy for the computer models, showing such things as the heights of the atmosphere at 1,000 millibars for the northern hemisphere in the month of March, and and so on (Postel, 2012).

The question of what it means for a weather forecast to be *accurate* does not lead to any simple answer. The problem from a verification standpoint is that translating the many, varying considerations (context, audience, predictability, and so forth) into some sort of catch-all metric really can't be done.

The question of what it means for a weather forecast to be *useful* or actionable also does not lead to any simple answer. For some weather events such as tornados, an alert that a tornado has been detected in the past 10 minutes in a particular region and is moving in such-and-such a direction at such-and-such a speed is a clear, sufficiently precise, and actionable forecast, although it would be great if the warning might have been posted earlier. In contrast, a forecast that says there is a 50% likelihood of rain showers somewhere within a large metropolitan area at some time across a given day is likely to affect people's attitudes about weather forecasts more than their umbrella-carrying behavior.

The literature of forecast verification not only confirms the view that forecaster reasoning depends on the key cognitive factors of expertise (rich knowledge base, reasoning skill, and perceptual skill), but also lends credence to the view that highly experienced weather forecasters can indeed qualify as experts, as defined in the literature of cognitive science (e.g., Ericsson et al., 2006; Hoffman, 1982; Norman, 1982; Shanteau, 1992a, 1992b). In terms of performance, forecasters can manifest levels of performance which approximate that of the consensus forecast, whether consensus is defined in terms of agreement among forecasters, agreement among forecasters and the computer models, or agreement among the computer models (more on this in chapter 12). There is evidence that forecasters are able to improve on the computer-generated forecasts even while they rely on those model outputs in addition to all the other available data and data types that they examine to make forecasts. In strictest terms, this suggests that forecasters are doing some form of reasoning that helps them go beyond the purely mathematical models. Forecasters are able to integrate vast amounts of data to create forecasts that are frequently as good as or better than computer-generated predictions (Carter and Polger, 1986; Roebber et al., 1996; Stewart et al., 1997; Swets et al., 2000).

How are forecasters able to perform this feat? Our view is that they create a mental model of the situation that combines a great deal of integrated information, and then they use that mental model to make forecasts (Morss et al., 2015). Their mental model allows them to go beyond the presented information (or at least determine when data- or computer-generated forecasts are misleading and in what ways). In-depth discussions of this process will be presented in chapters 8, 9, and 10.

This chapter establishes, at least initially, that some forecasters are genuine experts. How do they get to be expert? This is the topic of the next two chapters.

6 What Characterizes Expertise?

Who has the wisdom to count the clouds, And who can pour out the bottles of heaven?
—Book of Job 38:37

In the traditional craft guilds, a "journeyman" is a person who can perform a day's labor unsupervised (i.e., go on a journey), although working under orders. A journeyman is an experienced and a reliable worker, or one who has achieved a level of competence. An expert is a distinguished or brilliant authority, highly regarded by peers, whose judgments are accurate and reliable, whose performance shows consummate skill and economy of effort, and who can deal effectively with rare or tough cases. Also, experts tend to have special skills or knowledge derived from extensive experience with subdomains (Hoffman, 1998; see also chapter 7). To answer questions about how forecasters get to become expert forecasters, and then drill down on the question of how expert forecasters reason, we must first be clear about who is and is not an expert.

Although our focus is on expert cognition, we should make note of the sociological aspects of expertise. There are a number of excellent and provocative treatments and reviews of the sociology of expertise (Collins, 1992; Collins and Evans, 2007; Evetts, Mieg, and Felt, 2006; Kurz-Milcke and Gigerenzer, 2004; Mieg, 2006; Stein; 1997). Society as a whole has mixed attitudes about expertise. We touch on this matter in order to make our own stance clear.

> In the course of history the complexity of societies increased, and the skills needed for adequate governance and economic success grew more and more demanding and specialized ... the principle of heredity in elites became increasingly inappropriate and has largely been replaced by a principle of merit, mainly based on expertise. (Evetts, Mieg, and Felt, 2006, p. 1118; after Elias, 1982)

It is not uncommon to hear individuals with a medical condition say they are going to a certain clinic because they want to see *the* experts. Certainly, we all hope that an expert is piloting the plane on which we are passengers. However, some see the concept of expertise as elitist. In addition, experts are lampooned for their glorious failures. A case in point is the humorous motivational poster saying that economics is "The science of explaining tomorrow why the predictions you made yesterday didn't come true today." We do not want to single out economics, but it does afford many excellent examples of our point. A study conducted by *The Economist* has been widely cited:

> Economists are famous for their forecasting failures, but can anybody else do better? A Christmas quiz in *The Economist* in December 1984 allows us to put this to the test. At the time we sent a questionnaire to four ex-finance ministers ... four chairmen of multinational firms, four students at Oxford University and four London dustmen. They were each asked to predict economic prospects over the next decade. Final figures for 1994 now allow us to pick the winners. ... Nobody expected the price of oil ($29 a barrel at the time of the quiz) to drop below $25; seven of the 16 forecast a price of over $40. It is currently around $17. The dustmen came closest as a group. (*The Economist*, 1995, p. 70)

This sort of condemnation is often launched in the domain of politics: Editorials assert that "Pundits are clueless." More broadly, there is a large literature of books, both scientific and popular, arguing that essentially all human judgment is necessarily flawed and biased, that experts are no better than anyone (see, e.g., Ariley, 2008; Brafman and Brafman, 2009; Dawes, 2001; Kaplan and Kaplan, 2010). Psychological research has demonstrated limitations in people's ability to engage in logical thinking and reasoning about probabilities (Johnson-Laird, Khemlani, and Goodwin, 2015). Dozens of "cognitive biases" have been demonstrated, and new ones appear in the scientific journals on a regular basis. People get "anchored" by previous evidence or knowledge and do not try to disconfirm hypotheses. They tend to "project" their feelings onto others. People ignore "base rates" or frequency of occurrence when predicting events; people believe that small samples represent the whole; people miscalibrate their own understanding and confidence, and so on. In summary, the broad swath of humanity is not capable of sound critical thinking, and this extends to the claim that human reasoning, including expert reasoning, is inherently and necessarily limited (see, e.g., Evans, 1989; Fischoff and Beyth, 1975; Gilovich, Griffin, and Kahneman, 2002).

However, it has been pointed out that much of the research in which generalizations are made about "people" comes from academic laboratories, using college freshmen as "subjects" and simplistic, artificial, puzzle-like tasks (Klein et al., 2003). On the whole, the evidence for bias on the part of experts is mixed. Some experts seem overly confident, but who would want a surgeon who expresses uncertainty as he or she raises the scalpel? It is relatively easy to find example studies showing that so-called experts are not really experts by looking at domains in which the practitioner's ostensive main task involves the attempt to predict individual or aggregate human behavior (e.g., financial forecasting, parole counselors, clinical psychologists, etc.) (Dawes et al., 1989; Hoch, 1988; Shanteau, 1992a, 1992b). Some researchers have expressed doubt as to the general frequency of occurrence and magnitude of biases outside of the contrived academic laboratory (Christensen-Szalanski and Beach, 1984; Dreyfus and Dreyfus, 1986; Jungerman, 1983; Wallsten, 1983). Evidence for the occurrence and magnitude of bias is mixed in the literature on medicine (Schwartz and Griffin, 1986), and the literature on auditing (Shields, Solomon & Waller, 1987).

For some decades now, researchers have argued that "previous laboratory research on decision heuristics and biases is not applicable to real world, information-rich, interactive estimation and decision contexts" (Cohen, 1993a, 1993b; Northcraft and Neale, 1987, p. 96; Payne, 1982; Zakay and Wooler, 1984). "Much of the decision-making research seems to have a relatively obvious applied orientation, and yet, little concern about ecological validity issues has been expressed" (Ebbesen and Konecni, 1980, p. 39). One distinct possibility is that bias effects are merely the reflection of task-induced strategies. For example, Norman et al. (1989) were able to induce a bias in expert dermatologists by first presenting a series of test cases (slides) along with diagnoses and then obtaining subsequent judgments of the plausibility of the correct diagnosis of some additional cases. The experts could be biased to opt for an incorrect diagnosis even when the correct alternative was known. But it should be noted that this task was *not* the experts' usual or familiar task.

Experts show a range of proficiencies, depending on a variety of domain, task, contextual, and personal characteristics. The level of performance depends on domain and situational constraints that influence opportunities for feedback and interactions with others (Shanteau, 1984). Indeed, we could redefine expertise so as to limit it to certain domains, but this would short-change people's ability to improve on their methods. Experts certainly evolve ways to avoid at least some of the errors that are traditionally associated with bias. We do not doubt that critical thinking is a skill and that it takes practice (compare Heuer, 1999, with Hoffman et al., 2011). However, the occurrence of bias in expert judgment, to whatever extent it may manifest, need not be taken as a wholesale devaluation of expertise.

The History of Expertise Studies

Dating back to work by Francis Galton in the 1800s (1869, 1874), psychometricians and historiometricians have asked questions about the relations of intelligence, talent, genius, and high achievement. Much of this work has focused on the "nature versus nurture" issue, attempting to identify the relative contributions of environmental, social, cultural, and individual variables. Good reviews and theoretical integrations can be found in the works of Dean Keith Simonton (1988, 1999, 2000, 2006). Opinions have ranged from the view that genius (or talent) is born to the view that genius (or talent) is taught. It was with the emergence of Expertise Studies in the 1980s that another sort of question was asked about experts: What is the nature of their knowledge and reasoning skills?

Expertise Studies emerged in the 1980s, linked to the development of "expert systems" in the field of Artificial Intelligence (see chapter 5; also Amirault and Branson, 2006; Bereiter and Scardamalia, 1993; Buchanan, Davis, and Feigenbaum, 2006; Hoffman and Militello, 2009, chapter 8). Since the days of expert systems, psychological research on expertise has burgeoned. Diverse domains have been studied, and much has been learned (see, e.g., Ericsson et al., 2006; Hoffman, 1992; Hoffman et al., 2014). Expertise Studies has become a professional scientific field of inquiry, represented by researchers from many communities of practice and traditional academic areas, such as sociology of science, psychology of science, applied psychology, and others (see Ericsson, 2009; Ericsson et al., 2006; Feltovich, Ford and Hoffman, 1997; Hoffman, 1992; Hoffman and Militello, 2009 chapter 8).

To contextualize the idea of expert systems, perhaps the most well- known example is the software that guides the user through the process of preparing an income tax return. The software, it might be said, is an "expert in a box." To build an expert system, one must have a "knowledge base" of domain concepts and an "inference engine" of reasoning rules. Both of these key components are derived from cognitive interviews with domain experts (see Hoffman, 1992). Expert systems were developed for a great many domains, ranging from geological prospecting to the diagnosis of bacterial infections.

One traditional way of eliciting expert knowledge was to interview the experts. This interview process brought psychologists into collaboration with computer scientists. The computer scientists had encountered what was called the "knowledge acquisition bottleneck"—that it took too much time to elicit the experts' knowledge and reasoning strategies. Thus, there was a drive to develop efficient and effective methods of knowledge capture (see Hoffman et al., 1995). But there was also a dawning awareness that the psychology of human cognition had focused too much on experiments conducted in an academic laboratory using artificial and relatively simplistic problems and tasks, as well as using (mostly) college freshmen as the "subjects." The study of experts was a window onto human cognition at its highest levels of capacity and proficiency. Many researchers argued that a complete theory of human reasoning and intelligence would depend on the study of experts, in "real-world" domains (vs. the academic laboratory) and working "real-world" problems. This came to be called the study of "cognition in the wild" (Hutchins, 1995).

In hindsight, many academic journals had published studies about experts. For example, in the areas of organizational behavior and decision making, a great deal of effort had gone into the study of individuals who could be called experts, in such diverse domains as accounting, auditing, management, livestock judging, and finance. Today, expertise as a topic has its place in cognitive psychology textbooks and is a frequent focus of research in experimental psychology. A number of journals highlight experimental studies of expertise, such as the *Journal of Cognitive Engineering and Decision Making* and *Cognition, Technology and Work*. Expertise Studies has clear implications for basic theories of learning, skill, and cognition, as well as significant applications in training and education (se Proctor and Vu, 2006). Cognitive scientists have studied the thinking of highly proficient individuals in many domains, including science, public policy and administration, English composition, mathematics, jurisprudence, electronics, computer programming, medicine, aviation, and many others. For a broad review of the psychology of expertise in diverse domains, see Ericsson, Charness, Feltovich, and Hoffman (2006).

In what follows, we present some key points, defining and characterizing expertise.

Methods for Peering Into the Black Box

Whenever a skill (e.g., reading, bicycle riding, medical diagnosis) is highly practiced, knowledge that is initially taught explicitly can become tacit, and skills that were once deliberative can become "automatic" (Lesgold et al., 1988; Sanderson, 1989). This "declarative-to-procedural shift" leads to a potential paradox: As experts learn more, do they lose awareness of what they know? Do they become incapable of expressing what they know and describing how they think? Some developers of expert systems observed that highly skilled experts can carry out tasks without being aware of how or why they do what they do (Kidd and Welbank, 1984). Donald Broadbent and his colleagues (Berry, 1987; Berry and Broadbent, 1984; Broadbent and Aston, 1978; Broadbent, Fitzgerald, and Broadbent, 1986) demonstrated that when learning to control a complex system (e.g., a simulated model for a transportation system or an economic system), an individual may achieve a high level of performance through a great deal of practice, yet be unable to correctly answer questions about the system being controlled. There can be an increase in task performance without an increase in verbalizable knowledge, or there can be a drastic change in conceptual understanding and verbalizable knowledge without a corresponding major change in task performance.

It is not clear what it means to assert that there is a kind of knowledge that is "not verbalizable in principle." Setting this philosophical matter aside, the existence of "knowledge that is verbalizable but only with some difficulty" has not presented any insurmountable obstacle to either the task of identifying experts or the task of developing knowledge-based systems. "Ericsson and Smith (1991) found evidence that experts maintain an ability to control their performance, and they are able to give detailed accounts of their thought processes that can be validated against other observable performance and process date" (Feltovich, Prietula and Ericsson, 2006, p. 59). It is clear, nevertheless, that processes of identifying experts and probing their reasoning should never rely on a single method for eliciting knowledge or demonstrating that an individual possesses expertise (Hartley, 1981). Different knowledge elicitation methods may yield different types of forms of knowledge, and special methods may be needed to elicit tacit (so-called nonverbalizable) knowledge (Cooke, 1992; Hoffman, 1987b). A number of expert systems developers have claimed that special tasks are needed to reveal experts' tacit or procedural knowledge, that is, their reasoning strategies (Hoffman et al., 1993). Research in the fields of Human Factors and Cognitive Systems Engineering has led to the development of a number of methods of "cognitive task analysis" that can be used to reveal the knowledge and reasoning of experts, including expert weather forecasters. Detailed discussions of the methods appear in Crandall, Klein, and Hoffman (2006) and Hoffman and Militello (2009).

Although our focus is on how experts solve problems in their domain, it should be noted that there are important pragmatic components to expertise. Specifically, experts are typically not just good at working problems in their domain; they are also experts with regard to their organization. They know its history, how it works, why it works the way it does, the best ways for working with others, and so forth. This has been referred to as "franchise expertise" (Hoffman et al., 2011). This shows up in the ethnographic studies of forecasters discussed in chapter 3: recognition of the importance of collaborative analysis, recognition of the value of mentoring and on-the-job learning, and recognition of the need to acquire knowledge of and sensitivity to customer needs.

With regard to expertise specifically at forecasting, research using cognitive task analysis methodologies has yielded results about expert memory extent and organization, reasoning strategies, and perceptual skills (Feltovich, Prietula, and Ericsson, 2006).

Expert Knowledge

A straight forward way of scaling expertise is in terms of the extent of experience. The widely cited benchmark for the achievement of expertise is 10 years of full-time job experience. This benchmark was an approximation asserted by William Chase and Herbert Simon (1973) based on their studies of chess players. This amounts to about 16,000 hours (10 years x 40 weeks per year x 40 hours per week). Another benchmark, 10,000 hours of experience, was asserted by Herbert Simon and Kevin Gilmartin (1973) in an article on computer simulation of chess. This benchmark was subsequently verified by John R. Hayes (1985), who charted the careers of famous composers. Since then, the "ten years-10,000 hours" rule has been popularized by journalists. However, there is abundant evidence that the rule should not be generalized. As for all "pop psychology," this rule is open to question, and in particular the failure to consider factors other than practice, especially inherent ability and motivation (see Hambrick et al., 2014a, 2014b; Macnamara et al., 2014; Wai, 2014). It is certain that practice and experience alone do not necessarily lead to expertise. In the early 1900s, educational psychologist Edward Lee Thorndike (1912) argued that to get good at any skill, one must "practice with zeal." Ericsson, Krampe, and Tesch-Römer (1993) showed that the achievement of expertise depends on what they call "deliberate practice," where the performer works hard on hard problems.

As we discuss in chapter 7, Pliske et al., (1997) and Hoffman, Coffey, and Ford (2000), in their studies of weather forecasters, showed how some highly proficient individuals lack the job interest and motivation to achieve and extend their expertise, becoming "disengaged proceduralists." Given this disconnect of raw experience with the achievement of expertise, expertise is sometimes defined in terms of memory extent and organization (Ericsson et al., 2006; Hoffman, 1992; Hoffman and Militello, 2007). Estimates of the extent of expert knowledge put it anywhere from tens to hundreds of thousands of individual propositions and somewhere in the neighborhood of 50,000 concepts or "chunks" (Lenat and Feigenbaum, 1987; Simon and Gilmartin, 1973). For example, it has been claimed that chess masters can recognize tens of thousands of meaningful game patterns (Chase and Simon, 1973).

By the time one has become an expert, one's knowledge is both specific to the domain and extensive (Chase, 1983; Chiesi et al., 1979; Glaser, 1987; Lajoie, 2003; Scribner, 1984). It has been shown empirically that experts draw more complex conceptual distinctions than novices. For experts, the level of "basic objects" within their domain of expertise is more highly differentiated (Murphy and Wright, 1984).

Expert knowledge differs from novice knowledge in its organization as well as its extent. That is, concepts are interrelated in meaningful ways and memories are concept-, context-, and content-addressable (Chi et al., 1982; Feltovich, Prietula, and Ericsson, 2006; Glaser, 1987; Lajoie, 2003; Lesgold, 1984; Lesgold et al., 1988; Mandler, 1967). Experts also rely on conceptual categories that are principled (or more "abstract") (Voss, Tyler, and Yengo, 1983), and they know that conceptually different problem types may nonetheless manifest the same features (Murphy and Wright, 1984). Experts understand problems in terms of the main concepts and principles that are needed to accomplish their goals (Adelson, 1981; Jeffries, Turner, Polson, and Atwood, 1981; McKeithen, Reitman, Reuter, and Hirtle, 1981).

Studies of medical experts illustrate these ideas. Groen and Patel (1988) had expert medical diagnosticians and medical students read and recall descriptions of clinical cases. The results showed that experts tend to remember the underlying "gist" or meaning of the case, including the expert's own inferences, rather than just a verbatim recitation. This finding of "gist recall" has its analog in normal memory for short stories (Bartlett, 1932). A number of studies of experts' recall have shown this effect; furthermore, experts in diverse domains are better able to reorganize and make sense of scrambled information (Chase and Simon, 1973).

Empirical studies of pilots, expert chemists, social policymakers, electronic circuit designers, clinical psychologists, nurses, mathematicians, radiologists, telephone system operators, and musicians have all produced similar demonstrations of the expert–novice difference in memory organization.

Expertise must also be understood with reference to reasoning processes.

Expert Reasoning

Experts often rely on "case-based reasoning," where they refer to illustrative or prototypical examples of past cases when asked to justify or explain their decisions or actions. (Weather forecasting is a good case in point.) In domains where decisions have to be made under time pressure, such as firefighting, experts often engage in "Recognition Primed Decision Making." Experts do not engage in a process of listing alternative courses of action and evaluating them for such things as costs, risks, and benefits. Rather, they see a situation, immediately recognize what is going on, and then directly engage actions that will resolve the problem (see Klein, 1993; Klein, Moon, and Hoffman, 2006).

Experts like to "tell stories." Sometimes it seems as if a great deal of an expert's knowledge is remembered in the form of previously encountered cases. (Klein and

Hoffman, 1992). Indeed, there are domains of expertise wherein the primary method of reasoning involves explicitly comparing each given case to past cases. A clear example comes from avionics engineering (Weitzenfeld, 1984). The task of some avionics engineers is to predict the reliability and maintainability of new aircraft components or systems, and they do so on the basis of historical data about functionally or structurally similar components on older aircraft. Case-based reasoning is manifest in weather forecasting in the detailed analyses of specific weather events, as in the AMS's *Monthly Weather Review* and the NWA's *Journal of Operational Meteorology*. In these and other venues, forecasters share their knowledge and reasoning by telling stories of past forecasting experiences.

Expert reasoning can also be understood as a form of problem solving. Pioneering psychologist Karl Duncker (1945) presented word problems and mathematics problems to college students and had them "think aloud" during their attempts to solve the problems. Duncker observed that there were two general kinds of strategy or search, which he referred to as "suggestion from below" and "suggestion from above." In the modern literature on problem solving, these two particular concepts are referred to as top-down and bottom-up reasoning. According to Duncker's specification of reasoning, problem solving often involves a cycle:

1. Inspect available data →
2. Form an understanding and related hypotheses →
3. Seek information to test alternative hypotheses →
4. Cycle back to Step 2, that is, refine the understanding →
5. Produce a judgment.

In modern times, a number of studies of expertise and knowledge acquisition have revealed this refinement cycle or some variation of it (e.g., Anderson, 1982; Chi et al., 1982; Hoffman and Militello, 2009, ch. 8). For example, the Duncker cycle appeared explicitly in Lederberg and Feigenbaum's (1968) description of the goal for their expert system for the analysis of organic molecules:

> Data somehow suggest a hypothesis, and deductive algorithms are applied to the hypothesis to make logically necessary predictions; these are then matched with the data in a search for contradictions. (p. 187)

The Duncker refinement cycle appears explicitly in the field of weather forecasting, where it has been referred to as "scientific forecasting" (Bosart, 2003; Doswell, 1986a, 1986b; Doswell and Maddox, 1986).

Although expert and "everyday" reasoning may rely on the same fundamental cognitive operations, the flow of expert reasoning is definitely shaped by the tasks that are involved in the domain (Greeno, 1978; Scribner, 1984, 1986). Experts are, by definition, adept at their usual or familiar tasks (Hoffman, Shadbolt, Burton, and Klein, 1995). Hence, disruption of their familiar task can cause experts' superior performance to decline markedly: In a disrupted task, they cannot form meaningful representations or solutions. For example, chess masters' memory for game positions is disrupted for scrambled games (Chase and Simon, 1973). Expert bridge players are disrupted by meaningful rule changes (e.g., who leads each round) more than by superficial changes (e.g., the names of the suits) (Sternberg and Frensch, 1992).

Making sense of problem situations by reference to concepts and principles is a manifestation of comprehension in general, and of itself does not distinguish experts from nonexperts. But a number of studies have shown that experts and novices do reason differently. Larkin (1983) asked physics students and experienced physicists to solve mechanics problems (involving levers, weights, inclined planes, pulleys, forces, etc.) while thinking aloud. In the initial stages of problem solving, experts spend proportionately more time than novices in forming a conceptual understanding of the problem. Furthermore, experts generate representations that are conceptually richer and more organized than those of the novices. Novices tend to use hastily formed "concrete" (i.e., superficial) problem representations whereas experts use "abstract" representations that rely on "deep" knowledge." These more abstract representations provide the imaginal and conceptual understanding of functional relations and physical principles that relate concept (in the case of the research on experts at mechanics, principles such as conservation of energy). Furthermore, experts are better able to gauge the difficulty of problems and know the conditions for the use of particular knowledge and procedures (e.g., if there is acceleration, use Newton's second law) (Chi et al., 1982).

Learning and Cognitive Flexibility

In his seminal work on "the reflective practitioner," Donald Schön (1983) asserted that applied science problems tend to be muddy and indeterminate, requiring the practitioner to frame a relevant, manageable context while also considering how the larger situation might impact the problem at hand. In a professional setting, cognition must be freed for a focus on metacognitive strategies, including reflection or reasoning about one's own reasoning strategies (Anderson, 2005). Professionals do not necessarily use metacognition in this way and are not necessarily encouraged to do so. But for learning in a complex domain, such as weather forecasting, capacity once needed for thinking through tasks, such as how to display and understand data, becomes freed for thinking through potential weather scenarios.

Learning

Applied psychologists Stevens and Collins (1978) assumed that the understanding of complex phenomena hinges on the formation of "mental models," a concept we first mentioned in chapter 1. Stevens and Collins explored how school children form mental models for weather, with the goal of improving on the teaching of strategies and skills by examining protocols of teacher–student dialogs. In addition, they asked the teachers to comment on whether they could tell what a student knew or did not know. They asked about how they as teachers went about correcting student misconceptions. They asked the children questions about the causes of heavy rainfall (e.g., "How is the moisture content of the air related to heavy rainfall?"). Stevens and Collins identified numerous "bugs" (simplifications and distortions) that arise in student understanding. For example, many students fall prey to the "cooling by contact bug," which was illustrated in such statements as "cold air masses cool warm air masses when they collide." This type of cooling is not critical in causing heavy rainfall.

Many of the errors in student understanding can be related to their adoption of one or another rudimentary metaphor through which the weather is understood, and on the basis of which their mental model of weather is constructed. Metaphors include regarding the atmosphere as a mass of billiard balls or as a process like molecular attraction. Such models differ in the degree to which they adequately explain weather phenomena (i.e., why cold fronts usually herald the arrival of dry weather). What stemmed from the Stevens and Collins analysis was a set of rules for productive teacher–student dialogs (e.g., "if a student gives as an explanation a factor that is not an immediate cause in a causal chain, ask the student to describe the intermediate steps").

The work of Collins and Stevens serves as a clear case of the utility of adopting the mental model approach in the analysis of meteorological reasoning: (1) The progression of learning can be conceived of as involving such operations as adding,

Box 6.1
Heavy Rainfall

Heavy rainfall is related more to the rising of air of differing relative temperatures and moisture content. Air cools as it rises, owing to adiabatic expansion of the volume of air, which is the expansion of volume without any transfer of heat. When cooling causes air to reach its dew point or condensation temperature, clouds and precipitation can then develop. The condensation process also allows for the release of latent heat. This further warms the air, which enhances the rising process.

differentiating, and replacing components or factors to a mental model, and (2) there is value in encouraging students to use multiple conceptual (i.e., metaphorical) and functional (i.e., principled) models.

Advanced learning is promoted by emphasizing the interconnectedness of multiple concepts along multiple dimensions, and the use of multiple, organized representations. Learners need to overcome their simplifying mental models, and so instructional methods should be based on the ideas of "debugging" and progressive complexity (Stevens and Collins, 1978).

Cognitive Flexibility and Cognitive Transformation

Two main hypotheses about the acquisition of reasoning skill that have emerged in the literature on expertise are the "cognitive flexibility" hypothesis of Paul Feltovich and his colleagues (Coulson et al., 1997; Feltovich, Spiro and Coulson, 1989, 1993, 1997; Spiro et al., 1988, 1992) and Gary Klein's "cognitive transformation" hypothesis (Klein, 1997; Klein & Baxter, 2009). Both of these hypotheses regard learning as the active construction of knowledge (as opposed to memorization).

The reasoning of experts is special in terms of its flexibility. As one expert soil judge pointed out when confronted with a classification error, "I helped set up the methods for classifying soils, and the methods may need to be changed for this type of case." He then went on to suggest precisely how the new judgments might be made (cited in Shanteau, 1989). The performance change that occurs over development includes an increased ability to form multiple alternative interpretations or representations of problems, revise old strategies, and create new strateges as problem solving proceeds (Alexander, 2003; Shanteau and Phelps, 1977). Experts possess the ability to generate scenarios or frameworks for reinterpreting novel or difficult decisions.

One might expect that wholesale delivery truck drivers, warehouse inventory managers, office clerks, and product assemblers would mostly rely on routine procedures and tasks, rather like the airline pilot's checklist. Scribner (1984, 1986) presented highly experienced workers with some test cases and revealed a great deal of reasoning flexibility in service of efficiency and economy (see also Kusterer, 1978; Schön, 1983). Shanteau and Phelps (1977) demonstrated that when unique situations are encountered, expert livestock judges are able to generate meaningful special-case strategies. In contrast, novices persisted in following the well-learned decision rules, even when they were inappropriate. A well-regarded livestock judge noted that one of the biggest difficulties in teaching students "is their persistence in using inflexible and outdated standards" (cited in Shanteau, 1989). When confronted with an error, novices frequently appear more interested in rationalizing or defending their past decisions than in learning from them (Groen and Patel, 1988).

The cognitive transformation notion is that training must support the learner in overcoming "reductive explanations," which are explanations that mislead because they oversimplify. Reductive explanation reinforces and preserves itself through misconceptions and "knowledge shields," in which the learner attempts to preserve their simplistic understandings. For example, people seek simple cause–effect relations rather than multiple causation, people regard factors that interact as separable, principles that are context dependent are thought of as being universal, and so forth. We note that the "reductive tendency" is not a cognitive bias; it is a necessary consequence of the fact that knowledge is, at any one time, incomplete. It would hardly do to accuse someone of being biased simply on the grounds that there are things they do not know (Feltovich et al., 2004).

Because simplified mental models and knowledge shields lead to wrong diagnoses and enable the discounting of evidence, learning must also involve *unlearning*. A traditional learning hierarchy approach that compartmentalizes material and assumes fixed tasks (e.g., Gagné, 1968) can detract from advanced knowledge acquisition because the decomposition limits opportunities to learn about interactions. Feltovich, Klein, and their colleagues emphasize case-based learning because the development of expertise relies on experience working with especially "tough" ones that stretch knowledge and skill. The emphasis is less on trying to teach people to think like experts as it is on teaching people to learn like experts (Klein, 1997).

There is truth to the old saw that people learn more from their mistakes than from what they get right. When everything works the way it is supposed to, one is less likely to receive feedback about what did not work or what might have been done better. Experts seek out corrective feedback, especially feedback pointing out targets for improvement. A popular saying is that apprentices make the same mistake twice, journeymen make the same mistake once, and experts work until they never make mistakes. Although this point is well taken, domain specialists who are intrinsically motivated often seek out corrective feedback that allows them to perceive their errors. Sonnentag, Niessen, and Volmer (2006) showed that the more experienced problem solvers (in the domain of software engineering) sought out corrective feedback from coworkers. Likewise, Deakin and Cobley (2003) showed that advanced skaters fall in practice more often than the less skilled—because they attempt moves they have yet to master in order to receive corrective feedback.

The cognitive transformation hypothesis implies a paradox: the better the mental model, the harder it is to move past it because the model does a better job and is easier to protect using knowledge shields to explain away contrary data. Thus, at higher levels of proficiency, workers benefit from tough cases—experiences that force them to lose faith in their mental models so that they can move to the next level. The implication is that high levels of proficiency are achieved when the practitioner has the ability to "confidently lose confidence" in an existing mental model or concept.

In addition to knowledge structure and reasoning strategies, experts also differ from novices in terms of their perceptual skills.

Expert Perceptual Skill

The reasoning of experts often manifests itself as perceptual skill (Hoffman and Fiore, 2007; Klein and Hoffman, 1992; Lesgold et al., 1988; Proctor and Vu, 2006). To illustrate, upon witnessing the performance of a high diver, the expert sports commentator asserts that the diver's legs were not neatly tucked during a particular portion of the dive. Sure enough, the slow motion replay shows us novices that the expert was right. A more esoteric example is the sexing of newly hatched chicks, but examples can also be found in "ordinary" skills such as speech perception, face perception, and reading—which are actually remarkable accomplishments.

At least since the pioneering work on the reasoning of chess masters (deGroot, 1948/1978), perceptual skill has been regarded as a key to the advantage of experts (Charness et al., 2001; Glaser, 1987). Perceptual learning is an interesting notion because it cuts across a Cartesian distinction between two psychological processes that are believed to be distinct and fundamental to how the mind works (Fahle and Poggio, 2002; Gibson and Gibson, 1955; Goldstone, 2000). Perceptual learning has been shown in studies of expert–novice differences in a variety of domains including neonatal critical care nursing, radiology, baggage screening, commercial fishing, learning to discriminate flavors of beer, and learning how to determine the sex of newborn chicks (Biederman and Shiffrar, 1987; Fiore et al., 2000, 2004; Goldstone, 1998; Lesgold et al., 1988; Peron and Allen, 1998; Tanaka, Curran, and Sheinberg, 2005).

Perceptual learning results in the phenomenon, referred to earlier, of recognition-primed decision making, which enables experts to rapidly evaluate a situation and determine an appropriate plan of action (Klein, 1993). Within the first second of exposure to a novel chess position, chess experts can extract important information about the relations of the chess pieces' positions and begin identifying promising moves (Charness et al., 2001). In many domains of expertise, such as firefighting, power plant operation, jurisprudence, and design engineering, experts often make decisions through rapid recognition of causal factors and goals, rather than through any explicit process of generating and evaluating solutions (Klein and Hoffman, 1992).

Myles-Worsley et al. (1988) had expert radiologists and medical students observe and then attempt to recognize a series of chest X-rays. Reaction time data showed that experts allocated their attention more efficiently, focusing on abnormal features that distinguished the images. Similarly, Norman et al. (1989) demonstrated the effect of perceptual learning in a study of student, intern, and resident dermatologists, in which the participants were shown 100 slides for diagnosis. The researchers measured reaction time and error rates in a subsequent recognition task. The reaction time data showed that experts engaged in a rapid perceptual process that does not rely on independent analyses of separate cues. In other words, reaction times and errors were not predictable on the basis of the simple presence or absence of isolated features that are presumed to be definitive of each diagnostic category (Kundel and Nodine, 1978).

Research has shown that experts in domains as diverse as architecture, nursing, and electronic circuit design can indeed "see" things that novices cannot (Klein and Hoffman, 1992). When novice livestock judges confront the judgment task, they can miss seeing important features of livestock that experts readily detect (Phelps and Shanteau, 1978); the eye movements of radiologists while they scan X-ray films are quite different from those of novices—the experts can selectively search for abnormalities (Kundel and Nodine, 1978); expert cardiologists directly comprehend cardiovascular biomechanical events as perceived through a stethoscope (Jenkins, 1985).

Perceptual learning is not just about the perception of cues or the reckoning of variables; it is about their meaningful integration (Hoffman and Fiore, 2007). The knowledge-based integration of cues is illustrated in research in psychology on cue utilization. Expert decision makers do not always seem to rely on all of the relevant cues. What makes experts unique is "their ability to evaluate what is relevant in specific contexts. It is the study of that skill, not the number of cues used, that should guide future research on experts" (Phelps and Shanteau, 1978; Shanteau, 1992a, p. 84). Here's an example story. An expert examined the photos that had been used to support the claim that Iraq was developing biological weapons. After his own analysis, he said:

> I don't see any other decontamination vehicles down there that I would recognize ... the standard vehicle was a Soviet-made box-body van. This truck was too long. If you are an expert you can tell one hell of a lot from pictures like this. (Gladwell, 2004, p. 81)

From this we see that it was not only the detection of cues that were *absent* but also the meaning and significance of these factors when coupled with what was perceived.

Often the patterns that bear meaning cannot be defined in terms of the simple presence or absence (or the values) of separable cues. Meaningful patterns are sometimes defined by the relations among functionally integral cues. Wittgenstein (1953) was getting at this with his notion of "featureless family resemblances." Meaningful patterns are also defined by dynamics. When expert radiologists look at mammograms, they see patterns of shades of white, gray, and black. But they *perceive* processes such as calcification. The patterns that experts perceive, *even in static images*, are dynamic. The expert firefighter can tell the location and cause of a fire by the *movement* of the flame and smoke; the expert bird watcher can identify a species even when all there is to see is a fleeting shadow of movement in flight. Meaningful patterns sometimes exist *only* over time.

Complicating things further, the patterns that experts perceive sometimes do not exist in individual data types. Indeed, the really critical information often is "trans-modal"—it exists *only* across data types (Hoffman and Fiore, 2007). For instance, in weather forecasting, the radar images are not the only thing guiding sensemaking activity and shaping the forecaster's formation of a mental model. A great many other data types are involved, such as satellite images, ground-based weather data, storm spotter reports, and so on.

Integrated Models of Expert "Macrocognition"

A long-standing goal for cognitive science is to understand through controlled laboratory experimentation the causal sequences of mental events down to the level of basic mental operations. Modeling of cognition at the micro level relies on such notions as sensory encoding of stimuli, memory storage and retrieval, shifts of attention, and the execution of motor commands. Such operations occur in miliseconds. They are so-called "atomic components of thought."

Research over the last few decades in the emerging fields of Expertise Studies and Naturalistic Decision Making have converged on a different approach called Macrocognition, based largely on results from research on the cognition of experts, which we have summarized in this chapter (Hoffman and Militello, 2009; Klein, Hoffman, and Schraagen, 2008; Klein, Moon, and Hoffman, 2006). Macrocognition is distinguished from microcognition in a number of ways. The most important of these differences is clear in the key research findings:

- Cognitive work in complex contexts involves certain primary, goal-directed functions, including decision making, sensemaking, (re)planning, adapting, detecting problems, and coordinating.
- Supporting these are high-level cognitive and social processes, including maintaining common ground, developing mental models, managing uncertainty, identifying leverage points, and managing attention.

• Operating in combination, different primary functions and supporting processes are critical to cognitive work depending on the domain, particular task, and context.
• In cognitive work, the sorts of things that we might point to and call "mental operations" include attention management during multitasking situations, replanning triggered by problem recognition, and goal and decision modification based on situation awareness. These processes are all always parallel and highly interacting.

Micro- and macrocognition can be thought of as complementary paradigms or methodologies for the study of human cognition. These differences are described in table 6.1. Microcognition and macrocognition paradigms are both necessary for a complete science of mind. Together they provide a broader and more comprehensive view than either by itself. The microcognitive modeling approach has met with considerable success in identifying usability problems with new software tools and new interfaces, estimating the cost of training, evaluating alternative designs for interfaces and determining their usability, suggesting ways of improving on software to decrease task execution times, suggesting ways of measuring situation awareness, and informing the design of training aids and intelligent tutoring systems that can predict the errors students are likely to make given the stage of their skill development.

One of the intended functions of a macrocognition framework is to encourage the development of descriptive models of processes such as decision making, sensemaking, and problem detection. Macrocognitive modeling is informative for envisioning new technologies in terms of *usefulness*. For example, a macrocognitive approach could be used to generate creative ideas in terms of the requirements and desirements of workers to inform the design of new technologies and decision aids (e.g., Gray and Salzman, 1998; Hoffman and McCloskey, 2013).

In complex sociotechnical work contexts, one must consider adaptation, opportunism, dynamicism, and the unexpected—rather than routine, well-learned, separable, tasks. When a human who is working on a tough decision problem in context has to deviate from known task sequences to engage in problem solving, collaborative problem solving, or similar activities, microcognitive models become less applicable. In complex cognitive work, activities are rarely tasks in the sense of fixed action sequences (Simon, 1973). Instead, domain practitioners engage in knowledge-driven, context-sensitive choice among activities. The research findings we have discussed in this chapter have been integrated in the form of two models of expertise. One is a model of sensemaking and the other is a model of flexecution (flexible execution).

Gary Klein's Sensemaking model (Klein et al., 2006a, 2000b) is presented in figure 6.1. The purpose of this model is to explain what happens when experts try to understand

Table 6.1
The distinctions between Micro- and Macrocognition research paradigms

	Microcognition	Macrocognition
Methodology	Reductionist Experimentalism: Controlled laboratory experimentation to isolate cause–effect relations.	Wholistic Naturalism: Field studies and cognitive task analysis to go beyond the actual work and reveal the nature of the true work.
Methods	Traditional methods in cognitive psychology (puzzle solving, recall, recognition, reaction time) using simple, artificial tasks and simple, artificial materials.	Structured interviews, observations, simulations, constrained processing tasks, "tough case" tasks, and other methods using rich and realistic cases.
Participants	Typically, college students or naïve individuals are the "subjects." They are, by definition, domain-naive. Cognition is examined in brief experiments, looking at scales of minutes to weeks.	Experienced domain practitioners are the participants. Cognition is studied over scales ranging to entire careers. The full proficiency continuum is examined.
Phenomena Studied	Typically, phenomena that generally only appear in the laboratory, such as phenomena in the solving of preformulated puzzles and problems.	Phenomena that are not likely to occur in the laboratory and that laboratory studies would be highly unlikely to discover, such as problem detection.
Explanatory Goal	Reductionist causal chain theories of cognition at the scale of milliseconds (e.g., memory access, attention shifts, etc.) to minutes (puzzle solving).	Understanding expert knowledge and reasoning, and understanding how cognition adapts to complexity.
Stance	Cognition is characterized by limitations (e.g., memory, perceptual, temporal, etc.) and biases (e.g., reasoning). The goal of decision aids is to mitigate those limitations and biases.	Cognition is characterized by flexibility and adaptability. Bias is not typical in expert reasoning. The goal of decision aids is to contribute to the true work and not assume that experts need tools that help them mitigate bias.
Applied Goal	General applications of psychology to such areas as training, instructional design, behavior modification, and so on.	Technologies that amplify and extend the human abilities to learn, know, perceive, reason, and collaborate.

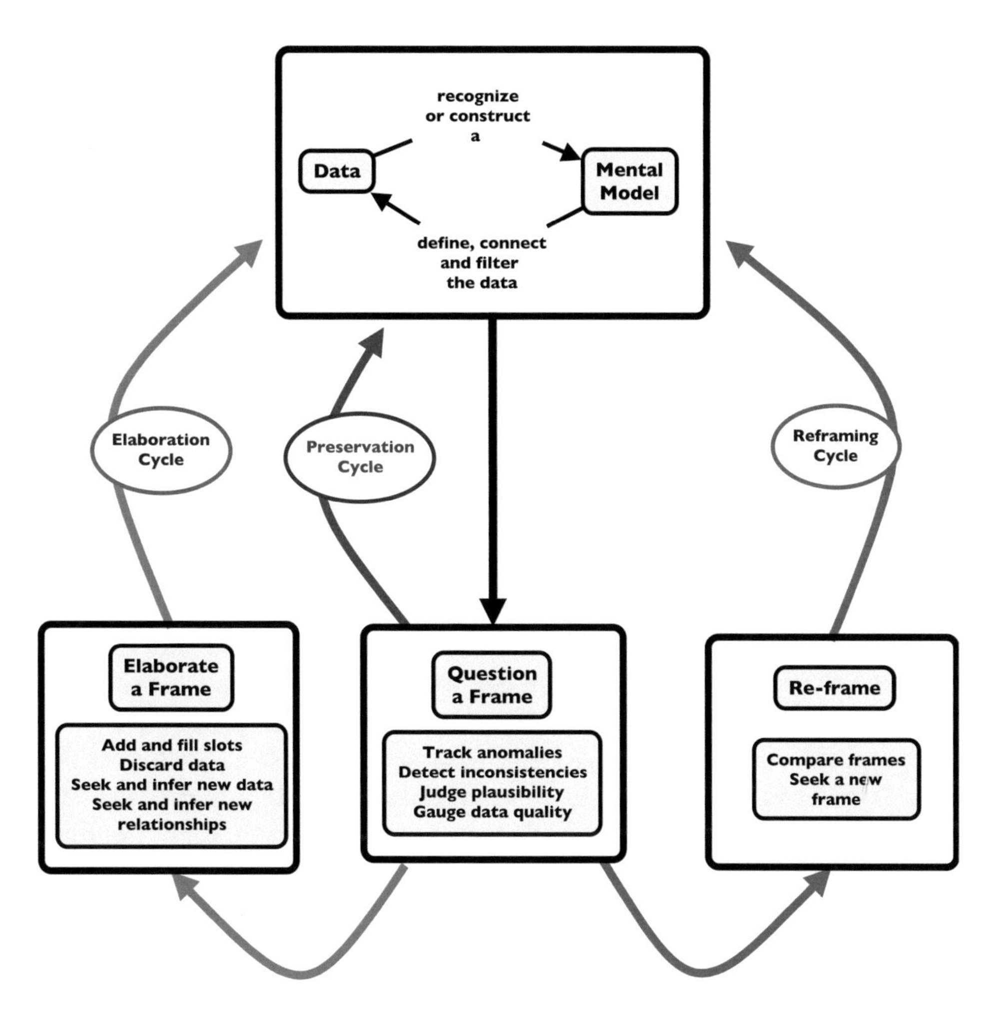

Figure 6.1
The Klein Sensemaking model (copyright R. Hoffman 2016, all rights reserved).

complex situations and continually work to refine and improve on that understanding. The model depicts sequences of activities or kinds of deliberative reasoning.

When one looks retrospectively at an instance of reasoning, one can describe it as a causal chain, as in: (1) an initial "frame" (or conceptual model) is formed, (2) it is used to seek confirming and disconfirming evidence, and (3) the frame is elaborated on, discarded, and so on. But before the fact, as a generic process model, all of the sequences are possible. Hence, the Sensemaking model consists entirely of closed loops. The three-step causal chain we describe above is Karl Duncker's (1945) classic model of hypothesis testing. The Sensemaking model includes Duncker's notion of a "mental model refinement cycle" and also integrates psychological models of expert reasoning, such as that developed by Einhorn and Hogarth (1981) in the era of expert systems (for more examples, see Hoffman and Militello, 2009). These previous models are basically all causal chain, input–output models. A major premise of the Sensemaking model is that the high-level functions of cognition are parallel, continuous, and interacting. Sensemaking often proceeds in fits and starts; there can be gaps, distractions, and multitaskings. Beginnings and endings can be anything but clear-cut. Although we can retrospectively describe instances in which reasoning seems to follow a sequence with a clear-cut starting point ("surprise" is often the trigger to problem solving) and an apparent stopping point (a decision is "made"), such causal chains are the exception and not the rule (see Hoffman and Yates, 2005). Thus, there are no input and output arrows in figure 6.1; it is all closed loops. The sensemaking process can commence anywhere, although as we just noted, it is often triggered by surprise (Trickett and Trafton, 2007). But even in that case, some sort of mental model must have "been in mind" earlier, otherwise there would be no anomaly to notice.

Activity involves implementing plans and routines, but it also involves changing plans and routines (Koopman and Hoffman, 2003). The nominal case is that a plan is made de novo and then followed lock-step. This is taken as the prototype in linear, causal chain theories. The more typical case is continual, adaptive replanning. Numerous oft-cited quotations attest to the empirical experience and hence the descriptive necessity of a model of replanning, such as, "No plan survives first contact with the enemy" (Field Marshall von Moltke) and "The best-laid schemes of mice and men often go awry" (Robert Burns). Gary Klein's Flexecution model of planning is aligned with Dwight Eisenhower's comment, "In preparing for battle I have always found that plans are useless, but planning is indispensable," and Publius Syrus's comment, "It is a bad plan that admits of no modification" (~100 BC).

The Flexecution model of replanning inherits significantly from the Sensemaking model and is depicted in figure 6.2. The central notion of the Flexecution model is that the execution of predefined plans is actually the exception rather than the rule. More often, cognitive work involves replanning. Furthermore, goals have to change while they are being pursued. This distinguishes the flexecution model from other models of replanning, which generally assume fixed goals. An example of flexible execution in weather forecasting would be the situation in which severe storms are emerging but the forecasting office has lost its link to the weather radar. Although the primary goal of issuing good forecasts does not change, a crucial subgoal has to change: The forecaster

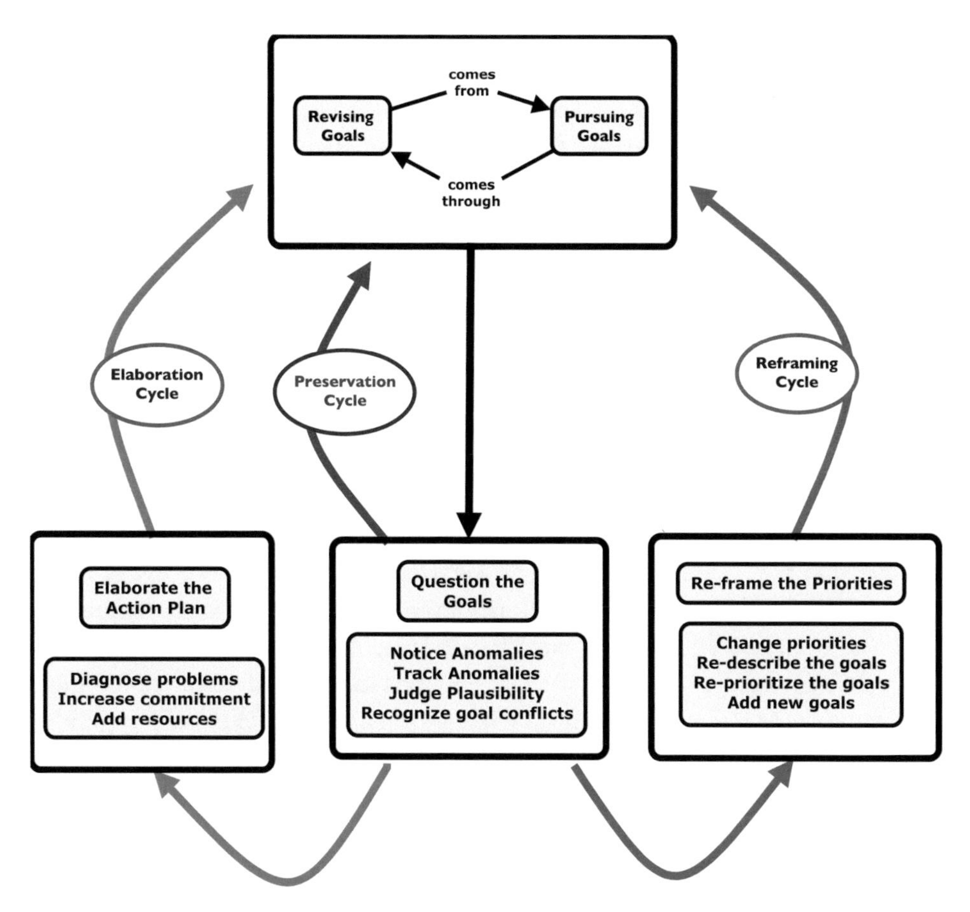

Figure 6.2
Klein's Flexecution model of (re)planning (copyright R. Hoffman 2016, all rights reserved).

cannot investigate storms with the radar and instead must get crucial information some other way, say, by making phone calls to other regional forecasting offices.

The closed loop at the top of the Flexecution model is the counterpart to the topmost closed loop in the Sensemaking model (figure 6.1). Likewise, the other loops in the Flexecution model are counterparts to those in the Sensemaking model. The Flexecution model does not assume that there are "start" and "stop" points, which is consistent with our understanding of flexible planning (Hoffman, Klein, and Miller, 2011).

We will rely on these descriptions of macrocognition as we address the question of the reasoning of expert weather forecasters.

Conclusions

The field of Expertise Studies emerged in the 1980s and has become a major scientific field with significant applications, including applications to education, training, and workforce issues. Significant findings that pertain to our analysis of forecasting expertise include the following:

- Although expertise can be defined in terms of proficient performance ("hit rate" on representative tasks), experts are also distinguished by their flexibility, and in particular their ability to cope with rare, tough, and challenging cases that require adaptation "on the fly."
- Expertise relies on the formation of an extensive and highly organized knowledge base, and the use of knowledge to understand situations by forming rich mental models.
- Expertise relies heavily on perceptual skill and perceptual learning. With experience and practice, new distinctive features are discovered, permitting the active, strategic search for critical information. As proficiency grows from the journeyman to the expert level, there emerges an ability to rapidly detect and discriminate, but perception is not of isolated features or cues. Rather, attention is drawn to invariant patterns and dynamic patterns. Over the course of development, perceptual skills change, permitting the rapid search, discrimination, recognition, and comprehension of complex informational patterns. This certainly applies to the domain of weather forecasting, especially the perception of patterns in satellite images and other graphical data displays.
- We have good, general descrptions of expert reasoning, including the Cognitive Flexibility and Cognitive Transformation theories, and we have good, general models of expert reasoning, such as the Sensemaking and Flexecution models. We will rely on these ideas from the field of Expertise Studies in subsequent chapters that ask specific questions about the cognition of weather forecasters.

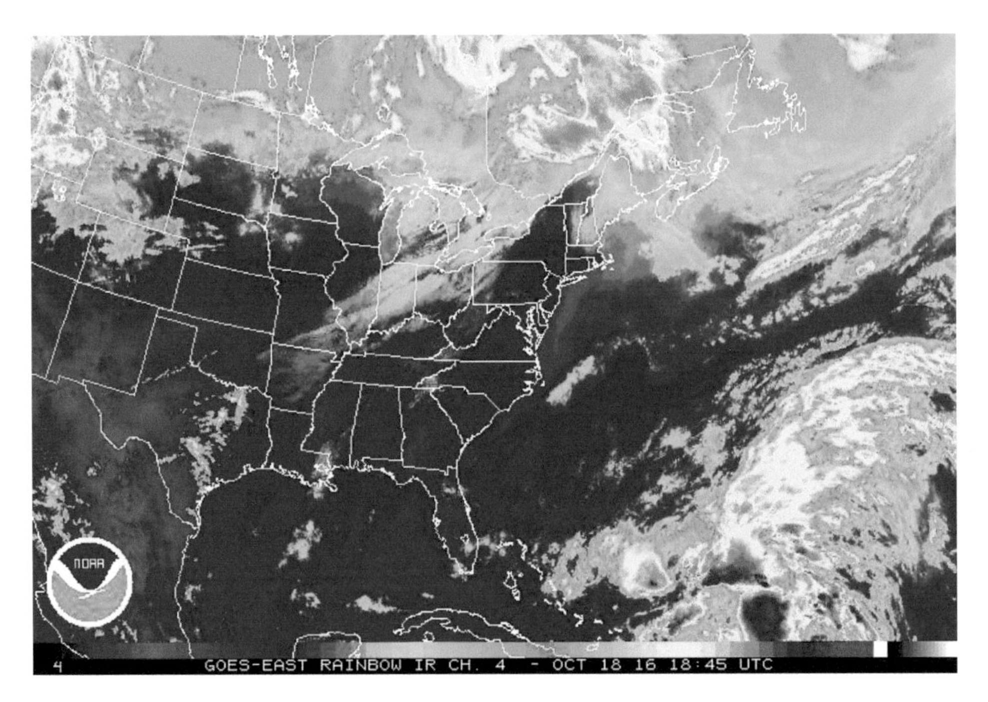

Plate 1

[figure 1.2] A colorized infrared GOES satellite image [downloaded 18 October 2016 from http://www.goes.noaa.gov/goes-e.html].

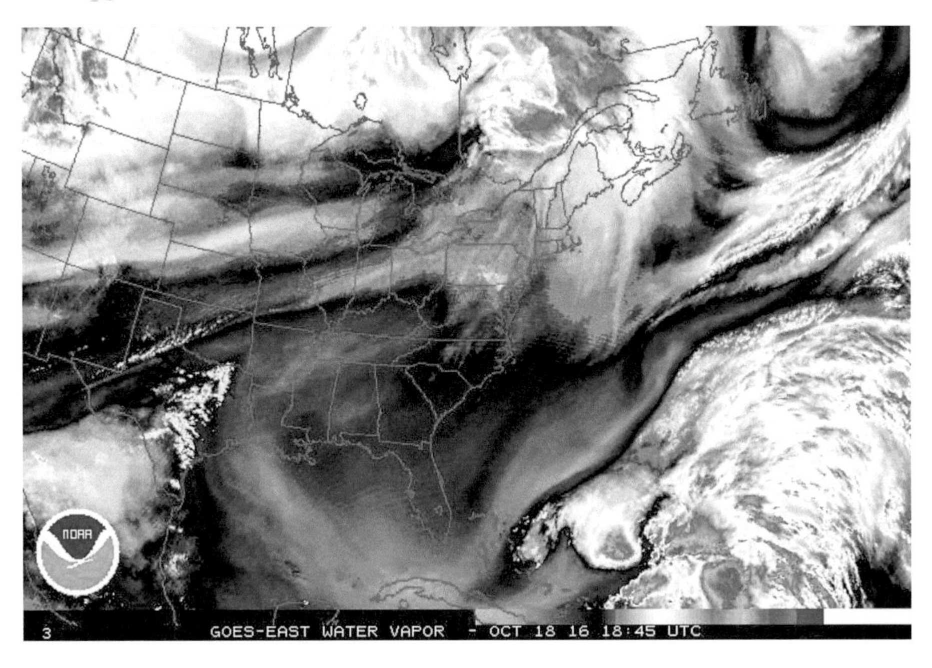

Plate 2

[figure 1.3] A colorized GOES water vapor satellite image [downloaded 18 October 2016 from http://www.goes.noaa.gov/goes-e.html].

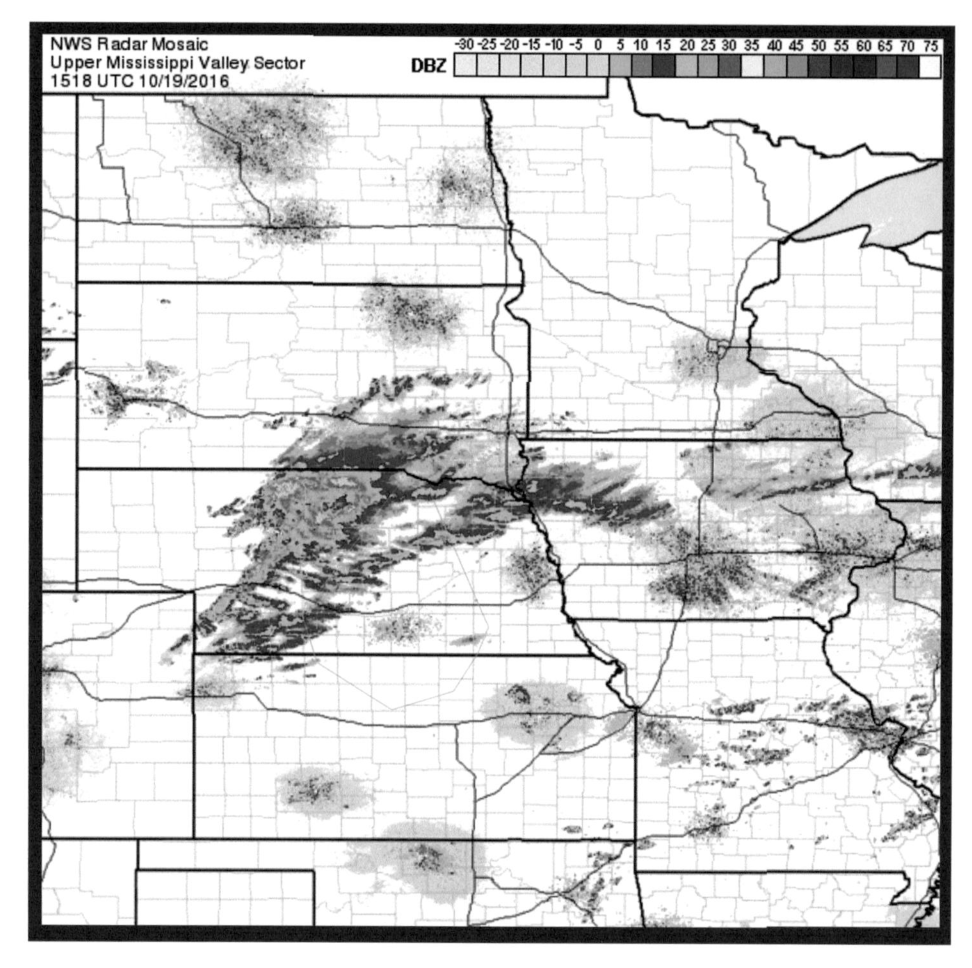

Plate 3

[figure 1.4] An example of a NEXRAD Base Reflectivity product [downloaded 3 January 2015 from http://radar.weather.gov/ridge/Conus/full.php/].

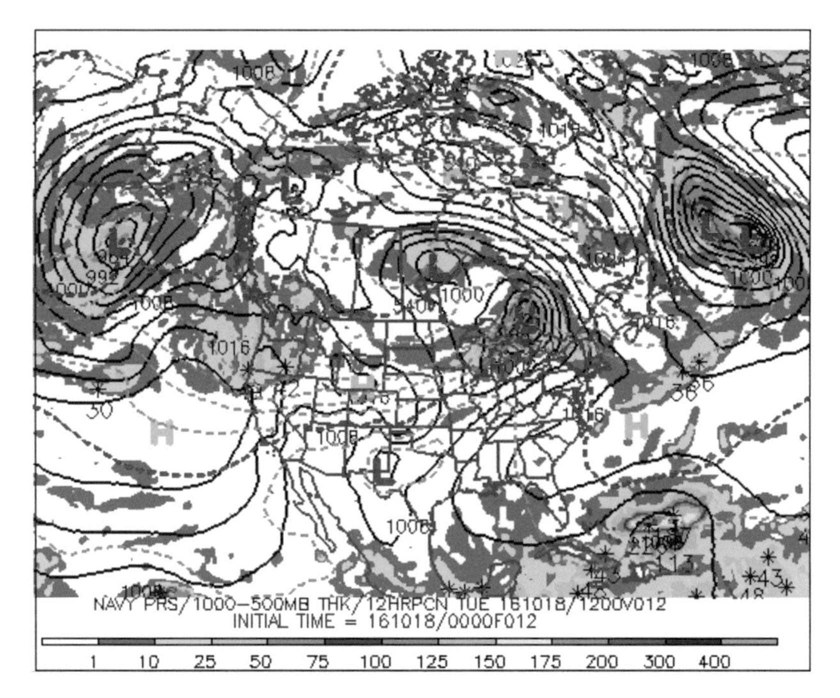

Plate 4

[figure 1.5] This display shows the 12-hour forecast for upper level winds generated on 18 October 2016 by the U.S. Navy's Navy Operational Global Atmospheric Prediction System (NOGAPS). [http://mp1.met.psu.edu/~fxg1/NOGAPS_0z/nogapsloopw.html]

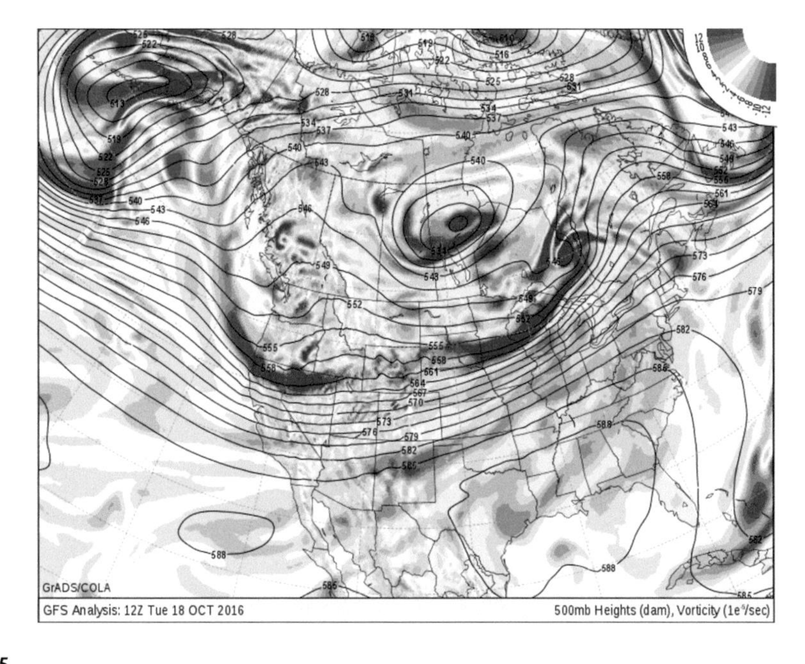

Plate 5

[figure 1.7] An example 500-millibar product [downloaded 18 October 2016 from http://radar.weather.gov/Conus/full_lite.php].

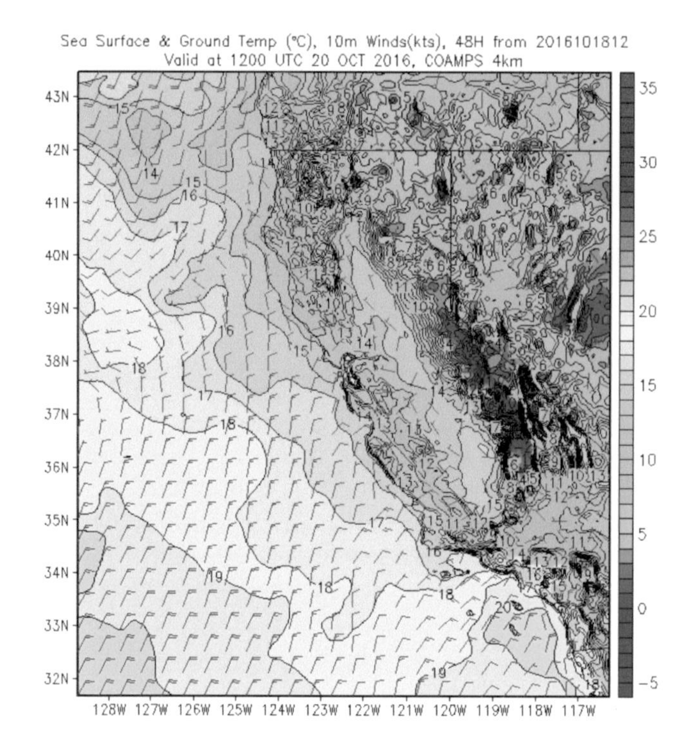

Plate 6

[figure 1.8] An image from the COAMPS web site showing the 2-day forecast for surface temperatures and winds. [Downloaded 18 October 2016 from http://www.nrlmry.navy.mil/coamps-web/web/home]

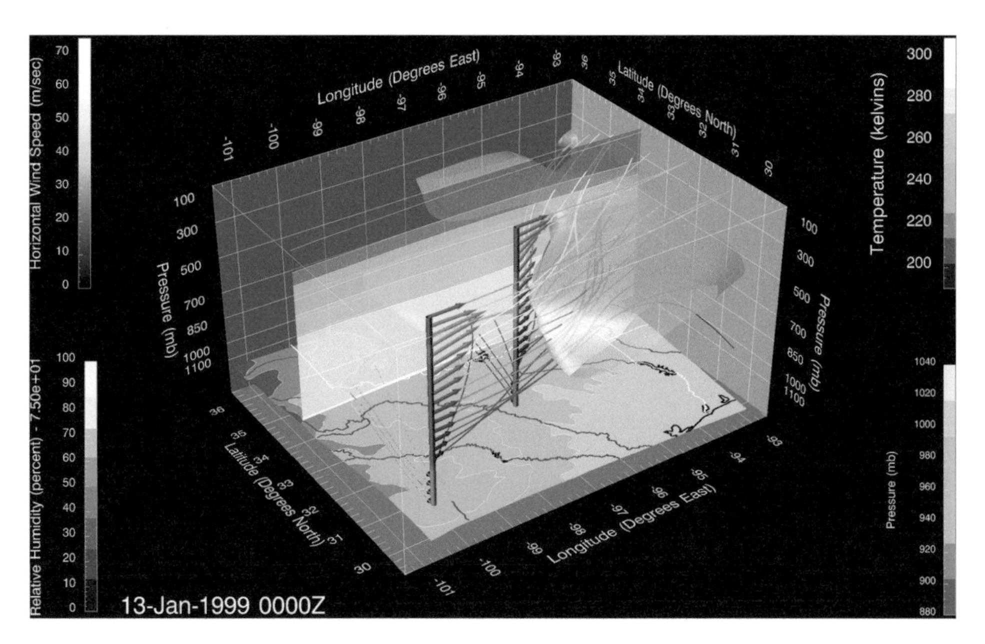

Plate 7

[figure 2.9] An example PRAVDA display (courtesy of Lloyd Treinish, IBM) [http://www.research.ibm.com/dx/bonuspak/html/bonuspak295.html].

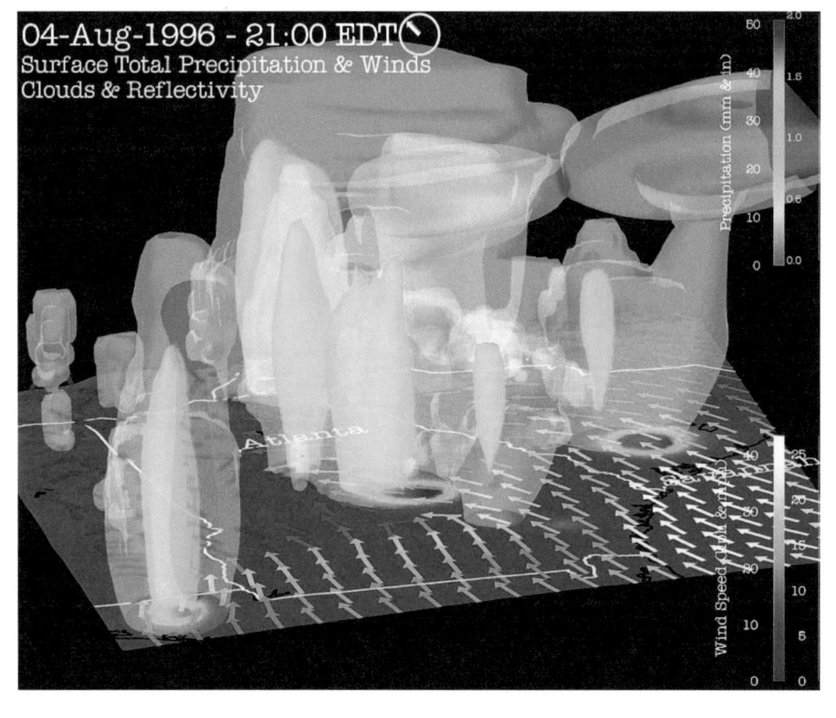

Plate 8

[figure 2.10] An example PRAVDA display (courtesy of Lloyd Treinish, IBM) [http://www.research.ibm.com/dx/bonuspak/html/bonuspak295.html].

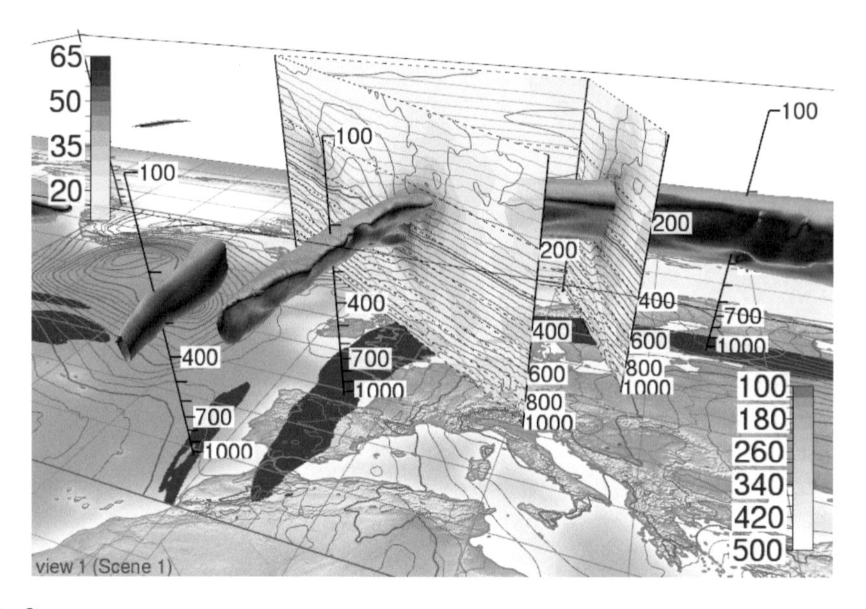

Plate 9

[figure 2.11] An example display in MET.3D (courtesy of Marc Rautenhaus, Technische Universität München).

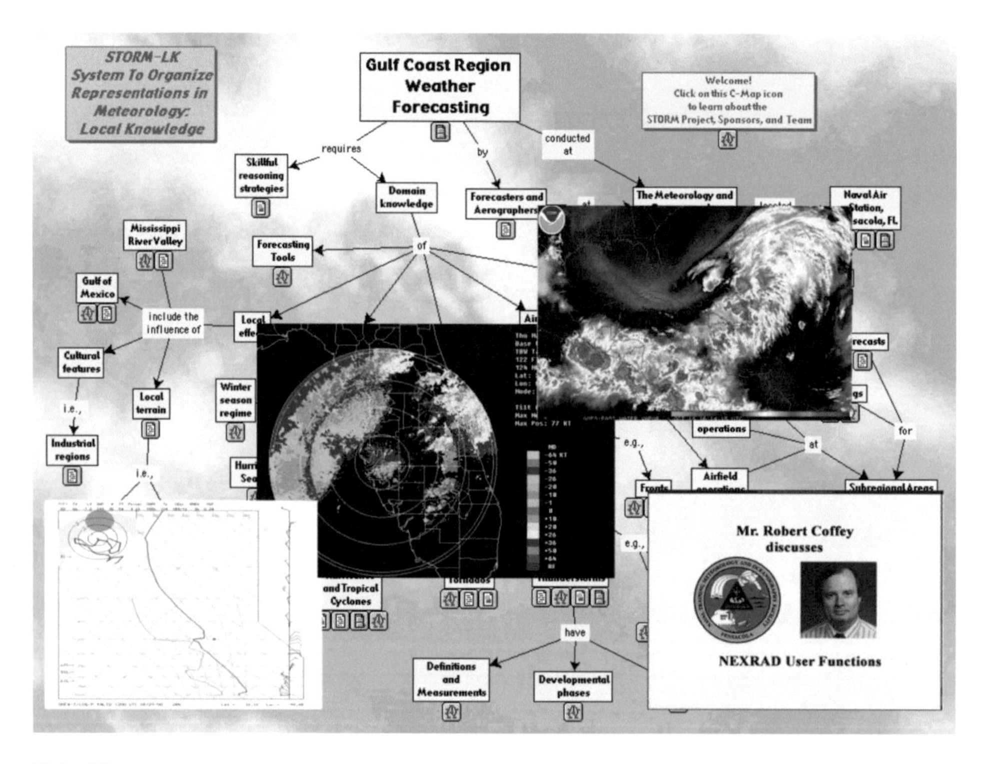

Plate 10

[figure 8.6] A "presentation" version of the top map in the STORM-LK knowledge model, illustrating some of the multimedia resources that are appended to concept nodes: tutorial videos, data charts, satellite images, and radar composites.

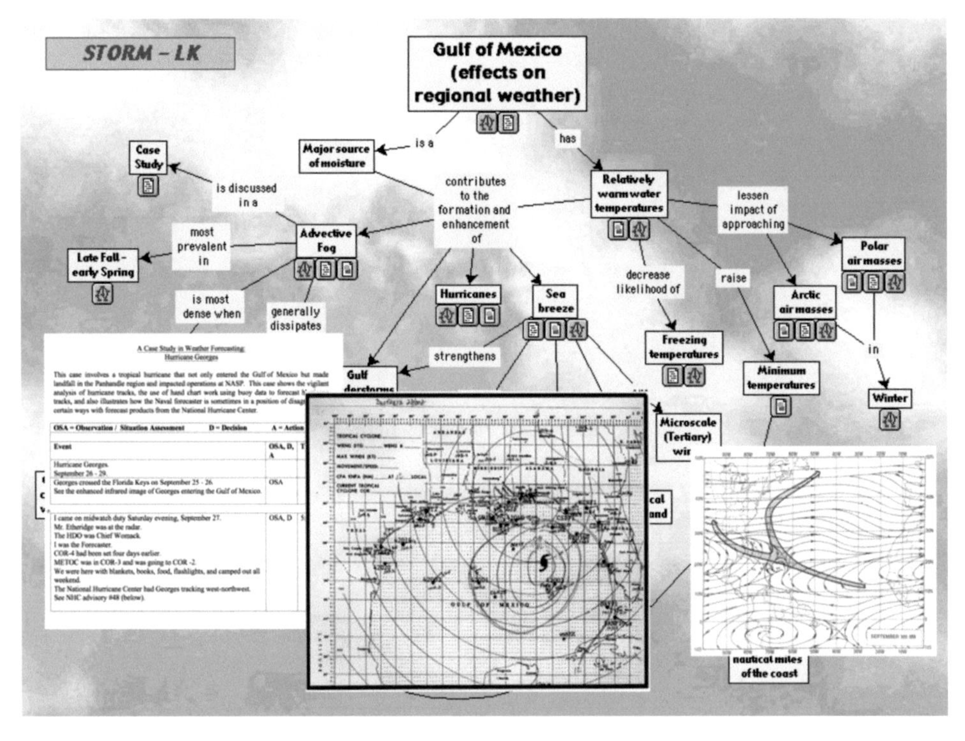

Plate 11

[figure 8.7] The concept map about Gulf of Mexico effects with resources from the CDM procedure that covered a case of hurricane track forecasting.

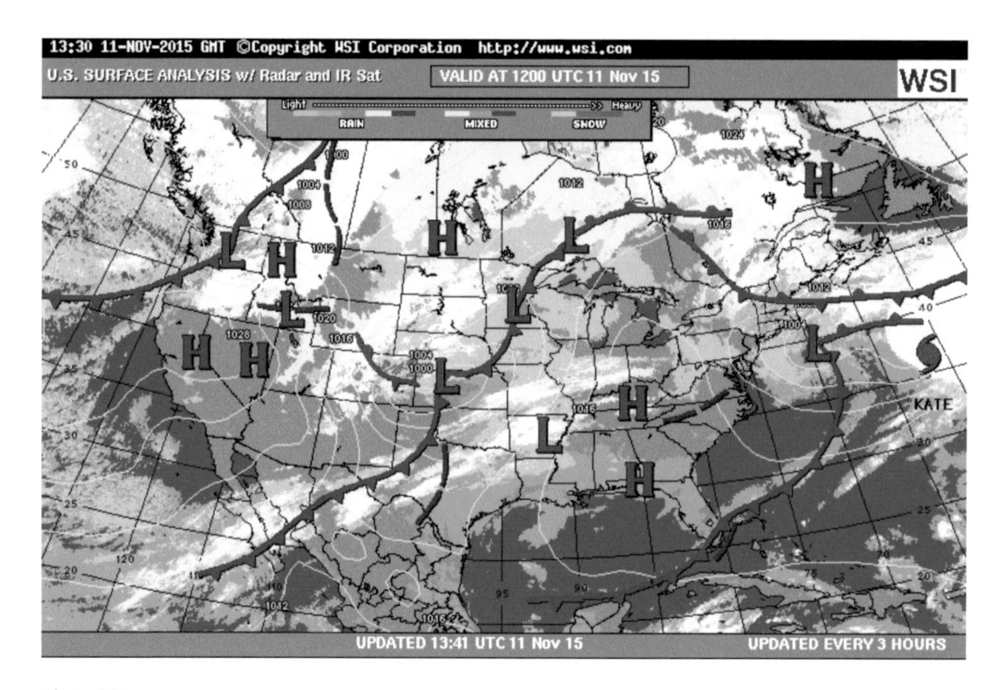

Plate 12

[figure 9.4] A "Sat-Rad" weather map produced by WSI Corporation.
Reproduced with permission from WSI Corporation [downloaded 11 November 2015, from http://www.intellicast.com/National/Surface/Mixed.aspx/].

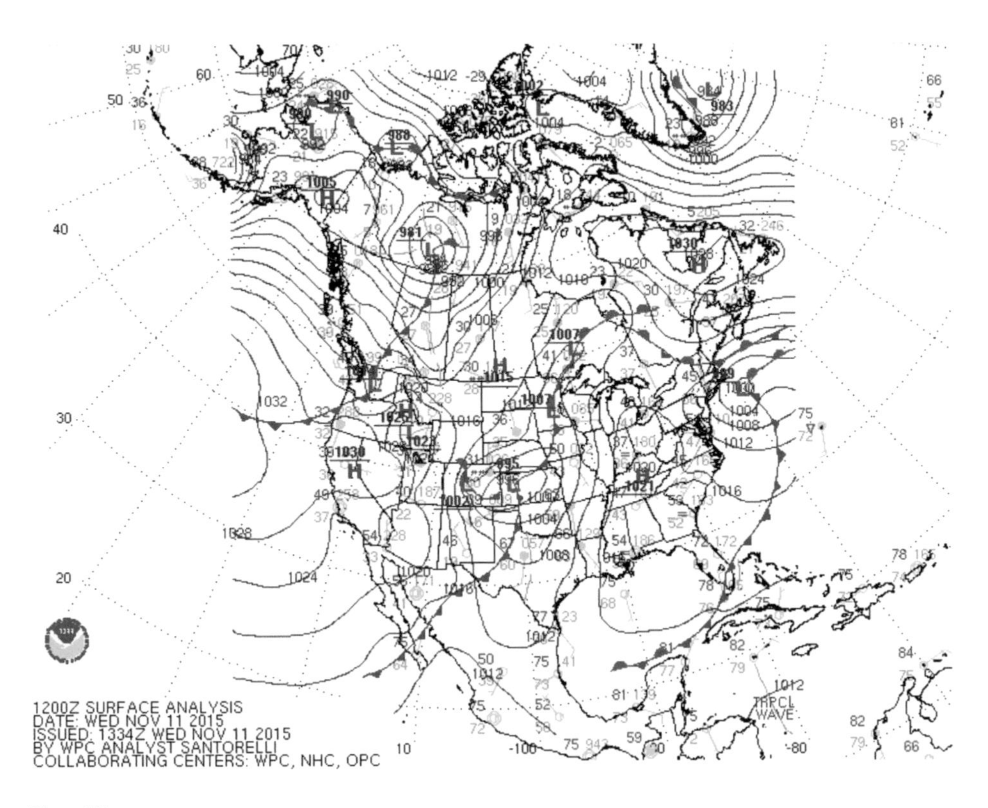

Plate 13

[figure 9.5] A "surface analysis" weather map [downloaded 10 November 2015, from http://www.wpc.ncep.noaa.gov/sfc/90fwbg.gif].

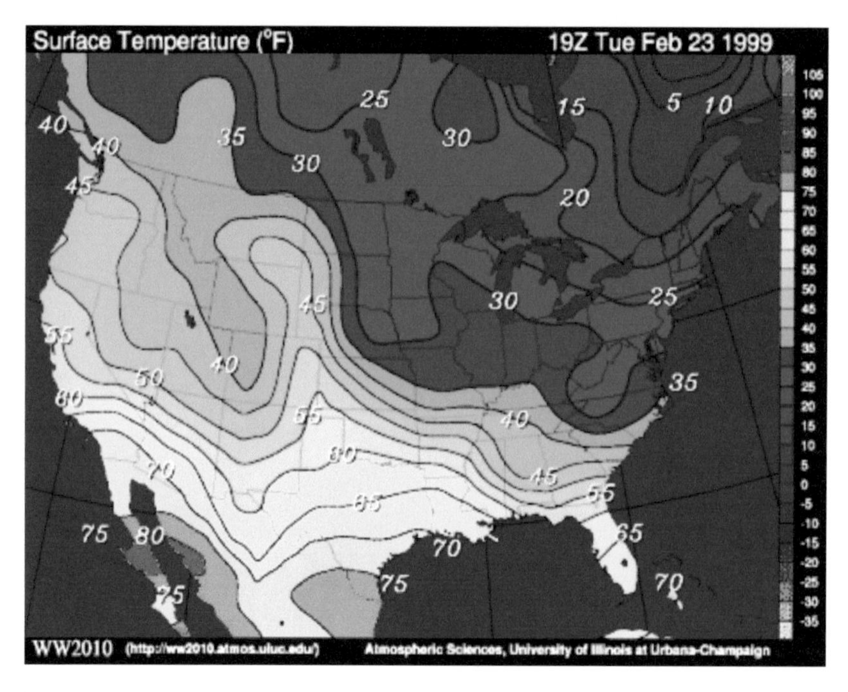

Plate 14

[figure 9.6] A surface temperature weather map. Reproduced by permission from the Department of Atmospheric Sciences, University of Illinois, Champaign–Urbana.

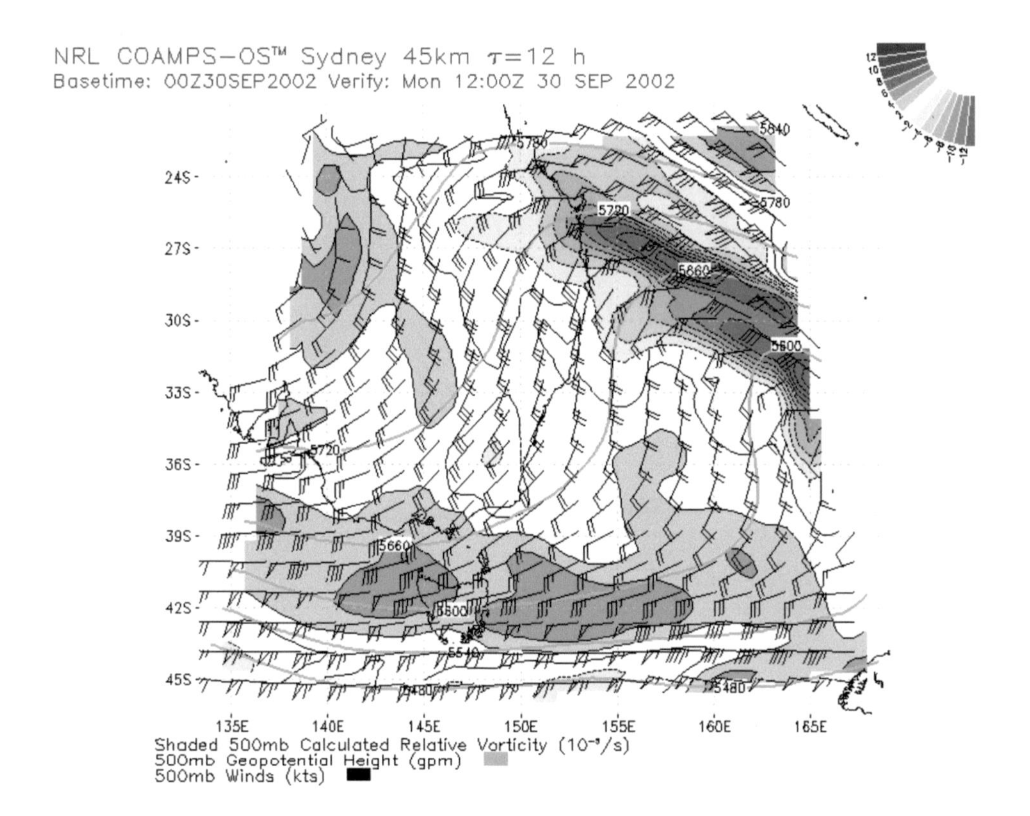

Plate 15

[figure 9.7] An example COAMPS 500-mb height/wind product [downloaded 11 March 2016, from https://cavu.nrlmry.navy.mil]. See [http://www.nrlmry.navy.mil/coamps-web/web/home].

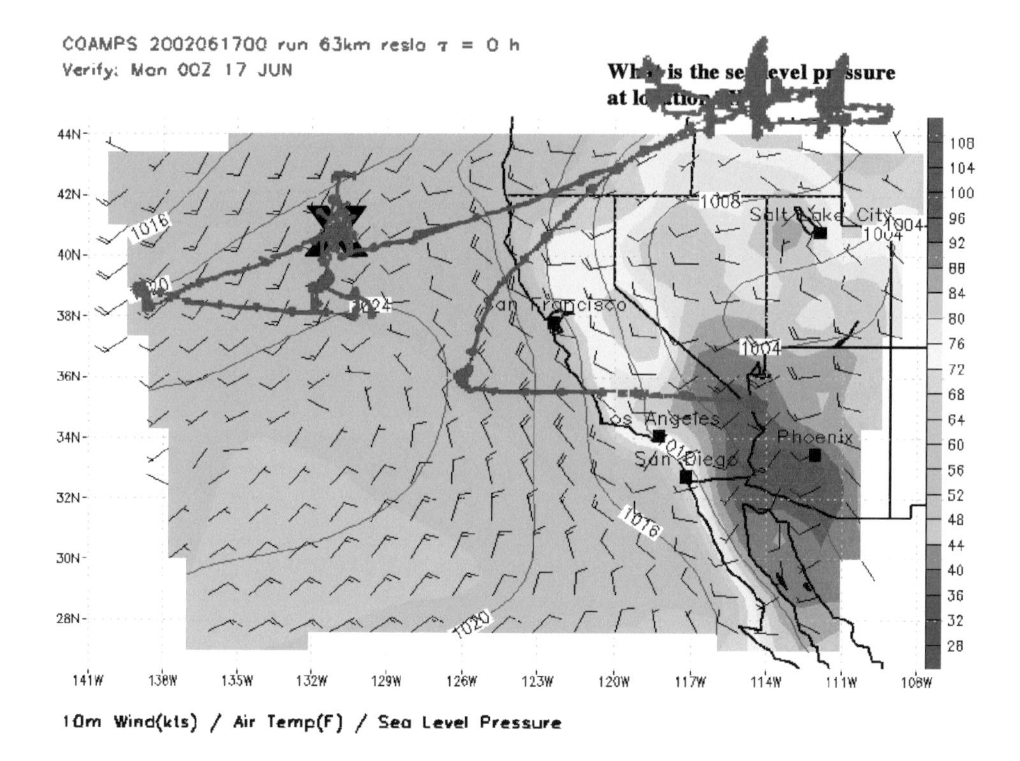

Plate 16

[figure 9.9] A COAMPS product with a re-creation of eye movement data from the experiments by Trafton et al.

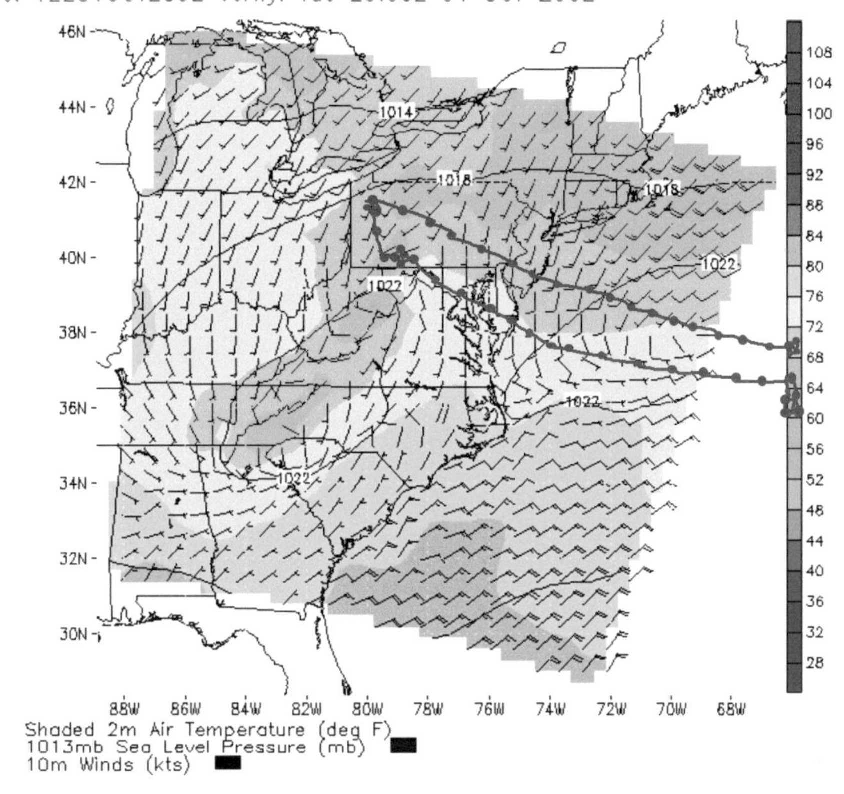

Plate 17

[figure 9.10] A COAMPS product with a re-creation of eye movement data from the experiments by Trafton et al.

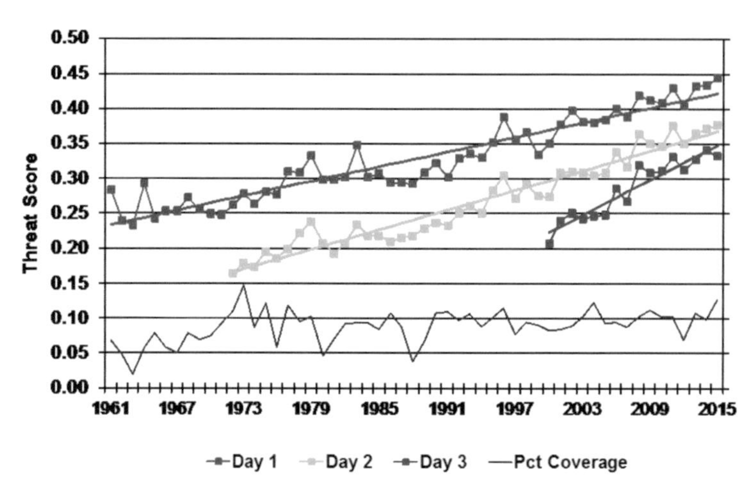

Plate 18

[figure 12.1] "Threat" (performance) scores for computer models across the years 1961 through 2015 [downloaded March 28, 2016, from http://www.wpc.ncep.noaa.gov/images/hpcvrf/wpc05yr.gif].

Plate 19

[figure B.8] This figure shows a GOES satellite image of Hurricane George just after it crossed the Florida Keys.

Plate 20

[figure B.15] This figure shows an infrared image at the time of landfall of the center of Hurricane George the morning of 28 September 1998.

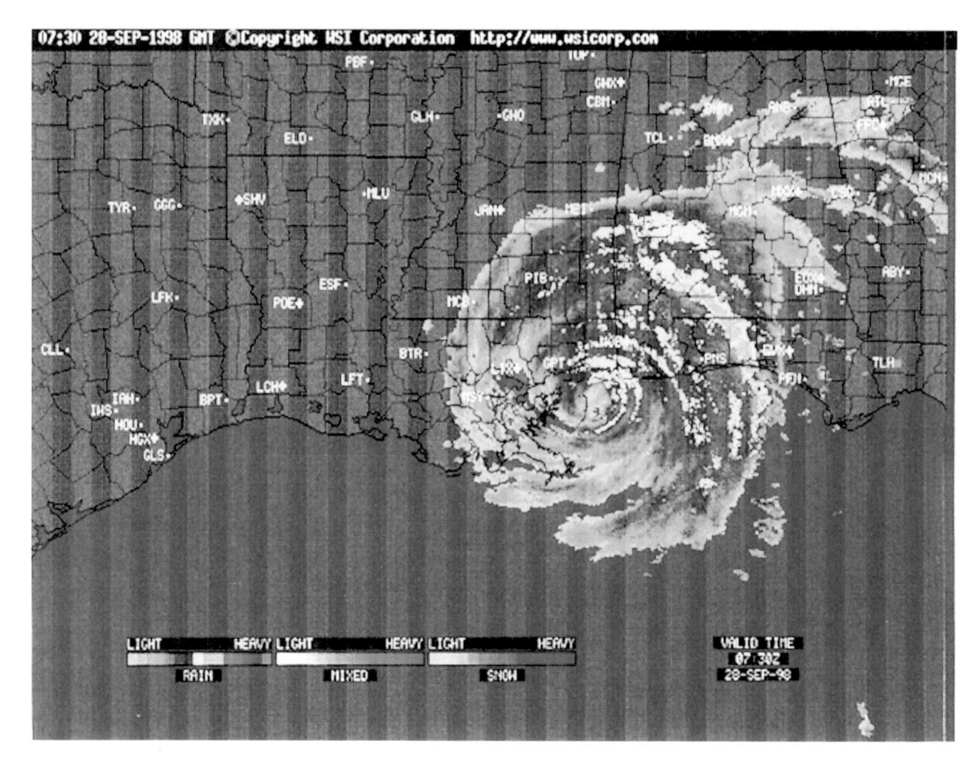

Plate 21

[figure B.16] This figure shows the radar image for Hurricane George at the time the center made landfall.

7 How Do Forecasters Get to Be Expert Forecasters?

In chapter 6, we reviewed the research on the question of what makes someone an expert. That review provides a framework for understanding expert cognition as we look at weather forecasting in particular. Beginning in this chapter, we converge on the concept of expertise at weather forecasting.

The literature on expert systems, broadly, includes many discussions of the differences between experts and novices (see Hoffman and Militello, 2007). Most research reports bifurcate people into only those two categories. Most of the people who are called novices are in fact not novices, and it is rare indeed to find a study that includes convincing evidence that the so-called experts actually qualify as experts in terms of their measured performance. Almost always, what is said is that the "experts" have more years of experience (sometimes just a few years, relative to trainees). Rarely considered is the fact that a proficiency scale can and should be more refined than a simple bifurcation (see Hoffman, 1998).

As we will show in this chapter, there is more to proficiency scaling than simply binning people as being either expert or novice. The core notion is that in order to make claims about expertise in any domain, you have to be able to demonstrate that the individuals referenced are in fact experts. To our knowledge, there have been almost no studies that specifically addressed the question of whether individuals who are called experts at forecasting in fact perform better than individuals who are not called experts (say, journeyman). To be sure, there is something of a circularity here. To show that someone is an expert, you have to show that their performance is superior, but to do that you have to show that their performance is superior to that of others who are less proficient. As we discussed in chapter 6, there are multiple ways to ask and answer the question of whether an individual is an expert because there is more to expertise than raw performance on a primary task. For example, experts can adapt to rare and tough problems, their knowledge is extensive, they can explain and justify their reasoning, and so forth.

In the study of experts specifically at weather forecasting, there are reports on the forecasting competitions and contests in which the performance of instructors (i.e., meteorology professors) is compared to that of the students (see chapter 3 and also later in this chapter). To be sure, many of the professors in the forecasting contests likely would qualify as experts at forecasting even though they were not day-to-day operational forecasters.

The literature on expert systems in weather forecasting (see chapter 11) includes many projects that involved the creation of expert systems based on knowledge acquisition from people who are said to be experts. There, too, we are provided no empirical evidence that the so-called experts were experts, except to say that they were professional meteorologists or professional meteorologists with so many years of experience. To be sure, many of the forecasters who help make expert systems would qualify as experts. "With very few exceptions, the highest [skill scores] go to the more experienced forecasters with lengthy, continuous service at [a] particular location" (Dyer, 1987, p. 20).

Of the many studies we cite in this book—especially chapters 9 and 10 on forecaster perception and reasoning and chapter 12 on forecasters versus computer models—we see discussions of studies that involved "forecasters" or "expert forecasters" accompanied by no information whatsoever about their actual proficiency level except to say that they were forecasters, operational forecasters, or forecasters having so many years of experience. To be sure, many of the forecasters whose performance was studied would qualify as experts. The superior skill scores found in studies of forecaster performance (chapter 5) can be taken as prima facie evidence that at least some of the forecasters *were* experts. Nevertheless, it is necessary to have a strong empirical base on this matter. Thus, in this chapter, we address the methodology of proficiency scaling and the means for identifying experts, which includes and goes beyond measures of performance solely at the forecasting task.

A first question to ask, and set aside, is whether forecasters get to be proficient forecasters simply because they are more intelligent.

Expertise versus Intelligence

Experimental psychologist (and former Air Weather Service forecaster) James J. Jenkins presented 90 Air Force forecasters with a number of standardized psychological tests of mental ability (Jenkins, 1953). On these tests, the forecasters surpassed a sample of college students (including engineering students), a sample of nonmeteorologist adults, and a sample of professional clerical workers. Next, performance on the psychological tests was correlated with forecast skill. The measure of forecast skill was based on the method used then by the Air Weather Service—precision and accuracy in predicting the probabilities of values (i.e., pressure, temperature, visibility, and ceiling) and events (i.e., precipitation). This score had been standardized on the basis of forecasts obtained three times per week for every forecaster serving in the Air Weather Service during the period 1943–1945. Based on the skill scores of his sample of forecasters, Jenkins divided them into two groups: those with higher and those with lower scores. There resulted no difference between these two subgroups in terms of their results on standard psychological ability tests. Forecast skill correlated with only one of the measures from one of the psychological aptitude tests—the Names section of the Minnesota Clerical Aptitude Test, which was basically a test of speed and accuracy of perception. Other than this relation, the other psychological measures could only be used for negative selection, that is, those with the lowest percentile forecast skill scores tended to be those with the lowest percentile scores on one or more of the psychological ability tests. Given that entrance requirements for training in meteorology in both academia and the military include mathematical ability and other skills usually associated with high intelligence, it not surprising that a high versus low division of forecasters on the basis of intelligence test scores did not go clearly hand-in-hand with high versus low skill scores. In other words, there was a ceiling effect because the forecasters Jenkins studied were all pretty smart.

Jenkins' results fit with more recent observations suggesting that forecasters rely heavily on perceptual skill and judgment (Hoffman, 1991; Klein and Hoffman, 1992). The results also fit recent research showing that proficiency and expertise are not necessarily highly correlated with general intelligence (e.g., Ceci and Liker, 1986), but instead are more related to educational background and especially degree of experience and motivated practice (see Ericsson, Krampe, and Tesch-Römer, 1993; Ericsson and Lehman, 1996).

Thus, we can explore the question of how expertise at forecasting develops while setting aside the notion that general intelligence will distinguish expert from proficient forecasters. We start by referencing again the studies that have been conducted on the "weather forecasting competitions" because those studies looked at the forecasting performance of experts as well as meteorology students.

Forecasting Competitions

Roebber and Bosart (1996b) examined data from a forecasting contest conducted at the State University of New York–Albany over the years 1988–1992, in which groups of

10 to 20 students and meteorology faculty made forecasts of daily high and low temperatures and POP for about 66 days in each of the two academic semesters each year. Experience at forecasting was taken into account as well as education level.

One set of results confirmed the notion that experience was critical in the development of expertise (see Ericsson, Krampe, and Tesch-Römer, 1993). Many undergraduates who enter college as meteorology majors have already had considerable forecasting experience. Roebber and Bosart (1996b) placed each participant into either a high- or low-experienced group according to background experience (i.e., interest in weather, familiarity with weather data types). There was a statistically significant difference in the forecast skill scores comparing participants when grouped according to high versus low experience but not when grouped by education level (i.e., students vs. faculty).

The results also confirmed previous findings that the skill scores of forecasts made in the academic setting by experienced students and meteorology faculty are uniformly high (in the range of 0.98 to 0.91) and do not suffer from conditional bias, that is, any tendency for forecasts to take on extreme values when one or more individual data values (e.g., surface dewpoint, windspeed) takes on an extreme value. The results also confirmed the finding that skill scores of consensus forecasts made in the academic setting can approximate those of the NWS forecasts (see also Bosart, 1983; Sanders, 1986), and that the development of proficiency at forecasting precipitation involves a longer time frame than that for forecasting temperatures. Finally, the results confirmed the expected finding that beginning in the late 1960s, when the first computer models came out, and across the years, during which the numerical models were developed, improved, and operationalized, there was a continual loss of skill for the human forecasts. That is, as the computer models got better, less value was added by the human's adjustments of the computer model guidance. This trend continues today, although it should not be interpreted to mean that forecasters provide no value added.

This study also showed that forecasting skill is determined by experience that happens beyond the baseline provided by meteorological training, that is, experience that occurs after the apprentice stage but before the journeyman stage. The first ten or so forecasts showed high errors relative to the consensus forecast, a "break-in" period when the forecaster learns the basic logistics of the forecasting operation. A second stage lasts up to about 100 forecasts. The skill advantage that develops over this time seems related to developing a consistent procedure, but also on the ability to recognize when to adjust the computer model outputs based on the weather pattern and know when deviations from a standard forecasting procedure are needed (see also Roebber, 1998).

Roebber and Bosart (1996b) looked in detail at those weather situations in which the more highly experienced forecasters created forecasts that deviated significantly from those of the computer models (e.g., temperature forecasts differed by five or more degrees Fahrenheit). Many of these were situations in which the meaning or importance of a particular cue depended on the value of some other cue in a way that computer models did not accommodate (e.g., temperature at a particular height in the atmosphere depends on whether there are thunderstorms in the vicinity; dry air warms more rapidly, but this depends on cloud cover; etc.). Thus, there is a rapid progression in forecast skill for initially inexperienced forecasters—across the first year of experience at a full seasonal cycle for their forecast region, they come to recognize the patterns that are played out in the atmosphere and, especially when supported by feedback and good mentoring, the trainee can come to understand the rules and develop strategies to accommodate the patterns (e.g., in the Albany region, winds off the Great Lakes influence the rate of night time radiational cooling and thus affect the minimum low temperatures).

Roebber and Bosart (1996b) approximated the effect on skill score of moving a forecaster to a new location—a drop in skill at forecasting temperatures by 5% to 10% relative to a consensus forecast and a drop in skill at forecasting precipitation by 10% or more. This result conforms to a view of meteorologists (e.g., Doswell, 1986c) and a conclusion from first-generation expert systems work (Elio, de Haan, and Strong, 1987) that local knowledge is critical in forecasting.

> It is my experience that most forecasters indeed develop a personal set of rules (heuristics) for accomplishing the tasks of diagnosis and forecasting with the available information. The development of personal heuristics begins during the education and early operational practice phases of a forecaster's career and, in some cases, proceeds indefinitely. (Doswell, 2004, p. 1120)

In the clearest demonstration of this, Roebber, Bosart, and Forbes (1996) examined data from the 1992–1993 National Collegiate Weather Forecasting Contest. In this contest, teams of student forecasters from a number of North American colleges made forecasts for a range of sites. Roebber et al. (1996) examined errors in daily temperature forecasts as a function of distance from the team home site. Although high-experience-level participants (faculty and graduate students) suffered less from moving their forecast locations to a distant site, for both high- and low-experience (undergraduate) groups, distance from the familiar site significantly impacted forecast accuracy—the differences in errors comparing familiar to distant sites were half again as much as the differences in errors comparing the high- and low-experience groups. Roebber et al. (1996) concluded that greater experience is reflected in:

1. a greater ability to take weather conditions into account in understanding the causation of precipitation (e.g., precipitation due to fronts, troughs, upslope winds, intensifying cyclonic activity), and
2. a greater ability to adjust computer model guidance in light of the weather situation, that is, a greater ability to take computer model biases into account.

The results of the research by Roebber et al. (1996) also speak to the importance of the formation of mental models that are rich with knowledge down to the level of particular local effects—prevailing seasonal winds, land and sea breezes, terrain effects, and the effects of nearby warm lakes. Using local knowledge, the more experienced forecasters can make better use of the data cues in forming mental models that link observations to the forecasted events via causal explanation.

Recognition of the Need for Training to High Proficiency

In chapter 3, we explained how actual forecasting skill is acquired—especially by on-the-job learning. In her ethnographic study, LaDue (2011) found that newly hired forecasters are stymied by what seems to be the ill-structured nature of forecasting, on top of the inherent uncertainty, amount of data, and complexity of tools. This can be overwhelming. Those exhibiting qualities predictive of the eventual achievement of expertise are deeply engaged in their forecasting activity, relating ongoing weather and their forecasts to their mental models. They keep up with the state of the art, and some of them push the state by doing research. Those who have opportunities for social interaction generally take advantage of it, although it was clear that not all around them were people from whom they wish to learn. They learned a great deal by experiencing the weather and reviewing their bad forecasts for the weather that they experience. When particular types of weather are rare (for the region, season, or climate), they have trouble learning to forecast it well.

Clearly, it is from on-the-job experience that expertise is achieved. A wave of modernization and restructuring of the NWS in the early 1990s was motivated by considerable concern about training issues for new staff (including forecasters, science officers, operations officers, and warning coordinators) (Rothfusz et al., 1992). Some new staff were able to spend "a few weeks" learning from veterans, but the modernization effort mandated the development of a more extensive and formal training program, which was instituted at three WFOs. Training included seminars, co-forecasting with veterans, and lots of practice forecasting. Seminars focused on the use of new radar products, the use of the AWIPS workstation system (see chapter 2), and local climatologies. Seminars also focused on WFO operations (e.g., shift duties, WFO organization, and staff duties). Forecast accuracy was assessed at each WFO for a number of weeks following the initial seminars and forecasting practice.

The training methods and their emphases differed across the WFOs. For instance, forecast practice at Amarillo involved providing feedback from veterans, whereas at Tulsa, apparently, it did not. Therefore, conclusions about training efficacy are of limited generality. The researchers reported only that performance was "quite good." The researchers concluded that:

> By the end of the [practice] period, the forecasters were able to focus on the meteorological aspects of their products rather than be distracted by learning the "mechanics" of their jobs. ... A certain psychological aspect was also noted. Issuing forecasts can be stressful for a new forecaster. (p. 82)

They also concluded that four months was the minimum time to adequately train a forecasting staff. This training program had to be developed quickly, and the approach to assessment might have been more systematic. The conclusions were based largely on the researchers' and forecaster's poststudy subjective impressions and judgments. Although the researchers acknowledged that the forecaster had to "gain extensive knowledge" (p. 82) and "trainees must demonstrate their knowledge" (p. 83), the cognitive elements of forecasting were not systematically investigated. It was claimed that "effective" seminars needed to be presented by experts and include such things as a clear set of objectives, copies of all the transparencies, case studies, and so on—but these are known to educational psychologists as being important features of any instruction design. This project showed that there was a need for studies of the development of meteorological reasoning and perceptual skill in the midrange of the novice–expert scale going from educated meteorologist to expert forecaster.

Meteorologists involved in training NWS personnel contracted a cognitive task analysis of the warning forecaster task. Hahn, Rall, and Klinger (2002) interviewed seven forecasters from Alabama, Oklahoma, Texas, and Missouri. Six of these forecasters had extensive experience (12 to 20 years), four were currently in Science and Operations Officer positions, and two were meteorologists in charge of their office. The seventh forecaster was a journeywoman with three months of experience, including two severe weather warning events. The interviews used the Critical Decision Method (see Crandall, Klein, and Hoffman, 2006; Hoffman, Crandall, and Shadbolt, 1998), in which the participant is guided in the detailed recall of recently experienced "tough cases." In successive waves of retelling, the participant goes back over the event to verify the timeline, identify decision points, and add details about the cues, expectations, options evaluated, and options chosen.

Hahn et al. (2003) reported a number of findings concerning the achievement of forecasting expertise:

• Forecasters talked about a strong social component to their work, specifically the importance of on-the-job training. They take the opportunity to train others in their office, get feedback on their conceptual models or ideas of how weather would evolve, and gain confidence in their decisions. Forecasters interact with others to learn and get feedback to gain confidence in their decisions.
• Forecasters use weather events, particularly unusual ones, to build their knowledge. The forecasters said it was important to gain direct feedback on how the weather impacted people so they could relate those effects to the data they had to work with as the event unfolded. They would later analyze an event in detail, particularly after failing to issue a warning prior to an event. They strongly desired immediate feedback, or ground truth, about the impacts of the weather as it evolved. Warning forecasters often conducted postevent analyses of weather events to relate the information they had with postevent damage and other reports, especially if they had failed to issue a timely warning in advance of a severe weather event (see the discussion of snowfall forecasting in chapter 5).

The goal for the National Weather Service was to use these findings about forecaster skill acquisition to inform the design of new training programs. But what does it mean when Hahn et al. (2003) assert that their interviewees were "highly experienced"? Does that mean they were experts?

A General Proficiency Scale

Both scientists and laypersons are generally rather fast and loose in their use of the word "expert" and rely on a simplistic distinction between experts and novices, as if humanity could be so neatly bifurcated. In a study on the classification of dinosaurs (Chi and Koeske, 1983), the "experts" were actually children (ardent dinosaur hobbyists), yet their knowledge showed some of the features of expertise—depth of comprehension as evidenced in their differential classification of meat-eaters versus plant-eaters. In some research, college students have served as experts because of their knowledge of particular domains (e.g., football, wedding apparel, regional geography) (see Chi, Glaser, and Farr, 1988; Ericsson and Smith, 1991). In some studies of mechanics problem solving, graduate students have been the "experts" (Chi, Feltovich, and Glaser, 1981). In some domains, people who are regarded as "novices" can actually be highly experienced. In the judging of livestock, for instance, one might remain a "novice" even after as much as 10 years of experience at school and club training; "experts" typically have 20 to 30 years of experience (Phelps and Shanteau, 1978). Workers in some domains (and many organizations) naturally distinguish their peers who are good, even very good, from the "real" experts or what might be called the "grand masters" (Hoffman et al., 2011; Shanteau, 1984, 1988, 1989). Clearly, the terms "novice" and "expert" leave much to be specified.

As we discussed in chapter 6, the widely cited benchmark for the achievement of expertise is 10 years of full-time job experience. The studies of weather forecasting competitions (see chapter 3) suggest that proficient performance can be achieved in fewer than 10 years. Even in chess, players differ widely in the amount of practice needed to achieve the highest levels of proficiency; a 13-year-old once beat a Grand Master (see Gobet and Campitelli, 2007). Studies of world-class athletes have shown similarly that some individuals achieve high proficiency in fewer than 10 years (see Hambrick et al., 2014a, 2014b).

Benchmark generalizations aside, it does take practice to achieve high proficiency, and practice takes time. Also, it takes practice at difficult, rare, and tough cases or problem instances, rather than just more and more practice at common or routine cases (see Hoffman et al., 2014). Even more important is the fact that the world should not be bifurcated into people who are experts and people who are novices. Despite its maturation as a scientific field (Ericsson et al., 2006), Expertise Studies is still frequented by papers which say that the research involved the comparison of experts versus novices, assuming there is nothing in between. Worse, research reports often provide no evidence, let alone convincing evidence, that the so-called experts actually qualified as experts.

A refined proficiency scale is presented in table 7.1. This is based on concepts from the traditional craft guilds of the Middle Ages.

This provides better concepts for studying the development of proficiency. In proficiency scaling, one attempts to benchmark all of the classes listed in table 7.1 (see Hoffman and Militello, 2007). If research is predicated on any notion of expertise (e.g., a need to build a new decision support system that will help experts but can also be used to teach apprentices), then it is necessary to have some sort of empirical anchor on what it really means for a person to be an "expert," say, or an "apprentice." Ideally, a proficiency scale will be based on more than one method, from the following general classes:

1. Professional criteria (e.g., graduate degrees, training experience, publication record, memberships in professional societies, and licensing).

Table 7.1
Basic proficiency categories based on the traditional craft guild terminology

Naïve	One who is ignorant of a domain.
Novice	Literally, someone who is new—a probationary member. There has been some ("minimal") exposure to the domain.
Initiate	Literally, someone who has been through an initiation ceremony—a novice who has begun introductory instruction.
Apprentice	Literally, one who is learning—a student undergoing a program of instruction beyond the introductory level. Traditionally, the apprentice is immersed in the domain by living with and assisting someone at a higher level. The length of an apprenticeship depends on the domain, ranging from about 1 to 12 years in the craft guilds.
Journeyman	Literally, a person who can perform a day's labor unsupervised, although working under orders. An experienced and reliable worker, or one who has achieved a level of competence. It is possible to remain at this level for life.
Expert	The distinguished or brilliant journeyman, highly regarded by peers, whose judgments are uncommonly accurate and reliable, whose performance shows consummate skill and economy of effort, and who can deal effectively with certain types of rare or "tough" cases. Also, an expert is one who has special skills or knowledge derived from extensive experience with subdomains.
Master	Traditionally, any journeyman or expert who is also qualified to teach those at a lower level. A master is one of an elite group of experts whose judgments set the regulations, standards, or ideals. Also, a master can be that expert who is regarded by the other experts as being "the" expert, or the "real" expert, especially with regard to subdomain knowledge.

Source: Adapted from Hoffman (1998).

2. Estimates of experience extent, breadth, and depth based on results from some sort of Career Interview.

3. Evaluations of performance based on clearly defined performance measures (see chapter 5).

4. Interviewing workers in an organization to identify the experts (e.g., who talks to whom for advice), a process called sociogrammetry (Stein, 1992, 1997).

With regard to professional criteria in meteorology, the AMS and the NWA offer ways to earn "master's" status. The AMS offers a Certified Consulting Meteorologist Seal and a TV Broadcast Seal. The NWA offers a similar TV Seal and now offers an online blogging certification. (The latter is intended to mitigate some of the poor quality information that gets posted online by nonexperts.) The certifications require real-life experience periods, plus certification exams and/or reviews. CCMs are highly regarded by their peers. They are considered experts in the application of weather information to a host of practical challenges ranging from specialized forecasts to engineering design support and expert testimony on weather-related court cases (https://www.ametsoc.org/ams/index.cfm/education-careers/ams-professional-certification-programs/certified-consulting-meteorologist-program-ccm/). Hence, applicants must have a minimum of five years of work at the professional level in meteorology or a related field. Substitution of some advanced degree work for professional-level work is possible. Although the "10-year rule" doesn't apply, there is clear recognition that experience matters.

The Proficiency Scaling Process

Proficiency scaling is the attempt to forge a domain- and organizationally appropriate scale for distinguishing levels of proficiency. Weather forecasting affords an interesting opportunity to compare experience measures with actual performance measures. At most weather forecasting facilities, forecast verification records are kept, and thus there is a ready-made performance measure (see chapter 5). Presumably, forecasters with more experience should produce forecasts that "verify" more often.

Studies of chess players, musicians, and professionals in various domains typically attempted to estimate how much practice individuals have had. Most often, the claim that research participants are experts is based on a rough calculation of years or hours of experience. Given that this scaling method is common, relatively easy to conduct, and established in such studies as those by Simon and his colleagues, it deserves particular consideration. It is prudent to rely on what is called the "Gilbreth Correction Factor."

Frank and Lillian Gilbreth (Gilbreth, 1911, 1934; Gilbreth and Gilbreth, 1917) helped pioneer the study of work efficiency using the method of "time and motion study." Studies were conducted of scores of jobs, including bricklaying, handkerchief folding, manual typesetting, radio assembly, cigar wrapping, and envelope stuffing. Results consistently showed that simple redesign of the workspace could often result in a doubling of output. "[Lillian Gilbreth] observed that while all her motions were made with great rapidity, about half of them would be unnecessary if she arranged her work a little differently" (from the Introduction by R. T. Kent). "The workman's output can always be doubled and oftentimes more than tripled by scientific motion study" (Gilbreth, 1911, p. 93). Likewise, studies of the causes of fatigue (Gilbreth and Gilbreth, 1919), based on results from fatigue surveys and the analysis of activity records, showed that during the course of a work day, manual laborers would spend roughly half their time engaged in motions that could be completely eliminated by redesign of the workspace, workbenches, and chairs. Redesign often resulted in a doubling of output and a reduction in fatigue. From these findings came the Gilbreth Correction Factor of 2, that is, one takes the number of hours during which workers are *at* their workplace and halves that number to obtain an estimate of the amount of time spent actually working at the primary job tasks. This is certainly a conservative estimate: It may underestimate the amount of time workers spend conducting their primary job tasks. But the need to have confidence when saying that an individual qualifies as an expert means that it is better to error on the side of disqualification.

We next present a case study in proficiency scaling for U.S. Navy forecasting. Following the presentation of the U.S. Navy case, we present a second proficiency scaling case study of U.S. Air Force forecasting. The two case studies highlight the different ways in which proficiency scaling can be conducted.

U.S. Navy Weather Forecasting Case Study

Proficiency scaling involved multiple procedures including career interviews and performance analysis. The combined results permitted the construction of a proficiency scale.

Hours of Experience

Hoffman, Coffey, and Ford (2000) began this project by interviewing 22 individuals at the weather forecasting facility at the Pensacola Naval Air Station. All were forecasters (civilians) or "aerographers" (enlisted personnel or officers). The group included senior forecasters (i.e., individuals in their 50s) senior officers (i.e., individuals in their 30s and 40s), and individuals with only a few years of experience in the U.S. Navy weather forecasting community. Those individuals were assigned to this particular facility precisely because it was a facility for training aerographers.

The primary mission of forecasting for aviation, and aviation training, lends some unique qualities to the cognitive work of the forecasters. For example, it might seem ironic that the forecasters sometimes feel that they need their clients (the pilots and pilot trainers) to leave them alone. Because there are multiple training flights on any given day, the forecaster ends up having to repeat the same answer to the same questions as pilots or pilot trainers come to the forecasting facility. Through their own limited understanding of the forecaster's activities, their inquiries lead to an increase in forecaster workload. The emotional impact on the forecasters becomes salient.

Despite the uniqueness of Navy-Pensacola as a facility for both pilot training and forecaster training, the proficiency scale results that Hoffman et al. obtained are not unrepresentative, given that the forecasters reason about weather using the same information and processes as utilized in other forecasting contexts and given what we know about how people get to be forecasters (see chapter 3). This said, our focus here is on the process proficiency scaling, which we illustrate by laying out the results.

The forecasters work duty shifts, from which one can derive reasonably accurate estimates on total time on shift as a forecaster across any given period of time or duty assignment. While on watch in a forecasting facility, a forecaster will be engaged in many activities other than actual forecasting. These include: forecasting-related activities (e.g., occasionally looking at a satellite or radar display to maintain situational awareness); job-related activities (e.g., studying meteorology materials to prepare for a qualifying exam, shift management); collateral activities (e.g., facility security check), and non-job-related activities (e.g., watching a cable news channel or engaging in casual discussions). This illustrated why time at work is not a good measure of who has the experience necessary to be rated as expert, and hence the value of applying the Gilbreth Correction Factor.

The Hoffman et al. (2000) analysis involved comparing depth of experience (number of hours scaled to the Gilbreth Correction Factor) to breadth of experience (as indicated by the variety of experiences possible—onship forecasting, experience in more than one climate area, etc.). The expectation was that depth of experience (number of hours) would be verified by breadth (variety of experiences), if only because more experience implies more opportunities to have a greater variety of experiences.

In the participant interviews, participants were asked about their general education, their training in weather forecasting or meteorology, and how they developed an interest in weather. Next, they were supported in laying out their career experience and attempting to estimate the amount of time they had spent engaged in weather analysis or related forecasting activities. A sample data record from the participant interviews appears in table 7.2. This participant was subsequently determined to be a Junior Journeyman. For comparison, a data record for Junior Expert is presented in table 7.3, and a data record for a Senior Expert is presented in table 7.4. In all three tables, certain information has been removed to ensure anonymity, and the data record has been slightly modified to ensure ease of reading. (A school and C school are the two main training experiences, in between which is usually a deployment. School training can last between three and nine months. Although the interviewees were all either forecasters or aerographers, references are made to "Observers." These are junior aerographers who make weather observations on ship or at airfields, and they are not qualified to issue forecasts.)

Table 7.2
Data record from the Participant Interview with a Junior Journeyman

Age	26	
Current Position	E-6, 1st Class Petty Officer	
Current Rank	Aerographer-1	
Education	High school and one year college (aeronautical engineering, physics)	
Early Interest in Weather?	Always enjoyed weather—watching storms, smelling rain, counting lightning. Interest in weather arose when recruited. Is clearly a weather lover now; plans to continue as a career.	
Meteorology Training	1. Boot camp, 8 weeks 2. Basic Aerography school, 16 weeks (observing, clouds, observation codes, plotting and analyzing charts, skew-t, basic oceanography; tropical and aviation forecasting, radar). 3. On-the-job training on ship. On the first tour, was an Airman (E-2). On third tour, did weather observations, sea heights, ran satellite downloads; made oceanographic observations for anti-submarine warfare (pressure, temperature, salinity at depths). Learned chart analysis, satellite interpretation, assisted senior forecaster in preparing airfield forecasts. Was promoted to Aerographer 2nd class (E-5). 4. Tested out of Aerography C school. 5. Self-directed distance learning: Astronomy, weather models, computer programming.	
EXPERIENCE		
Where and When	**What**	**Time**
USS XXXX December 19XX–October 19XX	Observer/forecaster assistant. With 2–3 forecasters and 8–10 observers (2 sections). Lots of on-the-job training—satellite image interpretation, chart analysis and interpretation. A lot of collateral duties—supplies ordering, training record-keeping, safety officer, anti-submarine warfare liaison, equipment maintenance, security officer.	5,628 hours (12 hours per day x 7 days per week x a number of cruises of varying durations totaling 67 weeks)
A Naval Air Station October 19XX–August 19XX	Aerographer-forecaster with three other forecasters and 4 observers. Did pilot briefings, local forecasts, harbor forecasts, warnings, airfield forecasts; supervised and trained Observers. Little on-the-job training—two forecasters were senior but were disengaged. Some collaterals—Network administration, administration record-keeping, Local Area Network system administration (took up a lot of time).	2,160 hours (12 hour watch per day x 15 days per month x 12 months = 2,160 hours)

Table 7.2 (continued)

EXPERIENCE		
Where and When	**What**	**Time**
Naval Air Station September 19XX-19XX	Aerographer. On-the-job training from all the lead forecasters. Some collaterals—fitness coordinator, forecasting software administrator, operations training	3,960 hours (12 hour watch per day x 15 days per month x 22 months)
TOTAL	11,748 hours Gilbreth Correction Factor = 5,874	
Notes		
Determination: This participant qualified to issue forecasts, but the correction suggests that the appropriate proficiency category is junior journeyman. Was always mathematically inclined; took physics, calculus, and chemistry in high school, got all As and Bs. Was promoted at the fastest possible rate. Had lots of on-the-job training. Tested out of Aerography C school.		

Referencing the 10,000–16,000 hours benchmark, military shift duty is more than 40 hours per week and more than 40 weeks per year. But even with that, and *even* taking the Gilbreth Correction Factor into account, the U.S. Navy personnel who Hoffman et al. (2000) felt were best considered journeymen would ordinarily be considered experts. Therefore, Hoffman et al. (2000) crafted a scale that seemed more appropriate to this domain and this organization, one having junior and senior grades within the major proficiency categories. To accomplish this goal, there had to be a reasonable mapping between skill levels and degree of experience, taking into account the conceptual definitions from the craft guilds (table 7.1, above). This is presented in table 7.5. How all of the participants were finally sorted is described in table 7.6.

The participant interviews confirm the findings from LaDue's (2011) ethnographic study (see chapter 3), specifically that people who go into weather forecasting often have a background as a "weather lover" beginning in childhood and generally acquire most of their actual skill at the forecasting process on-the-job.

The sorting in table 7.6 includes some uncertainties (e.g., whether a particular forecaster qualified as a journeyman or a senior apprentice). Although this scale seemed to be appropriate to the domain and the organization, the research project was designed to get an additional handle on proficiency for this domain and organization. Expertise may be only partially correlated with experience (e.g., a highly experienced forecaster

Table 7.3
Data record from a Participant Interview with a Junior Expert

Age	37	
Current Position	E-6, 1st Class Petty Officer	
Current Rank	Aerographer-1	
Education	High school, some meteorology in a physics class.	
Early Interest in Weather?	Always like to observe weather, and astronomy—general sky watching. Got a telescope when age 12 or 13. Visited weather observatory on family trips twice, age 13, 15. Participant was impressed by a TV forecaster. He had excellent delivery. His forecasts included bits of meteorology knowledge, quizzes. He often visited schools and did outreach—this impressed this participant greatly. Most forecasters gave data from airport, but he always mentioned the names of local observers, usually young people, who phoned in their observations. Participant was interested to learn that people could do this sort of thing.	
Meteorology Training	1. Boot camp 2. Aerographer A School 3. Aerographer C School. Took the AG-1 test his second time at the end of this. 4. Tactical Environmental Support System (TESS) training on software and use, 3 months. TESS was an older system for forecasting operations on board ships.	
EXPERIENCE		
Where and When	**What**	**Time**
USS XXXXX 19XX–mid 19XX	Observer. Ship had a section leader, a forecaster, and one or two observers. Did observations, updated observations during flight operations, maintained uplink to satellite and radio fax. Weekly on-the-job training from senior forecaster. One collateral (Quality Control Petty Officer).	4,032 hours (1 major 6-month cruise and several 3-week work-ups; 12 months x 4 weeks per month x 7 days x 12 hours per day-shift)
A Naval Air Station June 19XX–March 19XX	Worked with 10–12 personnel. 1. P qualified as an observer; was Forecast Verification Quality Control Petty Officer, 2. Assistant forecaster. Received a great deal of on-the-job training by listening to lead forecaster's experiences 3. Site waiver for forecaster ~1986 One collateral—Equipment Petty Officer.	4,752 hours (33 months x 12 watches per month x 12 hours per watch)

Table 7.3 (continued)

EXPERIENCE		
Where and When	**What**	**Time**
USS XXXX December 19XX–January 19XX	Forecaster (made Aerographer-1 right after getting there). Aviation support for helicopters and a Marine detachment, so was low-level weather (up to ~10,000 feet), and near shore oceanography—coastal environment weather qualified. Daily on-the-job training from senior forecaster. One collateral—Maintenance Petty Officer.	6,552 hours [(2 deployments of 6 months + two 3-month cruises + two work-ups of 3 weeks = 78 weeks) x 7 days per week x 12 per shift]
A Naval Air Station February 19XX–December 19XX	1. Forecaster—did aviation, weather forecasts for pilots, weather, weather briefings. 2. Subregional forecaster—also did warnings. 3. Forecast Duty Officer—also did warnings. Was training others. Three collaterals—Career Counselor Petty Officer, Administration Petty Officer, Liaison Representative.	8,280 hours (46 months x 15 days per month x 12 hours per shift)
Mobile Environment Team Beginning 19XX–mid 19XX	Forecaster. Deployed three times. Several ships in the Caribbean; two frigates for NATO exercises; Weather forecaster for the Commander of Joint Operations—shore assignment—mostly tropical support. Trained others when not deployed; was Training Petty Officer. One administrative collateral.	12,706 hours Onship deployments: 60% of time was at sea, and therefore P was doing weather. Haiti: 3 months of constant weather work. (706 days x 16 hours per day = 11,296 hours) + (470 days x 3 hours per day)
A Naval Air Station Mid-19XX–19XX	AG-1 (Forecaster Under Instruction). Passed test for subregional forecaster and then Forecast Duty Officer. Trained others. Career Counseling Petty Officer, Training Petty Officer.	Just began current assignment
TOTAL	36,898 Gilbreth Correction Factor = 18,499	
Notes		
Determination: This individual is one of the facility's senior forecasters and corrected hours of experience supports the assignment to the expert level of proficiency. However, other forecasters at the facility were older and had many more hours of experience (see table 7.4). Thus, a decision was made to distinguish levels of proficiency *within* the expert category, and this participant was deemed a junior expert. Clearly a weather lover; seems to have had a great deal of on-the-job training.		

Table 7.4
Data record from the Participant Interview with a Senior Expert

Age	59	
Current Position	Forecaster	
Current Rank	GS-9	
Education	High school, college Claims to have had a good science education Associates degree in Forecasting and BS in Management (ABD)	
Early Interest in Weather?	Saw many storms while growing up. Remembers hurricanes, floods. Evacuated a lot.	
Meteorology Training	Associates degree in Forecasting Air Force Observer's school, 3 months Forecasting school, September, 9 months AFB 5 week NEXRAD course Satellite school, 2 weeks	
EXPERIENCE		
WHERE and WHEN	**WHAT**	**TIME**
An Army Base 19XX–19XX	Observer. Little opportunity for on-the-job training. No collaterals.	560 hours [8 hours per shift x 6 shifts every nine days x 21 months (=70 shifts)]
An Air Force Base November 19XX– September 19XX	Observer, but was in a forward area so really functioned as a forecaster, forecasting for Agent Orange. Little opportunity for on-the-job training. No collaterals.	7,920 hours [8 hours per day x 5 days per week x 33 months (on average, 990 days x 8)]
An Air Force Base 19XX–January 19XX	Forecaster, Strategic Air Command. Daily on-the-job training with senior forecaster. Few collaterals.	12,640 hours [8 hours per shift x 6 shifts every 9 days x 6 years and 6 months = (2370 days)]
An Air Force Base 19XX–19XX	U.S. Air Force and USN aerographer schools were consolidated; participant was selected to be an instructor.	11,040 hours (8 hours per day x 5 days per week x 276 weeks)

Table 7.4 (continued)

EXPERIENCE		
WHERE and WHEN	**WHAT**	**TIME**
An Air Force Base April 19XX –19XX	Forecaster for missions. On-the-job training from officer who had a degree in meteorology. Lots of collaterals—had to "run the shop" (administrative).	8,200 hours (5 hours per day x 7 days per week x 1,640 days)
A Naval Air Station March 19XX– October 19XX	Forecaster. Was training others.	3,360 hours (40 hours per week x 84 weeks)
Naval Air Station October 19XX–19XX	Forecaster. Trained others.	10,416 hours (seven 12-hour shifts every two weeks = 36 hours every 2 weeks x 124 weeks)
TOTAL		54,136 hours Gilbreth Correction Factor = 27,068
Notes		
Determination: Senior Expert, Master (qualified to teach others). Demonstrated high aptitude and high levels of achievement from his school years onward. Remembers growing up "in weather"—remembers hurricanes, floods, evacuating. 19XX–19XX spent a great deal of time (sometimes 100+ miles a day) riding bicycle in the countryside and would watch clouds and storms grow, develop, and so on.		

Table 7.5

Adaptation of a standard proficiency scale (from table 7.1) to the domain of U.S. Navy Weather Forecasting

Apprentice	Minimal skill; unable to issue a competent forecast without supervision.
Journeyman	Has qualified as a forecaster. Can issue a competent forecast without supervision but has difficulty with tough or unfamiliar weather scenarios.
Expert	Can issue a proficient forecast for tough as well as easy weather scenarios; proficiency may be limited to particular regions and local effects.
Senior Expert	Can issue proficient forecasts for tough as well as easy weather scenarios; proficiency extends beyond particular regions or climates; they play a role in establishing standards and procedures.

Table 7.6
Sorting of the U.S. Navy Forecasters according to the refined proficiency scale

Determination	Hours of Experience	Age	Hours of Experience (uncorrected)	Notes and Qualifications
Senior Expert	50,000–up	59	55,412	Weather lover
Senior Expert		59	54,136	Weather lover
Expert	40,000–49,999	41	47,064	High aptitude for math and public speaking
Expert		42	43,908	High aptitude for math; a biology inclination
Junior Expert	30,000–39,999	37	36,898	Weather lover
Junior Expert		37	31,032	
Sr. Journeyman	25,000–29,999	48	29,756	Indications of high ability
Sr. Journeyman		42	29,127	
Sr. Journeyman		32	25,720	Indications of high ability
Journeyman	15,000–24,999	30	20,940	Science lover
Journeyman		32	19,758	Biology lover
Jr. Journeyman	10,000–14,999	44	14,464	High science aptitude
Jr. Journeyman		28	14,460	
Jr. Journeyman		33	13,320	High science aptitude
Jr. Journeyman		26	11,748	Weather lover; high math aptitude
Jr. Journeyman		24	11,067	Indications of high ability; science lover
Sr. Apprentice	5,000–9,999	25	8,448	Qualified as a forecaster, therefore a Journeyman?
Sr. Apprentice		36	8,232	Weather lover
Sr. Apprentice		23	6,384	Weather lover
Jr. Apprentice	–4,999	26	2,016	
Jr. Apprentice		27	1, 848	Science lover
Jr. Apprentice		24	1,848	Oceanography lover; tested out of A School; Indications of high aptitude

may not generate expert-level forecasts). For this reason, it can be valuable, if not necessary, to approach proficiency scaling by assessing actual performance as well as hours of experience.

Breadth of Experience

The scaling of skill in terms of degree of experience does not necessarily reflect the richness or variety of experience (breadth). Based on the data from the participant interviews, an attempt was made to scale breadth of experience and compare this to hours of experience. The U.S. Navy forecasters' careers involved the following kinds of experiences:

- onship forecasting experience,
- experience on a Mobile Environmental Team (a small group of forecasters and observers who are assigned temporary duties at facilities that do not have a weather forecasting operation or to some other form of deployed unit),
- experience in more than one climate,
- on-the-job training with a senior forecaster,
- a site waiver (allowing one to serve as a forecaster but only at the selected site),
- experience doing hand chart work, and
- experience at instructing others.

Data on these experiences for each participant came from the participant interviews. For each experience, Hoffman et al. (2000) determined a simple yes = 1, 0 = no scale and added up the values for each participant. This simple measure assumes that each of the possible experiences counts equally, and that is almost certainly not true. However, this simple scale avoids breadth bias. For instance, Participant X might have had a site waiver whereas Participant Y might have had extensive experience on deployments. Which counts more toward the achievement of proficiency? If more scale points were to be awarded to Participant X than Participant Y, then the breadth scale might merely be a reflection of the shear amount of time that the individual had been working in the U.S. Navy weather community.

The purpose of this analysis using a simple scale was to see whether one could "cut to the chase" by counting Yes's and No's *without* differentially weighting the experiences. However, Hoffman et al. (2000) did use a 0, 1, 2 scale for on-the-job training because it is certain that training from a senior forecaster is an important experience contributing to the achievement of proficiency (see chapter 3). It would be obvious to say that the more hours of experience people had, the greater the likelihood that they could have had a broader range of experiences. One cannot take this as a foregone conclusion, however. The participant interviews had revealed cases where individuals had a number of different kinds of experiences (e.g., multiple duty stations, duty aboard ship) but had never come under the wing of a senior forecaster who provided them with on-the-job training.

A plot of the results comparing the breadth score to rank for hours of experience forms a nice 45-degree angle. The implication is that one need not grapple with the apples–oranges comparison involved in creating a breadth of experience score, but this conclusion may not hold for other domains of expertise.

Analysis of Performance

U.S. Naval Air Station weather forecasting facilities, like most weather forecasting facilities, compare the daily forecasts to the actual weather data in a process called forecast verification. This involves the following weather parameters: average wind direction, average wind speed, maximum winds, sky conditions, visibility, ceiling, maximum temperature, and minimum temperature. For each of these parameters, there is a threshold—a range of values within which a forecast is said to have verified. Every day, each forecaster produces a number of "terminal aerodrome forecasts" (TAFs), each of which specifies the predicted values of the forecast parameters. On a monthly basis, the Quality Assurance Officer compares the forecasts to the actual observations in terms of the verification thresholds and calculates an average percent correct for each forecaster for the month in question. Hoffman et al. (2000) obtained verification data sets for the U.S. Naval Air Station for the months of March 1995 through July 1999, and culled from those data sets were the data for those forecasters who had completed the participant interview (N = 8).

A representative data set from one of the forecasters appears in table 7.7. Missing from this table, for sake of clarity, is the number of amendments. An amendment is an update made to a forecast before the valid interval of the forecast has expired. Although these are not considered "errors," they are taken into account in the calculations of percent correct.

Hoffman et al. (2000) reduced each of the participant data tables to the form shown in table 7.8 (showing only the months of May through September for the years covered). The number of forecasts was multiplied by the number of parameters that were forecast in each TAF (which changed once during the time span of months covered in these data sets). Thus, for example, from the first line in the table 7.8:

374 forecasts – 75 errors = 299
299/364 = 79.95% correct.

Table 7.7
Representative data set for the Naval Air Station Terminal Aerodrome Forecasts, showing the total number of forecast values, the number of errors, and the percent correct

Date	Total Number of Forecasts	Total Number of Errors	Percent Correct
Mar-95	280	42	85.00%
Apr-95	30	7	76.67%
May-95	240	44	81.67%
Jun-95	190	24	87.37%
Aug-95	250	42	83.20%
Sep-95	40	9	77.50%
Oct-95	270	43	84.07%
Nov-95	200	31	84.50%
Dec-95	20	2	90.00%
Jan-96	180	22	87.78%
Feb-96	250	53	78.80%
Mar-96	20	2	90.00%
Apr-96	180	22	87.78%
May-96	280	32	88.57%
Jun-96	260	28	89.23%
Jul-96	300	55	81.67%
Aug-96	300	55	81.67%
Sep-96	200	37	81.50%
Oct-96	280	52	81.43%
Nov-96	220	37	83.18%
Dec-96	280	51	81.79%
Jan-97	310	48	84.52%
Mar-97	40	10	75.00%
Apr-97	10	0	100.00%
Jun-97	110	20	81.82%
Jul-97	420	51	87.86%
Aug-97	340	56	83.53%
Sep-97	396	41	89.65%
Oct-97	165	22	86.67%
Nov-97	495	34	93.13%
Dec-97	462	29	93.72%
Jan-98	400	37	90.75%
Feb-98	649	35	94.61%

Table 7.7 (continued)

Date	Total Number of Forecasts	Total Number of Errors	Percent Correct
Mar-98	77	2	97.40%
Apr-98	11	0	100.00%
May-98	77	6	92.21%
Jun-98	11	2	81.82%
Jul-98	88	2	97.73%
Aug-98	22	3	86.36%
Sep-98	66	1	98.48%
Oct-98	55	3	94.55%
Nov-98	22	4	81.82%
Dec-98	176	19	88.92%
Feb-99	33	1	96.97%
Mar-99	44	5	88.64%
Apr-99	44	1	97.73%
May-99	55	14	73.64%

Table 7.8
An example of the data reduction performed on the TAF verification data

Date	# AMD/ COR	# of FCST	Total FCST	# of Errors	% Correct
May-98	5	29	374	75	79.95%
Jun-98	0	15	165	28	83.03%
Jul-98	5	30	385	71	81.56%
Aug-98	0	22	242	28.5	88.20%
Sep-98	0	4	44	14.5	67.00%
May-99	3	5	88	15	82.95%
Jun-99	0	2	22	4	81.80%
					Average = 80.64%

Ideally, the verification data might show some relation to hours of experience, but this hope has to be qualified by the understanding that there would be a significant "ceiling effect." Most of the forecasts verified in the 70% to 90% correct range, meaning that there might not be enough variance to allow a relation to hours of experience to appear. Hoffman et al. (2000) performed two analyses on the data. The first compared forecasting performance to proficiency levels as determined by the participant interviews for all of the months of available data. Results are shown in table 7.9.

Table 7.9
Relation of forecast verification score to experience level

Participant Proficiency Level	Proficiency Rank	Hours of Experience (uncorrected)	Mean Percent Correct	Standard Deviation	Average (all months)	Average (May–September)
Sr. Expert	1	55,412	82.45	0.069	84.8	83.8
Sr. Expert	2	54,136	87.14	0.067		
Expert	3	47,064	85.88	0.036		
Jr. Expert	5	31,032	83.84	0.072		
Journeyman	9	19,758	78.11	0.102	81.4	80.2
Jr. Journeyman	11	14,460	82.8	0.06		
Jr. Journeyman	12	13,320	80.99	0.075		
Jr. Journeyman	13	11,748	83.83	0.04		

There was a trend for average verification scores to decrease as hours of experience decreased, but this is only within the range of about 81% to 87%. The rightmost column in table 7.9 shows the averages for the months of May through September—months that weather typical of the Gulf region (thunderstorms, hurricanes) is more difficult to forecast and requires greater local skill. For the May to September data, higher verification scores were accompanied by a somewhat smaller range. One thing that this analysis of averages hides is the important finding that the range of scores across the months was considerably greater for the journeymen. Indeed, for some journeymen, the monthly score was in the 50% to 55% range. The range for experts never fell below 70%. Journeymen are often about as good as the experts, but when they err, they can really err. Errors for the journeymen were within a range of values that were lower than that of experts.

Overall, the results demonstrate how proficiency scaling from interview data on hours of experience can be integrated with the analysis of actual performance data, and the two methods can converge. However, the data suggest that there is not that much improvement in skill comparing average performance of the journeymen to that of the experts. The score for the Junior Journeyman ranked 13th in terms of hours of experience was actually 1.4 percentage points greater than that for the expert ranked first in terms of hours of experience. One interpretation is that this confirms the notion that the journeymen qualify as a journeyman: They were issuing valid forecasts. A journeyman can perform nearly as well as an expert. For this reason, the proficiency scale distinguished sublevels (junior, senior). Clarification on the expert–journeyman differences matter must come from further studies that attempt to develop a proficiency scale. The data Hoffman et al. (2000) obtained were for verifications that included amendments, which can be thought of as sort of a "cheat" (i.e., a forecast could be corrected just before its valid interval expired). Furthermore, experts might be better distinguished in terms of their performance on especially rare or difficult forecasting events.

Most broadly, the data confirm an alternative benchmark to the "10,000-hours rule." Specifically, in terms of performance measures:

- Journeymen "get it right" about 85% of the time for routine forecasting situations *on average* and show a greater range in performance than experts, and
- Experts "get it right" 85% or more of the time for both routine and more difficult forecasting situations and show a smaller range in performance.

However, we are not advocating *any* decontextualized and simplistic rule. We do advocate for domain and organization appropriate scales, and for scaling procedures that rely on more than one method. In addition to scaling proficiency in terms of extent of experience and performance, Hoffman et al. (2000) also investigated breadth of experience.

Proficiency Scaling Results

The U.S. Navy forecasters were asked to describe the process they had gone through in their most recent forecast. One Junior Journeyman forecaster reported that he began with the climatological norm. Another reported that he relied most heavily on computer models. Both of these forecasters depended on guidance (climatology, computer models), as opposed to using guidance in service of the formation of a mental model. Hoffman et al. (2000) summarized their findings by integrating evidence about the forecasting process and strategy. The proficiency scaling findings are presented in table 7.10.

Hoffman et al. (2000) are not the only researchers to have studied the development of proficiency in weather forecasting. Additional evidence comes from a study of U.S. Air Force weather forecasters.

U.S. Air Force Weather Forecasting Case Study

A project conducted by Rebecca Pliske and her colleagues (Pliske, Crandall, and Klein, 2004; Pliske et al., 1997) focused on revealing the differing reasoning styles and skill

Table 7.10
A summary proficiency scale appropriate to U.S. Navy forecasting

	TRAINEE	JOURNEYMAN	EXPERT (Proficient)	EXPERT (Senior)
Forecasting Process	Forecasting by extrapolation from the previous weather and forecast and by reliance on computer models.	Begins by formulating the problem of the day but focuses on forecasting by extrapolation from the previous weather and forecast and by reliance on computer models.	Begins by formulating the problem of the day and then building a mental model to guide further information search.	Begins by formulating the problem of the day and then building a mental model to guide further information search.
Reasoning Strategy	Reasoning is at the level of individual cues within data types.	Reasoning is mostly at the level of individual cues, some ability to recognize cue configurations within and across data types.	Reasoning is in terms of both cues and cue configurations, both within and across data types. Some recognition-primed decision-making occurs.	Process of mental model formation and refinement is more likely to be short-circuited by recognition-primed decision-making skill.

sets that distinguish expert forecasters. Pliske et al. (1997) utilized a cognitive task analysis method called the Knowledge Audit. This method invited the participant to recount recently experienced "tough cases" that involved making a difficult decision that challenged their expertise. Probe questions focus on the recall of specific, lived experiences. Each participant was asked to reflect on generic knowledge or skills and also to recall specific instances in which this knowledge was used and/or skills were implemented. Example interview probes for the forecasting study appear in table 7.11.

Pliske et al. (1997) interviewed 65 U.S. Air Force and NWS forecasters with varying degrees of experience. To analyze the interview data, the researchers conducted a multi-trial sorting task in which they reached a consensus on categories of the reasoning styles they had observed. These categories focused on the forecasters' overall strategic approach to the task of forecasting, their strategy in the use of computer weather models, their process in creating forecasts, their means for coping with data or mental overload, and their metacognition. The categories they identified were dubbed "Scientist," "Proceduralist," "Mechanic," and "Disengaged." Each of these was further categorized in terms of their Affect, Skills, and Activities.

Table 7.11
Probes used in the studies of weather forecasting by Pliske et al. (1997, 2004)

Probe	Knowledge/Skill of Interest
Can you recall a time when you made a forecast that others thought was wrong, but it turned out you were right?	Ability to think critically about one's own reasoning.
Can you recall and discuss some experiences where part of a situation just "popped" out at you, where you noticed things that others did not catch?	Skill at perceiving cues and patterns.
Have there been times when you walked into a situation and knew exactly how things got there and where they were headed?	Skill at situation assessment.
Can you recall past experiences in which you: • Found ways of accomplishing more with less? • Noticed opportunities to do things better? • Relied on experience to avoid being led astray by the equipment?	Flexibility, ability to think critically about one's own thinking.

Scientists tend to have had a wide range of experience in the domain, including experience at a variety of scenarios.

• In terms of *Affect*, they are typically "lovers" of the domain. They like to experience domain events and see patterns develop. They are motivated to improve their understanding of the domain.
• In terms of *Skills*, they possess a high level of pattern-recognition skill. They possess a high level of skill at mental simulation. They understand domain events as a dynamic system. Their reasoning is deliberative, analytical, and critical. They possess an extensive knowledge base of domain concepts, principles, and reasoning rules. They are likely to act like a mechanic when stressed or when problems are easy. They can be slowed down by difficult or unusual problems.
• In terms of *Activities*, they show a high level of flexibility. They spend proportionately more time trying to understand the weather problem of the day and building and refining a mental model of the weather. They possess skill at using a wide variety of tools. They are most likely to be able to engage in recognition-primed decision making. They spend relatively little time generating products because this is done so efficiently.

Proceduralists are typically younger and less experienced than scientists. Some reach the advanced journeyman and even the expert level (i.e., they can have had years of experience).

- In terms of *Affect*, they sometimes love the domain. They like to experience domain events and see patterns develop. They are motivated to improve their understanding of the domain.
- In terms of *Skills*, they are less likely than scientists to understand domain events as a complex dynamic system. They see their job as having the goal of completing a fixed set of procedures, but these are often reliant on a knowledge base. They sometimes rely on a knowledge base of principles of rules, but this tends to be limited to types of events they have worked on in the past.
- In terms of *Activities*, they spend proportionately less time than scientists building a mental model and proportionately more time examining the guidance. They can engage in recognition-primed decision making only some of the time. They are proficient with the tools they have been taught to use.

Mechanics sometimes have years of experience, but they are distinguished especially by their affect.

- In terms of *Affect*, they are not interested in knowing more than what it takes to do the job; they are not highly motivated to improve.
- In terms of *Skills*, they see their job as having the goal of completing a fixed set of procedures, and these are often not knowledge-based. They possess a limited ability to describe their reasoning. They are likely to be unaware of factors that make problems difficult.
- In terms of *Activities*, they spend proportionately less time building a mental model than both scientists and mechanics and proportionately more time examining the guidance. Mechanics cannot engage in recognition-primed decision making. They are skilled at using tools with which they are familiar, but changes in the tools can be disruptive.

Disengaged Forecasters are also distinguished especially by affect. Job assignment or tour-of-duty contexts can result in a "disengaged" style at the medium- and (especially) low-skill levels. The Disengaged style can be manifest in individuals with little experience or individuals having many years of experience.

- In terms of *Affect*, disengaged forecasters dislike their job. They are not interested in knowing more than what it takes to do the job. They are not motivated to improve.
- In terms of *Skills*, they possess a limited knowledge base of domain concepts, principles, and reasoning rules. Their knowledge and skill are limited to scenarios they have worked in the past.
- In terms of *Activities*, their products are of minimally acceptable quality. They are likely to be unaware of factors that make problems difficult. They spend most of the

time generating routine products or filling out routine forms. They spend almost no time building a mental model and proportionately much more time examining the guidance. They cannot engage in recognition-primed decision making.

Pliske et al. (1997) did not claim that this set of styles categories is exhaustive, that all practitioners will fall neatly into one or another of the categories. However, their findings are suggestive because the categories are partly related to proficiency levels. The affect and motivational aspects of the Pliske et al. (1997) styles accord with the findings from LaDue's (2011) ethnographic study. At the mid-career stage, some forecasters go beyond what is nominally expected of them to engage outside partnerships, collaborate with researchers, and identify and solve reasons for communication issues in recent events. They will research challenging forecasts, serve as reviewers for journal papers, and pursue their own research.

Elements of the Pliske et al. (1997) findings were replicated in a study of U.S. Navy forecasters by Josslyn and Jones (2008; Jones and Josslyn, 2004), who invited forecasters to think out loud as they produced Terminal Aerodrome Forecasts (TAFs). Josslyn and Jones found that less experienced forecasters relied more on rules of thumb (i.e., they would be Procedualists in the Pliske et al. scheme). Also in concert with the other studies, Josslyn and Jones (2008) found that the less experienced forecasters relied on just one or two of the available computer models, they overrelied on the outputs of the computer models, and they only looked at a half dozen of the statistics that the models produce and did not take time to evaluate trends in the models' errors. The least experienced forecasters engaged in routine, fixed procedures, relied the most on the model predictions, and were more focused on the weather problem of the day and producing the particulars to complete a TAF every six hours than they were on building a mental model (i.e., they were Proceduralists). The most experienced forecaster looked at model outputs later in his process and compared the model outputs.

Josslyn and Jones (2008) concluded that the least experienced forecasters were merely adapting (successfully) to the pressures of their job: the requirement to issue TAFs on schedule in difficult circumstances characterized by frequent interruptions. Hence, they relied on procedural rules and a routine process. But the more experienced forecasters were working under the same circumstances and yet manifested a rather different approach. Thus, it is likely that we see here an interaction between work context (or circumstance) and proficiency. In general, the findings of Josslyn and Jones (2008) fit with the findings of Pliske et al. (2004) and Hoffman, Coffey, and Ford (2000). Thus, we can merge the findings of the Hoffman et al., Pliske et al., and Josslyn and Jones studies into a description of the cognition of forecasters at the different proficiency levels. Experts gain something from experience and constitution that enables them to transcend circumstance.

Integration of the Findings

Now we present an integrated scheme for analyzing forecaster proficiency based on the findings from the forecasting competitions and the two proficiency scaling studies. We rely on the characterizations of expertise and models of expert reasoning that we find in the literature of Expertise Studies and the concepts of Macrocognition (see chapter 6). Tables 7.12, 7.13, and 7.14 present this integration. For ease of exposition, we group into Lower Levels of Proficiency (Naive, Initiate, Novice), Medium Levels of Proficiency (Apprentice, Journeyman), and Higher Levels of Proficiency (Expert, Master) (see the definitions of these in table 7.1).

The Forecasting Process

These case studies provide a rich description of what it means for weather forecasters to be experts, expressed in terms of their reasoning skills and other elements of cognition as they generate forecasts.

The "Quick Size-Up"

Experts can do a quick size-up and sometimes do go with their first impression, but they are more likely to be correct in their size-up. Even when they go with a first impression, they remain skeptical and open to disconfirming evidence. They can deliberately undertake a quick size-up, but then they second-guess themselves and can also consider that second-guessing themselves can lead them astray, in what Sanders (1973) calls "agonizing reappraisal." As one forecaster described it:

> Yeah—I've been burned enough times to know how quickly things can go the other way! Also, I think of the forecast as considering a set of possible outcomes, with corresponding probabilities attached to each of those outcomes. You hope to be in a situation where the probabilities peak on one particular outcome (high confidence) rather than equally probable outcomes (low confidence). But as new evidence in the form of observations comes in, the probabilities may shift and a less likely outcome will become more likely, perhaps even the most likely.

The experts' quick size-up is never a mere default process. Experts' richer experience base enables them to take local effects into account even when doing their quick size-up. They ascribe greater importance to local effects than do journeymen. They

Table 7.12

A description of the cognition of forecasters falling at lower levels of proficiency

Forecasting Strategy

- They rely heavily on guidance (e.g., outputs of computer models), and not enough time is spent in service of forming a mental model.
- They use a single strategy or a fixed set of procedures—they conduct tasks in a highly proceduralized manner using the specific methods taught in the schoolhouse.
- Their data-gathering efforts are limited, but they tend to say that they look at everything.
- They are insensitive to contextual factors or local effects.
- They are unaware of situations in which data can be misleading and do not adjust their judgments appropriately.
- They tend to make wholesale judgments (i.e., that guidance is either "right" or "wrong").
- They are less adroit; things that are difficult for them to do are easy for the expert.
- They are not strategic opportunists.
- They sometimes use a "quick size-up" strategy, in which they move quickly from an initial diagnosis to a judgment—but they only skim the surface of the data and go with their first impression.
- Their reasoning is not analytical or critical.
- They sometimes second-guess themselves or generate alternative scenarios, with no awareness of whether that is a positive strategy, and are unclear as to how they handle situations in which there is conflicting evidence.

Knowledge

- They possess a limited knowledge base of domain rules and causal principles.
- Their knowledge is limited to the types of scenarios they have experienced.

Pattern Recognition Skill

- They possess a low level of pattern recognition skill.
- Recognition-primed decision making rarely occurs, if at all.

Causal Reasoning and Mental Modeling

- Their meaning-making activities reflect little depth of understanding.
- Their inferences involve memorized rules rather than an attempt at causal explanation.
- They are unlikely to be able to build a "mental picture," although they may profess the idea that it is important to attempt to understand the "big picture."
- The mental representations that they do report are most likely to be images that resemble the displays, forms, and charts that are used in the operational context—rather than causal models of the domain events.

Affect and Effort

- They can be "lovers" of the domain, with early life experiences motivating them.
- Their performance standards focus on satisfying literal job requirements.
- Their process of providing information to clients is laborious and does not rely on deep understanding.
- They report difficulty in handling the uncertainty inherent in the task.
- They are uncomfortable at any need to improvise.
- They can be fearful of having to adopt strategies that deviate from standard operating procedures.
- They cannot do multitasking without suffering from overload; become less efficient and more prone to error.
- They can be unaware of problem difficulty (i.e., that subtle cues or situational factors have made a seemingly routine case into a tough case).

Goal Orientation

- Their performance standards focus on satisfying literal job requirements.
- A primary goal is to satisfy literal job requirements and avoid error or criticism.
- They are not fully aware of the client's true needs.

Table 7.13

A description of the cognition of forecasters falling at medium levels of proficiency

Forecasting Strategy

• Despite motivation, it is possible to remain at this level for life and not progress to the expert level.
• They begin by formulating the problem, but then like individuals at the low skill levels, those at the medium skill levels tend to follow routine procedures—they tend to seek information in a proceduralized or routine way, following the same steps every time.
• Some time is spent forming a mental model.
• They tend to place great reliance on the available guidance, but they are often skeptical of it.
• They tend to treat guidance as a whole and not look to subelements—they focus on determining whether the guidance is "right" or "wrong."
• They are proficient at using those tools that they rely on most heavily or routinely.
• Journeymen can issue a competent product without supervision, but both apprentices and journeymen have difficulty with tough or unfamiliar cases.
• As a group, they are most likely to show a variety of individual differences in their standard procedure:
✓ A majority use those strategies taught in school, but they can be somewhat adaptable and are interested in learning about new effective strategies.
✓ Some will begin their deliberation with an awareness of a standard or typical scenario and then search for data that would entail a modification of the standard scenario.
✓ Some begin their deliberation by examining the guidance and then searching for data that would entail a modification of the guidance.
✓ Some will follow a procedure that is more like that of experts.

Knowledge

• Their experience involves work in a few contexts and experience with a few subdomains of practice and knowledge.
• Their knowledge base is formed more around rules than around conceptual or causal principles.
• Their knowledge and skill are somewhat limited by the types and range of experiences they have had.

Pattern Recognition Skill

• Any recognition-primed decision making that occurs is likely to be cue-dependent, that is, dependent on isolated cues or data of a single type.
• The journeyman can detect cues that the novice misses.

Causal Reasoning and Mental Modeling

• Their sensemaking activities involve some causal explanation, but often they must consult with or defer to a practitioner with a greater range of experience to fully understand the problem situation.
• They are somewhat more likely to report building a "mental picture" or attempt to understand the "big picture."
• The mental representations that they do report are most likely to be images that resemble the displays, forms, and charts that are used in the operational context—rather than causal models of the domain events.

Table 7.13 (continued)

Affect and Effort

- They are sometimes "lovers" of their domain and are motivated to learn more and improve.
- They can improvise if necessary, but they lack confidence and lose efficiency when doing so.
- When under time pressure, they are like individuals at lower skill levels, that is, they "chase the observations."
- Although they are aware that overload can occur, they are limited in their ability to anticipate overload situations. They are able to cope with overload, but they are at a loss in efficiency and an increase in the likelihood of error.
- May tend to fall back on routine or normative procedures when confronted with high levels of mental workload.

Goal Orientation

- Their performance standards focus on literal job requirements, but the higher standard is seen as an achievable goal.
- They seek to create high-quality products.
- They try to provide information of a type and format that is useful to the client.

Metacognition

- They manifest some conscious, deliberate management of cognitive resources, especially in situations of time pressure or conflicting data.
- They clearly understand as well as recognize the data overload problem.

sometimes do chart work, looking across data types and time to get a sense of emerging patterns. Some forecasters report that they often (or always) feel that their most skilled forecasts were issued toward the end of a shift because in spending a shift working with the data, they got a better feel for how things were evolving.

Apprentices and some journeymen also sometimes use a "quick size-up" strategy, in which they move quickly from an initial diagnosis to a forecast. But they do this in ways that distinguish them from experts: They just skim the surface of the data and go with their first impression. Their limited experience base makes them less able to take local knowledge into account. They are overly reliant on persistence—when the weather today will probably be like yesterday's weather. In persistence situations, they make persistence forecasts. They show limited or no ability to do traditional hand chart work to determine the future course of weather events. This is largely because technology has to a great extent obviated the requirement to make forecasts by conducting hand chart work. Journeymen sometimes second-guess themselves with no feeling about whether or when that is a positive strategy. Apprentices and some journeymen are unclear as to how to handle situations in which there is disconfirming evidence. When creating a forecast, they provide information to clients based largely or entirely on the requirements of forms and/or standard operating procedures.

Table 7.14

A description of the cognition of forecasters falling at the high levels of proficiency

Forecasting Style

• Their first goal is to rapidly come to a detailed understanding of the problem at hand. They will spend proportionately more time doing this than individuals of lower skill levels.
• They know, based on their mental model, which guidance to look at for confirming or disconfirming evidence—which guidance gives good versus poor information depending on the nature of the situation.
• They look at agreement of the various types of guidance—features or developments on which the various guidance agrees. They do not make wholesale judgments that guidance is either "right" or "wrong." They can tell when equipment, data, or data type is misleading; they know when to be skeptical.
• They are able to take into account data source, location, and age. They are always skeptical of guidance, and they know the biases and weakness of each of the various types of guidance.
• They can conduct tasks using traditional "by hand" methods without relying on the guidance.
• They are flexible and inventive in the use of tools and procedures. They show a high level of flexibility and adaptability, including an ability to create procedures ad hoc. They use and create strategies not taught in school.
• They are able to take context and local effects into account.
• They can shift direction or attention to take advantage of opportunities.
• Their reasoning is analytical and critical; they know when to be skeptical.
• Their process then focuses on building a rich and coherent mental model to guide further information search and analysis.
• They examine guidance only after an initial mental model has been formulated. They are aware that after an understanding of the problem at hand, the task is to go to specific data types to determine what to do about the problem.
• If they do not like the guidance, they rely on their own strategy—either "mental simulation" or continuity of trends.
• When predicting by continuity, they can take local effects and context into account.
• They recognize the need to verify/disconfirm hypotheses, but they know that second-guessing can lead one astray.
• They sometimes assert that the proceduralized method prescribed in the operational forms does not work well for them; they prefer to "work like a detective."
• They can shift direction or attention to take advantage of opportunities. They can tell when equipment, data, or a data type is misleading.
• When it comes to the final step—generating products—they will spend proportionately less time than individuals of lower skill levels.
• They can issue high-quality products for tough as well as easy scenarios.

Knowledge

• Their reasoning relies on an extensive and coherent knowledge base of conceptual or causal principles and domain rules.
• They understand the domain to be a complex, dynamic one.
• They have had extensive experience with diverse situations or scenarios.
• They have extensive knowledge of subdomains, particular contexts, and local effects.

Table 7.14 (continued)

Pattern Recognition Skill

• They possess a high level of pattern recognition skill. Their data search is in terms of both cues and cue configurations, both within and across data types.
• The expert can detect cues and cue configurations that novices and journeymen miss.
• They are able to rapidly recognize a data pattern or cue configuration that triggers a mental schema or awareness that the case at hand fits a commonly occurring or familiar pattern.
• Rather than engaging in a hypothesis-testing cycle, the practitioner can sometimes go directly from the act of recognition to some sort of judgment or decision.
• Their recognition-primed decision making is more likely to be pattern-dependent, that is, dependent on relations of cues across data types (cue configurations) rather than being dependent on isolated cues.
• Because cue configuration-dependent, recognition-primed decision making necessitates the inspection and perception of more than one data type, followed by an integrative process, the expert's pattern-dependent, recognition-primed decision making can take longer that the journeyman's cue-dependent decision making.
• Except in cases where recognition priming occurs, they will spend proportionately more time inspecting data than individuals of lower skill levels.

Causal Reasoning and Mental Modeling

• They engage in more "meaning-making" activities: They generate more complete causal explanations, including linkages of all the pertinent data.
• They reason ahead of the data.
• There is a greater likelihood of reasoning in terms of multiple potential causes and the interactions of causal forces.
• There is a greater likelihood of anticipating possible interactions among complex factors.
• They possess a high level of skill at mental simulation of domain events.
• They will spend proportionately more time forming and refining a mental model than individuals of lower skill levels.
• They express a "need" to build a mental picture. Their descriptions of their thinking involve references to mental picture-building, vivid imagery, and three- or four-dimensional mental simulations of events, grounded in causal principles.
• They will describe perceptible patterns using visual metaphors (e.g., the weather forecaster describes a cloud pattern in a satellite image as looking like a "wobbling turkey" or a "sneering face").
• They possess an ability to predict using a mental simulation—running a mental model forward in time in accordance with domain principles and rules. They can use their mental model to quickly provide information they are asked for.

Affect and Effort

• They are often "lovers" of their domain and the experiences it entails (e.g., "weather lovers" enjoy experiencing weather phenomena directly).
• They can use their mental model to quickly provide the information they are asked for.
• Things that are difficult for them to do may be either difficult for the novice to do or may be cases in which the difficulty or subtleties go totally unrecognized by the novice.
• They are comfortable when improvising.
• They can do multitasking and still conduct tasks efficiently without wasting time or resources, although they will slow down under conditions of high mental workload.
• They can explicitly recognize when overload, or the potential for data overload, is occurring.

Table 7.14 (continued)

- They recognize that practitioners of lower skill levels are especially prone to being drowned by the data when experiencing high levels of mental workload.
- They regard the strategy for focused information search as being critical in coping with data overload.
- They are comfortable improvising even under conditions of high mental workload, and sometimes especially under those conditions.
- They have high personal standards for performance.
- They seek to create definitive solutions and products of the highest quality.
- They provide information of a type and format that is useful to the client.

Metacognition

- Typically, they are highly verbal and are able to describe and talk about their own reasoning (e.g., "You have to think about what you have and figure out how to fill in what you don't have").
- Except in cases of recognition-primed decision making, they engage in conscious, deliberate management of cognitive resources and conscious evaluation of what works best.
- They know that accurate products are more likely to result from detailed understanding.
- They can have strongly held views about technology's impact on information management.
- Data overload makes the job more difficult and forces a need to create ways of managing information.
- The need for getting "the big picture" (i.e., situation awareness) is not well supported by the technology—"You have to hold various elements in your head."
- They bemoan the need to do "ad hoc work," "make-work," or "work-around" activities that are forced on them by the technology.

Use of Computer Models

Apprentices and journeymen differ from experts in terms of how they use the computer models as they create forecasts. Apprentices and journeymen are likely to approach the forecasting process as a prescribed sequence of steps. They are uncomfortable at any need to improvise and use only those strategies taught in school. The focus for their forecast is often on determining whether the computer model is "right" or "wrong." Apprentices and even some journeymen rely uncritically on the computer models. They tend to treat computer model outputs as a whole and not to regard the computer outputs as guidance. They can be over-reliant on computer model outputs: What the computer models generate, they put out as their forecasts. They look at guidance from individual computer models (i.e., the outputs of each of the various computer models) and are less likely to engage in comparison or integration across the computer model outputs. They chose which computer model outputs to look at depending on things they are told about what model works well ("the model of the day") and what model does not. A noteworthy finding in the Pliske et al. (1997) study of U.S. Air Force forecasters was that forecasting skill was degraded when training methods emphasized automated products rather than sensemaking and conceptual models.

Senior journeymen and experts do not make wholesale judgments that computer model output is either "right" or "wrong." When looking at the outputs of each particular computer model, they examine sub-elements for particular data types (e.g., wind speeds). They look for convergence of computer models' outputs—features or developments on which various models agree. Experts are almost always skeptical of computer model guidance and know the biases and weaknesses of each of the various computer model products. They know, based on their experience and understanding of the weather situation, which models to look at for a confirming match—which models give good versus poor information depending on the nature of the weather situation. As expert forecasters sometimes say, "You can't predict the weather well by using the computer models unless you can predict the weather well without using the computer models." Apprentices and even journeymen sometimes find themselves in situations where they would be unable to make forecasts at all without using the models.

Sensemaking

Forecasters at all skill levels report some sensemaking activity in which they attempt to organize and explain data. They frequently report experiencing "pictures in the mind" as they try to understand weather. Many forecasters express a deeply felt need to build a mental simulation, and they rely heavily on visual metaphors for weather: the atmosphere as a fluid, "ripple" effects, "rock in a pond," and frontal patterns have "waves." For forecasters of low and medium skill levels (apprentices, junior journeymen, and some journeymen), sensemaking is usually an attempt to determine how quickly the weather will change. They are less likely than experts to put in the time and effort to achieve a deep understanding of the causes of weather events and indeed are less able to do so. Their sensemaking activities reflect little depth of understanding of meteorology. For example, the assertion that "If the K index (a measure of instability) is in the 30s, there's a chance of thunderstorms" is sensemaking, in that it involves an inference but reflects a memorized rule rather than an attempt at causal explanation. The "mental pictures" of apprentices and some journeymen are more likely to seem as static images, stacked features, or stacked static charts, not dynamic mental simulations. These individuals do not form or attempt to refine rich mental models. They do not think like a detective.

Experts, senior journeymen, and some journeymen engage in more meaning-making activities than individuals at other skill levels. These individuals are more likely to spend time seeking causal relations, causal linkages of data elements, and causal explanations of weather events. They are more likely to reason in terms of multiple potential causes and the interactions of causal forces. They are more able to anticipate possible interactions among complex factors—how complex systems develop across the Earth's surface. They can tell when equipment, data, or a data type is misleading. They know when to be skeptical. They can tell when an apparent incorrect data observation is the key to evolving events.

Experts and senior journeymen form rich mental models that, by definition, depict causal forces. They speak of "mental picture-building," involving vivid imagery in four-dimensional mental simulations. These are used to predict weather using a mental simulation—running a mental model forward in time (e.g., can "see thunderstorms redeveloping ahead of a squall line").

After building a tentative mental model, they seek missing data on causal factors or seek data to test a hypothesis that derives from their mental model. When critical pieces of information are missing, highly skilled forecasters (experts, senior journeymen) return to the diagnosis step to obtain additional information. The forecasters sometimes describe this as hypothesis testing or working on a premise. They can use their mental model to quickly provide information they are asked for.

Conclusion

Many researchers initially presumed that forecasting expertise followed the 10-year, 10,000 hours of experience rule-of-thumb. Most realized in the course of their research that forecasting does not follow that rule (e.g., Hahn et al., 2002; Hoffman et al., 2006; Pliske et al., 1997; Trickett, Trafton, and Schunn, 2009). Several researchers observed that years of experience in forecasting do not directly correspond to expertise (Hahn et al., 2003; Hoffman et al., 2006; Pliske et al., 1997). That being said, experience and on-the-job training were shown to have a great effect on skill at generating weather forecasts, perhaps even greater than knowledge of the underlying science gained from classroom education (Roebber and Bosart, 1996a, 1996b).

Forecasters of low and medium proficiency levels rely on computer models, use a fixed set of procedures and rules-of-thumb, have a narrow focus, fail to consider larger scale weather features, and often end up reactive to unexpected weather (Bosart, 2003; Hahn et al., 2002; Joslyn and Jones, 2008; Pliske et al., 1997). In contrast, expert forecasters might be identified with the "scientist" style identified by Pliske et al. (1997, 2004). They can also be regarded as what Donald Schön (1983, 1987) called "reflective practitioners." They are also typified by the intrinsic motive to learn, which is manifested as self-directed learning (see Hammond and Collins, 1991; Knowles, 1975; LaDue, 2011).

The two case studies reviewed in this chapter involved some form of proficiency scaling but then applied different methods to understand what it means for a weather forecaster to be an expert. As is the case in all cognitive task analysis projects, methods are adapted and combined, as appropriate, to the needs of the project and other practical constraints and goals. Proficiency scaling can be accomplished by a number of means and methods and should generally be accomplished through the use of more than one converging measure (for more details, see Crandall, Klein and Hoffman, 2006). A protocol for conducting multimethod proficiency scaling, including template forms for interviewing and data collection, is included in the "Protocols for Cognitive Task Analysis" (Hoffman, Crandall, and Klein, 2008), which is available for download at [www.dtic.mil/cgi-bin/GetTRDoc?AD=ADA475456]

The following chapters go into more detail concerning the cognition of expert forecasters by focusing on results from research concerning forecaster knowledge (chapter 8) and concerning forecaster perception (chapter 9).

8 What Does Research on Forecaster Knowledge Tell Us?

As we pointed out in chapter 6, a defining feature of expertise is the depth, extent, and organization of domain knowledge. Experts' knowledge involves finer gradations of functional categories, that is, "basic object-level" categories fall at a finer level than for non-experts. For example, limestone is simply a kind of rock to most people, but to the expert, there are many variants that inform of geological dynamics (e.g., tilted thinly interbedded limestone-shale with limestone predominating). Experts' categories fall at a functional level rather than a literal surface feature level (Glaser, 1987). Thus, experts can rapidly evaluate a situation and determine an appropriate plan of action, a phenomenon called "recognition-primed decision making" (Klein, 1989, 1993, 1997a). Within the first second of exposure to a novel chess position, chess experts can extract important information about the relations of the chess pieces' positions and begin identifying promising moves (Charness et al., 2001).

A research and development project illustrating this in weather forecasting (Hoffman et al., 2000, 2001, 2006) resulted in a knowledge model called "STORM-LK," standing for System to Organize Representations in Meteorology-Local Knowledge. The broad goal of the STORM project was to illustrate a full process of developing models of expert knowledge based on cognitive task analysis. The research was conducted at the U.S. Navy Meteorology and Oceanography Command (METOC) weather forecasting and training facility at Pensacola Naval Air Station. (Another aspect of this project was the proficiency scaling effort that was described in chapter 7.) The facility produced forecasts for pilots and pilot trainers and provided training on aviation weather to trainees. With the additional support of the National Technology Alliance, the STORM project was a valuable opportunity to empirically compare and evaluate a greater variety of methods of knowledge elicitation and modeling than had ever been brought to bear in a study of a single domain of expertise. Methods included documentation analysis, workspace analysis, work patterns observations, and two knowledge elicitation interview methods: the Critical Decision Method (CDM) and Concept Mapping.

The pool of participants (n = 22) included civilian weather forecasters and military aerographers and observers. Some of the participants were involved in the CDM interviews and some in the Concept Mapping.

Preparation

Prior to conducting the knowledge elicitation procedures, the researchers had to familiarize themselves with the facility, organization, and forecasting procedural guides, even though the researchers were already conversant in the domain of weather forecasting (Hoffman, 1991; Hoffman and Conway, 1990; Hoffman et al., 1993). This familiarization process involved:

- *Work Space Analysis.* The researchers and a qualified aerographer visited each workstation/work area and discussed the activities conducted, resources needed, communication and collaboration patterns, and so on.
- *Observations of Weather Briefings.* The researchers observed and audio recorded nine briefings to pilots, pilot trainers, and the forecasting staff. The transcripts were analyzed for propositional content concerning weather concepts and phenomena.
- *Documentation Analysis.* Two of the researchers and each of three aerographers (who had qualified as Forecast Duty Officers or Subregional Forecast Officers) reviewed the facility's Standard Operating Procedure (SOP) documents. The facility had 58 SOP documents, and the interviews went into greatest detail concerning those that were most pertinent to forecasting. Examples are the procedure for selecting and displaying products on the Satellite, AlphaNumeric, and NEXRAD Displays (SAND); and the procedure for adding annotations about significant weather events (SIGMETS, from NWS Products) to the forecasting facility's Home Page. For each SOP, the participants indicated which designated officer conducted each procedure, when, and why, and for each procedure the things that made it easy and the things that made it difficult.

Having identified the individuals who could be designated as proficient (i.e., journeymen and experts; see chapter 7), and having familiarized themselves with the organization and its SOPs, the researchers were prepared to conduct the knowledge elicitation procedures.

Critical Decision Method

Background

In the mid-1980s, Gary Klein and his colleagues conducted a number of studies on domains such as firefighting and neonatal critical care, using a cognitive task analysis method they called the Critical Decision Method. This was a variation of the Critical Incident technique developed by the U.S. military for after-action review to determine the causes of incidents in which there was loss of life or materiel (e.g., Flanagan, 1954). But the focus of the CDM is on incidents that involved difficult decisions in time-pressured, high-risk settings (Calderwood, Crandall, and Klein, 1987; Klein, Calderwood, and Clinton-Cirocco, 1986; Klein, Calderwood, and MacGregor, 1989). The CDM works by avoiding generic questions of the kind, "Tell me everything you know about x" or "Can you describe your typical procedure?" The CDM leverages the fact that domain experts often retain detailed memories of previously encountered cases, especially ones that were unusual or challenging—cases where one might expect differences between the decisions and actions of an expert and those of someone with less experience, and cases where elements of expertise are likely to emerge. Participants describe events in terms of timelines and answer probe questions about each decision point on the timeline (e.g., "What information did you need at that point?", "What were you seeing at that point?", "What were your options at that point?"). The results included information about the experts' actions, goals, and plans. The probe questioning yielded information about the cues to which the experts perceive and information about how the cues were linked to causal relations, actions, and plans.

Klein and his colleagues were able to specify many of the important cues in various situations. Some of the cues and cue patterns that were revealed were ones that the expert has never explicitly deliberated or specified. As an example, from the Klein et al. study of firefighters, in the initial description of one of his experiences, a firefighter explained that he had a "sixth sense" for judging the safety of a fire ground (i.e., a burning roof). Upon the subsequent sweep through the retrospective recall, using the probe questions, the expert "discovered" the perceptual pattern that he relied on, involving such things as smoke color and the feel of a "spongy" roof. Another finding was that the experts did not spend much time generating and evaluating options. Indeed, in this high-pressure decision-making situation, the deliberation of options is not an option: There's no time. Yet the experts were able to make good decisions, many of them at scales including small scale (e.g., the location of the seat of the fire) and larger scale (e.g., when to call in extra tanker trucks). This phenomenon came to be called "recognition-primed decision making" (see chapter 6).

The CDM was envisioned as a technique to leverage descriptions of expert knowledge in training and instructional design, and has proven to be a useful technique in many studies (see Hoffman and Militello, 2007). "Organizations suffer when they do not properly value their own expertise and when they lose skilled personnel without a chance to retain, share or preserve the knowledge of people who retire or leave" (Klein et al., 1989, p. 471). Klein's (1992) seminal paper on "preserving corporate memory" helped usher in a wave of interest in what came to be called knowledge management (cf. Brooking, 1999; O'Dell and Grayson, 1998). Hoffman, Crandall, and Shadbolt (1998) and Crandall, Klein and Hoffman (2006) provided a detailed protocol for conducting the CDM procedure and performing validity checks on the results.

Method

The CDM procedure is composed of the following "sweeps" through an incident account:

1. *Incident selection.* The expert is supported in listing a handful of past tough cases, with probe questions inviting the recall of cases that challenged the expert's knowledge, strategies, or skills. One of these is selected for further analysis if it involved tough decisions and spoke directly to the nature of forecaster experience.
2. *Incident recall.* Through the use of probe questions, the expert is supported in telling the story from beginning to end.
3. *Re-telling.* The elicitor then tells the story back, matching as closely as possible the expert's own phrasing and terminology, and inviting the participant to offer additional details, clarifications, and corrections.
4. *Timeline construction.* The interviewer and expert work together to create a timeline, laid out in terms of the sequence of events and the activities of the practitioner (e.g., observations, actions, decisions, judgments, etc.).
5. *Deepening.* The timeline is used to scaffold yet another re-telling, which also invariably results in the recall of more details. The elicitor asks probe questions, such as, "What were you seeing at this point?", "What were your goals and objectives at this particular time?", "What other courses of action were considered or were available?", "Did you imagine the events that would unfold?", "What mistakes were likely at this point?", and "How might a novice have acted differently?"

The results of the CDM are rich case studies, including timelines and indications of critical cues, and decisions. The narratives can be accompanied by supporting media, such as images, videos, or documents. Thus, the results can be applied directly in training as well as contribute to a library or repository of expert knowledge and experience.

Results

The eight CDM procedures conducted with the U.S. Navy forecasters took an average of 6.5 hours, ranging from 52 minutes to 10 hours. More than half of the procedures were broken up, with steps conducted over two days. This was surprising because previous studies of experience using the CDM reported that CDM procedures took at most two hours. The main reason that these procedures took so long is the fact of expert knowledge: its depth and detail. The senior experts in particular had clear and remarkably detailed memories of specific weather events, especially those that were particularly challenging or for which their forecasts "busted." Indeed, the workplace observations revealed that one of the senior experts kept a file cabinet with images and other records for a great many past cases that he had found particularly interesting and difficult.

The CDM procedures focused on forecasting the weather events having the greatest impact on aviation (severe storms and fog) and on the Gulf Coast region (hurricanes). Stories included cases of forecasting severe thunderstorms, storms associated with frontal systems, tornadoes associated with supercells, hurricane tracks, and fog. In some cases, the forecaster created diagrams during the CDM interview, showing such things as the dynamics of frontal systems and the dynamics inside supercells. For some cases, the forecaster had kept charts and satellite or radar images.

The CDM procedures identified decision requirements, information requirements, hypotheses, options, and other aspects of forecasting. The forecasters described their reasoning and also provided advice for less experienced forecasters. The highly experienced forecaster engages in a great deal of hypothetical reasoning, sometimes despite the fact that incoming data cut against a favored hypothesis or make one or another scenario less likely. Forecasting involves not just determining what will happen and when, but sometimes involves determining why something that is expected in a given typical scenario is not happening. Table 8.1 presents the CDM results for the decision requirements for severe weather (i.e., thunderstorms).

Appendix B presents the "Final Integration" of information gained across all of the steps of the CDM for two cases. One is a case in which an expert described an experience at forecasting the track of a hurricane. The other is a case in which an expert described an experience at forecasting a severe storm. These cases illustrate the depth and richness of forecaster knowledge. They are also a window on the actual process of forecasting evolving severe weather.

The CDM resulted in rich case studies in which "lessons learned" are conveyed in the expert's own words, laid out in a format that is populated by data and other records, such as forecaster notes, weather bulletins, satellite images, charts, and

Table 8.1

Results from the Critical Decision Method for forecasting severe weather

Information Requirements

• Forecasting relies on a number of data types, including radar reflectivity (i.e., precipitation), radar echo tops (showing the altitude of cloud tops), lightning network data, GOES (satellite) imagery and loops, pilot reports, and observations on wind speed and direction.
• Forecasters need to have radar data, frame by frame, to be able to track storms.
• Forecasters need to have data from buoys.
• Forecasters need to determine the direction in which fronts and storm cells are moving and the rate at which they are moving.
• Forecasters need upper air data on troughs and vorticity, which can be critical in forecasting rapidly developing severe weather.
• Forecasters need to identify the causes of lifting: moisture at the low levels, the amount of lifting, convergence-divergence across levels, and a "trigger," such as an upper level trough or vorticity advection.

Cautionary Tales About the Information Requirements

• Localized severe weather events are not always noticed or discussed in products from the NWS. Overreliance on individual sources can lead to errors.
• Computer models sometimes gloss over smaller scale event. The forecaster needs to be alert to weather events that the models can miss and then look to data types that can be informative.
• The experienced forecaster never relies exclusively on the data provided through the observing and information-processing systems. They always directly observe the atmosphere and clouds at the airfield.

Hypothetical Reasoning

• Sometimes the decision to explore a possibility hinges on seeing a single key clue.
• It can be easy to miss a flare-up or strengthening in storm cells unless one keeps examining the data.
• It is not a good idea to deliberately overforecast for fear of misjudging storm severity. This can lead to loss of credibility (i.e., overwarning).
• Rules-of-thumb for hypothetical reasoning in standard scenarios can often be stated succinctly (e.g., stationary front over the Gulf with weak lows can be energized by upper level troughs overrunning them from the southwest). "You look out to the southwest and if you see any approaching trough ... the front will develop one or two storm systems."
• Forecasters need to be able to reason speculatively about what might cause intensification or dissipation of developing storm systems.
• Night shift affords the opportunity for more in-depth analysis of weather dynamics.
• Storms and supercells can sometimes turn in anomalous directions.

Options to Consider

• In some weather forecasting situations, even in severe weather, there are no alternative courses of action and no options.
• Even if severe weather seems to be dissipating, one still has to keep monitoring it.
• The forecaster is advised to always forecast for the worst possible situations (warnings, etc.). Issuing a thunderstorm warning sometimes constitutes justified over-forecasting. If severe weather does occur and a warning had not been put out, that would be a worse error than putting out a warning and then the bad weather does not happen. "It is better to have them laugh at you because you were wrong (forecast rain, but no rain) than complain to you because you were wrong (no rain forecast but it rained)."

Table 8.1 (continued)

• The less experienced forecaster is advised to reduce mental workload by extending the valid interval for a forecast so that attention can be distributed to other areas of concern.
• The forecaster must be willing to communicate with distant observers/forecasters to get timely information. Reliance among forecasters at various stations enables them to coordinate warnings and not just share information.
• Avoid deviating from NWS watches and warnings but be aware of clear cases when departure is warranted (e.g., an NWS warning box covers an entire line of storm cells, but it is clear that the cells will be severe only in one region). Be willing to take it case by case.
• It is better to extend a warning out for a longer rather than a shorter valid interval. Any need to amend a warning implies a lack of understanding of the weather situation. Warnings can always be cut short at the watch change.

Goals to Consider

• The less experienced forecaster is advised to avoid overestimating storm severity and resist the temptation to inflate the forecast just to "cover your back."
• Reasoning about determining the valid interval for warnings depends on thorough understanding of client needs and the activities in which clients engage as a result of issued warnings.
• It is important to coordinate the warnings among responsible forecasting offices.
• The forecaster needs to be aware of the needs of the various clients and the circumstances in which they experience weather impacts (e.g., night-time supercell would not influence training flights but would impact Search and Rescue operations if they were needed).

Situation Assessment

• Satellite data can be critical in supporting ongoing situation assessment through long-term monitoring of the animations. Using a loop, one can readily calculate the direction and rate of motion of developing systems and thereby forecast the onset time of severe weather. Sometimes this can be the only way to determine onsets for smaller scale weather events that are sometimes glossed over by the computer models.
• Ponder what might cause intensification or dissipation in a developing storm system?
• It is important to keep watching for outflow boundaries and secondary storm development.
• In forecasting severe weather, events can sometimes transpire rapidly, and forecasters must be prepared to sometimes "go out on a limb" in anticipating severe weather outbreaks.

Time/Effort Considerations

• It is important to possess the willingness and fortitude to inspect incoming data stream frame by frame for long periods of time (many hours). Needed information may be obtained only after long periods of data monitoring (e.g., NEXRAD).
• In severe weather situations, hand chart-work and hand charting skills can be of critical importance to both understanding and forecasting.
• Supercells can last for a long time, many hours.
• Severe weather situations involve a need to monitor data for prolonged periods of time (e.g., manual plotting of storm tracks over a period of many hours).

Table 8.1 (continued)

Knowledge, Training, and Skill Level

• Knowledge of local forecasting rules is critical.
• It is important to have a thorough knowledge of typical scenarios, but also enough lived experience so as to have had the chance to learn from errors during the typical scenarios. Training in the school house on the standard scenarios should include gaming in which the students are set up to "get burned."
• It is critical to be familiar with the dynamics of frontal systems and their manifestations in the local climate regime.
• The less experienced forecaster needs to try and explain why expected events do NOT occur and why the unexpected CAN occur.
• Novices may fail to notice small features that are "upstream" and not salient but that can develop into severe weather that impacts operations.
• Novices are said to be weak in terms of picking up subtle clues for small-scale events that the computer models gloss over.
• Novices are said to fall prey to mindsets, such as that which says that after frontal passage the weather will be clear. Hence, they expect nothing to happen and do not look to see whether anything is happening. Schoolhouse exercises should help them break through such mindsets. "They are trained in school but tend to not do what they're taught, and they get caught with their pants down."

other information that was involved in the forecasting event and was retrieved from archives. These narratives were included in a "knowledge model" based on the use of concept maps.

Concept Mapping

Whereas the CDM is good for eliciting expert knowledge based on their experiences, concept mapping is good for eliciting and representing expert knowledge about the domain concepts and phenomena (Moon, Hoffman, Cañas, and Novak, 2011; Novak and Gowin, 1984). Concept maps are a form of meaningful diagram, showing concepts (enclosed in boxes that are called "nodes") and relationships among concepts. Relationships are indicated by linking lines bearing labels. Each node-link-node triad forms a meaningful proposition. Figure 8.1 is an example of a relatively simple concept map.

The literature on diagrammatic reasoning (research in education, cognitive science, computer science, and geography) includes reports on studies of how people understand a great many types of diagrams, ranging from topographic maps, to matrices, to schematic diagrams of machines, to semantic networks (a comprehensive review appears in Vekirl, 2002; see also Ausubel, Novak and Hanesian, 1978; Day, Arthur, and Gettman, 2001; Glasgow, Narayanan, and Chandrasekaran 1995; Mandl and Levin

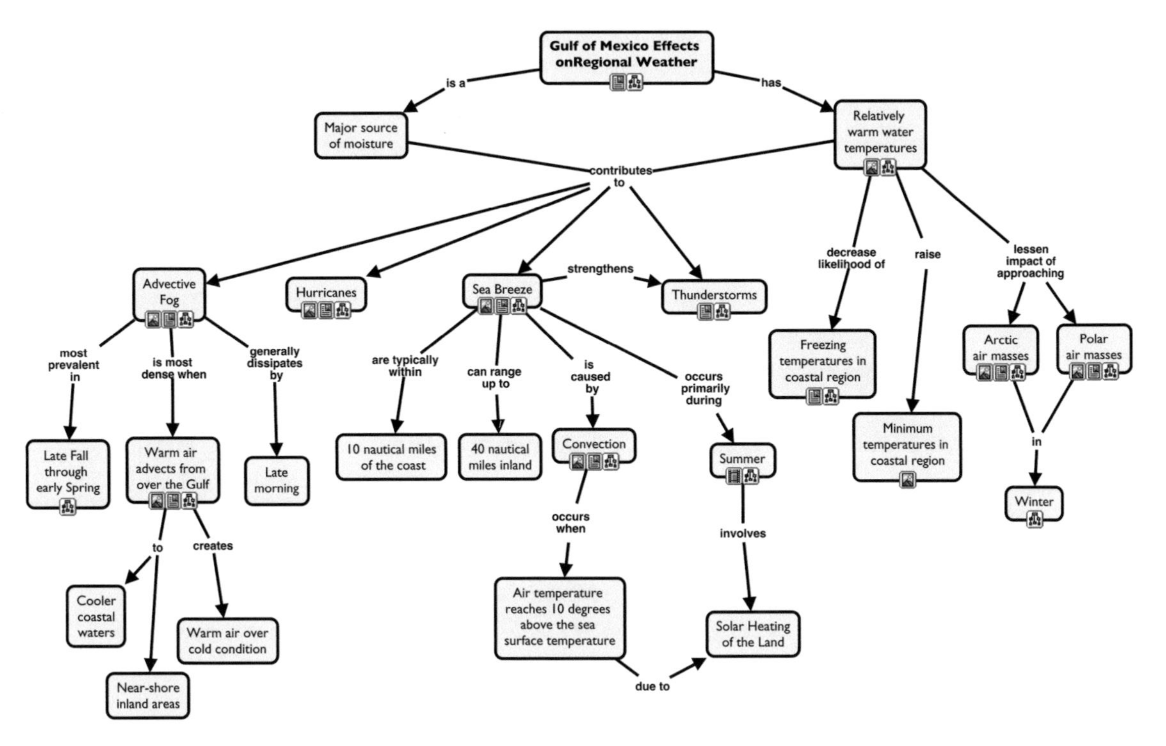

Figure 8.1
A concept map about the effects of the Gulf of Mexico on Gulf Coast region weather.

1989). Good diagrams are effective because they "externalize" cognition; they guide/constrain and facilitate cognition by supporting inference making. They have mnemonic value and reduce cognitive demands by enabling information integration at a glance (as opposed to overloading working memory) by shifting some of the burden of text processing onto the visual perception system. Diagrams that work well are ones that rely on proximity. The spatial organization or connection of information units induces people to see the units as being related and makes people likely to attempt to draw inferences about the relation.

These are all features of a concept maps. Concept mapping has foundations in the theory of meaningful learning (Ausubel, Novak, and Hanesian, 1978) and a background of decades of research and application, primarily in education (Novak, 1998). Concept maps are being used by groups as disparate as school children across South America, astrobiologists at NASA, curriculum designers in the U.S. Navy, university professors preparing distance learning-based courses, trainers in the electric power utility industry, and businesses where focus groups create concept maps for brainstorming (Cañas et al., 2003, 2004; Gaines and Shaw, 1995; Hanes and Gross, 2002). Concept maps have been used as knowledge representations in cognitive science (Dorsey, Campbell, Foster, and Miles, 1999). Concept maps made by domain experts can be used to show agreements and disagreements (see Gordon, Schmierer, and Gill, 1993). In the field of human factors engineering, diagramming has been proven useful as a procedure whereby domain practitioners describe their knowledge and reasoning (Cooke and McDonald, 1986; McNeese, et al., 1990). Furthermore, concept maps have been used as the basis for the explanation component of knowledge-based systems and performance support systems (Cañas et al., 2003; Coffey and Hoffman, 2003; Dodson, 1989; Ford et al., 1992; McNeese et al., 1990; Sutcliffe, 1985).

Reviews of the literature on concept mapping, discussion of methods for making concept maps, and discussion of the differences between concept maps and other types of meaningful diagrams can be found in Cañas et al. (2003, 2004), Coffey and Hoffman (2003), and Crandall, Klein, and Hoffman (2006). Although concept mapping has the variety of applications we have detailed here, it is by no means a tool for all purposes. Its primary strength lies in the creation and representation of knowledge about domain concepts. That is the purpose to which it was applied in the weather forecasting project.

Participants

Participants in the concept mapping effort were those eight individuals at the Pensacola Naval Air Station weather forecasting facility who were qualified to produce forecasts. We did not engage the less experienced aerographers or any of the observers in the concept mapping sessions. The group of eight included four individuals who had been designated as experts on the basis of the career interview and analysis of performance data (see chapter 7)—two of the three senior civilian forecasters and the Command Master Chief Petty Officer. Although the Command Master Chief had only one year on-station at the Pensacola Naval Air Station, he had more than 14 years of experience forecasting tropical weather. The group of eight also included Chief Petty Officers and Petty Officers who had been designated as journeymen on the basis of the career interviews. As a group, the eight had been on-station in Pensacola for a range of one to six years, averaging about three years. Three had been on-station for fewer than three years. On the basis of the career interview (see chapter 7), we had designated the METOC Commanding Officer as a journeyman, but he had authored the Local Forecasting Handbook used by the Navy in the Pacific Oeean area of responsibility. His main role in the knowledge modeling project was to assist in the process of validating, finalizing, and approving the concept maps.

A final participant was a retired U.S. Navy forecaster who was designated a junior expert on our proficiency scale. He reviewed all of the concept maps, validated all of them one proposition at a time, and offered suggestions for refinements in their phrasing. Validation took about seven minutes per concept map, on average. About 10% of the propositions were modified. Some of the changes involved important subtleties (e.g., "X causes Y" vs. "X facilitates Y"). Some of the changes seemed like wordsmithing. For example, the proposition, "Dry line which acts like a frontal slope" was changed to "Dry line acts like a frontal slope" (the change in the scope of the qualification might be regarded as both subtle and important). The main point is that we found little in the way of what might be regarded as outright disagreement (i.e., statements to the effect that "This proposition is wrong").

Knowledge Elicitation Procedure

The knowledge elicitation interviews were supported by use of *CmapTools*, a freeware software suite created at the Florida Institute for Human and Machine Cognition [http://cmap.ihmc.us]. *CmapTools* has a simple interface that guides the user in the creation of concept maps using simple point-and-click and drag-and-drop operations.

In the knowledge elicitation interviews, one researcher stood at a screen and served as the facilitator while another researcher worked at a laptop computer and created the concept map, which was projected on a screen. Participants interacted with the researchers to create the concept maps. This process is illustrated in figure 8.2. Referring to the projected concept map as it developed, the facilitator helped the forecaster build up a representation of domain knowledge of weather concepts by suggesting alternative phrasings for concepts and propositions. The facilitator avoided imposing ideas or word choices.

Results

The concept mapping sessions were aimed at breadth rather than depth, resulting in more than 150 concept maps. Topics included advisories (storms, winds, tropical systems), warnings (storms, winds, tropical systems), communications equipment, forecast products, computer models (e.g., icing), aviation instrument flight rules, tropical conditions of readiness, and NEXRAD radar. It took about 1.5 to 2 hours to create, refine, and verify each of the concept maps.

For building an integrated knowledge model, we focused on capturing expertise at forecasting weather phenomena that are important in the Gulf Coast, including

Figure 8.2
Photographs of a knowledge elicitation session.

regional seasonal tendencies, fog, turbulence, tornadoes, thunderstorms, and hurricanes. A knowledge model is a set of concept maps that are hyperlinked together and organized by a "top map" that shows their cross-links hierarchically. This is achieved by hyperlinking concept maps onto nodes inside other concept maps. In figure 8.1, for example, one sees that many of the nodes have icons appended at their bottom. These nodes link to other concept maps. The node for hurricanes in figure 8.1 hyperlinks to a concept map that is just about hurricanes. The node for tornadoes in figure 8.1 hyperlinks to a concept map that is just about tornadoes. In this way, all of the concept maps get "stitched together". In addition, other forms of digital media can be appended to the nodes in concept maps. The STORM-LK knowledge model included a number of brief video clips in which experts talked about various aspects of forecasting. STORM-LK included hyperlinks to satellite images, the local radar, and other online NWS products. Additional concept maps had this organizing principle. Figure 8.3 is the concept map about the Gulf region climate. It hyperlinks to concept maps about the seasons and various seasonal trends and patterns.

The primary organizing concept map from the knowledge model is presented in figure 8.4. This is the "top map" in the knowledge model, that is, the first thing one sees when viewing STORM-LK. From this top map, one can navigate to (and from) all of the other concept maps in the knowledge model by clicking on the resource icons that appear below the concept nodes.

Concept maps on regional seasonal tendencies, fog forecasting, turbulence, tornadoes, thunderstorms, and tropical weather (e.g., hurricanes). formed the core of the

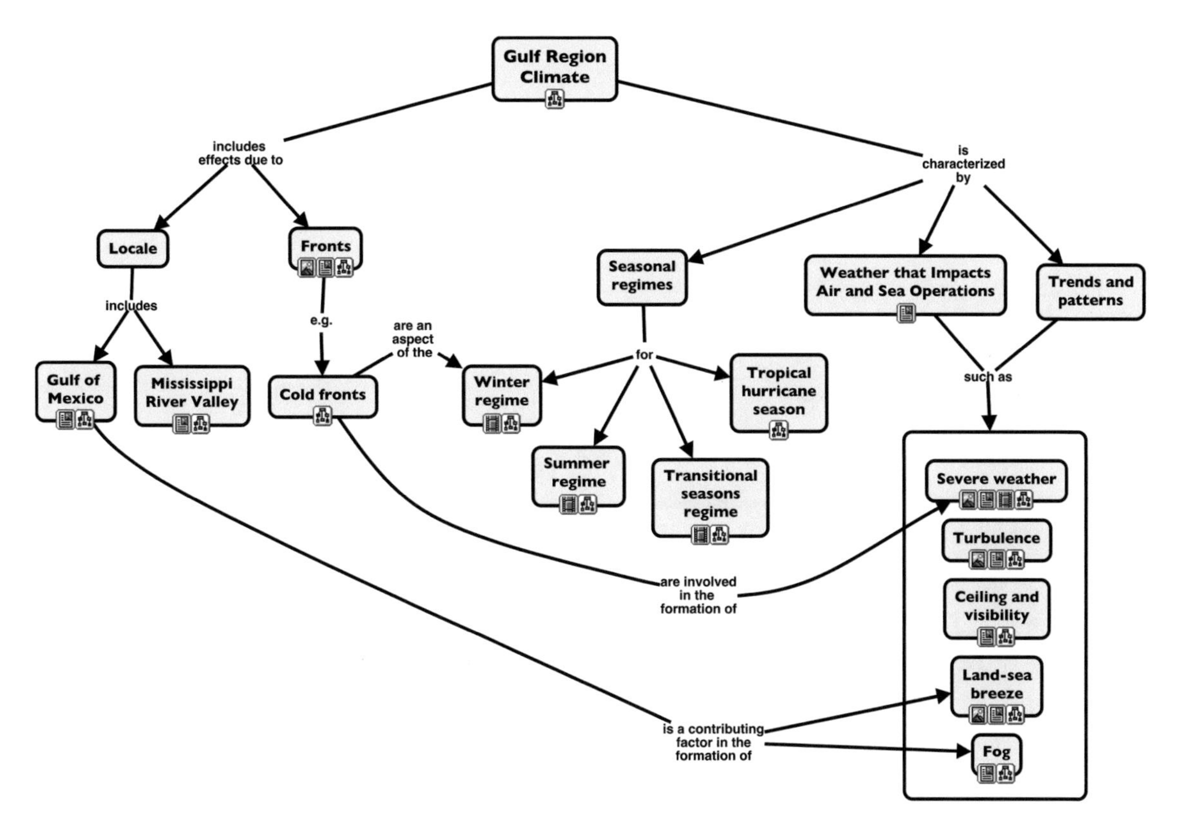

Figure 8.3

The concept map that organizes other concept maps about seasons and seasonal trends.

STORM-LK knowledge model. Additional models could easily have been made of other major topics, such as computer models and NEXRAD, but these were not regarded as central to region-specific expertise. The most elaborate concept map was for the dynamics of hurricanes. It is reproduced in figure 8.5.

The knowledge model and its various text and graphics resources in STORK-LK contain all of the information found in the Local Forecasting Handbook. Because the concept maps are web-enabled, they allow real-time access to data (radar satellite, computer forecasts, charts, etc.). Case studies from the CDM were appended to the appropriate nodes, as were digitized videos that allow the apprentice to "stand on the expert's shoulders" by viewing mini-tutorials about particular concepts. Figure 8.6 (plate 10) shows a "presentation" version of the top map, with open windows that illustrate the available resources.

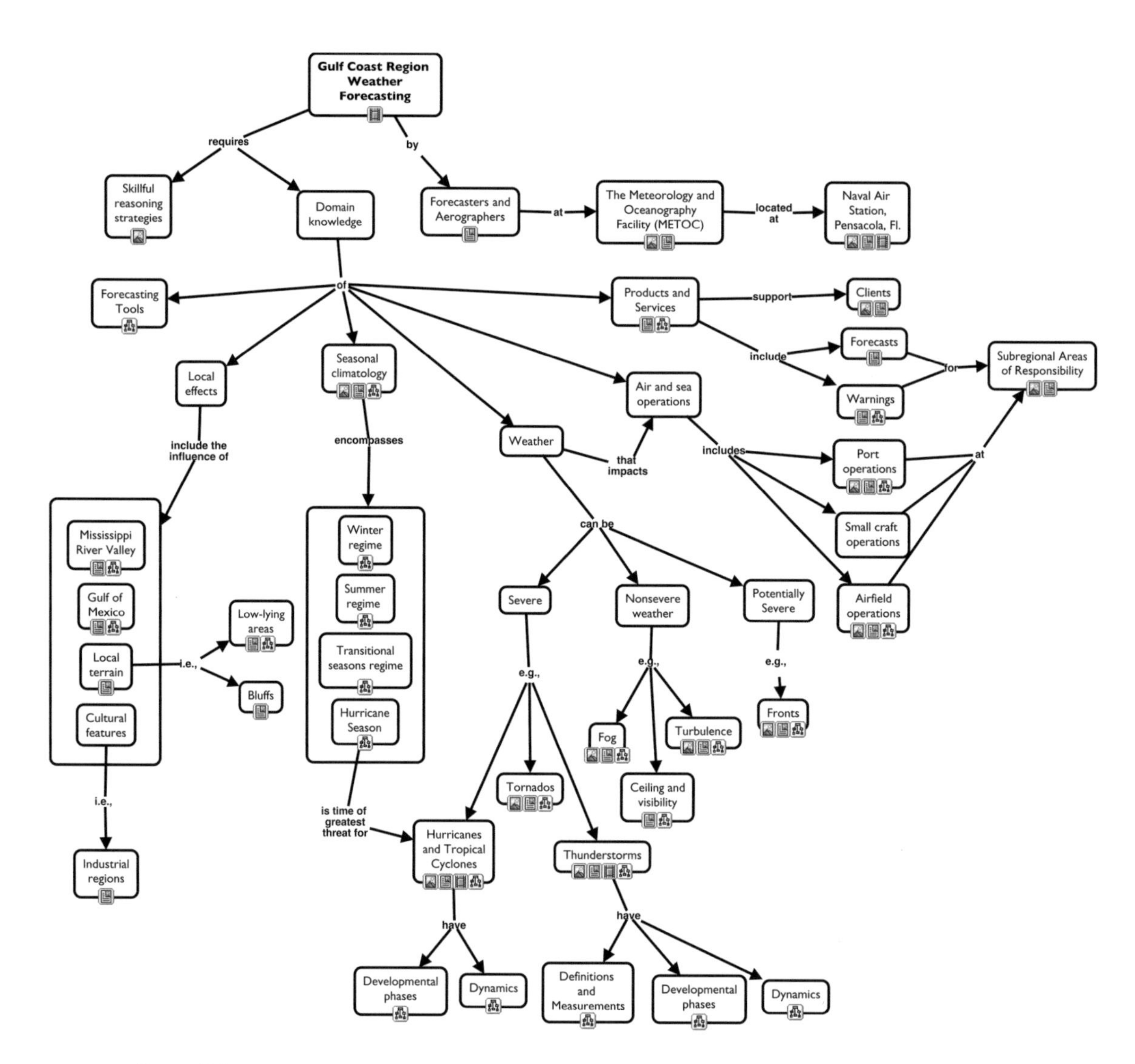

Figure 8.4

The "top map" in the STORM-LK knowledge model.

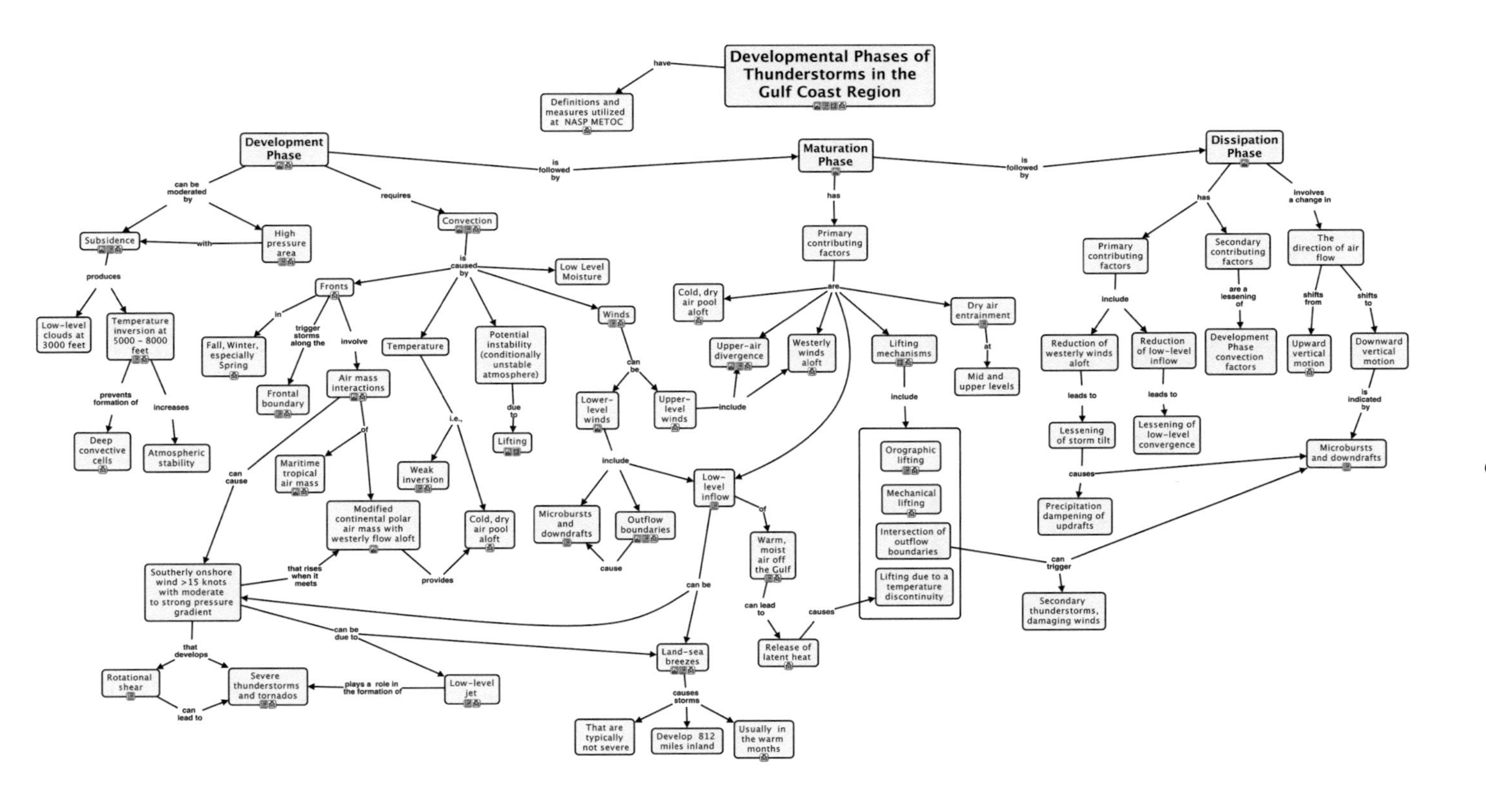

Figure 8.5
A concept map about the dynamics of hurricanes.

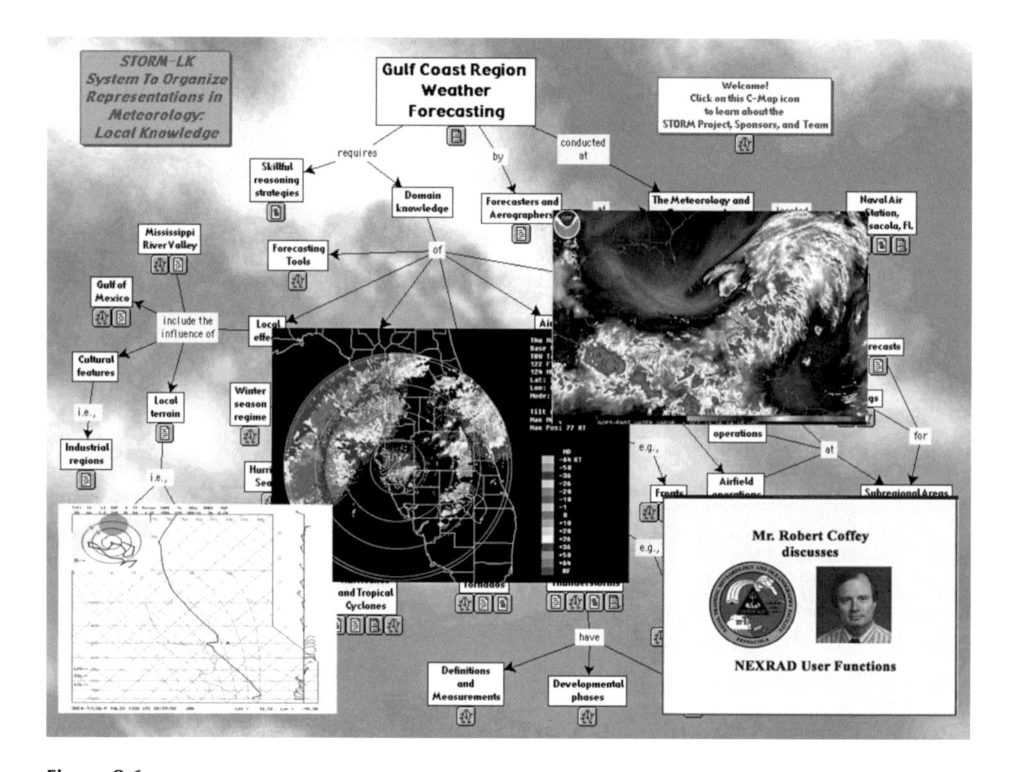

Figure 8.6

(plate 10) A "presentation" version of the top map in the STORM-LK knowledge model, illustrating some of the multimedia resources that are appended to concept nodes: tutorial videos, data charts, satellite images, and radar composites.

Figure 8.7 (plate 11) shows the concept map about Gulf of Mexico effects appended with open windows showing material from the CDM procedure that covered hurricane track forecasting (see appendix B).

The knowledge model uses the concept maps to be the interface to support trainees or practicing forecasters as they navigate through the domain knowledge. The user can also view real-time weather data, but within the context of the concept maps that provide the "explanatory glue" for the forecasting process. Weather forecasting expertise is often quite local, dependent on years of experience in a particular region. Hence, even an expert who is transferred to a new duty station needs to rebuild his or her local forecasting expertise (i.e., "re-qualifying"). In the military and the NWS, re-qualification involves learning from the Local Forecasting Handbook and also working forecasting shifts with supervision by the local veteran staff.

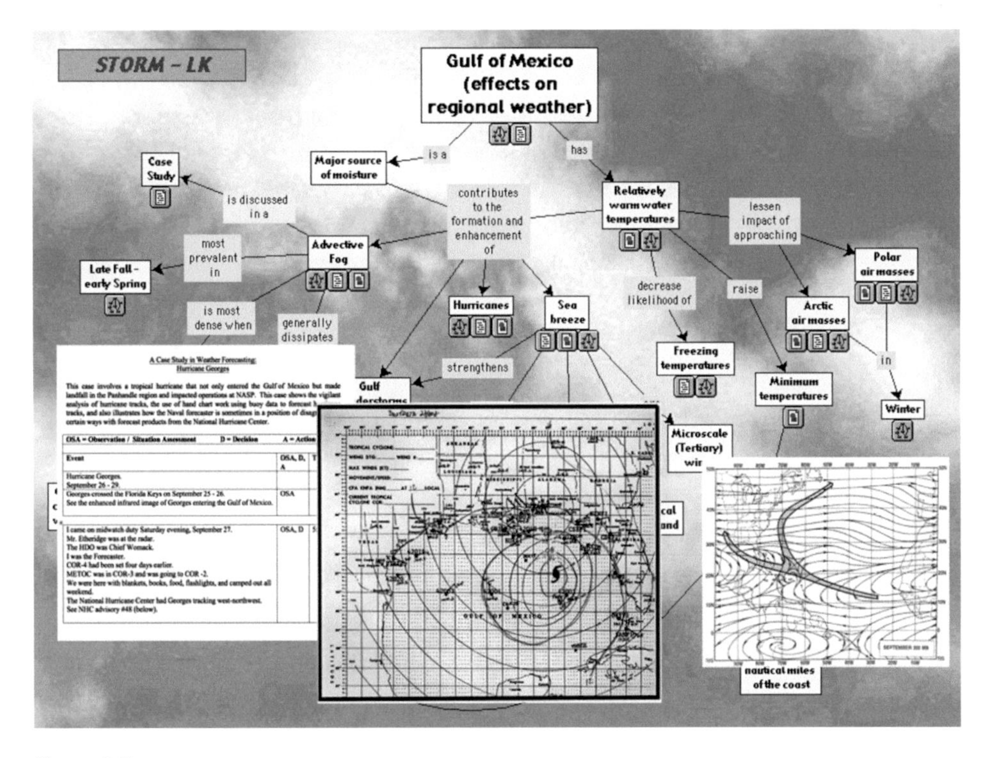

Figure 8.7
(plate 11) The concept map about Gulf of Mexico effects with resources from the CDM procedure that covered a case of hurricane track forecasting.

STORM-LK was made available on a compact disc to newly arriving aerographers and trainee aerographers at the METOC facility. They were invited to explore the concept maps and the appended resources and convey their judgments regarding the material's value as a learning aid as they prepared for re-qualification tests. We found an interesting bifurcation in the reactions of the aerographers. Those who were more advanced in their forecasting experience, and who had previously qualified at two or more duty stations using the Local Forecasting Handbook, felt more comfortable using that traditional document with its standardized format and linear organization. A question often asked about concept map knowledge models is, "Where do I begin?" The answer is, "Anywhere you want," but those who are accustomed to linear modes of thinking and rote learning strategies can find this perplexing. In contrast, those aerographers who were of the "digital native" generation found the concept maps interesting, almost like a video game, and regarded them as an invitation to follow the links and look at the resources. They found the digital videos with the experts' mini-tutorials to be especially memorable and believed that the exploration of the model assisted them in preparing for their qualification exams.

A human factors scientist who was interested in concept mapping and who heard about the STORM-LK project asked if he could see "the" concept map about weather forecasting. He was taken aback to learn that the project resulted more than 150 concept maps, not one. A rough estimate would be that it would take many hundreds of concept maps to capture the knowledge of expert forecasters. For this reason, subsequent knowledge modeling projects emphasized the need to focus on the expert's "tacit" knowledge. Tacit knowledge was defined as that which was unique to highly accomplished individuals, was critical to their organization, and had not yet been codified or recorded in any way (Hoffman and Hanes, 2003; Moon, Hoffman, and Ziebell, 2009).

It Is Not All Just in the Head

We should note one other important aspect of expert forecaster knowledge. Experts in diverse domains have been observed to keep "treasure maps" (Hoffman et al., 2008, 2011). These can take the form of a "go-to" list of other experts according to their sub-specializations or unique skill sets. It can take the form of a "cheat sheet" encapsulation of key procedures that consume many pages and are formally described in documents. In weather forecasting, the treasure maps take the form of records on previously encountered tough cases, unusual weather events, or cases where forecasts were a bust. Hoffman et al. discovered a senior expert's weather treasure map during a workspace analysis, when a lead forecaster had been asked the content of a particular filing cabinet, one that was kept close to the main workstation. "Oh, those are my old busts." Indeed, the cabinet was a treasure trove, including printouts of the actual forecasts and warnings, GOES and radar images, hand-drawn charts, and other records. These external memories serve multiple purposes. First, they remind the forecaster of their fallibilities, which they can feel painfully. Second, they preserve examples of how key data can be crucial in the proper interpretation of singular events:

> Experts recognize the importance of building a "lessons learned" knowledge base ... one forecaster was able to instantly recognize subtle signature in the radar data, an unusual cue that a less experienced person might have missed. The forecaster was able to make a nearly instantaneous decision because the pattern he saw on the radar screen matched a pattern he had seen in an investigation years before. In that previous case, he and other forecasters has poured over the data for hours, finding only this one persistent signature to explain the phenomenon. (Klinger, Hahn, and Rall, 2007, p. 369)

The other key use of the treasure is it is a key resource in the mentoring process. The senior forecaster serves as a sounding board as less experienced forecasters describe their developing mental model and will sometimes reference their records of previously encountered cases:

> The more experienced forecasters did not attempt to dominate difficult situations, but rather leveraged opportunities presented within those incidents to mentor and share insights with their colleagues, (Klinger, Hahn, and Rall, 2007, p. 367)

Conclusion

Together the CDM and concept mapping results are strongly supportive of the notion that expertise in weather forecasting involves rich, deep, and broad conceptual knowledge. STORM-LK is also suggestive of how similar efforts may be undertaken to preserve expert knowledge for other regions and climates. STORM-LK facilitates the process of preserving organizational expertise and case studies by substituting a new type of "living e-document" for the traditional hardcopy Local Forecasting Handbooks. Forecasting handbooks are updated every few years or so. The STORM-LK approach has the virtue of allowing the organization's knowledge base to always be up-to-date; it can be continually refined and expanded by the incorporation of new knowledge and models of situation-specific reasoning. This is particularly important because at the Navy facility and many other similar organizations, knowledge is "shared" in only such things as briefings and documents. Yet we know from ethnographic studies (see chapters 3 and 7) that most forecasting expertise is actually acquired by being passed from generation to generation by mentored on-the-job experiences. Both of those can be systematically employed and embedded on the organization's culture, and both are typically haphazard. Not all junior forecasters get good mentoring or can get good mentoring, and not all junior forecasters find themselves in knowledge-sharing organizations. As we will show in chapter 10, experienced forecasters can recognize the strategies and error tendencies of apprentices, but within a forecasting organization, forecasters might remain largely uniformed about the sensemaking and reasoning strategies of their fellow forecasters.

Weather forecasting, like numerous other domains and venues of professional practice, is experiencing the "grey tsunami." Individuals of the boomer generation are retiring, and when they retire, their expertise usually goes out the door with them (Hoffman et al., 2008, 2011). The forecasting facility at Pensacola Naval Air Station was downgraded to a Detachment, and the senior civilian forecasters took the opportunity to retire. Were it not for the knowledge that had been captured in the STORM-LK project, all of their experience, strategies, and case libraries would have been effectively lost to the broader Navy weather community—and that includes descriptions of domain knowledge and reasoning that were not discussed in the Local Forecasting Handbook.

It may be useful to explore the possibility that the creation of concept maps might be a valuable component of mentoring, and perhaps even a task to be included as part of qualification examinations (Hoffman et al., 2001, 2006). An outstanding need is for studies that use concept map knowledge models and concept mapping exercises in an educational intervention, comparing learning and performance for meteorology students who are given traditional instruction, with appropriate control for experimental demand characteristics (i.e., performance might improve because of the special treatment not because of the nature of that treatment). Such studies in other domains have shown significant and lasting gains for concept mappers (see Novak, 1998). Related to this, there is an outstanding need to conduct long-term longitudinal studies of the development of forecasting expertise. Perhaps the most significant open question is whether knowledge representation activities of this kind will ultimately help accelerate the achievement of expertise, that is, contribute to improved forecaster performance (Hoffman, et al, 2014).

9 What Does Research on Forecaster Perception Tell Us?

In weather forecasting, each day can be considered to be unique, and a day exactly like it will never be encountered again. Hence, it can be argued that weather forecasters are trapped into trying to deal with the smallest possible sample size—a sample of one—every day. Pattern recognition ... has long been a traditional tool in weather forecasting (Doswell, 2004, p. 1122).

A great deal of research has been conducted on diagrammatic reasoning, map interpretation, graph comprehension, and data visualization. Research has been conducted by geographers, cartographers, experimental psychologists, cognitive scientists, educators, and computer scientists, among others. Studies have looked at the design of maps and graphical representations of many kinds. Studies have investigated display effectiveness for knowledge expression, comprehension, learning, communication, and other functions. Educational and cognitive psychologists have conducted experiments on how college-age students learn from illustrations in scientific texts, and on how people interpret statistical graphs and charts (e.g., Barker-Plummer, Cox, and Swoboda, 2006; Fabrikant, Hespanha, and Hegarty, 2010; Hegarty, 2011; Kosslyn, 2006; Mayer, 1989, 1993; Mayer and Anderson, 1991; Mayer and Gallini, 1990). Cartographers and remote sensing scientists have conducted extensive research on map/display design (e.g., Bertin, 1967/1983; Davies, Fabrikant, and Hegarty 2015; Hoffman and Markman, 2001; MacEachren, 1995; Ooms, De Maeyer, and Fack, 2014; Tufte, 1990, 2001), and computer scientists have also extensively investigated visualization issues (e.g., Glasgow et al. 1995; Silva, Santos, and Madeira, 2011; also see appendix E). A great deal of research has been conducted on the design of weather information displays for pilots and air traffic controllers (Ahlstrom, 2003; Ghirardelli and Glahn, 2010; Krebs and Ahumada, 2001; O'Hare and Stenhouse, 2009). New designs are being tested to enable proper interpretation of probability forecasts. For example, a new visualization for the display of the probability of strong winds enables meteorology students to make warning decisions without bias in either high or low likelihood situations (i.e., to err on the side of caution) (e.g., Joslyn et al., 2007). The research results affirm the importance of visualization in the integration and interpretation processes, for laypersons as well as meteorologists and forecasters.

A great deal of research is now being conducted on how laypersons interpret (or misinterpret) forecasts and weather data visualizations (e.g., Martin et al., 2008; Savelli and Joslyn, 2013). Apparently even straightforward things such as daily high and low temperature forecasts can be misinterpreted. In this chapter, we focus on how forecasters use data displays and visualizations in their sensemaking activities.

How and Why Diagrams and Visualizations Aid Sensemaking

Mayer (1995; Mayer et al., 1995) extended the educational psychology findings to the domain of meteorology. College students who lacked knowledge of meteorology were presented with a series of illustrations showing a sequence of events in the formation of lightning: lifting of warm moist air, falling of rain drops and ice crystals, and build-up of electrical charges, followed by cloud-to-ground and then ground-to-cloud negative and positive discharges. In one condition the explanatory text was presented in the figure captions, whereas in another condition annotations of the key explanatory information were included in the diagrams. The participants' recall of the summary text and their performance on a transfer problem were facilitated by having seen the illustration. Recall of the summary text was not helped if the diagram was also accompanied by a fuller text or if the explanatory text material was included in the captions that accompanied the illustrations.

Thus, there are good and not-so-good ways of designing integrated diagrams and text explanations. When properly designed, an illustration can actually render lengthy verbal explanation unhelpful. Mayer's view is that dynamic illustrations containing integrated labels and succinct annotations assist in the formation of mental representations that combine principle-based understanding of causation with a dynamic, imagistic understanding. Mayer's research demonstrates the significance of mental modeling and the need for mental models to be based on a correct understanding of fundamental principles.

It is well established that diagrams can present information efficiently and guide reasoning, support memory, and support perception and attention because diagrams shift some of the burden of text processing over to the visual perception system (Kriz and Hegarty, 2007; Larkin and Simon, 1987; Lowe, 2015; Scaife and Rogers, 1996). It would be well beyond the scope of this or any single book to review all this literature. Reviews on the psychology and educational psychology research literature are presented by Anderson and Meyer (2013), Barker-Plummer, et al. (2006), Bauer and Johnson-Laird (1993), Mandl and Levin (1989), and Vekirl (2002). Researchers have relied on a variety of experimental tasks to explore hypotheses about display perception and understanding: accuracy and response times to detecting map features of data elements, eye movements during map reading, recall of features or data elements shown in briefly presented maps, the ability to detect changes in successively presented maps, the ability to make inferences from maps, and judgments of preferences based on map appearance and aesthetics.

We should note that diagrammatic visualizations do not always help (Rogowitz and Treinish, 1996). For example, Susan Joslyn and her colleagues at the University of Washington (Nadav-Greenberg, Joslyn, and Taing, 2008; Savelli and Joslyn, 2013) demonstrated that laypersons' understanding of temperature forecasts is significantly aided by presenting predictive intervals (80% confidence intervals), but only when the data are presented as text (e.g., "Today's high temperature will be 41 degrees but could be as high as 44 degrees or as low as 38 degrees"). When presented graphically in the form of a statistical histogram, the temperature range was sometimes interpreted to mean how the temperature would fluctuate over the day. It should be noted that the data were misinterpreted sometimes for all of the different ways in which the data were presented.

In another study, both meteorologists and laypersons were presented information about wind speed forecasts in three different ways: a map with overlaid colors indicating the predicted median wind speeds across the region, a similar diagram showing the range of speeds between the median and the upper bound, and a statistical graph, called a "box plot," which showed median, range, and upper bound. The participants were asked to assess the uncertainty expressed in the data and decide what the winds would be in an identified region on the map and whether they would post a high wind advisory (if winds were expected to be 20 knots or more somewhere in the mapped region). For the participants who were laypersons, the display showing only the upper bounds of the winds served as a biasing anchor, leading to higher estimates of wind speeds. The chart showing the range between the median and the upper bound was harder for participants to use, but it resulted in less bias. For both laypersons and meteorologists, the box plot was both easiest to use and resulted in the best performance. Thus, in this study, a statistical graph was better than a colorized map display. The researchers cautioned that, "There may be other situations in which visualizations are ill-advised" (Savelli and Josslyn, 2013, p. 538).

This being said, there is strong evidence from psychological research that there is a primary role of perception in knowledge acquisition and use, reasoning is often visual thinking, conceptual reasoning is often perceptual (Brown, Collins, and Duguid, 1989;), and reasoning is often in the form of "analog mental simulations" (Hegarty, 2004) that make inference "automatic" (Day and Goldstone, 2012; Glenberg and Robertson, 2000). This is a main theme of this chapter.

Display Design Principles: Some Work, Some Don't

A number of principles for the design of displays or data visualizations pertain to meteorology and forecasting. Some of these principles seem to be valid for this domain, some do not, and some need to be taken with a grain of salt.

The Principle of Visual Diagrams

Much of the research on the perception of graphs, charts, and maps has relied on the classical theories of cartographic representation. For example, Bertin's (1967/1983) theory asserted that different kinds of graphical elements (lines, shapes, symbols, colors, brightness values, etc.) would be differentially appropriate depending on the nature of the information they were intended to represent (e.g., continuous variables, qualitative variables); consequently, the important meanings that maps have to convey should be depicted by making graphical elements particularly salient. The difficulty here, especially in the case of meteorology, is that salience is not solely a property of diagrams or display visualizations: It is an interaction of the visualization with the perceiver. Many graphic aspects can stand out (e.g., color, contrast, or asymmetry), but there is also a significant "top-down" component to visual sensemaking, related to the ability of the trained meteorologist to see things that the untrained eye cannot (Klein and Hoffman, 1992). Perception can be guided top-down by knowledge, goals, and the task one has to perform as well as by bottom-up processes triggered by raw visual salience (Davies et al., 2015). Perceptual learning is a crucial element to sensemaking in weather forecasting, and we also discuss this in more detail below.

The Principle of Naturalness of Representation

One of the most widely cited principles for display design is Donald Norman's (1990, 1993) "naturalness of representation" principle. This principle states that the properties of the representation need to match the properties of the thing being represented:

> Perceptual and spatial representations are more natural and therefore to be preferred but only if the mapping between the representation and what it stands for is natural—analogous to the real perceptual and spatial environment (Norman, 1993, p. 72).

Thus, for example, it is better to represent population density in a map using brightness shades of a single color than one using multiple colors. Norman (1993) also stated, "It is easiest to present people with the same representations used by the machines: numbers. This is not the way it ought to be" (p. 226). Norman's emphasis was on the mapping, but the word "natural" and the references to the "real perceptual environment" imply that the representation has to look like the thing being represented. Whether realism or "naturalness" is desirable is a matter of some debate in human factors psychology (see Smallman and St. John, 2005). But meteorology, it is typically not possible. Indeed, a very "non-natural" representation is sometimes a better aid to perception and understanding. Figure 9.1 shows winds at a height of 700 millibars (700 mb—roughly 10,000 feet or 3,100 meters above ground level). This is a counterintuitive technique of measuring height in the atmosphere in terms of air pressure. The chart has a graphical map (familiar, if not "natural") and individual data points (i.e., altitudes and wind barbs; definitely not "natural"). Charts of this kind depict information that is clearly a step beyond direct or everyday experience.

A wind barb shows both wind speed (by the number of full and half lines on each barb) and direction (by the direction of the major line, blowing into the station location). On the one hand, collapsing information in a way that might be regarded as efficient comes at a cost of legibility and display clutter. On the other hand, because individual wind barbs tend to cluster in ways suggestive of atmospheric dynamics (see the circulation patterns in figure 9.1), the display allows the forecaster to see the qualitative aspects of the wind field (e.g., areas of increased wind speed and presence of high and low pressure systems). A wind barb has the added benefit of allowing a forecaster to extract quantitative information from the glyph (e.g., long barbs are 10 knots and short barbs are 5 knots). These ways of representing data were standardized internationally long ago and are well known and routinely used by today's forecasters. It has thus become easy to ignore the fact that glyphs such as wind barbs are not ideal in terms of how the visual perception system works. Research on how laypersons (college students) interpret wind barb charts (generally, vector fields) has shown that both wind speeds and directions are often misinterpreted (Martin et al., 2008). It has been argued that streamlines would be a better way of depicting wind fields (Ware, Kelley, and Pilar, 2014), but those too would not be "natural."

Another example of how the naturalness of representation principle does not quite work in meteorology is the Geostationary Operational Environmental Satellite (GOES) satellite imagery. The GOES system of satellites provides visible (light-reflected) and a series of infrared images. Fifteen GOES satellites have been launched since 1975, and three are currently active. From their orbits of more than 20,000 miles above the Earth,

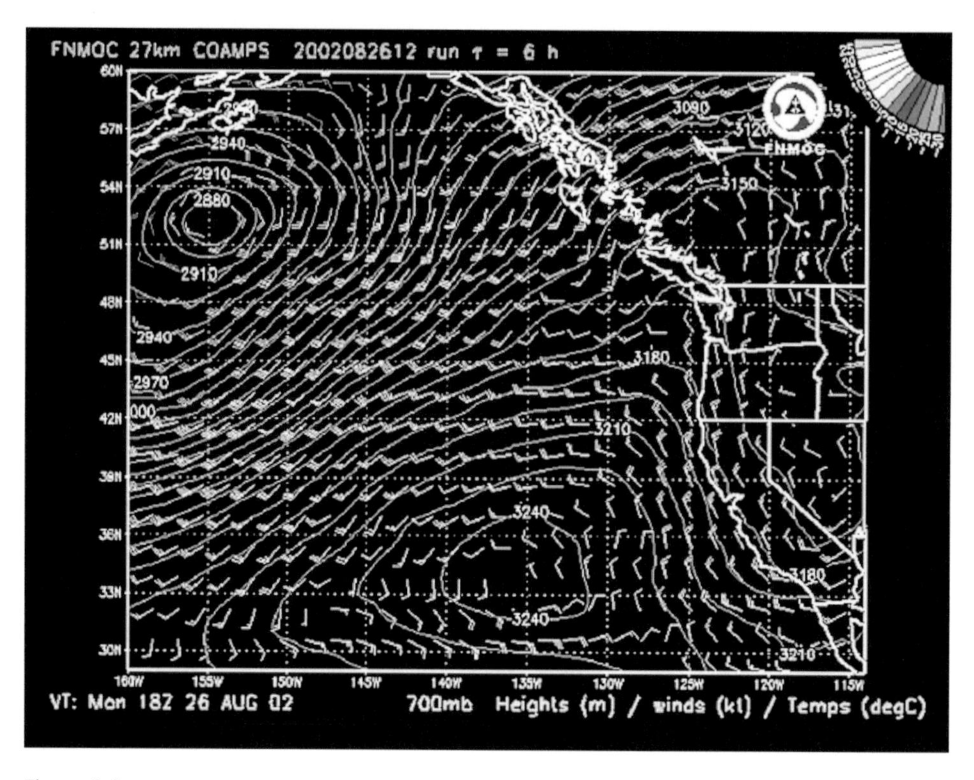

Figure 9.1

A constant pressure chart showing wind speed, wind direction using wind barbs, and isolines (representing height). Image courtesy of the U.S. Navy.

they can view large expanses of the Earth. Collectively, a network of GOES-type satellites operated by different countries or agencies covers the globe, providing a wealth of near real-time atmospheric and oceanic information.

Originally, GOES infrared imagery depicted cloud temperature using gray scales, in part, due to the reliance on difax (a now outdated printer system) and also due to the cost of color printing and cathode ray tube monitors. How should one display temperature? One way may be to display colder clouds using dark gray and black tones (using the "natural" association that cold is dark) and warmer clouds using lighter shades and white (hot is bright). However, if one looks at planet Earth using this hue palette for infrared images, it looks more like a surrealistic marble cake than a planet with an atmosphere (see figure 9.2).

The operational GOES images flip the color-coding, making clouds bright and the background or the Earth dark (figure 9.3). This color coding is the reverse of this

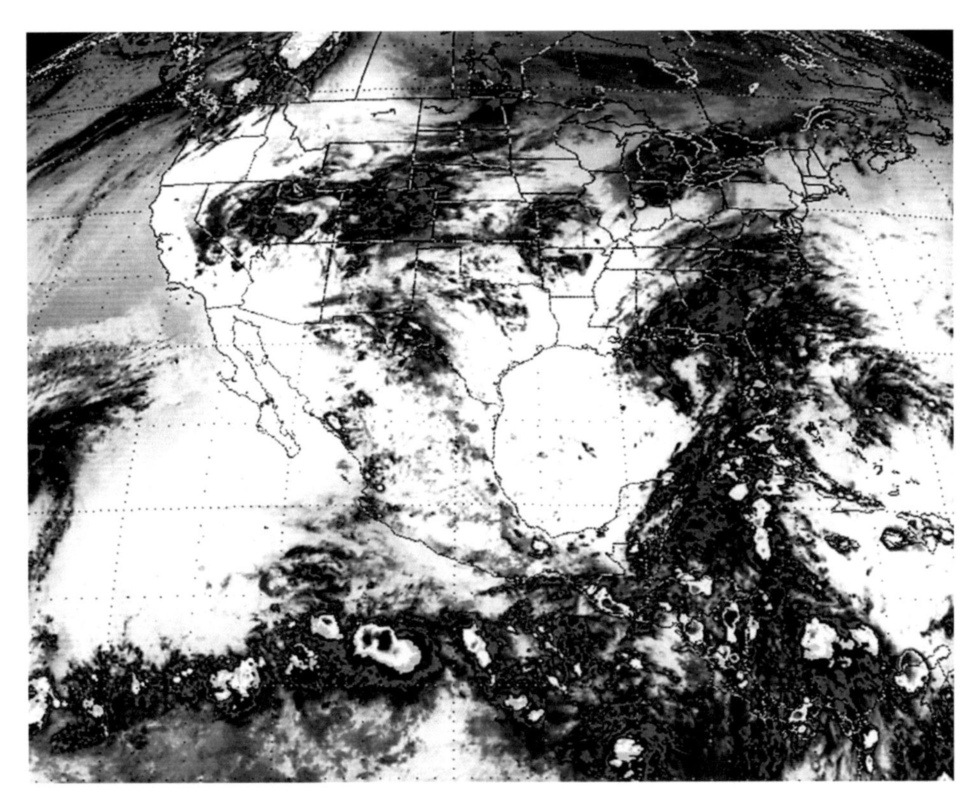

Figure 9.2
An infrared satellite image in which bright tones indicate warmer temperatures and dark tones indicate colder temperatures. Original image courtesy of Michael Mogil, NOAA/NESDIS. Post-processed by R. R. Hoffman.

"natural" scheme: Cold clouds are depicted in whiter shades and warmer ones in darker shades. But suddenly, the Earth looks like the Earth and the clouds in the atmosphere look like clouds (that one might see if viewing the scene with a naked eye; compare figure 9.3 with figure 9.2).

The Principle of Clutter

It is a generally accepted axiom in the human factors of visualization that "clutter" is to be avoided (see Moacdieh and Sarter, 2015). But in meteorology and forecasting, clutter cannot be avoided. Here, we use a standard definition of clutter: Clutter is the state in which excess items, or their representation or organization, lead to a degradation of performance at some task (Rosenholtz, Mansfield, and Jin, 2005). It is

Figure 9.3
A standard GOES infrared satellite image corresponding to the image in figure 9.2. Image courtesy of Michael Mogil, NOAA/NESDIS.

clear that for some tasks (e.g., finding a specific object on a map), high density can negatively impact performance. In fact, there are even automated systems for scaling the amount of clutter in an image (Lohrenz, Trafton, Beck, and Gendron, 2009; Rosenholtz et al., 2005; Zuschlag, 2004). However, one person's clutter is another person's critical data.

As figure 9.1 shows, the data displays that forecasters use not only portray multiple data types at multiple scales, but they have to do so. Even the simplest of weather maps is actually conceptually dense, and has to be because the meaning lies in the interrelations among the data types. Figure 9.4 (plate 12) illustrates this. It is the new "Sat-Rad" display, which is intended for public consumption. It overlays radar imagery (the green and blue colorations indicate precipitation) with infrared satellite imagery (gray tones showing cloud cover), in addition to the kinds of data shown on a

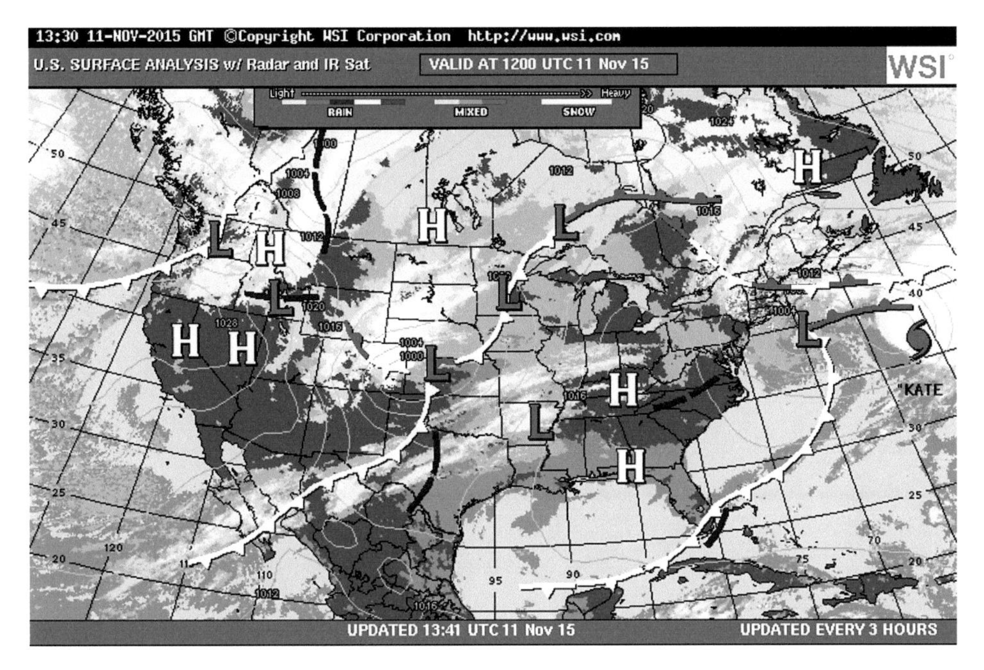

Figure 9.4
(plate 12) A "Sat-Rad" weather map produced by WSI Corporation.
Reproduced with permission from WSI Corporation [downloaded 11 November 2015, from http://www.intellicast.com/National/Surface/Mixed.aspx/].

traditional surface chart (symbols for lows, highs, fronts, hurricanes, and isobars—lines of equal pressure).

Figure 9.5 (plate 13) is a large-scale weather map of the sort used by meteorologists and forecasters. This map showcases the interrelationships among various pressure systems and air masses. To the untrained eye, it is definitely cluttered.

In both figures 9.4 (plate 12) and 9.5 (plate 13), most readers will recognize symbols of the sort used in televised weather broadcasts: H and L symbols for high- and low-pressure centers. Surface lows are regions where air is generally rising, whereas highs are regions where air is generally sinking. Blue lines depict cold fronts (relatively colder air is moving toward relatively warmer air), with the blue triangles showing the direction that the frontal boundary is moving. In contrast, red lines indicate warm fronts (relatively warmer air is moving toward relatively cooler air). Alternating red and blue warm and cold fronts are used to depict stationary fronts (two air masses experiencing little relative movement). There happen to be no occluded fronts in figures 9.4 (plate 12) and 9.5 (plate 13); these are typically shown in purple. Occluded

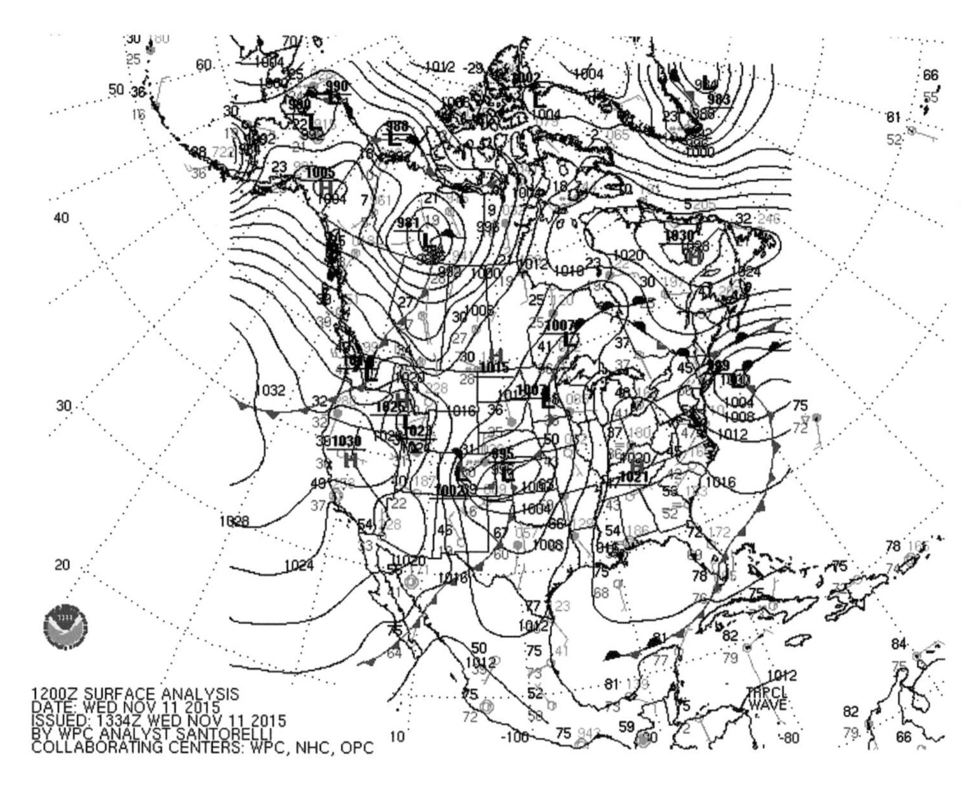

Figure 9.5
(plate 13) A "surface analysis" weather map [downloaded 10 November 2015, from http://www.wpc.ncep.noaa.gov/sfc/90fwbg.gif].

fronts indicate where one air mass overtakes another and replaces it near the surface of the Earth.

As mentioned earlier, the main reason that data displays need to be dense is that forecasters need to understand the relationships among the different types of data at a mesoscale as well as at a synoptic scale (e.g., how far a front extends). Sensemaking of atmospheric dynamics depends on integrating multiple data types so as to perceive patterns, as opposed to the cues that individual data types reveal (Mogil, 2001). Thus, "clutter" for meteorologists is not always a bad thing—rather it is a required high level of density of data.

The Principle of Color

The use of color in informational displays is one of the most heavily researched aspects of display design (for example, see Christ, 1975; Christ and Corso, 1983; Davidoff,

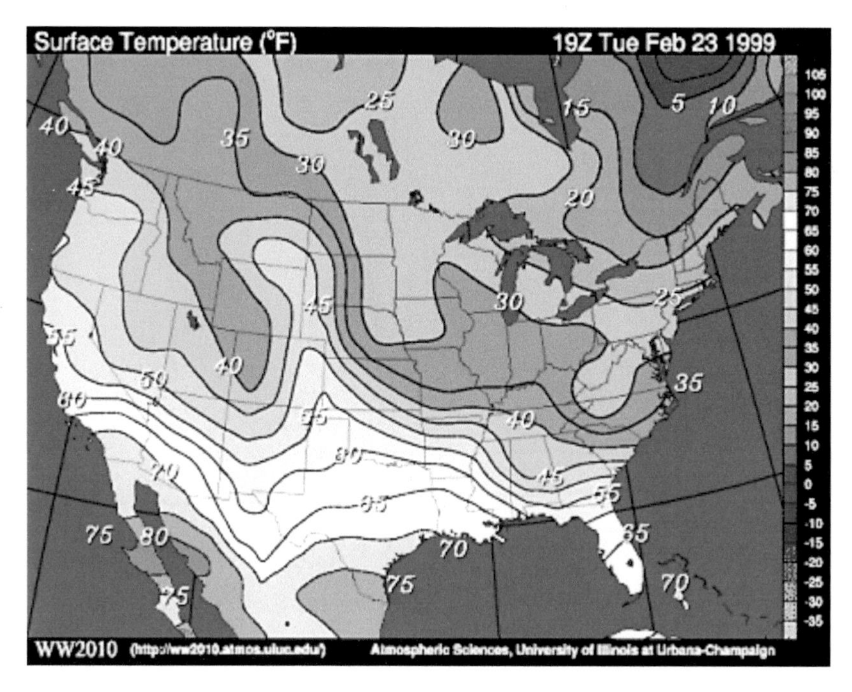

Figure 9.6
(plate 14) A surface temperature weather map. Reproduced by permission from the Department of Atmospheric Sciences, University of Illinois, Champaign–Urbana.

1987; Durett, 1987; Hoffman, Detweiler, Conway, and Lipton, 1993; Krebs and Wolf, 1979) In many displays of weather data, and remote sensing in general, it is assumed that (1) color has to be used to encode information (see Henson, 2010), (2) the colors should be highly saturated, and (3) the encoding should preserve the ordering of colors in the visible portion of the electromagnetic spectrum. The ROYGBIV "rainbow" coding scheme is the preferred or default scheme in many graphics applications (Borland and Taylor, 2007). An example is presented in figure 9.6 (plate 14). As with all of the other principles we discuss here, problems arise in the implementation.

Data displays for meteorology and forecasting do use other schemes for encoding information. Figure 9.7 (plate 15) shows a display of a 500-mb forecast map that shows winds and vorticity (i.e., spin) generated by COAMPS, the U.S. Navy's Combined Ocean/Atmosphere Mesoscale Prediction System, for a region in the Southern Hemisphere. This display uses a dual-hue color encoding and also relies on saturation shades of the two colors. Wind barbs show wind speed and direction. Isolines with numeric labels show 500-mb height values (see Hodur, 1997).

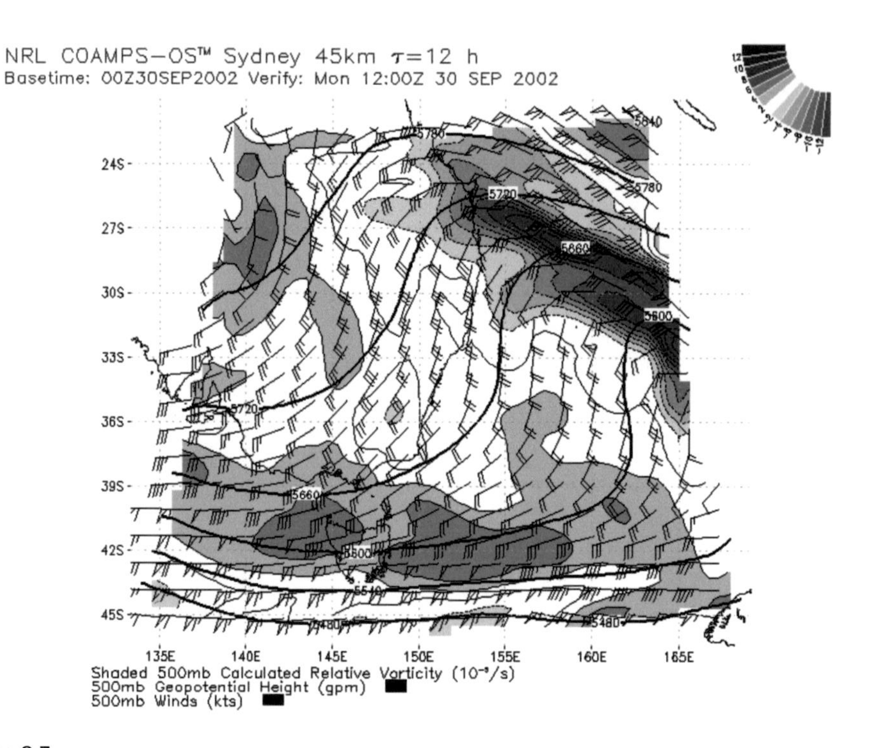

Figure 9.7

(plate 15) An example COAMPS 500-mb height/wind product [downloaded 11 March 2016, from https://cavu.nrlmry.navy.mil]. See [http://www.nrlmry.navy.mil/coamps-web/web/home]

Traditional GOES images use a brightness scale (as in figure 9.3). An enhancement of GOES images, called the MB enhancement, is more complicated than the encoding used in GOES infrared images (figure 9.2), but it actually makes the images more useful. An example is shown in figure 9.8. Decreasing cloud temperatures (i.e., increasing elevations) are depicted in shades going from white (lower, warmer clouds) to gray (mid-level cooler clouds) to dark gray (mid-level colder clouds), but then *back* to light gray (high-level cooler clouds), and up again to black (high-level cold clouds). This gray-shade paletting is quite clever. The upper gray scale in the legend at the bottom of figure 9.6 (plate 14) maps temperatures on a continuous tonal palette. The bottom gray scale is the enhancement curve. The upper right corner appears cropped but is the apparent temperature of space, coded as white (i.e., cold). Taken out of context as a tone-to-temperature mapping scale, it is somewhat mysterious to the outsider. But with practice comes the skill of being able to use the repeating ascending tone scales to perceive cloud height and thereby gain an awareness of atmospheric dynamics. Higher

Figure 9.8
An example GOES infrared image using the MB Enhancement.

cloud tops are relatively colder and more massive, thus representing the presence of greater amounts of moisture that can be associated with thunderstorms and heavy precipitation at ground levels.

In the 1980s, color was introduced in GOES imagery, as shown in figure 1.2 (plate 1). At the time, it was not entirely clear that adding color was a good thing, especially because meteorologists had learned to perceive depth using the MB enhancement. With the multi-hue palette, clouds got colored in green, tapping into a natural association with precipitation, whereas the presence of cloud cover does not necessarily mean that there is precipitation. Color indicates only the cloud height (via temperature). Because higher clouds contain more moisture, the highest clouds (tops of 40,000 feet or more) often entail the potential for precipitation, but those are colored red (see figure 1.2, plate 1), which is a color that does not naturally imply precipitation. Furthermore, the NEXRAD radar images were using green to denote actual measured precipitation (see figure 1.4, plate 3). This meant another potential source of misinterpretation, attributable to the assumption that color coding of data would *have* to use the multi-hue palette of bright, highly saturated colors (Hoffman et al., 1993; Stauffer et al., 2015).

Thus, the principles of color and naturalness are at loggerheads. As an example of the problems of naive association, if one uses red to mean hot and blue to mean cold, how do green or violet fit in? Violet would be hotter than hot, but naive associations of heat do not stretch from red to violet. If one uses green to depict anything, then people are inclined to assume that it depicts rain, when it does not necessarily mean that at all. There are also significant issues of display interaction. One display might use green to denote areas of rain, but another might use the same shade of green to depict something entirely different, such as dew point temperatures, relative humidity (weather maps), or winds moving toward a radar (in a Doppler display).

The principles of color and clutter are also at loggerheads. Many weather data fields are complex, such as arrays of wind barbs showing wind speeds and directions as a function of height in the atmosphere. The jumbles of tiny colored arrows can be nearly indiscriminable (see Hoffman et al., 1993). Interactions emerge in displays that combine data types. For instance, lightning strike data (traditionally depicted as yellow dots) might be displayed along with colored isolines showing wind speeds, resulting in more jumbles of colored graphical elements. Visualizations can be accurate but cluttered and more difficult to read (Moacdieh and Sarter, 2015). Meteorological visualizations need to display a huge amount of information, and that information needs to be constructed in layers with different information types (e.g., wind speed, humidity, temperature). Typically, only one of these layers can be colored.

The principle of color runs out of steam. Whatever data type one needs to show, the rainbow palette nearly always runs out of hues before one covers the full range of the variable to be depicted. Some remote sensing displays use dozens of colors, and geological maps use many dozens, compounded by such graphical additions as crosshatching (MacEachren, 1993). The Color Guidelines Subcommittee of the Committee on Interactive Information Processing Systems of the American Meteorological Society conducted a survey of the uses of color and the pertinent human factors considerations (Hoffman, Detweiler, Conway, and Lipton, 1993). It recommended that the color palette can be significantly expanded by utilizing desaturated hues, that is, pastels. It is interesting to note that PRAVDA data displays presented in chapter 2 utilize a palette that includes pastels (see figures 2.9 [plate 7] and 2.10 [plate 8]). PRAVDA can help meteorologists and forecasters generate advanced colorized displays. It includes a rule base and a library of color maps that together permit users to make decisions about the visualization of data without requiring them to become experts in human vision, data structures, visualization algorithms, or color theory. In other words, PRAVDA places the visualization design process in the hands of the end-user. The rule base ensures that data content is reflected in the image displays, and that perceptual artifacts are not erroneously interpreted as data features.

The principle of color is challenged in another way: how data are displayed interacts with the tasks to be performed.

The Principle of Task Dependence

The color coding can make some tasks easier and some more difficult depending on the viewer's task, which in turn determines how the legend and other graphical elements are used. J. G. Trafton and his colleagues at the U.S. Naval Research Laboratory (Breslow, Ratwani, and Trafton, 2009; Breslow, Trafton, and Ratwani, 2009) recorded U.S. Navy weather forecasters' eye movements as they examined COAMPS weather charts like that shown in figure 9.9 (plate 16). In this example, the forecasters' task was an identification task: "What is the sea level pressure at location 'X'?" The gaze pattern is indicated by the red gaze tracing in figure 9.9 and shows that in this task, the forecasters had to interpolate between the location of the "X" and the nearest numerical labels for the surface pressure values of the nearest isolines.

Figure 9.10 (plate 17) shows a similar tracing. The task was also an identification task: "What is the surface air temperature in Pittsburgh?" In this case, the color scale was the crucial thing, and forecasters had to interpolate between two similar shades of green, requiring repeated glances at the legend.

Trafton et al. found that when conducting an *identification* task, forecasters spent more time looking at the legend in displays having brightness scales than displays having multicolored scales. In the *comparison* task (e.g., Is it hotter in Pittsburgh than in Atlanta?), the forecasters would directly compare colors when the displays used a brightness scale, but they still had to rely on the legend a great deal for displays using a multi-hue scales. Multi-hue palettes are best because it is easy to do fast search of the legend, but for a comparison task, a brightness palette is best because it is easy to directly tell which target region is lighter or darker.

All of these design notions and principles—naturalness, salience, clutter, color, and task dependence—point in one direction: Perceptual learning is the key. So how do forecasters interpret displays and learn to perceive meaningful patterns? One way of addressing this is to compare how forecasters and non-forecasters interpret weather maps.

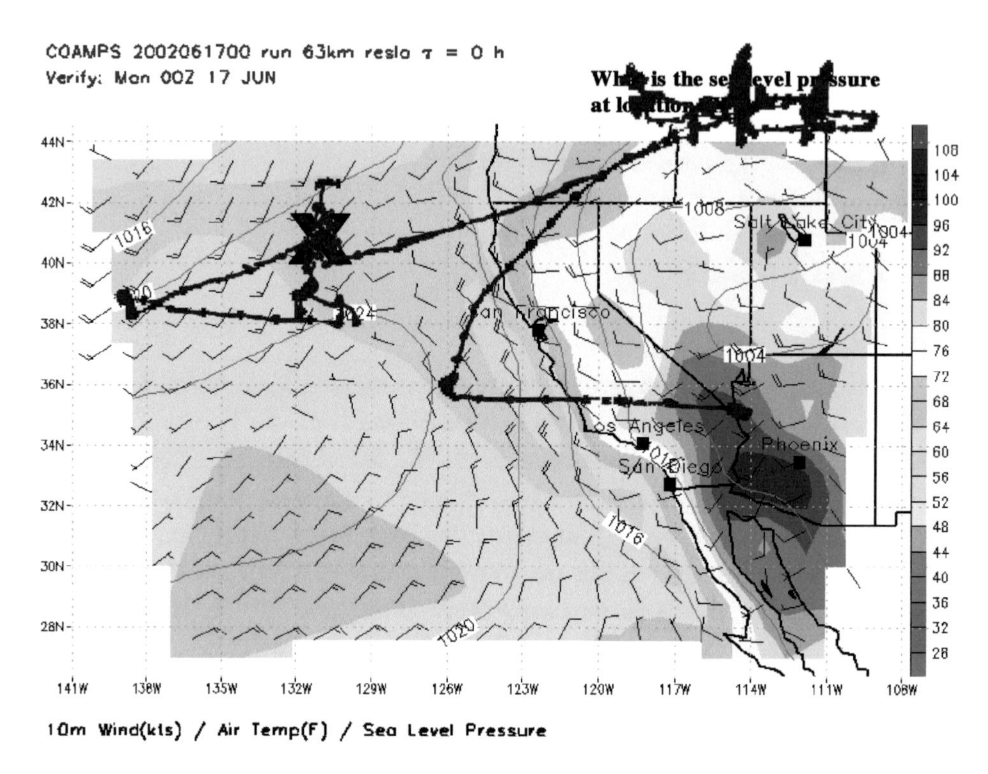

Figure 9.9
(plate 16) A COAMPS product with a re-creation of eye movement data from the experiments by Trafton et al.

Interpretation of Weather Maps by Forecasters and Non-Forecasters

Mary Hegarty of the University of California and her colleagues have conducted a number of experiments on how people interpret weather maps of different designs, and this research is especially pertinent because some of the studies involved tracking what happens as people learn to perceive the information that is presented (Canham and Hegarty, 2010; Davies et al., 2006; Hegarty, 2013; Hegarty et al., 2009, 2010, 2012, 2014; Smallman and Hegarty, 2007). Most of the participants were college students, although for comparison, some of the participants were aerographers in the U.S. Navy. Of those participants, some had begun courses of instruction at the U.S. Naval Postgraduate School. Aerographers take from between 6 and 12 months of training and then they deploy, after which they take qualifying exams to be allowed to create forecasts, but even then the forecasts that are issued by a facility are those that are approved by the forecaster-in-charge, not one of the aerographers. While deployed, aerographers

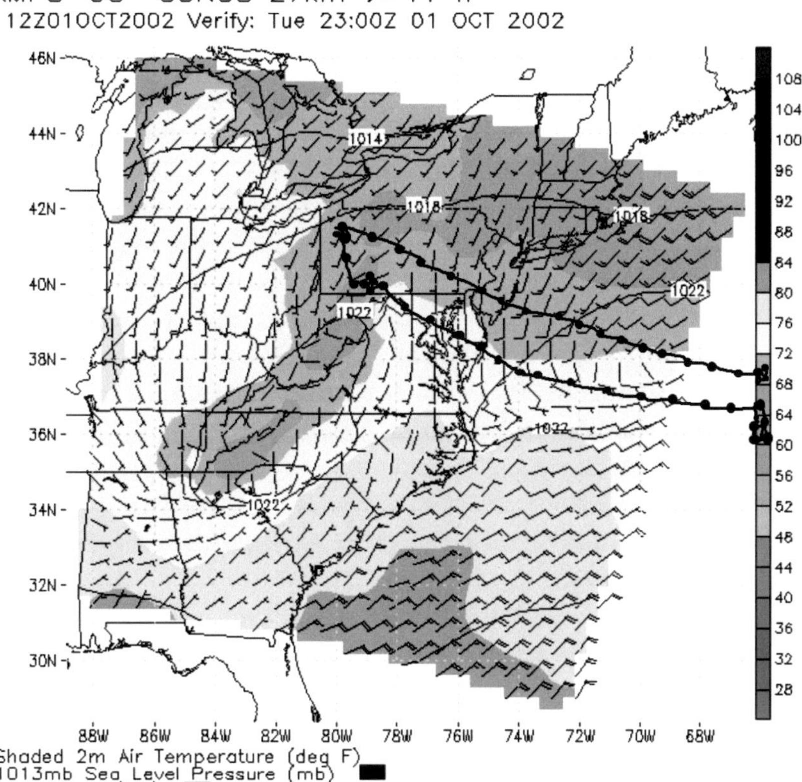

Figure 9.10
(plate 17) A COAMPS product with a re-creation of eye movement data from the experiments by Trafton et al.

Box 9.1
And Then There Is Color Vision Weakness

Roughly 10% of the U.S. population (mostly male) has a color vision weakness. Although it ranges all the way to complete color blindness, the common color vision weaknesses involve difficulty discriminating yellows and blues or discriminating reds and greens. The myriad of colors and color displays described in this chapter can wreak havoc on forecasters and the public when it comes to understanding weather imagery. Because red and green are often used together (and in overlays of different images), it is easy to see how important information can be lost (see Bolton and Blumberg, 2015; Hoffman et al., 1993). The NWS has recently taken this matter under consideration.

have to engage in many collateral duties. Thus, these aerographers could be considered to fall roughly at the journeymen level of proficiency, because to qualify as expert they would certainly have to have had more years of experience (see chapter 7).

Interviews with aerographers about their forecasting process revealed that some of them would inspect the outputs of computer models and then begin to examine data, whereas others would inspect data, form an initial conceptual model, and then begin inspecting the various computer model outputs. This reasoning approach or strategy is known to distinguish journeymen from experts (see Hoffman et al., 2000; Pliske et al., 1997, 2004; chapters 7 and 10, this volume).

The materials that Hegarty et al. used were a number of alternative forms of weather maps, all of which overlaid data fields on an outline of North America. Some charts showed isobars (not unlike figure 9.5 [plate 13] but simplified), some showed winds using wind barbs (not unlike figure 9.1), and some showed surface temperatures using swaths of vibrant colors (not unlike figure 9.6 [plate 14]). The primary task for participants was to determine the wind direction at some arbitrary point on the chart, indicated by an arrow. The researchers measured response time and response accuracy.

One of the main findings was that the vibrant temperature scale was distracting, in that the participants spent more time looking at the temperature fields than at the data that were directly pertinent to the wind detection task. However, after the students had been given some basic instruction in chart interpretation, their gaze patterns changed, showing more time looking at the task-relevant information. Looking at the results across the experiments, between 54% and 74% of the time the participants preferred the chart that most simply and directly presented the information that was relevant to the task (i.e., the isobars).

Although two-thirds of the participants preferred to use the chart that most simply and directly presented the task-relevant information (i.e., the charts showing the isobar lines), Hegarty et al. focused their discussions on the one-third of participants who preferred the more complex charts, arguing against human competence and presenting an overall dim view of forecasting expertise:

> Meteorologists were as likely as undergraduate students to prefer geographically complex (realistic) displays and more likely than undergraduates to opt for displays that added extraneous weather variables ... college students and experienced weather forecasters alike have a tendency to choose more realistic and complex maps over less realistic and simple ones, even though performance is more efficient with simple maps. (Hegarty, 2013, pp. 1, 6)

This paints a misleading picture in two respects. Response times increased by about a second for each task-irrelevant data field included in a display. It is doubtful that this magnitude of a difference of one to a few seconds would be of any practical significance in the forecasting context. Hegarty et al. highlighted the fact that even experienced forecasters showed slower response times when more task-irrelevant variables were presented in a chart, but this may be due to the very distinct likelihood that the forecasters were not merely engaging in the one task that the researchers presented to them. To the forecaster the "extraneous" variables would not be extraneous regardless of whether they were task-irrelevant. When shown a weather chart, no matter what data fields it presents, experienced forecasters would almost certainly attempt to develop a conceptual model of what was going on in the atmosphere. Thus, a longer response time for the specific wind detection task would be expected. It would also be expected that the experienced forecasters would choose to use the more complex charts *at least* about half the time (which is what the Hegarty et al. results showed). The reason is that the more complex charts provided information that supported the formation of an overall conceptual model. As an example, a map of surface temperatures (see figure 9.6 [plate 14]) is actually suggestive of the pattern of winds at higher levels of the atmosphere.

Hegarty (2013) did balance their stance, somewhat, in saying:

> ... the tasks we assigned to our expert participants were not necessarily the ones they did on a daily basis. The additional time taken to make a judgment using a more cluttered map was relatively short in our experiments, and this might be offset against the time it takes to find or create the most efficient map for a user's task. ... Furthermore, efficiency in the short term may not be the best criterion for evaluating the effectiveness of a map. For example, a weather forecaster might take more time to make a focused judgment about one variable if there are extraneous variables on a map, but at the same time, the more complex map may give him a deeper understanding that allows him to better anticipate tomorrow's weather. (p. 7)

This search for a deeper understanding was revealed in experiments by Rik Lowe of Curtin University in Australia, which we discuss next.

Forecasters' Understanding of Weather Charts: Filling in the Gaps

Australian educational psychologist Rik Lowe conducted a number of experiments (1989, 1992, 1993a, 1993b, 1994) on forecasters' understanding of weather charts, using a variety of innovative experimental tasks and procedures. The experimental materials were traditional weather charts showing the outlines of land masses, pressure contours (isobars), wind barbs depicting speed and direction, and the locations of fronts and pressure systems. Such conventional weather charts give a selective and somewhat decontextualized presentation of a meteorological situation, and they depict information that is beyond the realm of direct experience. Meteorologically, the importance of the visuospatial properties of chart symbology lies in the way symbols capture dynamic aspects of the atmosphere that reflect its nature as a gaseous fluid. Appropriate patterning of these individual graphic elements into higher levels of meteorological organization is not always readily apparent from a consideration of their literal visuospatial characteristics.

In one of Lowe's experiments, professional civilian forecasters were engaged in a sorting task, in which meteorological concepts depicted on surface analysis charts (highs, lows, frontal lines, etc.) were sorted into categories (i.e., entities, relations, qualifiers). Novices tended to sort in terms of salient visuospatial features. Forecasters sorted in terms of situation-specific relations (i.e., isobars might or might not be associated with fronts depending on whether there is an underlying causal dynamic in a particular weather situation). In a task in which people had to group map elements and explain the groupings, meteorologists' groupings involved the division of the map into a northern chunk and a southern chunk, which corresponds with the quite different meteorological influences that operate for these two major landforms of the Australian continent. Next, they would indicate large-scale patterns that corresponded to the location of zones of regional meteorological significance. In contrast, the novices' groupings divided the map into eastern and western chunks on the basis of groups of figurally similar elements that happened to be in close proximity. Such subdivision has no real meteorological foundation.

In another task, participants were shown a map with an unfilled perimeter and had to attempt to extend the markings in the map. In addition to producing significantly fewer markings in the extended region, the novices' markings derived directly from the graphics of the existing original markings by extrapolation (e.g., direct continuation of lines and curves). In contrast, the meteorologists were operating in accordance with superordinate constraints involving a variety of external relations that integrated the original map area with the wider meteorological context. The resulting patterns in markings suggested the progressive clustering of lower level weather map elements into high-level composite structures that correspond to meteorologically significant features and dynamics. Thus, a frontal line might be extrapolated beyond the border of the image, but become curved; or alternatively, a curved frontal line might straighten.

In another experiment, forecasters attempted to reproduce weather charts from memory. Their reproduction process involved two stages. They began by drawing the major meteorological features that they remembered. The second stage was then to go over the map again and fill in subsidiary elements around this framework. In contrast, the novices tended to make a single continuous pass around the map, filling in all elements they could remember in each region as they progressed, influenced primarily by the figural similarity of elements and their spatial proximity (e.g., a region of semicircular isobars). The forecasters' recall of the wind barbs along a frontal line was actually worse than that of the novices. This was because the meteorologists were concerned with the meteorologically important aspect of the cold front (the cold front line) while glossing over details such as the particular number of barbs on a frontal line. The forecasters' ability to reproduce the charts hinged on their understanding of weather dynamics—what the forecasters had perceived and remembered was the meteorological dynamics, as expressed in terms of the chart features that have meteorological significance. Hence, forecasters often did not accurately reproduce chart information that was not of meteorological significance. Furthermore, forecasters sometimes made errors that brought the chart closer to forecasters' understanding of the meteorological dynamics.

In another of Lowe's innovative tasks, forecasters were presented with a map outline of Australia gridded into 35 sectors. Weather map symbols were provided for only a few of the grid sectors. During the acquisition phase of the experiment, the participants were allowed to choose sectors of weather data to add onto the chart (using paste-on squares). This continued until 10 of the 35 sector squares had been filled. The task in the test phase was to attempt to reproduce the presented chart and then extrapolate by filling in the *unfilled* sectors, that is, attempt to produce a full weather chart. The novices showed far more variation among their choices. Often they would choose squares to add so that they could simply follow a frontal line square by square (a continuation that an expert might see as obvious). Some novices did a "checkerboard" pattern of sampling. Many of the novices were (inappropriately) biased toward filling in the squares that were more toward the center of the map. In selecting grid squares, novices tended to "travel" only short distances, in the attempt to extrapolate chart symbols directly from the given data, or they attempted to "sample" by selecting widely dispersed grid squares.

Forecasters, perhaps not surprisingly, showed consensus in the grid square selection task. They tended to initially select one of the squares in a particular region of the continent (i.e., the southeast) because of what it might tell them about the season of the given weather scenario. Next, they focused on a few localized groups of squares to identify key atmospheric dynamics. As they proceeded in the task, the forecasters also showed consensus in their inference-making, as revealed by a uniformity in the sequence of sector selections. Their choices of squares tended to form clusters—but of nonadjacent as well as adjacent squares that corresponded with their mental projection of the major weather dynamics (i.e., fronts, highs, lows, troughs).

In the reproduction-extrapolation task, novices created more idiosyncratic reproductions. Their charts tended to extend the literal features of the presented chart in meteorologically inappropriate ways (e.g., a curved isobar in the presented chart would be completed as a circle in the novices' reproduced map but would be extended along a frontal line in the expert's reproduced map). The novices appeared to construct limited mental models that were insufficiently constrained, that lacked conceptual structure, and therefore provided an ineffective basis for interpretation and extrapolation. In contrast, the forecasters were able to correctly fill in the entire map (e.g., create frontal lines that crossed a greater number of nonchosen squares). The forecasters' domain knowledge allowed them to develop a meteorologically coherent mental model of the depicted situation and hence successfully extrapolate/interpolate based on incomplete information. In the reproduction task, the forecasters were able to predict the weather well beyond the presented chart, based on their knowledge of latitudinal, global, and climactic patterns.

In another task, participants attempted to predict future weather on the basis of what was shown in a map. For the non-meteorologists, markings on the forecast maps could be largely accounted for as the results of simple graphic manipulations of the original markings (i.e., they tended to move markings *en masse* from west to east without regard to meteorological dynamics). In contrast, the meteorologists' predictions showed a much greater differentiation in the way the various markings on the map were treated. Rather than moving markings *en masse*, new markings were added. This shows that meteorologists' mental representation of a weather map extends into the surrounding atmospheric context or scale.

In general, novices construct limited mental models that are insufficiently constrained, that lack a principled hierarchical structure, and that provide an ineffective basis for interpretation or memory. A major weakness of their mental models was the apparent lack of information available regarding the dynamics of weather systems. "The expert's mental model would be of a particular meteorological situation in the real world, not merely a snapshot or image of a set of graphic elements arranged on a page" (pp. 187, 188).

The consistent pattern of findings suggested a training intervention based on animations that portrayed temporal changes that occur across a sequence of weather charts, the idea being that animations would empower novices to develop richer mental models that would include or provide necessary dynamic information. But when novices worked with the animations, Lowe (2001) got a surprise:

> ... animated material itself introduces perceptual and cognitive processing factors that may actually work against the development of a high quality mental model. ... When the information extracted by novices was examined, it was found that they were highly selective in their approach, tending to extract material that was perceptually conspicuous, rather than thematically relevant to the domain of meteorology. ... [For example] for highly mobile features such as high pressure cells, trajectory information was extracted while information about internal changes to the form of the feature tended to be lacking. There is clearly more research required to tease out the complexities involved in addressing ways to help meteorological novices become more adept at weather map interpretation. In particular, we need to know more about the ways in which they interact with both static and dynamic displays. (p. 205)

Lowe's research shows how mental models rely on that conceptual level of understanding by demonstrating something that forecasters experience all the time—the filling in of details. Forecasters do this, in part, because there are gaps in the observations. But they also do this because it is part of what it means to form a mental model of atmospheric dynamics (Roebber, 1996b, 1998). In summary, what novices perceive and remember is the literal stimulus. What the experts perceive and remember is the meaning, enabling them to not only remember more of the stimulus than the novices, but make appropriate inferences that enabled them to "remember" (i.e., correctly extrapolate) much more of what wasn't in the stimulus. This finding fits exactly to a great deal of psychological research showing that people remember gist and meaning, rather than literal cues or facts, that is, whether the acquisition task involves understanding at a deep or meaningful level (see, e.g., Bartlett, 1932; Bransford, 1979; Bransford and Franks, 1971).

From Cues to Patterns to Dynamics

We pointed out in chapter 6 that, in general, perceptual learning and perceptual skill are defining aspects of expertise. Experts learn not only what to perceive, but *how* to perceive. For example, research has shown that highly salient graphical elements such as color can distract from the interpretation task, but professional cartographers who are inspecting maps to discern particular meanings of information are *not* distracted by graphical salience (Davies et al., 2006). This is no doubt a consequence of extended practice and perceptual learning. Experts develop "scanning patterns" that are driven by their knowledge; they know what cues to look for, which Doswell (2004) calls "multiple fallible indicators." Going beyond individual cues or indicators, experts learn to perceive meaningful patterns. Although eye fixations may tell you what a person is looking at, they do not necessarily tell you what a person is apprehending (see Hoffman, 1990; Hoffman and Fiore, 2007).

As has been demonstrated in many domains of expertise where the primary tasks involve perceiving meaningful patterns, expert forecasters can indeed "see the invisible," finding meaningful patterns that escape the untrained eye (Klein and Hoffman, 1992). Mogil (2001) presents detailed yet easily understood analyses of example GOES images, showing how each image can include numerous cues and patterns to swirls, bands, and clusters of clouds suggestive of weather dynamics. Meaningful cues and patterns in GOES satellite images include:

- patterns that distinguish clouds from snow cover and fog,
- patterns that show the effects of mountains on cloud and storm formation,
- shapes of clouds (e.g., an expanse of "popcorn cumulus" clouds), which are indicative of different weather events,
- shapes of tropical cyclones, which are indicative of their winds and pressures, and
- patterns indicative of lake effect snows and snowfall gradients (see chapter 5).

Meaning is also found when cues are *absent*. Mogil (2001) gives the example of the Bermuda High, a high-pressure area that has a significant impact on storm formation in the southeastern United States. It is signaled by the *absence* of clouds. Experts can perceive many things that the novice can see but rarely pauses to look for and cannot appreciate. For instance, there are often meaningful patterns *inside* expanses of clouds:

> Altocumulus [clouds] usually have alternating bands of clouds and clear spaces. Stratocumulus are similar, but with larger individual cloud elements. Sometimes there are two patterns that seem to crisscross in these layered cumulus-type clouds. At the time I first noticed these subtle patterns, I did not appreciate the interactions taking place in these wave pattern clouds ... basically, transverse wave patterns ... air rises in one place and sinks in another ... ripple marks made in sand by moving water or wind show the presence of the same type of wave pattern. (Mogil, 2001, pp. 254–255)

As we pointed out in chapter 6, meaningful patterns are often defined by the relations among cues, their functional integrality. Returning to the example of the Bermuda High, where the ocean off the southeastern U.S. coast is largely cloud-free, indicating the suppression of cloud formation over warm water. However, there can be cloud bands off the U.S. coast, and their counter-clockwise curvature is suggestive of possible storm formation. Appendix C presents five GOES visible satellite images accompanied by synoptic analyses prepared by an NWS forecaster. In these examples, the reader will see things that novices can see and also read about things that experts can see that novices cannot (at least until they are pointed out).

As an example of integral cues in radar data, when weather forecasters look for tornadoes in a radar image, they *see* a pattern of colors and shapes, but that is not what they *perceive*. The "gate-to-gate" signature appears when winds are fast approaching the radar from one direction and fast receding in an opposing direction. In the radar, this looks like an owl's head (in its protoypical manifestation), but it rarely shows up in an individual radar scan. Signals preceding the gate-to-gate signature are often seen by examining a series of radar scans over time and height. This signature is a function of a *difference* (is a relation) in relative velocity (is a relation) of proximal (is a relation) winds, with strong (is a relation) winds moving toward the radar (is another relation) in close proximity (is yet another relation) to strong (is a relation) winds that are moving away from the radar (and, again, another relation) (Trafton and Hoffman, 2007). Clearly, this scenario involves a considerable nexus of relations. Seven of them, in fact.

The patterns that are meaningful to experts sometimes do involve individual cues, and sometimes sets of separable cues with some cues being necessary and some being sufficient. But sometimes the patterns can only be defined in terms of combinations of cues, sometimes in terms of relations among cues, and sometimes "featureless family resemblances" where cues are neither necessary nor sufficient when considered individually. For example, the patterns known as El Niño and La Niña were first discernable in integrations of satellite image data with sea surface temperature data from buoys and ship reports (Mogil, 2001). It is thus of no surprise that various "overlays" are used in forecasting. The Sat-Rad image in figure 9.3 is an example, but figures 1.5 (plate 4) and 1.6 are also integrations of multiple data types. Meteorologists and weather forecasters do not just like to see the overlays, they *need* them because they have to discern patterns across multiple data types (Mogil, 2001). A map showing "significant weather events" (e.g., storms) may have overlaid on it the data from the national lightning detection network, an image from the GOES satellite may have overlaid on it the data from the NEXRAD radar, and so on.

In addition, research on expertise in diverse domains has shown conclusively that experts do not just perceive cues and patterns, they perceive *processes and dynamics*. The expert firefighter can tell the location and cause of a fire by the *movement* of the flame and smoke, and the expert bird watcher can identify a species even when all there is to see is a fleeting shadow of movement in flight. Many meaningful patterns exist *only* over time.

When expert weather forecasters look for tornadoes in a radar image, they *see* a pattern of colors and shapes, but that is not what they *perceive*. Mogil (2001) highlights the example of the "dry line." On one side is an air mass of relatively dry, relatively hot air, and on the other side is an air mass that is relatively humid and relatively warm. Figure 9.11 is an example of what this looks like in a single visible GOES image, just at the onset of storm formation. Dry lines are common in the U.S. Great Plains (Texas–Oklahoma) region. A dry line is often "invisible," appearing at first only as a subtle difference in gray tone relative to the land, something the untrained eye is unlikely to discern. As the humid, warm air mass migrates toward the north and east, winds can blow up dust, and the line becomes visible in satellite images and can be tracked, but again it is hard for the untrained eye to discern. Then lines of small cumulus clouds develop, and these define the line more clearly (as in figure 9.11). Thunderstorms develop along the line, and in many instances clouds seem to explode from out of nowhere, but the trained forecaster can apprehend where, and when the storms will emerge. A dramatic animation of such storm formation can be seen at [http://cimss.ssec.wisc.edu/goes/blog/archives/15638].

It is because of the expert's understanding of the underlying causal dynamics that the expert can perceive patterns that the novice cannot. Fronts do not appear on satellite images, but the expert

Figure 9.11
A dry line and associated storm formation. Image courtesy of the Space Science and Engineering Center University of Wisconsin–Madison.

can tell you where the fronts are. High and low pressure systems, upper-air disturbances, the list goes on and on—these are things that can be *perceived* in images but cannot be *seen* in the images. (Mogil, 2001, p. 253)

Taking this yet another step, sometimes the cue configurations only exist *across* multiple data types (Mogil, 2001). So far we have talked about how experts perceive patterns in such things as GOES images and radar images. But the patterns that experts perceive sometimes do not exist in individual data types. Indeed, the really critical information is often "transmodal." For instance, the radar images are not the only thing guiding sensemaking activity and shaping the forecaster's formation of a mental model. A great many other data types are involved, such as satellite images, computer model outputs, wind fields, pressure data, and so on. For example, indications of emerging severe weather might lie in a combination of:

- Satellite image loops. These show at a spacetime scale on the order of continent/week the movement and interaction of air masses through visualization of cloud and movement of mid- and upper level moisture. The images can reveal the proximity of various cloud masses, movement of vorticity maxima that indicate lifting, cloud top temperatures, jet stream location, moisture content, and other weather variables.
- Wind fields as a function of height in the atmosphere. These show at a spacetime scale on the order of states/days the localized regions of jet stream winds, directional wind shear, and divergence-convergence regions.
- Surface observations. These show at a spacetime scale on the order of regions/hours the places where ground-based weather variables such as temperature, moisture, and pressure are changing, possibly helping to localize where storm formation is most likely to occur.

The "Aha!" moment might come when viewing radar, but the mind had been prepared in the sense that a mental model was formed on an integration of meaningful patterns that only exist across data types. Likewise, the expert terrain analyst makes determinations when viewing aerial photos but has engaged in systematic analysis of other data, such as maps.

As an example, the following cue configuration suggests that heavy precipitation would be likely to the northwest and west of a low pressure center:

- Computer models suggest the deepening upper level trough,
- The deepening coincides with the northern edge of a mid-latitude low, as shown in a GOES satellite image,
- There is a dry slot present to SE-SW of the low center, and

• Divergence is clearly evident at both the leading and trailing edges of the dry slot. Water vapor imagery suggests a strong inflow from over the eastern Gulf of Mexico ahead of the dry slot.

Countless cue configurations of this type could be gleaned from the literature on meteorology and elicited from the individuals who possess local expertise. Indeed, configurations of this kind form the core of some of the weather forecasting expert systems (discussed in chapter 11).

Weather data displays such as that shown in figure 9.5 (plate 13) are referred to as surface charts because they indicate the central low and high pressures as measured at the surface of the Earth. The actual centers of the air circulation for the lows (counterclockwise in the Northern Hemisphere) and highs (clockwise in the Northern Hemisphere) do not necessarily coincide with the low and high pressures measured at the surface. In addition, the fronts represent boundaries between air masses as projected down to the surface. As one moves upward in the atmosphere, frontal boundaries slant and curve in one direction or another and take on various shapes depending on the nature of their interactions, such as when one air mass overrides another, sometimes through the entire depth of the troposphere or about 10 kilometers. The general shape of the air mass boundaries can be determined from surface measurements (temperature, dew point, wind speed and direction, pressure and changes in these variables), which is a major reason that the notion of surface fronts was seminal to the emergence of the science of meteorology (Godske et al., 1957), although it should be noted that the researchers who developed the concept of the front did not have upper air data. In the case of stationary fronts and other types of air mass boundaries, two air masses can be abutting one another but only weakly interacting. They can be moving in opposing directions with no interaction at upper levels. In such cases, the frontal boundary can be projected upward through the atmosphere, to some extent, although it is usually somewhat curved and slanted. But in general, as one moves upward in the atmosphere, air masses can undercut one another, override each other, or neither depending on factors such as pressure differences, relative humidity, temperature, and wind. To make this imaginable, an analogy might be drawn to the blobs of hot wax that move around inside a "lava lamp." Fronts basically represent a transition zone between air masses and usually have above-surface projections. Indeed, air masses are also hypotheticals, to a large degree, even more so than fronts. An air mass is supposed to be homogenous in temperature, moisture, and other variables, but they are rarely close to that ideal. Some transitions exist within an overall uniform pattern.

Meteorologists and forecasters describe air mass shapes using such terms as "dome," "trough," and "ridge." These are not represented in the map symbology, but the trained forecaster can *perceive* them in the map isobar patterns. Isobar lines envelop regions with the same air pressure. In an area where the isobars are densely packed, as can be seen associated with the low-pressure center just north of Alberta in figure 9.5, one can expect to see strong winds, moving roughly parallel to the isobars in a counter-clockwise sense. Conversely, the lack of tight isobaric spacing (see high pressure over Kentucky) suggests lighter winds.

> In weather maps these [air masses and air mass interactions] are indicated not by isolated graphical features but rather by patterning. ... Minor local convolutions that are echoed across a series of adjacent isobars indicate the presence of a meteorologically significant feature. However, this subtle patterning of isobars can be obscured to a large extent by their visually distracting context, and so these features are likely to be overlooked unless given special attention. (Lowe, 2001, p. 189)

Perceptual Operations during the Forecasting Process

"What all forecasters do is image processing" (Dyer, 1987, p. 23). The deepening of understanding and the perception of dynamics are in full play when forecasters are engaged in the process of creating a forecast. Researchers have observed how expert forecasters create weather briefings and asking them questions about what data sources they are examining and why. One study found that forecasters refer to an average of three different data types/displays per minute (Hoffman, 1991). Another found that U.S. Navy forecasters refer to an average of eight displays of different data types during the creation of a forecast (Hegarty, Smallman, and Stull, 2012).

Trafton et al. (2000) observed U.S. Navy weather forecasters as they created a briefing and talked out loud as they created their briefing. The forecasters examined two data types per minute while trying to come to their basic understanding of the weather. Surprisingly, forecasters extracted primarily qualitative information from the visualizations: "the wind is fast" or "this low is going to make the temperature colder" rather than "the wind is 28 knots" or "the temperature will be 42 degrees." Interestingly, the quantitative information was explicitly available on the visualizations, so it was not just that the forecasters did not have access to it. Instead, forecasters seemed to prefer to extract and reason with the qualitative information more than the quantitative information. However, when the forecasters actually made a forecast, they generated quantitative information: They predicted that the temperature would be 72 degrees, that there would be four inches of rain, or that the wind speed would be 16 knots. Thus, forecasters extracted qualitative information from the visualizations and integrated that information into a coherent whole along with other data, their knowledge and expectations of local weather patterns, and what they knew about meteorology. From their qualitative representation, they generated numeric predictions.

Trafton et al. (2000) found that when forecasters look at data types and displays, by far the most frequent mental operation is comparison, either of two different data types or a comparison of their conceptual model to a data display. In addition, each data type that is compared to others is compared to as many as four others. This shows that comparisons are frequent and extremely important to the forecasting process. Forecasters compare computer model outputs to surface charts of observational data. They compare computer model outputs to other computer model outputs, they compare radar to satellite images, they compare charts (of various kinds) to other charts representing data across time, and so forth.

What sorts of mental operations do forecasters perform when comparing different data visualizations? The integration of multiple data types involves performing a number of different kinds of mental operations on a mental image (Bogacz and Trafton, 2002, 2005; Hegarty, 1992; Kosslyn, Sukel, and Bly, 1999; Trafton, Trickett, and Mintz, 2005):

- combining different perspectives,
- modifying a mental image by adding or deleting features,
- mentally moving or rotating an object (air mass, front, etc.), and
- mentally animating static images and projecting to the future.

Many data types are available as animations, satellite images and radar especially (see http://www.goes.noaa.gov). Most of the computer model outputs can be viewed as animations. Psychological research on diagrammatic reasoning shows that animations can have more explanatory value than a series of static images (Pane, Corbett, and John, 1996; Tversky and Morrison, 2002; see also Byrne, Catrambone, and Stasko, 1999; Mayer and Anderson, 1991; Rieber, Boyce, and Assad, 1990). Forecasters are especially focused on the dynamics of the weather—approximately one-third of their utterances in forecast discussions have a dynamic component (Bogacz and Trafton, 2002). We would therefore expect forecasters to rely on the animated data types. They sometimes do, especially, for example, in inspecting NEXRAD radar in severe weather situations in which warnings might have to be issued (i.e., the search for the emerging and dynamic signature indicating tornado formation). But in more routine forecasting situations, forecasters may not all prefer to view animated data. One study at a U.S. Navy forecasting facility found that forecasters did not rely much on animations, even though they talked about the dynamics they perceived. They preferred to look at sets of static images and then mentally animate their mental models to explore hypotheses about what the weather is going to do (Bogacz and Trafton, 2005). A likely explanation is that the forecasters prefer static images because they need to inspect details that are better determined by viewing static images. The experts apprehend the dynamics as part of developing their mental model, and they prefer to do it that way.

Conclusions

In his discussion of the issue of whether computer models would replace human forecasters, senior meteorologist Michael McIntyre (1988, 1994, 1999) of the United Kingdom came down squarely on the side of the human. His most vigorous reason for insisting on the necessary and continuing role of the human forecaster was human visual perception:

> that most powerful of data interfaces between computers and humans. For clear biological reasons, connected with the survival of the species, the human visual system has, for instance, a "four-dimensional intelligence" that even in terms of raw computing power still dwarfs, by many orders of magnitude, the power of today's largest electronic supercomputers. (McIntyre, 1999, p. 338)

While emphasizing the biological aspect of perception ("the eye-brain system"), McIntyre was drawing a bridge between the ability to apprehend meaningful patterns in data and using that understanding in the formation of conceptual or mental models, "the most remarkable cognitive skill of all" (p. 338).

The patterns that are meaningful to experts sometimes involve:

- individual cues,
- sets of separable cues with some cues being necessary and some being sufficient,
- patterns that can be defined in terms of combinations of cues,
- patterns defined in terms of relations among cues,
- featureless family resemblances where cues are neither necessary nor sufficient when considered individually,
- meaning that resides in the relations among cues that are integral cue configurations, and
- dynamic information defined over sets of integral cues that are transmodal (they exist over different data types).

Forecasting depends on the forecaster's ability to envision a dynamic mental model of the weather. Evidence for mental model building is clearest when weather forecasters use data visualizations to make their forecasts; imagistic and visual reasoning is a critical part of the forecasting process. This imagery is not merely "mental pictures":

it expresses qualitative/conceptual information (e.g., images of the internal structure and dynamics of storm clouds) and numerical/conceptual information (e.g., data about winds, air pressure changes, etc.). Additionally, forecasters use their mental model to generate quantitative forecasts. Indeed, forecasters of a previous generation, who grew up making weather charts by hand, have said that their conceptual models take the form of mental images that manifest as dynamic weather maps (Hoffman, Coffey, and Ford, 2000). Forecasters inspect and integrate a great number of complex visualizations and data sources. Estimates are in the range of eight or more different data type displays for forecasts in non-severe situations. Forecasters mentally manipulate visualizations to form a conceptual model of the atmosphere. Because the pertinent information is not displayed on any one visualization, forecasters must integrate that information into a coherent whole to make a prediction about what the future weather will be. Forecasters examine animated data types, but they also examine many static images and mentally animate them to make predictions.

As we have mentioned, forecasters refer to something on the order of eight or more different displays/data types as they generate their forecasts. This depends of course on the specific forecasting task and the weather situation, as will be elaborated in chapter 10.

10 What Does Research on Forecaster Reasoning Tell Us?

If meteorologists want to make [forecasting] more efficient, then they must become as concerned with the nature of [cognitive processes] as they are with the nature of the physical processes in the atmosphere (Murphy and Winkler, 1971, p. 239).

This chapter reviews research that highlights the strategies forecasters use to make sense of the weather. The research involves various manipulations of the tasks and materials, creating a "window on the mind." For example, in one of the earliest studies on this topic, Australian meteorologist Gordon Allen (1982) presented data representing a variety of weather situations to eight forecasters in the Australian Bureau of Meteorology. He then solicited their Probability of Precipitation (PoP) judgments for Melbourne during a specified forecast interval and, for those judgments, a corresponding rating of their confidence level. In one condition, the forecasters were presented only one type of data (satellite image, surface isobar chart, upper wind chart, or computer-generated rain forecast map). In another condition, all the data types were presented, but successively rather than all at once. Results indicated that the effect of limiting the information was to make forecasters overconfident; that is, they expressed high confidence but tended to be less accurate. For the second condition, there was a suggestion of an anchoring effect, in that the initial inspection of a certain data type served to bias subsequent judgments. Specifically, being presented with data that strongly suggested rain tended to evoke precipitation forecasts even though subsequently viewed data were less favorable toward precipitation. There appeared little improvement in judgment accuracy as more data were provided, suggesting that these forecasters did not optimally integrate the available data. That is, for situations in which one of the data types contained evidence that conflicted with that provided in the other data types, the forecasters tended to underutilize the information contained in the later map sets. Allen also asked the forecasters to evaluate the usefulness of each of the data types and to generate synoptic evaluations. "The verbal accounts of ... their judgmental processes rarely contain statistical frequencies and Bayesian algebra. They usually contain causal arguments" (G. Allen, 1982, p. 1). So if, in generating POP forecasts, forecasters do not mentally juggle probability numbers or make calculations based on the axioms and theorems of probability, what do they do?

Maja-Lisa Perby (1989) conducted a cognitive analysis of forecasting operations at the Malmö airport in Sturup, Sweden, over a period of several years, using methods of observation and interviewing. She noted that the forecasters spent most of their time with pen in hand at traditional work methods intended to "assimilate information" (drawing isobars, areas of fog, etc.). These map preparation activities were all in service of the formation of an "inner picture" that integrates background knowledge of principles and cause--effect relations, complications in the use of the principles when applied to specific dynamic circumstances, and some degree of "aesthetic consciousness" or a feeling of having achieved a coherent explanation.

> [This] inner weather picture gradually builds up in the minds of meteorologists and leads to understanding and the development of skill. ... To make a forecast is not a distinct step in the work of a meteorologist: forecasts are made continuously, as an integrated part of elaborating an inner weather picture. (Maja-Lisa Perby, 1989, pp. 39, 46)

One forecaster referred to "mental pictures":

> My first picture is quite abstract—I use a theoretical model of the strata of the atmosphere. During the work shift the abstractions disappear more and more. The picture is filled out by the weather as it actually is. (Maja-Lisa Perby, 1989, p. 46)

This "mental picture" (whether conceptual/abstract or imaginal/pictorial) is an important aspect of forecaster reasoning. As we explained in chapter 4, this has been quite familiar to the meteorology community for decades (see Chisholm et al., 1983; Doswell, 2004; Doswell and Maddox, 1986; Morss et al., 2015) because they have to distinguish forecaster understanding ("conceptual models") from partial glimpses of atmospheric processes that computer models and data provide. Forecasters do not reason solely, or even primarily, with numbers as they try to understand and predict the weather. In this chapter, we discuss this concept in more detail, along with other descriptions of forecaster reasoning strategies.

Mental Models

Cognitive psychologists, unlike behaviorally inclined psychologists, use the term "mental model" to refer to people's conceptual/imaginal understandings (Byrne, 2002, also see chapter 9). Mental models have been demonstrated in studies of reasoning, deduction, and problem solving (Johnson-Laird and Byrne, 1991). Donald Norman (1983, 1988) used the term "mental model" to explore the workings of devices such as computers, ATMs, and so on. Gentner and Stevens (1983) described a mental model as a representation of some domain or situation that supports understanding, reasoning, and prediction. Klein and Hoffman (2008) and Trafton (2004) defined mental models in this way:

- Mental models emerge in the interplay of perception, comprehension, and organized knowledge. Mental models are a phenomenon or "presentation" to consciousness, that is, they are "accessible" (retrievable from memory).
- Mental models combine imagery and propositions, expressing spatial, dynamic, qualitative, and quantitative concepts.
- Mental models are relatively enduring (i.e., they are not strictly static, not strictly "structures," and not strictly "stored" things). They interact with new information in a constructive process of dynamic problem representation, resulting in the phenomenon (comprehension or image) that is presented to consciousness.
- The ways in which a mental model emerges are shaped by the regularities, laws, principles, and dynamics that are known or believed to govern the "something in the world" that is being represented.
- Mental models are representations ("mappings") of something in the world (often some sort of dynamical system). Hence, mental models are not snapshots but are dynamic. Mental models often have a strong imagery component.
- A mental model is primarily a dynamic spatiotemporal representation, a conceptual simulation. However, the information is primarily represented qualitatively, although it sometimes includes metrical or quantitative information.
- Mental models allow the thinker to anticipate the future—"runnable" events that can be mentally inspected. Mental models include propositional reasoning in order to support inference-making and hypothesis testing (Altmann and Trafton, 2002; Anderson, Conrad, and Corbett, 1989). A mental model is a mix of images and propositions because it must be able to connect to our propositional and inferential reasoning mechanisms (Altmann and Trafton, 2002; Anderson, Conrad, and Corbett, 1989). Mental models often require a great deal of thought and mental effort to create.
- Mental models can be inferred from empirical data. People may not be able to tell you "everything" about their mental models, and they may not be able to tell it well. But with adequate scaffolding in a cognitive task analysis or knowledge elicitation procedure, people can tell you about their knowledge, or the knowledge can be manifested in the cognitive work in which people engage. In other words, mental models are

partially "declarative"—people can talk about their mental models. Mental models are almost always described in terms of images, concepts, and their relations, expressing states of affairs and dynamics.

Stevens and Collins (1978) assumed that the understanding of the weather, as with all complex phenomena, hinges on the formation of mental models. The researchers asked primary school teachers to comment on whether they could tell what a student knew or did not know, and how they as teachers went about correcting student misconceptions. They also asked students questions about phenomena such as the causes of heavy rainfall (e.g., "How is the moisture content of the air related to heavy rainfall?"). On the basis of the data, Stevens and Collins identified numerous "reasoning bugs" (simplifications and distortions) that arise in student understanding. For example, many students fall prey to the "cooling by contact bug," which was illustrated in such statements as "cold air masses cool warm air masses when they collide." This type of cooling is not critical in causing heavy rainfall. (See box 10.1.)

Many of the errors in student understanding could be related to their adoption of one or more rudimentary metaphors, through which the weather is understood and on the basis of which their mental model of weather is constructed. We know that metaphors play a key role in scientific thinking and contribute to the formation of conceptual models (Hoffman, 1980). But metaphors can lead to simplistic understandings. In the case of atmospheric dynamics, metaphors include regarding the atmosphere as a mass of billiard balls or as a process like molecular attraction. Such metaphors differ in the degree to which they adequately explain weather phenomena (i.e., why cold fronts usually herald the arrival of dry weather). What stemmed from the Stevens and Collins (1978) analysis was a set of rules for productive teacher–student dialogues (e.g., "If a student gives as an explanation a factor that is not an immediate cause in a causal chain, ask the student to describe the intermediate steps").

Box 10.1
What Causes Heavy Rainfall?

Heavy rainfall is related to the rising of masses of air of differing relative temperatures, pressures, and moisture content. Fronts, storm systems, terrain, and daily solar heating can all cause air to rise. Air cools as it rises, however, owing to expansion of the volume of air, without any transfer of heat. The cooling causes air to reach its dew point or condensation temperature, allowing clouds and precipitation to develop. The condensation process also allows for the release of latent heat. This further warms the air, which enhances the rising process. Air also cools through evaporation, as when rain falls from above into a dry layer.

The work of Collins and Stevens (1978) serves as a clear case of the utility of adopting the mental model approach in the analysis of forecaster reasoning:

1. the progression of learning can be conceived of as involving such operations as adding, differentiating, and replacing components or factors to a mental model; and
2. there is value in encouraging students to use multiple conceptual models (i.e., based on metaphors) and functional models (i.e., reliant on principles of atmospheric dynamics) (see Feltovich et al., 1997).

Despite its utility as a mental phenomenon, the mental model concept is a target for criticism and debate because it "wears its mentalism on its sleeve." The debate has occurred primarily in the United States, where we still feel a lingering hangover from the school of thought called behaviorism. The argument is that mental models cannot be observed objectively or directly in behavior, and so like all phenomena of mind, they are of dubious scientific status (Rouse and Morris, 1986). For Europeans, less impacted by behaviorism, the mental model concept emerged gracefully. But even some cognitive psychologists have wondered whether the imagery or pictorial aspect of mental models is merely "epiphenomenal" to some underlying cognitive calculus (e.g., Pylyshyn, 1973), although most psychologists now believe that mental imagery is functional (Kosslyn, 1994).

With regard to expertise, broadly, we recognize that experts' mental models are incomplete or even include some mistakes. This does not mean that expertise is to be discounted entirely, that mental models play no role in how people engage in sensemaking, or that the understanding of expert mental models is not crucial to a science of expertise and its applications:

> Our view is that the "mental model" designation/metaphor is invoked out of recognition of a phenomenon, not to be brushed aside for being subjective, or to be avoided because research on mental models fails to qualify as "hard" science. To assert that the mental models notion has no explanatory value is, to us, to merely choose to ignore the obvious. Those who do that are walking a different road than the one we walk. Our challenge is to empirically explore, understand, and explain things that phenomenology dishes up—and methodology makes difficult—not explain them away at some altar of methodolatry. (Hoffman and Klein, 2008, p. 59)

Our definition of mental model is consistent with definitions offered by many cognitive psychologists. Mental model formation and its role in scientific reasoning have been documented in diverse domains, ranging from mechanics to fluid dynamics to astronomy (e.g., Trickett, Trafton, and Schunn, 2009). It also seems to capture meteorologsts' notion of the "conceptual models," especially that the mental model has a significant qualitative aspect manifested as imagery (Pliske et al., 1997; Pliske, Crandall, and Klein, 2004). Senior forecaster Charles Doswell (2004) said:

> Weather forecasting has proceeded along a path that began with entirely intuitive cognition. ... A conceptual model, such as the Norwegian Cyclone Model [see Figure 4.1] is a nonmathematical form of abstraction—a so-called mental model. It is a prototypical example of a conceptual model in meteorology. (p. 1124)

The research literature on mental imagery shows that there are considerable individual differences in the extent to which people report imaginal experiences during reasoning (see Roe, 1951; Walkup, 1965). This applies to the reasoning of scientists, with some claiming they experience little in the way of imagery. Individual differences in imagery have not been empirically studied in the case of weather forecasters, but many forecasters do report that their conceptual model or understanding of the weather has a significant qualitative aspect manifested as imagery. In addition to the imagery aspect, forecasters' mental models are driven by an understanding of the principles of atmospheric dynamics. Some forecasters, those who grew up on the traditional methodology of hand chart work, report that their mental images are like animated charts populated with graphic elements such as frontal lines, pressure isolines, and wind barbs. Others report visualization of air masses and air mass interactions (fronts, jet streams, etc.).

> ... forecasters generally tend to follow [these] processes: acquires a mental three dimensional picture of the atmosphere and its properties ... [then] project that picture forward in time and incorporate dynamic changes that may occur ... [then] from this mental picture distill the weather parameters of interest. (Targett, 1994, p. 48)

An experienced weather forecaster is able to create a mental model of the atmospheric dynamics and project the likely future weather (Ballas, 2007; Doswell, 2004; Hahn, Rall, and Klinger 2002; Hoffman, Coffey, and Ford, 2000; Lowe, 1994; Morss et al., 2015; Perby, 1989; Trafton et al., 2000; Trickett, Trafton, and Schunn, 2009). As the research on forecaster perception shows (see chapter 9), expert forecasters do more than simply read off information from the charts or computer model outputs—they go through a process of apprehension (Hoffman, 1991; Lowe, 2000; Trafton et al., 2000). Forecasters form an initial mental model for an event before arriving at work (e.g., by viewing maps and charts on the Internet). They study the "big picture" of atmospheric dynamics, primarily by perceiving qualitative information (e.g., "The jet stream is further south today" or "The low seems to be developing along the Louisiana coast"). They project their model in time, and watch for signals indicating that an event might be unfolding or becoming unusual. They continuously refine their mental model, and they rely heavily on that model to generate a forecast, including numeric estimations (e.g., "The wind speed over San Diego at 500 mb will be 45 knots"). They use their mental model as their primary source of *information* (rather than copying *data* directly from the best or a favorite visualization or computer model output). Following a weather event, forecasters get feedback on how the weather impacted people so they can relate those effects to the data they had and to their mental model as the event unfolded (Hahn, Rall, and Klinger, 2002).

Some psychologists have argued that mental models, including those of experts, are incomplete and biased by misconceptions and false beliefs. For example, ecological psychologist Kim Vicente (1999, 2000) stated that in human factors research on systems design, there is a "widely held belief that it is always important to identify a worker's mental model of the work domain and then design a human-computer interface to be compatible with that model" (1999, p. 49). Thus, it is argued that mental models provide a weak foundation for scientific understanding of cognition (including expert cognition). However, incompleteness is an inevitable feature of human knowledge. At any given time, a person's understanding of a complex phenomenon can be seen as "reductive," that is, a simplification (Feltovich et al., 2004; see chapter 6).

In the case of weather forecasting, forecasters argue that their conceptual models are actually *more* complete than the data of the computer models outputs because the forecaster will mentally fill in phenomena or dynamics to form a complete and integrated understanding. Forecasting happens on many spacetime scales. For example, forecasters may simplify some things on the synoptic scale out of necessity: there is a lot going on in the weather at the scale of continents and weeks. But on the smaller spacetime scale, such as for a local downburst, their mental models are more complete than the data, and the forecaster scrutinizes data for hints that confirm the conceptual model. The data and computer model outputs do not provide much information about the processes involved in downbursts, hail formation, supercell formation, interactions of storms with boundaries that increase vorticity, and other events at the mesoscale and microscale. The forecasters' models have to be more complete, and they must fill in some blanks. Computer models routinely do well in their predictions at the synoptic scale (see chapter 12), but the algorithms cannot yet compete with humans on the smaller scales at which the forecaster is most concerned for localized warnings. Somewhere between these scales, it flips over to the human being best.

We need to understand forecaster reasoning at multiple spacetime scales because it differs at multiple scales. In other words, how forecasters reason depends on what it is that they are trying to understand. The conceptual models we present in this chapter will illustrate this point.

The formation of conceptual or mental models is not the only core element to forecaster reasoning. The literatures we have reviewed here and discussions in the previous chapters of this book include descriptions of other aspects and strategies of forecasting process by experienced forecasters who have looked at forecasting from a decision-making perspective. We can discern a consensus about how expert forecasters think.

Modeling Forecaster Reasoning

It is necessary to develop specific models of reasoning that are particular to weather problems (e.g., severe storms, tornadoes, etc.), specific to situations (e.g., regions, climates, etc.), and specific to the technologies available (Ballas, 2007). There is no doubt that processes involved in issuing general forecasts and processes involved in issuing warnings are different (Klinger, Hahn, and Rall, 2007), and thus we must adduce somewhat different models for these distinct main tasks. We pursue this notion in this chapter.

As we explained in chapter 4, there is a general consensus among forecasters in how experienced forecasters go about their general forecasting task, and their characterizations fit with descriptions of expert reasoning and problem solving (see chapter 6). In Leonard Snellman's (1982, 1991) "forecasting funnel," the forecaster starts with an attempt to get the "big picture" of what is happening in the hemisphere in terms of major forces and dynamics and then inspects data that focus the understanding down to the continental scale (synoptic scale), at which time one asks, "What will be the forecasting problem of the day? (see figure 4.5). Charles Doswell et al. (1996) proposed an "ingredients" approach, in which the forecaster looks at the ingredients necessary for certain types of weather events and assesses whether those ingredients are present in sufficient quantity or balance to cause those types of weather. Lance Bosart, a Senior meteorologist at the University of Albany, suggested that forecasting progresses through a series of questions: (1) What happened? (2) Why did it happen? (3) What is happening? And (4) Why is it happening? From the literature on the psychology of expertise, we get the sensemaking and flexecution models (see chapter 6). Both of these models can be thought of as generic.

A study of U.S. Navy forecasters by Hoffman, Ford, and Coffey (2000b) illustrates the modeling process. Based on the literature about the psychology of expertise, Hoffman et al. began with what they referred to as the "Base Model of Expertise." The model integrated the pioneering ideas from Karl Duncker (1945). According to Duncker's model of reasoning, problem solving involves a cycle in which hypotheses (or mental models) are refined:

1. Inspect available data →
2. Form an understanding and related hypotheses →

3. Seek information to test alternative hypotheses →
4. Cycle back to step 2, that is, refine the understanding →
5. Produce a judgment.

Many studies of expertise revealed this refinement cycle or some variation of it (e.g., Anderson, 1982; Chi et al., 1982; Hoffman and Militello, 2007, chapter 8; Newell, 1985). For example, the Duncker cycle appeared explicitly in Lederberg and Feigenbaum's (1968) description of the goal for their expert system on the analysis of organic molecules: "Data somehow suggest a hypothesis, and deductive algorithms are applied to the hypothesis to make logically necessary predictions; these are then matched with the data in a search for contradictions" (p.187). The Base Model is presented in figure 10.1.

The Base Model incorporates the notion of a mental model (Gentner and Stevens, 1983) and is consistent with ideas about the reasoning of experts in diverse domains (discussed in chapter 6). In fact, we have found no descriptions of domain-specific expertise that cannot be regarded as variations on this Base Model (Hoffman and Militello, 2007; Hoffman, Ford, and Coffey, 2000a). Finally, the Base Model integrates what forecasters have said about forecaster reasoning (e.g., Doswell and Maddox, 1986; Godske et al., 1957; Trafton et al., 2000; Snellman, 1982).

Hoffman, Coffey, and Ford (2000b) created a version of the Base Model specifically to capture the models proposed by forecasters—including Snellman's (1982) forecasting funnel and Bosart's (2003) progression of questions. This variation on the Base Model is generic, that is, it describes routine forecasting rather than reasoning for warning forecasts or reasoning for specialized forecasting situations (e.g., tornadoes). (More will be said about this matter of general vs. specific reasoning models later in this chapter.) This generic model is presented in figure 10.2.

The reasoning of proficient forecasters always involves multiple, parallel, and partially overlapping refinement cycles or iterations. Thus, the "Situation Awareness Cycle" along with the "Mental Model Refinement Cycle" can be regarded as an early version of Klein's sensemaking model, which we presented in chapter 6 as a generalized characterization of expert reasoning. Initial acts of recognition based on some data type (frequently the first data type inspected is a satellite image; Trafton et al., 2000) lead to the formation of an initial mental model. This then suggests which data types should be inspected, which leads to more acts of recognition, hypothesis tests, model refinements, and so on.

Furthermore, the loop in figure 10.2, which links "Mental Model of Atmospheric Dynamics" to "Judgments, Predictions" and then up to the "Action Queue," is an early

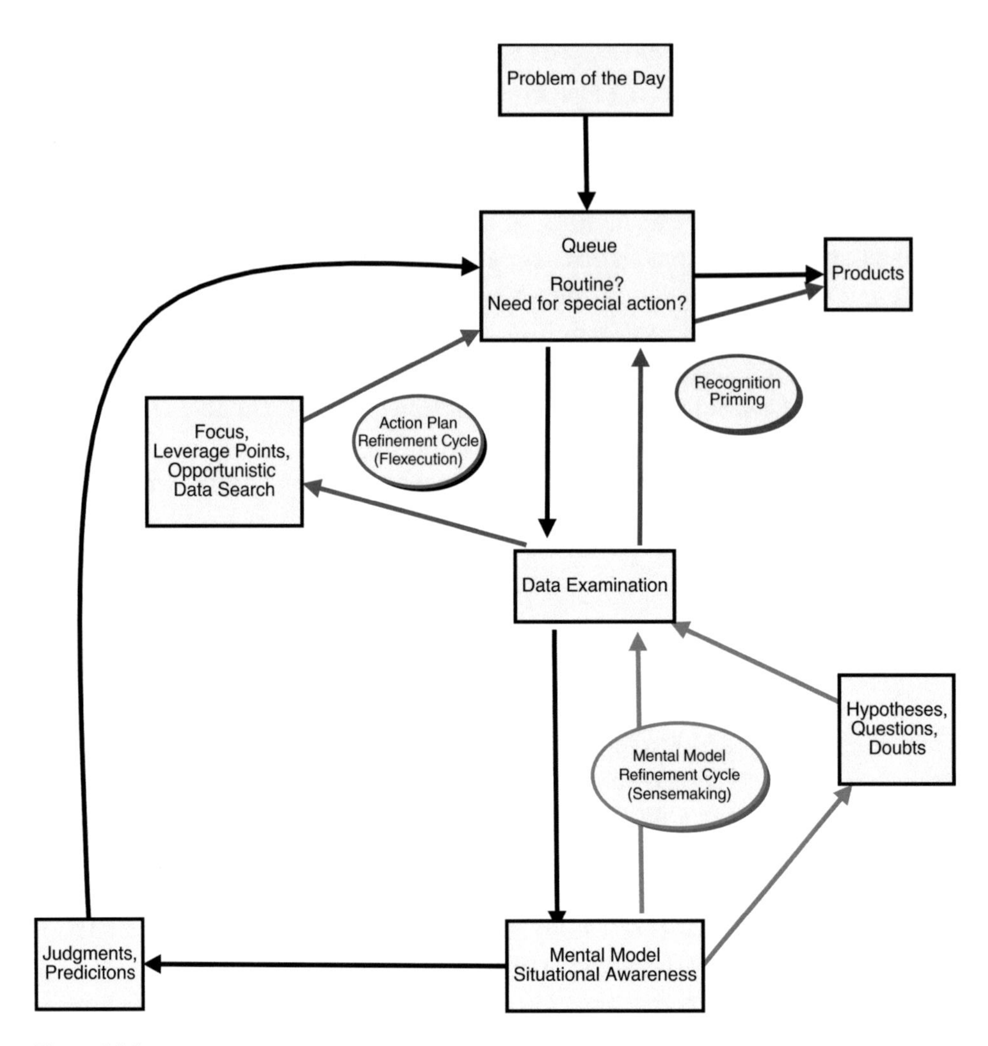

Figure 10.1
The Base Model of expert reasoning.

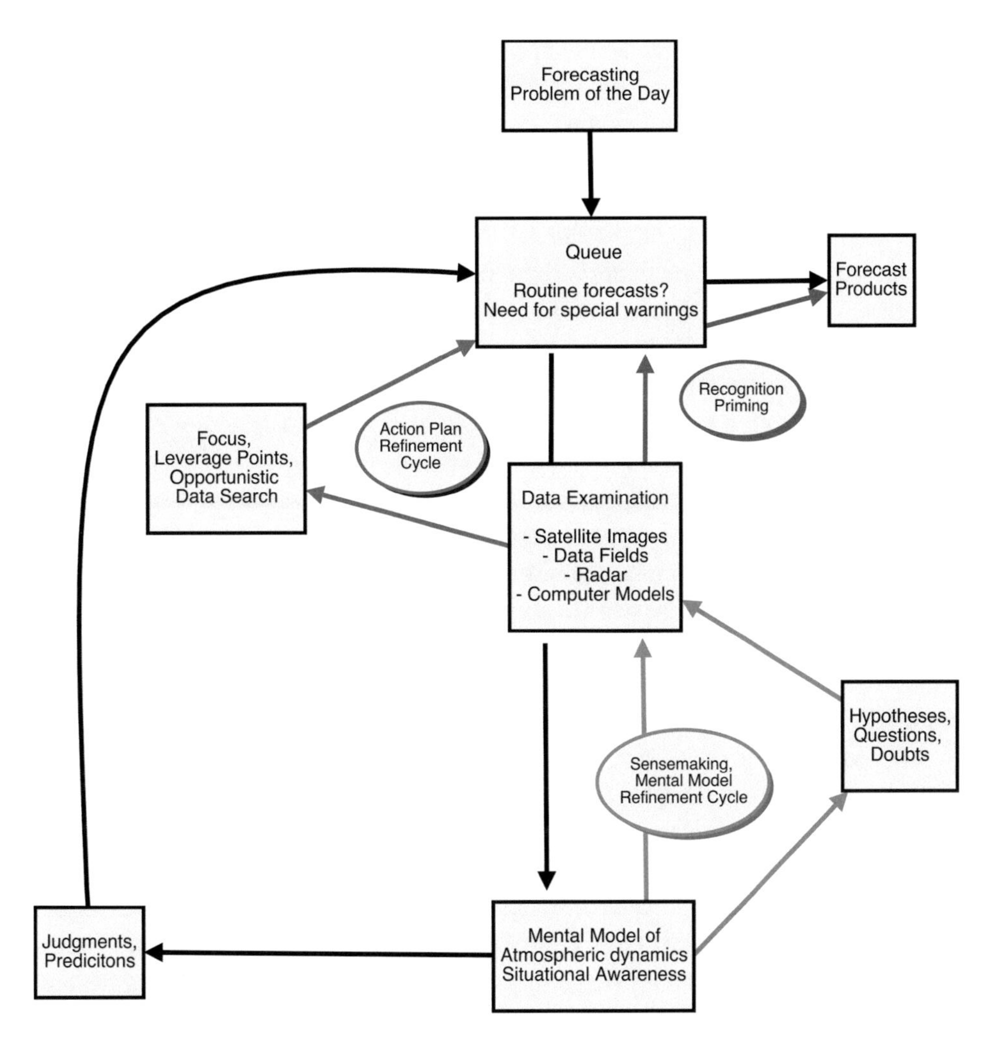

Figure 10.2
The Base Model of expertise adapted to the domain of weather forecasting.

version of the flexecution model of expert performance, which was also presented in chapter 6 as a characterization of expert reasoning.

Forecasters rarely come into a forecasting task without already having some idea of what is going on, especially if they have been doing daily shift work. They might look at television broadcasts or online weather data before going to work, and of course they watch the sky. As a result, they already have some idea of what the weather will be before going to work (Hahn et al., 2002; Klinger, Hahn, and Rall, 2007). At the beginning of a forecasting procedure, two things must be accomplished. The forecaster must develop a "big picture." Weather data are scanned to form an initial mental model and derive a ballpark assessment of the current weather situation and the relevant atmospheric dynamics at a large scale (i.e., hemispheric scale and a period of many days). Satellite imagery is important for this initial assessment at synoptic and hemispheric scales. At that point, reasoning becomes top-down, with the mental model suggesting both predictions and testable hypotheses. The forecaster then examines a subset of data and products, such as radar images and surface observations. Evidence on the development of forecasting expertise (see chapter 7) suggests that less experienced forecasters are less likely, and often less able, to begin their forecasting process with a clear focus on understanding the overall current weather situation at a synoptic scale, and this can contribute to data overload as they proceed.

A related and immediate task is to deepen their understanding of the "problem of the day" (Curtis, 1998; Daipha, 2007; Joslyn and Jones, 2008; Pliske et al., 1997; Targett, 1994). This has two aspects. One is the literal directive or task, which can be to create a regional forecast over a period of watch or to create a forecast tailored to a specific end-user or operation (e.g., aviation forecasting, port operations). The second aspect is that "the" weather problem of the day depends on the weather (i.e., a severe storm seems to be rapidly developing, suggesting a need to issue a warning). The weather and problem of the day means that particular data types must be examined (e.g., computer model outputs, surface temperatures, surface dew points). If the immediate problem is to predict the daily high temperature, the forecaster is likely to examine the current and forecasted low-altitude temperature profile and cloud cover, whereas, if the immediate task is to forecast precipitation, the forecaster is more likely to inspect multichannel GOES water vapor data, radar imagery, and local observations of humidity as a function of height in the atmosphere (see Roebber and Bosart, 1996b).

The inspection of data, especially for experienced forecasters, can sometimes trigger recognition-primed decision making. Inspection of data (e.g., a computer model output or a satellite image) representing a familiar or frequently occurring situation, represented as a cue configuration, can lead directly to an act of recognition of concepts or categories (e.g., a low-pressure center is forming off the North Carolina coast), which can "prime" a course of action (e.g., anticipate possible warnings of a "Nor' Easter"). Recognition-primed decision making has been observed in a number of domains of expertise, including firefighting, military command and control, design engineering, and clinical nursing (Klein, 1989, 1993; Lipshitz and Ben Saul, 1993; see chapter 6). Especially when working under time pressure, experts in diverse domains may not engage in any deliberative, analytical problem solving in the sense that they evaluate all of the possible alternatives and courses of action as prescribed by normative models of decision making. Rather, they can go from recognition of critical cues or cue configurations directly to ideas about the one most appropriate course of action.

Forecasters have observed that some forecasters have an intuitive (pattern recognition) style, whereas others have a more logical-analytical approach to their forecast decision (Curtis, 1998; Joslyn and Jones, 2008). Senior meteorologist Charles Doswell (2004) suggested that forecasters fall along a spectrum between those two types. He also cited historical references suggesting that the earliest forecasters used a wholly intuitive process. In his book about the National Weather Service, author Gary Fine (2007) argued that the development of intuition is key to weather forecasting: "gut feeling" is more important than knowledge of meteorology.

The "mental model refinement" cycle (see figure 10.2 above) begins with a detailed inspection of particular data types (e.g., weather maps, satellite data, and the output of computer forecasting models), with an eye toward refining or disconfirming the initial mental model and converging on local weather (mesoscale and microscale) in a shorter time frame (i.e., days to hours to minutes). For example, if the mental model suggests that thunderstorms are likely to occur, then the forecaster will almost certainly inspect visible and infrared satellite images. They are generally produced at intervals of 30 minutes, but they can sometimes be obtained more frequently in critical weather situations. (NOTE: When the new GOES-R satellites are operational, images will be obtained in shorter time invervals.) The forecaster will also inspect radar data.

There are variations on the mental model refinement cycle, of course. For example, model refinement can involve comparisons of the particular tendencies of different computer models. Also, the cycle is not just engaged a single time; it can occur repeatedly, especially for challenging forecasting situations (Hoffman, Ford, and Coffey, 2000a, 2000b).

As the weather forecasting and warning process proceeds, more looping of satellite imagery often occurs. Forecasters project their model in time and watch for signals indicating that an event might be becoming unusual. Receiving reports of weather impacts in real or near-real time has been shown to be critical to confirm forecasters' mental model. The mental model is checked and rechecked frequently as the weather evolves (Hahn et al., 2002). Inspection of certain data or data types may result in immediate recognition of cues, configurations, or patterns. Nevertheless, the forecasting procedure continues with inspection of more data and data types, in a sequence of attention shifts that depend on both the forecasting problem of the day and the forecaster's situation awareness. For instance, in circumstances in which it is known or suspected that the computer models may be biased (e.g., overly conservative in predicting storms under certain conditions), the forecaster may go on to inspect additional data or another computer model to test a hypothesis (e.g., about whether a storm might intensify or might dissipate because of shearing winds at upper levels) (see Roebber, 1998; Roebber and Bosart, 1996b; Roebber, Bosart, and Forbes, 1996; Roebber, Schultz, and Romero 2002). The forecaster works with the available data to develop a mental model of the current weather—the "kinematically and dynamically most probable system" (Godske et al., 1957, p. 651).

For some forecasters, the mental representation is less depictive and more analytical. That is, their mental model seems to be imaginal, but it also seems more essentially conceptual, in that it takes the form of images that are phenomenally like traditional weather charts and computer model outputs charts, on which forecasters are weaned, showing the forces and events as represented in physical/mathematical relationships—isobars, the location of low pressure centers, skew-T/log p diagrams showing temperature as a function of height in the atmosphere, and so on (see Perby, 1989). Some forecasters argue that they will not develop confident understanding of the weather unless they actually hand draw a surface chart (Doswell, 1986c, Pliske et al. 1997; Pliske, Crandall, and Klein, 2004).

Next, a forecast or warning is generated along with the required products. The form and format of the forecast is determined by the problem of the day (i.e., the current dynamics, such as a deep low-pressure system approaching the forecast area) and other constraints (i.e., the need to complete particular forms, the need to tailor a forecast to a particular type of operation). There is ample evidence that forecasters frequently generate forecasting predictions right out of their mental model. This is shown as the arrow at the bottom of figure 10.2 (above), going from "Mental Model" to "Judgments and Predictions." Those judgments and predictions must be conformed to the action queue (arrow leading to "Action Queue" from "Judgments and Predictions"). This step is brief and straightforward: The forecasters write or produce their product or forecast.

Sometimes a formal forecasting procedure comes to an end (or a pause) once a particular forecast product has been prepared and distributed (e.g., a weather forecast for a particular region for a particular period of time). However, the cognitive work of the forecaster does not stop after the forecast is completed. Over the period of their watch or shift, meteorologists continue the processes depicted in figure 10.2 (above). The mental model is not just refined, but it is modified and ammended in light of the dynamic changes that are occurring in the atmosphere. In this way, the mental model directs attention and supports a continuing state of situation awareness and a process of updating, adjusting, or otherwise modifying the forecasts.

Following a weather event, it is important for forecasters to get feedback on how the weather impacted people so they can relate those effects to the data they had to work with as the event unfolded. Forecasters analyze a difficult or an unusual event in detail, particularly after failing to issue a warning prior to an event, with a mindset of improving their product generation going forward.

The Base Model was derived from studies of expert reasoning in diverse domains, and it integrated the various stage models that researchers have adduced. Confirmation that the Base Model applies to the domain of meteorology comes from a study by Trafton et al. (2000; discussed in chapter 9), in which they observed U.S. Navy weather forecasters as they created a briefing and talked out loud as they created their briefing. Their step model of reasoning is entirely consistent with the Base Model: form an initial mental model at a large scale, refine the model for a more local scale, validate the model by an adjustment process, and then create a final product.

The Base Model was also verified in a study by Susan Kirschenbaum (2004), in which she took advantage of a unique opportunity to compare the reasoning of U.S. Navy and Royal Australian Navy (RAN) forecasters. In the Southern Hemisphere, it gets warmer as one travels north, and low-pressure systems circulate clockwise rather than counterclockwise, as they do in the Northern Hemisphere. Would these major context differences entail different forecasting procedures and reasoning threads? Experienced forecasters might predict that these manifestly significant differences would not make a difference in forecasters' reasoning process.

Kirschenbaum (2004) observed actual forecasting operations and forecasting in an experiment-like context using preplanned scenarios. Given that the forecasters were all qualified to issue forecasts, they would be considered journeyman, minimally, but some had many as 16 years of experience, suggesting that some of them might qualify as junior experts or experts. Taking into account the fact that the forecasting in the operational context was time-constrained, Kirschenbaum (2004) looked at the proportion of time spent in different activities (inspection of satellite image loops, comparison of different data types and fields, formation of hypothesis formation, etc.). She relied on a process model developed by Trafton et al. (2000), which is essentially similar to the steps or stages in the Base Model (figures 10.1 and 10.2 above) and its specific forecasting activities of: (1) forming the "big picture," (2) developing a more local mental model, (3) testing or adjusting that mental model, and then (4) forming a forecast. Kirschenbaum (2004) found that the USN and RAN forecasters spent proportionately the same amount of time in each of these activities. Consistent with other research findings, the formation of the "big picture" (usually relying on satellite images) takes proportionately little time. Verifying, refining, and adjusting the mental model by comparing different kinds of data and deriving hypotheses takes proportionately the most time. The RAN forecasters relied far less on computer model outputs than the USN forecasters, and the USN forecasters spent less time looking at different websites because all the pertinent websites were available at their one workstation. The comparison of the outputs of various computer models was a primary activity in the stage of validating and refining a mental model.

Joslyn and Jones (2008) also conducted cognitive task analysis with the participation of U.S. Navy forecasters, who thought out loud as they produced Terminal Aerodrome Forecasts (TAFs). The model of reasoning that the researchers developed conforms to those adduced in the other research on forecaster reasoning: develop the big picture (synoptic scale), form an initial mental model, check the mental model, and then produce a forecast.

Modeling Forecaster Macrocognition

In chapter 6, we introduced the concept of macrocognition: the high-level cognitive functions that are invoked as humans try to adapt to complexity. Hoffman, Coffey, and Ford (2000; Klein and Hoffman, 2008) used the Base Model of expertise to form a new procedure called the Macrocognitive Modeling Procedure (MMP) as a "fast track into the black box."

In experimental psychology, the development of models of reasoning has traditionally relied on the think-aloud problem solving task, in which participants are given test cases and are required to think out loud as they work through problems. Then the audio record is transcribed, and each statement is coded in terms of microcognitive categories, such as the millisecond scale of shifts of attention, inferencing, and accessing short-term memory. This method for studying thinking is called protocol analysis (Ericsson and Simon, 1993). Protocol analysis is known to be laborious and time-consuming (see Hoffman and Militello, 2007; Hoffman et al., 1995), hence the desire for a faster method of generating models of reasoning at somewhat greater time scales.

The MMP was first conceived when a weather forecaster was shown the figure 10.2 (above) model and was asked whether it seemed appropriate to the domain. Ordinarily, a theoretical concoction such as this, from an academic experimental psychologist, would be foreign language to a domain practitioner. In this case, it was felt that discussing the model would be sensible because the weather forecasting community has relied on a distinction between conceptual (mental) models versus computational models of the weather (see chapter 4). Not only did the model immediately resonate for the forecaster, but he spontaneously took the diagram as an opportunity to add more domain-specific details to the process description and modify some of the relations among the diagram elements. With this experience as a flash point, a more formal procedure was created consisting of three basic steps.

Step 1: Preparation

The researcher adapts the Base Model (figure 10.2 above) to make it directly pertinent to the domain. For example, comparing figures 10.1 and 10.2, the "Problem of the Day" would be specified as "The Forecasting Problem of the Day," and "Data Examination" would be specified as "Examination of images, data, or radar." Next, two alternative "bogus models" are created. At least one of these models includes some sort of loop, and both include a number of Base Model elements. Taken together, the bogus models include core macrocognitive functions (e.g., recognition priming, hypothesis testing). Ideally, bogus models are not too unrealistic; yet, the researcher would not expect the practitioner to be entirely satisfied with either of them. Examples appear in figure 10.3.

Step 2: Model Making

Participants are presented with two "bogus" models and are encouraged to select the one that best fits their own reasoning, knowing that neither one is totally faithful to how domain practitioners reason. Participants are asked to describe ways in which the selected model would have to be altered to be more faithful to their own reasoning. Using the bogus models and their elements as a scaffold, practitioners are invited to concoct their own reasoning diagram. Examples of experts' self-crafted reasoning models appear in figures 10.4, 10.5, and 10.6. Examples of journeymen's models appear in figures 10.7 and 10.8.

Step 3: Verification

The best opportunity to observe forecaster activity in a context that might afford evidence pertinent to the evaluation of the participants' models was to observe them when

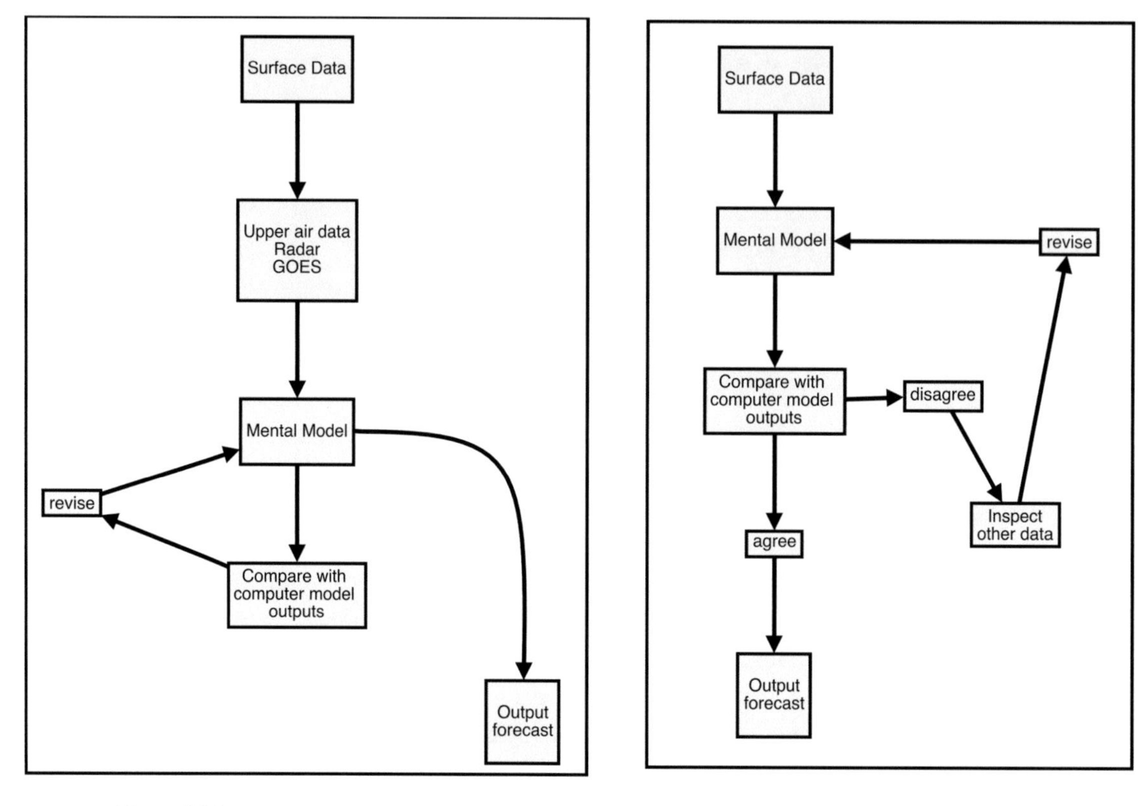

Figure 10.3
Example "bogus" models used in the study of weather forecasting.

they first came on watch after a period of several nonworking days. On such days, they would have to develop a mental model of the current overall weather situation. There were opportunities to observe five of the participants under these circumstances, after a delay of some days from when step 2 was conducted. The participants were observed from the time they arrived at the forecasting facility to begin a duty shift. Some aspects of each forecaster's conceptual model could be validated by observing their activities. For example, if forecasters asserted in step 2 that they began their process by examining satellite images, then that should have been verifiable by observing actual activity. Other aspects that are not so readily validated were the subject of probe questions. In addition, it was possible to validate elements of the Base Model (e.g., "examine satellite images"). Aspects of a reasoning model that would not be directly observable were verified using probe questions. Examples are:

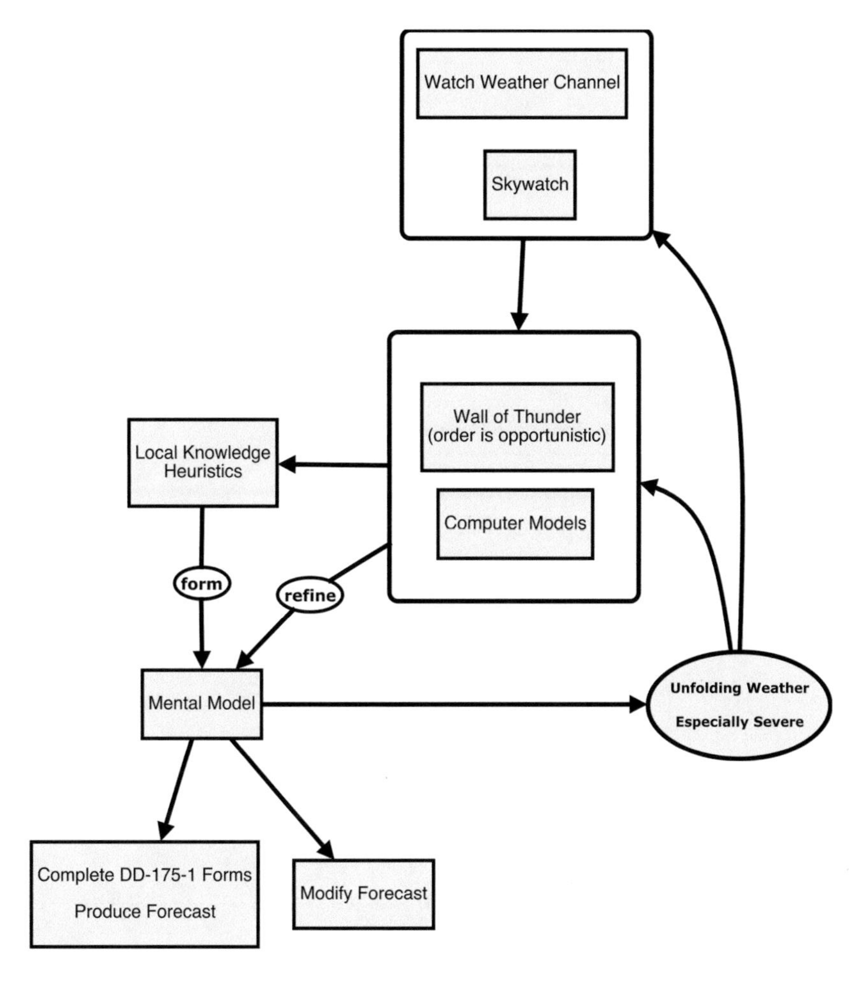

Figure 10.4
An example of a reasoning model created by an expert forecaster.

- "What are you doing now? Are you trying to understand the current weather situation?
- "Does what you're seeing fit with persistence?"
- "Are the computer models agreeing?"
- "Did you look at weather data or a forecast before you came in?"
- "What are you going to do now?

As a result of the observation-based verifications and the responses to probe questions, the reasoning models could be refined. An example record from the step 3

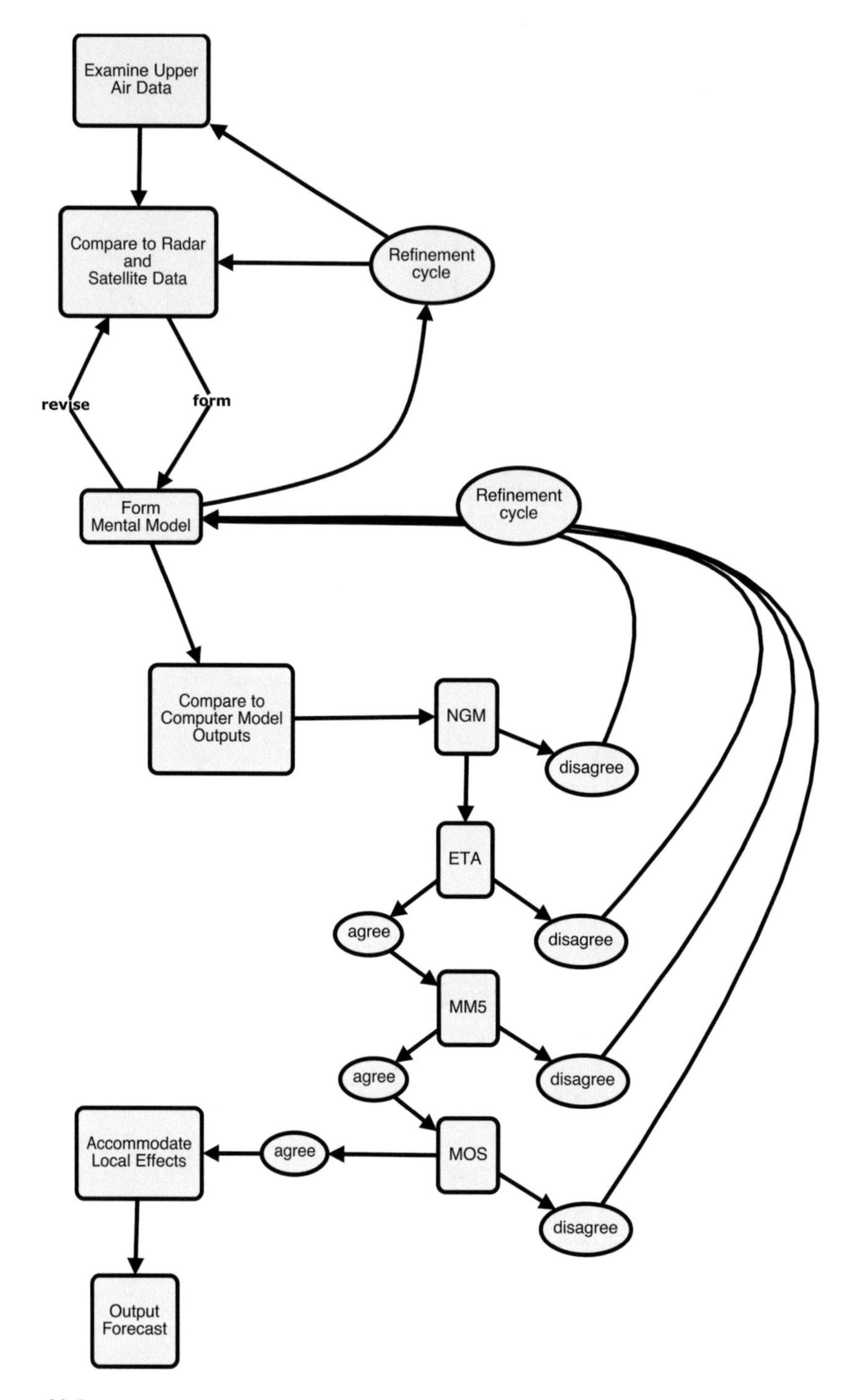

Figure 10.5

An example of a reasoning model created by an expert forecaster. The boxes labeled NDM, ETA, MM5, and MOS are names of computer models or model outputs.

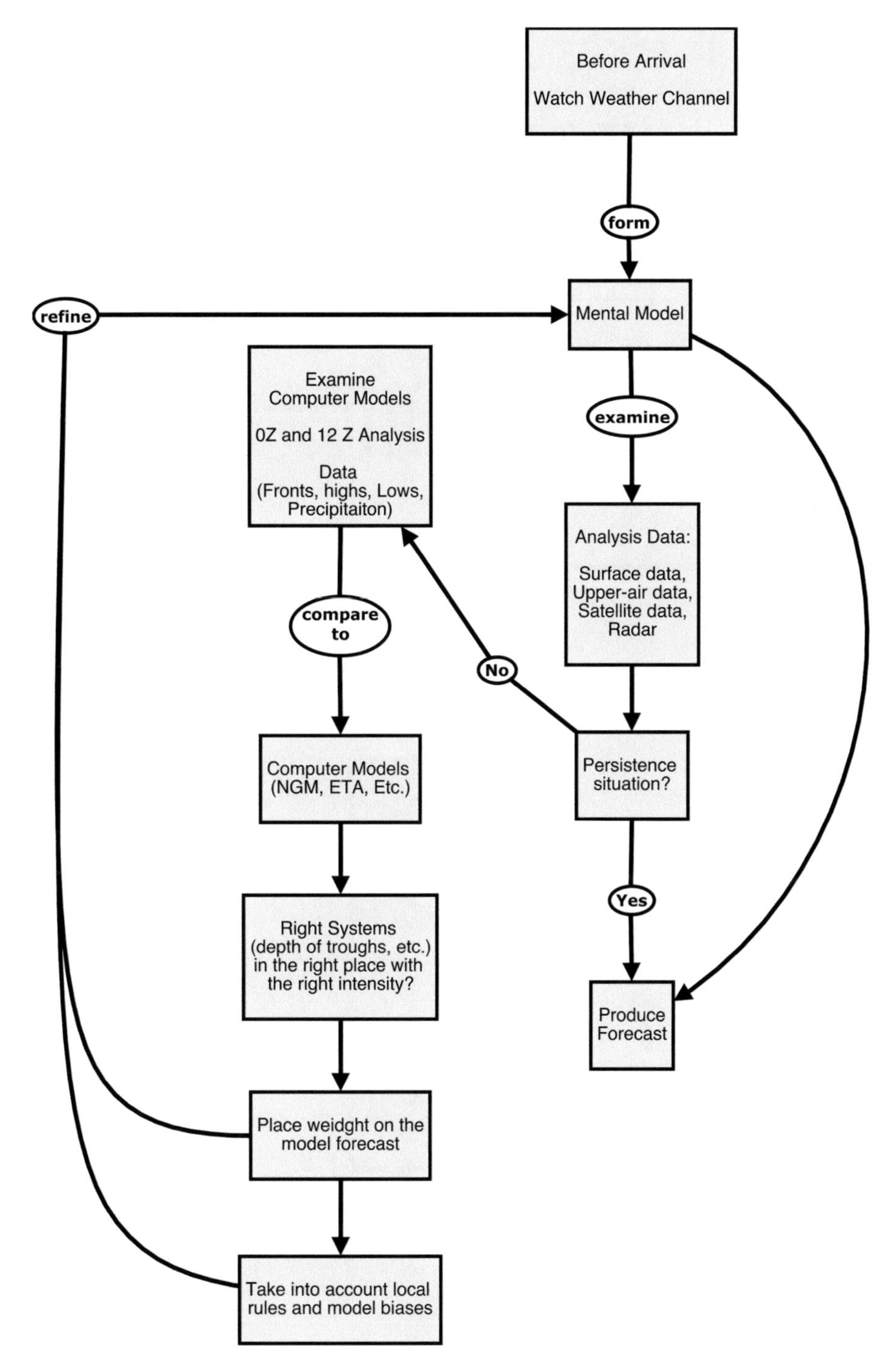

Figure 10.6

An example of a reasoning model created by an expert forecaster.

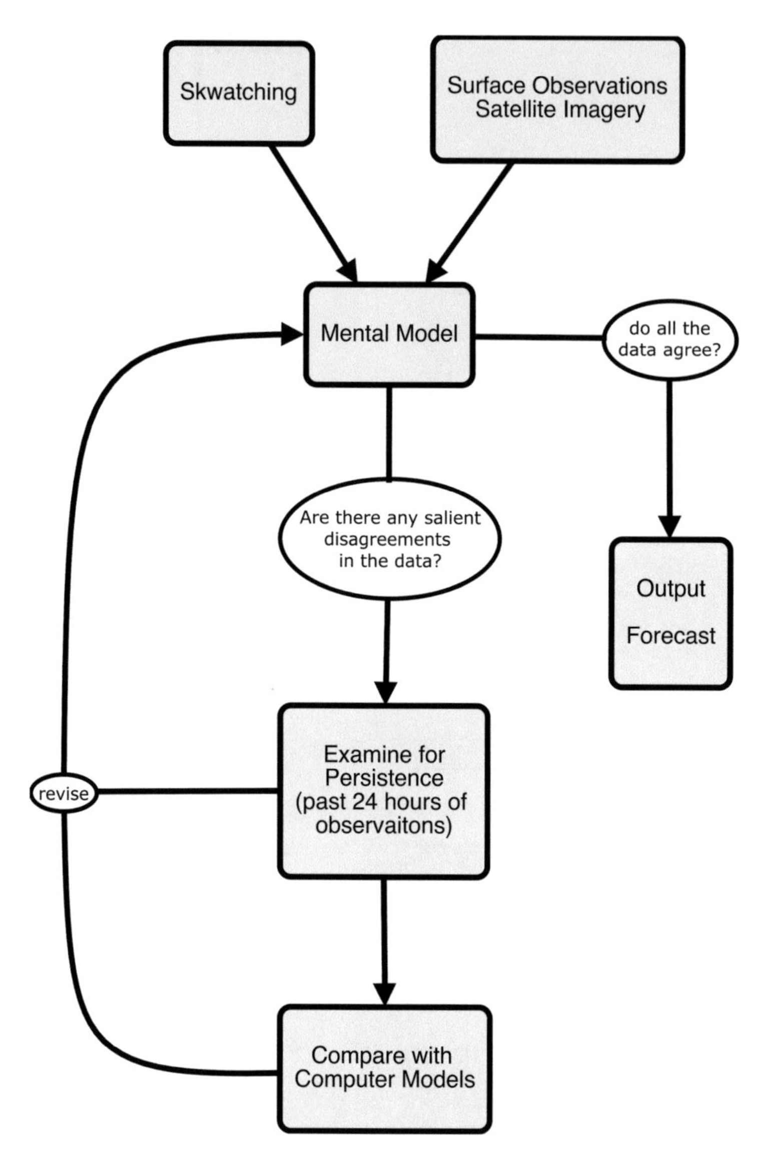

Figure 10.7
An example of a reasoning model created by a journeyman forecaster.

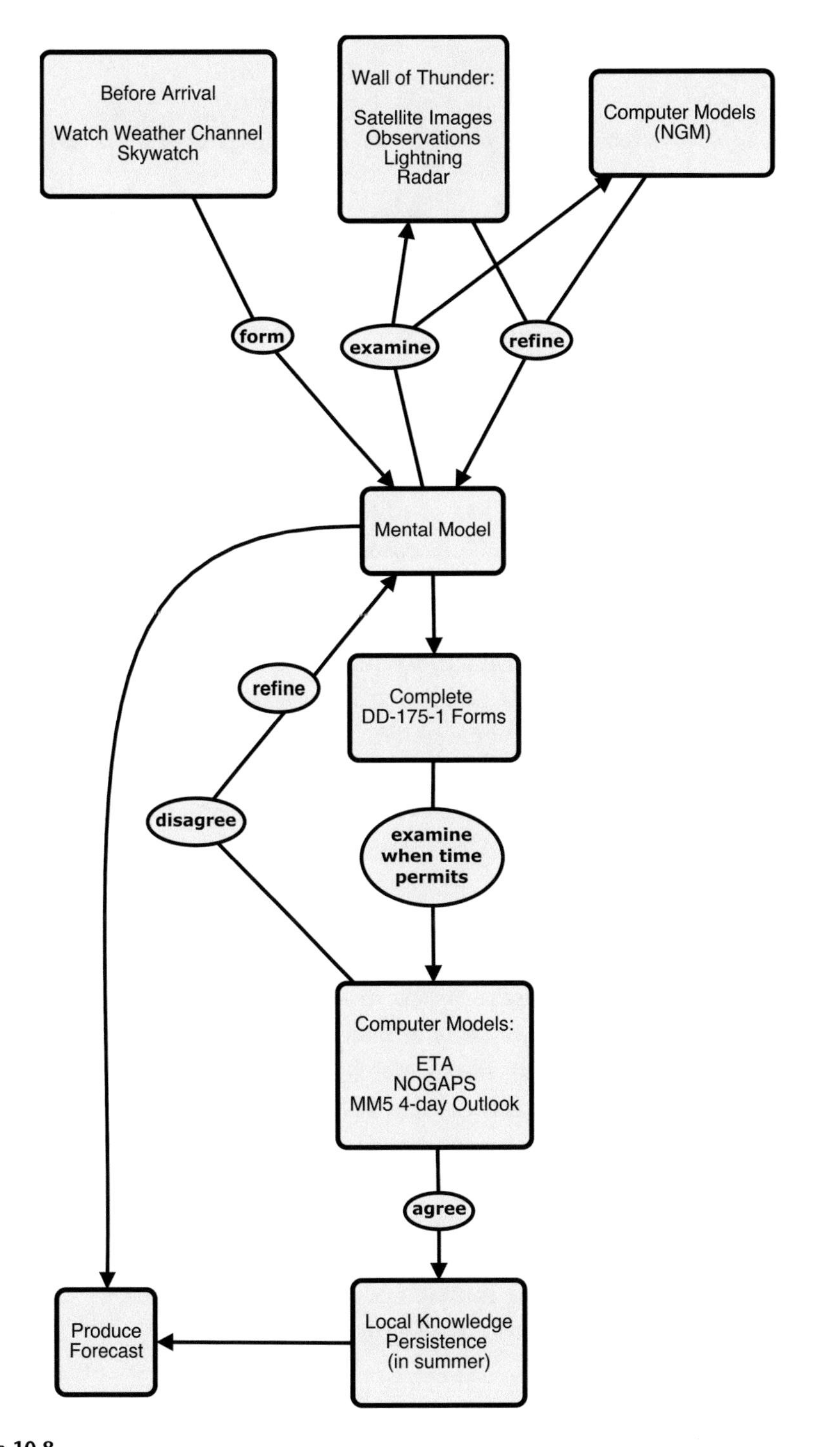

Figure 10.8

An example of a reasoning model created by a journeyman forecaster. ETA, NOGAPS, and MM5 are computer models. The "DD-175-1" forms provide the information for pilots about the weather expected for a given flight plan.

procedure appears in table 10.1. The researcher arrived just before the scheduled shift change.

Figure 10.9 graphically illustrates the verification results for the expert's reasoning model shown in figure 10.4 (above). Figure 10.10 graphically illustrates the verification results for the expert's reasoning model shown in figure 10.5 (above). Figure 10.11 graphically illustrates the verification results for the journeyman's reasoning model shown in figure 10.7 (above). Figure 10.12 graphically shows the verification results for the journeyman's reasoning model shown in figure 10.8 (above). Table 10.2 shows the verification results for all five forecasters.

Results from the procedure were models of the reasoning of seven proficient forecasters (experts and journeyman), five of which were validated by the step 3 observations of actual forecasting work. The procedure averaged out to 52 minutes total task time to develop and validate a reasoning model. This is without doubt less than the time taken in other widely used methods to reveal and verify reasoning models—preparing and coding a transcript of a think-aloud problem-solving protocol can take many hours (Hoffman et al., 1995). The results also validated the Base Model of expert reasoning. The results also clearly showed differences in proficiency, with less experienced forecasters tending to rely uncritically on computer forecasts and less likely to think hypothetically and counterfactually. Also, the less experienced forecasters were more likely to rely on a fixed sequence for inspecting the outputs of various computer models of the weather.

It is clear from these results that senior weather experts are constantly maintaining a state of situational awareness. However, they also understand that sometimes the weather situation can be one in which there is no need to actively maintain a heightened state of situational awareness. Situational awareness might be more critical in the following watch period, for instance. It is clear that the sequence of reasoning operations/strategies that the expert engages in is a function of the watch period. At times when forecasts do not have to be issued, situational awareness can lapse (but usually not completely). At times when there is a persistent situation, awareness can lapse and seems more likely to lapse for the less senior forecasters and journeymen. Furthermore, in a persistence situation, the forecaster's initial mental modeling can be superficial, including just what is pertinent to support forecasting for the facility's clients (i.e., pilots, in the case of this Navy facility).

Step 4: The "Guess Who" Game

This step is an optional procedure, intended to probe the extent to which the forecasters within the organization share information concerning their own reasoning and

Table 10.1
A representative data record from step 3 of the Cognitive Modeling Procedure

Time	Observation	Explanation
4:20	Participant (P) #2 is at the Forecast Duty Officer (FDO) station.	FDO workstation
4:49	P#1 arrives	
4:51	P#1 and P#2 discuss weather situation. GOES water vapor is shown on the FDO workstation display. P#2: "Little sea breeze, shifted at 3PM. East of us is getting clobbered." P#1: "They got good stuff last night." P#2: "A trough came through and lit everything off. All we got was cirrus aloft. Tomorrow the trough through our region will kick off some convection, but it will need some strong mechanical lifting to beat the ridge. The 00Z sounding will be interesting—there is +2 degrees up to 20,000 feet. The Senior Chief passed this information to you."	This verifies an aspect of the participant's reasoning model 00Z is midnight, Greenwich Mean Time
4:55	P#2 puts GOES visible image on the FDO. P#2 to P#1: "No cross-countries to do. Some thunder about noon. A T-1 is out until 8:00 PM. You might not need it. Other than that—a crazy day."	"Cross-country" is a flight plan. T-1 is a training aircraft.
	P#1: "Any Blue Angel special briefing?"	Performances of the Blue Angels require specialized weather briefings, which would be prepared the afternoon of this day.
	P#2: "No... The weather is interesting."	
	P#1: "That high ... 15,000 foot freezing level. Wow! That big cut-off ridge came right around it."	
	P#2: "And nothing's coming to change it. ... "	
	P#1: "You know, Europe is having a heat wave, too."	
4:57	P#2: "There was a potential severe, T2 for Charlie AOR 2 a while ago, but ... moving southeast but did not get strong enough. Tops at 55 to 60 (thousand feet), then as it moved it started to die. Tops in the 30s. I don't see it now. ... "	T2 is a thunderstorm, AOR2 is a particular Area of Responsibility.
	P#1: "Everything is sitting over the ridge—on the periphery."	

Table 10.1 (continued)

Time	Observation	Explanation
5:03	P#1 leaves operations floor.	
5:06	P#1 returns to operations floor and logs in.	
5:07	P#1 sits at FDO workstation and examines the TAF verifications.	TAF means "Terminal Aerodrome Forecast."
5:09	P#1 goes over to the NEXRAD Principal User Processor. P#1 to Researcher: "Stuff is coming around the ridge and following I-10 east to west. You can see the sea breeze, but the westerlies are overpowering it."	
5:11	P#1 conducts the security round.	
5:12	P#1 returns and sits at FDO station.	
5:13	Researcher: "Did you skywatch or look at the Weather Channel before you came in today? This is your first watch after two days off." P#1: "I was bike riding. Too hot. Like a blast furnace. I didn't look at the Weather Channel, but I stopped by last midshift. Big ridge with inversion. The air sets up and blankets you. I've lived on the Gulf all my life."	This probe was to validate an aspect of the participant's reasoning model.
5:31	P#1: "I usually stop by every day of the week (even when not on watch). When I go out to bike ride or drive around."	
5:41	Researcher: "Do you need to do TAFs or Dash-1s [DD-175-1 form]?" P#1: "No."	

strategies, or forecaster reasoning in general. This procedure is conducted some weeks or months after the previous steps. In the Hoffman et al. (2000b) study, the delay was about four months (two months in the case of one of the participants who joined the facility staff during the course of the investigation). In the game, each participant is presented with all of the forecasters' reasoning models plus additional bogus reasoning models and reasoning models generated in step 1 by apprentices. Participants are invited to guess which reasoning model goes with which forecaster.

Step 4 yielded some interesting and useful results, although it only took each participant about 15 minutes to complete. Most participants found the task to be quite interesting, if not fun. Some found it confusing. Most reported a "divide-and-conquer"

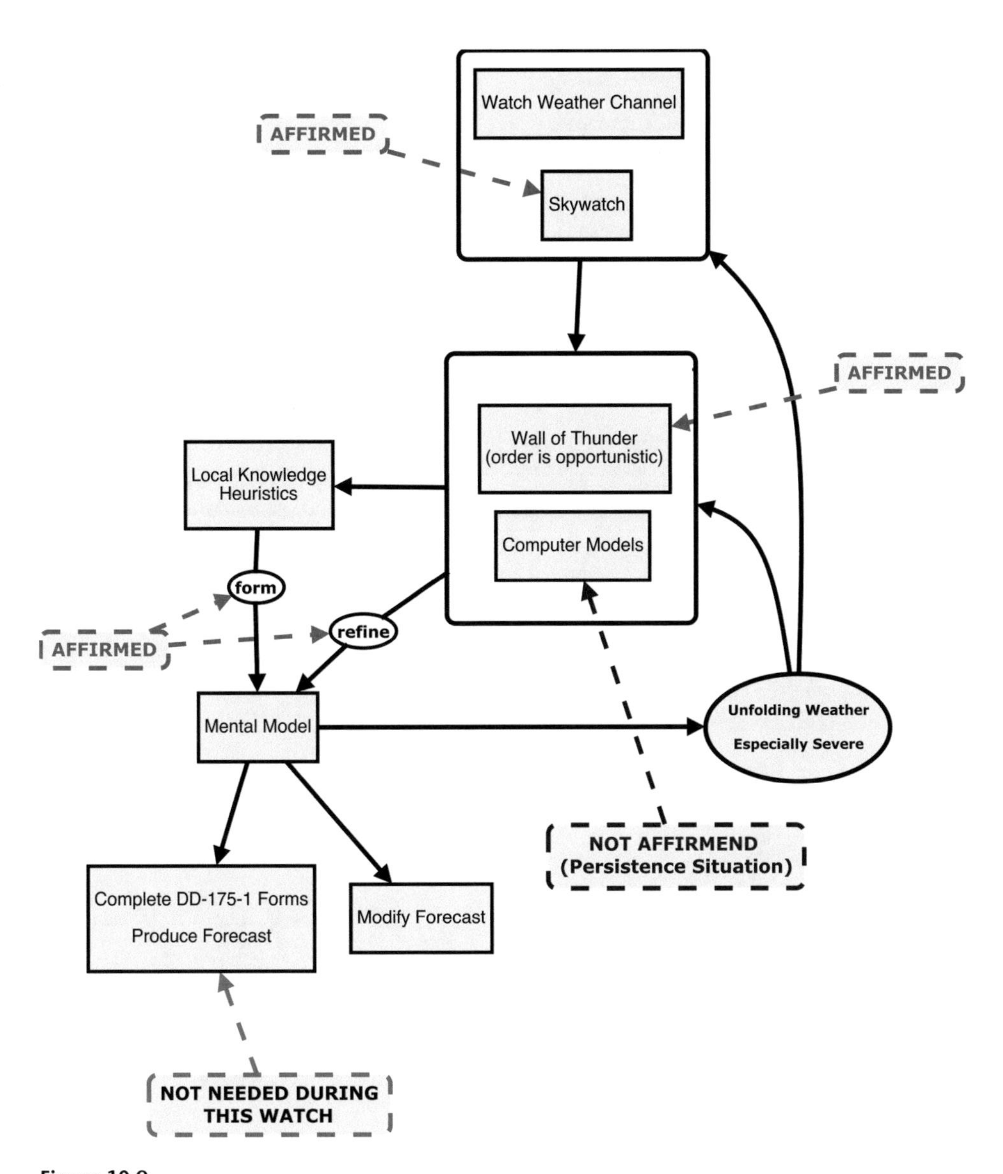

Figure 10.9

Results from an observational validation of the forecaster's model shown in figure 10.4. The "DD-175-1" forms provide the information for pilots about the weather expected for a given flight plan.

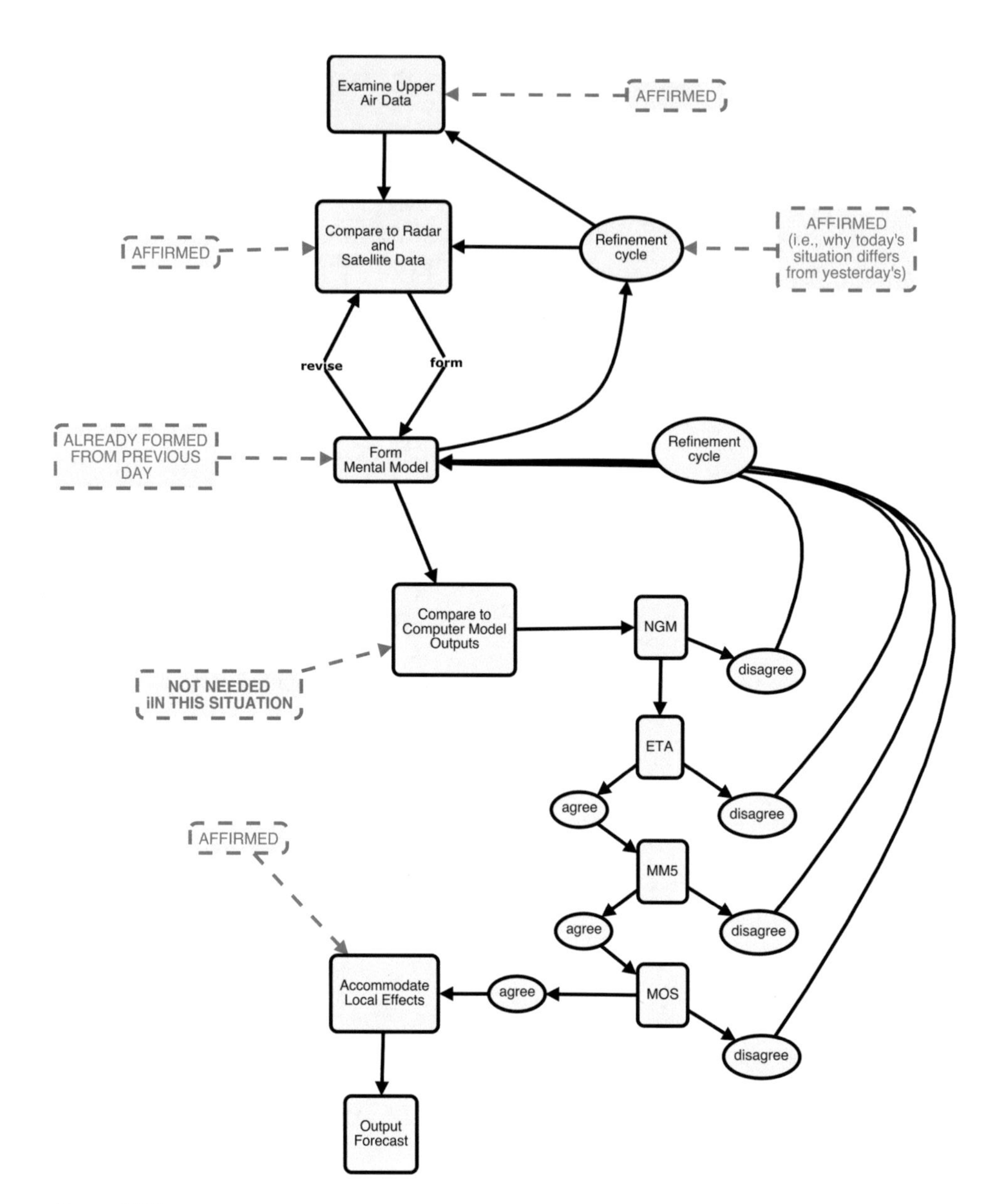

Figure 10.10
Results from an observational validation for the forecaster's reasoning model shown in figure 10.5.

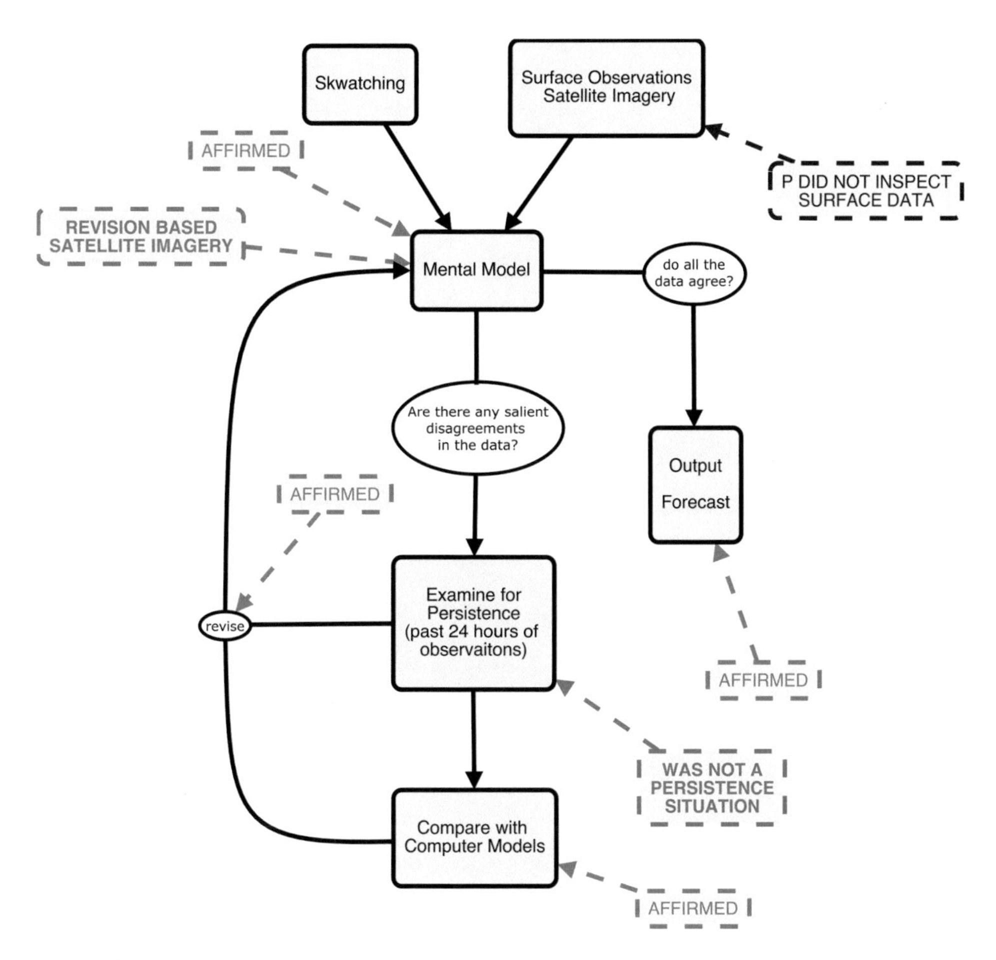

Figure 10.11
Results from an observational validation of the forecaster's model shown in figure 10.7.

strategy of first trying to identify the reasoning models of senior experts or forecasters with whom they were more familiar and then identify the reasoning models they thought were bogus and the reasoning models they thought were those of apprentices. Of all the identification judgments (N = 60), only 13 (or 22%) were correct. One of the bogus reasoning models was correctly identified as being bogus by three of the seven participants. As one participant described it, "Neither path in this model can yield any outcome to the forecast. There is not enough data in it, not much time being put into the forecasting according to this model." The fact that the bogus reasoning models were not uniformly correctly identified as being bogus suggests that they were successfully crafted so as to not be too unrealistic.

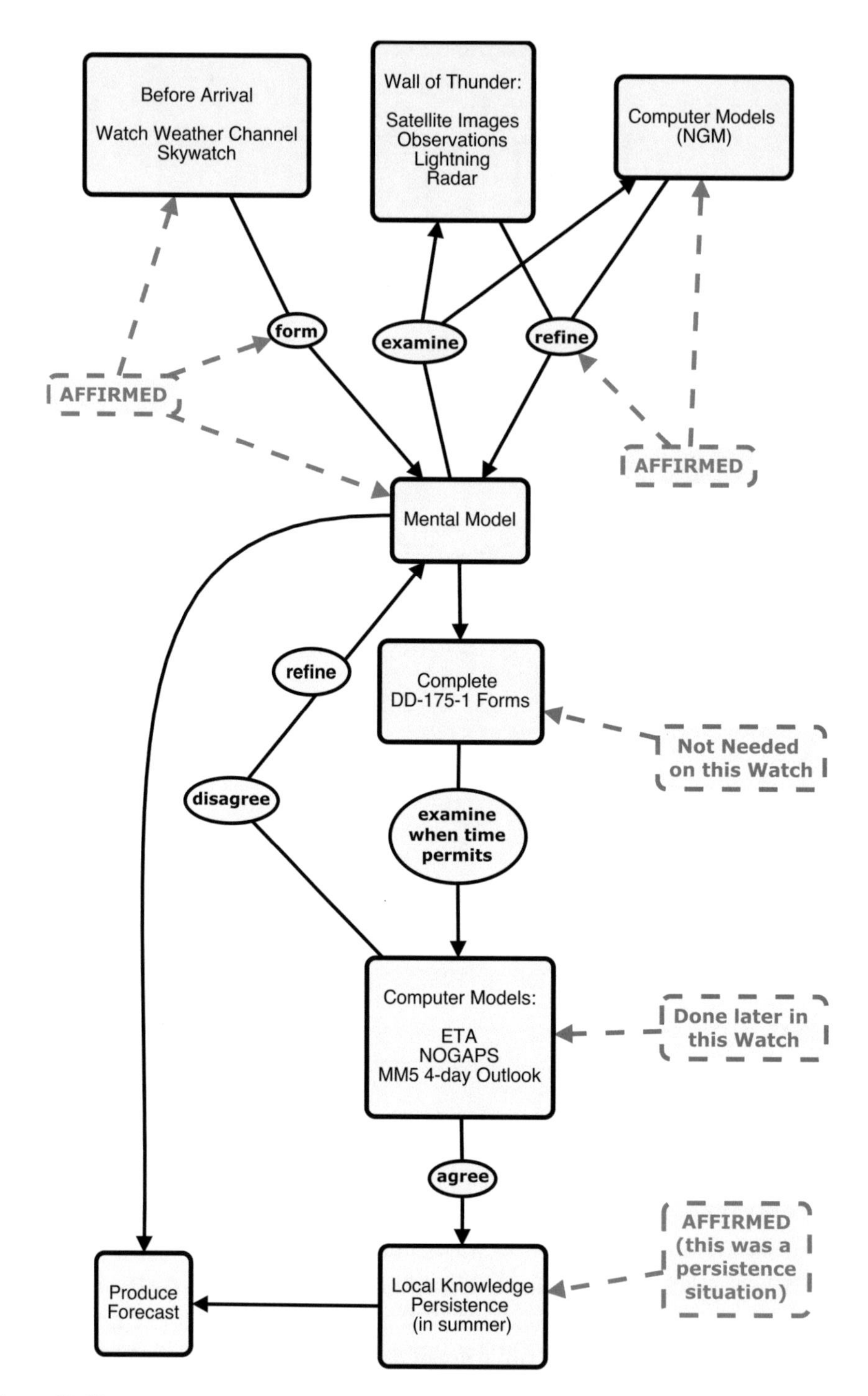

Figure 10.12

Results from an observational validation of the forecaster's model shown in figure 10.8. The "DD-175-1" forms provide the information for pilots about the weather expected for a given flight plan.

Table 10.2
Summary of the step 3 results for the expert forecasters

Elements of the Original Reasoning Model That Were Validated	Elements of the Original Reasoning Model That Were Qualified	Elements of Base Model That Were Affirmed
Expert 1		
• Skywatching and Weather Channel watching before arrival. • Examination of Wall of Thunder data. • Mental model formation. • Mental model refinement. • Maintenance of situational awareness. • Reliance on local knowledge.	• Forms completion time depends on watch period. • Comparison of models not needed in persistence situations.	• Mental model formation. • The refinement cycle. • Maintenance of situation awareness. • Reliance on local knowledge.
Expert 2		
Examination of upper air data. Examination of radar and satellite data. Mental model formation. Mental model refinement. Comparison to computer model guidance (with qualification). Reliance on local knowledge.	Forms completion time depends on watch period. Comparison of models not needed in this situation Maintenance of situation awareness—mental model already formed on previous day.	Data examination. Mental model formation. The refinement cycle. Reliance on local knowledge.
Expert 3		
Skywatching before arrival. Examination of Wall of Thunder data. Mental model formation. Mental model refinement. Comparison to computer model guidance. Reliance on persistence.	Forms completion time depends on watch period. Mental model formation, but only a superficial model is needed at times. Comparison of models only if there is a reason. Mental model refinement only if there is a reason. Preparing of forecasts is done on demand.	Mental model formation (with qualification). Mental model refinement (with qualification). Reliance on local knowledge (i.e., persistence).

Table 10.2 (continued)

Elements of the Original Reasoning Model That Were Validated	Elements of the Original Reasoning Model That Were Qualified	Elements of Base Model That Were Affirmed
Journeyman 1		
Examination of satellite imagery. Mental model formation. Mental model refinement. Comparison to computer model guidance. Reliance on local knowledge.	He did not begin by examining surface data; the key data were not to be found in surface observations, entailing a change in the initial mental modeling loop. It was not a persistence regime, eliminating one of the loops from his model Refinement focused on developing a coherent mental model.	Data examination. Mental model formation. The refinement cycle. Reliance on local knowledge (i.e., this was *not* a persistence situation).
Journeyman 2		
Skywatching and Weather Channel watching before arrival. Examination of Wall of Thunder data. Mental model formation. Mental model refinement. Comparison to computer model guidance. Reliance on local knowledge.	Forms completion time depends on watch period. Comparison of models not needed in persistence situations.	Mental model formation. The refinement cycle. Reliance on local knowledge.

Most of the time, the forecasters correctly identified the reasoning models of the apprentices: "This looks like somebody who has not had much experience. They start with the upper-air data, but that data could be biased. You should do satellite first since that is real data." Another forecaster said of one of the apprentice's reasoning models, "There is no mental modeling here, no thoughts of what's going on. This is just data-in/data-out."

None of the participants correctly attributed the reasoning models of two of the senior forecasters. But participants were sometimes able to identify other reasoning models:

[That forecaster] is methodical. This is his logic. From the Weather Channel he gets a mental model. If the data agree, OK. If they disagree he examines the models. Though I've never seen him work on the floor except in hurricane season.

Surprisingly, only two of the participants correctly identified their own reasoning model. The reasoning model of one of the experts was the one most often correctly identified. This forecaster had a particular strategy for forecasting severe weather and had presented that strategy to all of the personnel in a recent series of technical training briefings. Another forecaster who did not correctly identify his own reasoning model explained later that the model of his that had been generated in step 1 and was used in the Guess Who Game had been created during the previous season. The seasonal change entailed a forecasting strategy and process that differed from the strategy he had been using at the time the Guess Who Game had been played. We will have more to say about this later in this chapter.

One would expect that as individuals achieve proficiency their overall reasoning strategies might converge, causing them to experience confusion in this task. As one participant put it, "None of the last three models looks totally bogus. [These two models] look similar. [One] could be me as easily as [the other]." Another participant said:

> This looks like [Forecaster]. [He] is big on examining the past 24 hours of observations. But I don't do a lot of that. There is no other model left that could fit me. I don't look at NGM and ETA. This seems closest to me than any of the others.

On occasion, the participants pointed to reasoning convergence as a cause of their confusion at this task:

> This is not bogus. I could just about give it to myself. Start off with the big picture, compare to the numerical models. The sequence is OK for a pretty good forecast.

Most of the participants reported that they had to guess. However, they were sometimes confident, even when incorrect:

> [Forecaster X] puts care into it. [He] does stuff before he comes in. [He] applies his local knowledge to get a picture in his head. [He] is cognizant of the Wall of Thunder. [He] does stuff quickly since [he] already knows what's going on when he comes in.

In fact, the reasoning model in question was not the model of the participant being referred to. Another participant waffled, even about a reasoning model that was in fact his own:

> Skywatching and the Wall of Thunder when I come in, and I do use NGM [computer model]. But I don't think this is me since I don't like the MM5 [computer model]. I do a lot of the same things as [Forecaster B]. It could be the reasoning model [of Forecaster C]; it could also be the model of [Forecaster D]. He goes through everything.

Apart from reasoning convergence as a possible cause of the confusion, these results spoke clearly to the fact that this particular organization was not much engaged in sharing reasoning strategies, despite their avowed adoption of a divide-and-conquer approach to this task. Participants may have had opportunities to *see* what each another does, but they did not share much information about their actual strategies for data search, mental model formation, and hypothesis testing. As one forecaster reflected afterward:

You never really talk to others to see how they reason. Funny how you work with them but never systematically see what they do to form opinions. Even when you work side-by-side with someone you do not hear their thinking. When you are on the floor you don't have time, but it would be good to see how others do what they do. I'd be surprised if anyone knows how anyone else thinks.

A few times, expert reasoning models were incorrectly identified as being models of apprentices and bogus models. This finding was the most striking. An explanation came in this post-task interview:

When this Bermuda High set up early a few years ago like now, the ETA and MM5 [computer] models did not handle it well but NGM did. It is the same now but we have COAMPS as well, and it does well, too. This [conceptual] model of mine does not fit my reasoning now since I am not using the [computer] models in the same way as I did when we made my [conceptual] model. I do a different preferred order now due to the set-up of the cut-off low preventing the high from ridging into the Gulf and preventing the Pacific influence. This kills our weather.

Generic versus Specific Models

Figure 10.10 (above) shows the step 3 results for the participant whose model was presented in figure 10.5. This reasoning model included a sequence of inspections of the outputs of certain computer models. At the time the step 1 model was created, the forecaster utilized this preferred order. However, by the time step 3 was conducted, not only had the season changed, but the weather regime had changed to one that was seasonally atypical. A computer forecasting model that had been preferred was no longer preferred. As a result, in step 4 the forecaster rejected his own model as being his own. When this "error" was pointed out, he said:

This model of mine does not fit my reasoning now since I am not using the models in the same way as I did when we made my model. I do a different preferred order now due to the set-up of the cut-off low preventing the high from ridging into the Gulf and preventing the Pacific influence.

Earlier in this chapter, in discussing the question of whether the expert's mental model is or can be complete, we mentioned that the forecasting process differs considerably depending on the goal: The process for generating a daily forecast differs from that for creating and issuing a warning, in which the forecaster has to focus attention to be able to "stay ahead of the storm." In other chapters of this book, we have referred to a variety of additional factors that forecasters must take into consideration when understanding different kinds of weather events. In the studies by Lusk et al. (1990; see chapter 5), structured interviews were held with a group of experienced forecasters concerning the prediction of microbursts. The forecasters described the cues they would look for, including descending reflectivity core, a collapsing storm top, convergence about the cloud base, and a reflectivity notch. Next, the forecasters described the important cue combinations. For instance, collapsing storms usually involve descending cores. The interviews allowed the researchers to develop a prototypical scenario (i.e., moderate storms in the vicinity, including some microbursts and the maturing of an isolated storm, temperature near the convective temperature). The results also revealed a reasoning sequence particular to microbursts:

> The [forecasters] indicated both in discussion and in writing that they used a two-stage process. ... If descending core and collapsing storm variables were low, then the probability of a microburst would be low, regardless of the other cues. On the other hand, high values of descending core and collapsing storm would indicate a downdraft, and the forecasters would look at the other cues to determine the strength of the downdraft. (Lusk et al., 1990, p. 633)

This situational dependence notwithstanding, forecasting severe weather and issuing warnings does to some extent piggyback on the general forecasting reasoning process. The forecaster will still begin by getting the big picture and forming an initial mental model; after that there is some divergence, as described here by David Klinger and his colleagues who conducted cognitive task analyses in WFOs in Alabama, Oklahoma, Missouri and Texas:

> [Forecasters] spend a lot of time projecting what will happen in the near future. This allows the forecaster to stay ahead of the storm and to be proactive regarding warnings. [They] generate hypotheses concerning the worst-case scenario and play it out in their minds. ... Season after season of observing weather patterns has instilled a "sensitivity for severity." They have developed an ability to identify that a certain weather event is going to be much more intense ... they realize at some point that this storm would be larger, more destructive, or faster developing than the usual storms. ... It is not unusual for our expert forecasters are the first ones in their office to anticipate the extraordinary level of storm development. (Klinger Hahn, and Rall, 2007, p. 363)

Furthermore,

> ... it goes far beyond just anticipating "big storms." We would say that our experts were also able to pinpoint the exact microstorm within he larger storm front ... they were able to identify the exact data elements within the larger storm that needed to be monitored closely. (Klinger Hahn, and Rall, 2007, p. 363)

The procedures and strategies used in forecasting tornadoes tomorrow in Kansas are different from those used in trying to predict whether it will be sunny tomorrow in Chicago. If one were to attempt completeness of macrocognitive modeling to describe forecaster reasoning, one would need quite a great many models to cover reasoning as a function of season, general weather regime, and other factors (see, e.g., Ralph et al., 2005). One would have to describe expert forecaster reasoning for each site-specific situation (e.g., fog formation in the Gulf Coast in the winter and thunderstorm development in the Caribbean in the summer). In addition, it would be a "moving target" problem because the technologies are always changing and improving (Ballas, 2007). By the time one had developed even a small fraction of the models, the world of forecasting will have passed by because of advances in technology, forecasting procedures, and new methods and tools.

Modeling of forecaster reasoning must consider the many differences among apprentices, journeymen, and expert forecasters in terms of their reasoning skills, styles, and strategies (see chapter 7). Apprentices and journeymen tend to do limited data gathering; they tend to follow prescribed steps, and they use computer model outputs in ways different from how exerts use them. Experts are better at talking local effects into account, and so forth. Time well spent would be aimed at expanding our models of reasoning across the proficiency scale. Many dozens of macrocognitive models of reasoning strategies would be needed to present a rich and fair picture of practitioner reasoning across the proficiency scale and across climates and seasons. For example, consider the task of predicting tornadoes. Figure 10.13 presents a reasoning model for tornado forecasting based on discussions of this problem in the meteorology literature (i.e., Andra, Quoetone, and Bunting, 2002; Heinselman et al., 2015; Moller et al., 1994; Roebber et al., 2002). This is quite different from the reasoning models presented in figures 10.4 through 10.7 (above), but it represents systematic looping between data and mental models (for situational awareness and flexecution), in which different data types are inspected to look for different possibilities and patterns.

Another consideration in the modeling of forecaster reasoning has to do with the fact that forecasters often have to flexecute.

Flexecution in Forecasting

Forecasters have to be adaptive in many respects (Joslyn and Jones, 2008; Pliske, et al., 1997), especially regarding severe weather. A representative case study is the May 3, 1999, tornado outbreak in central Oklahoma that was associated with a potent storm system. Forecasters at the Norman Oklahoma Weather Forecast Office were wary of

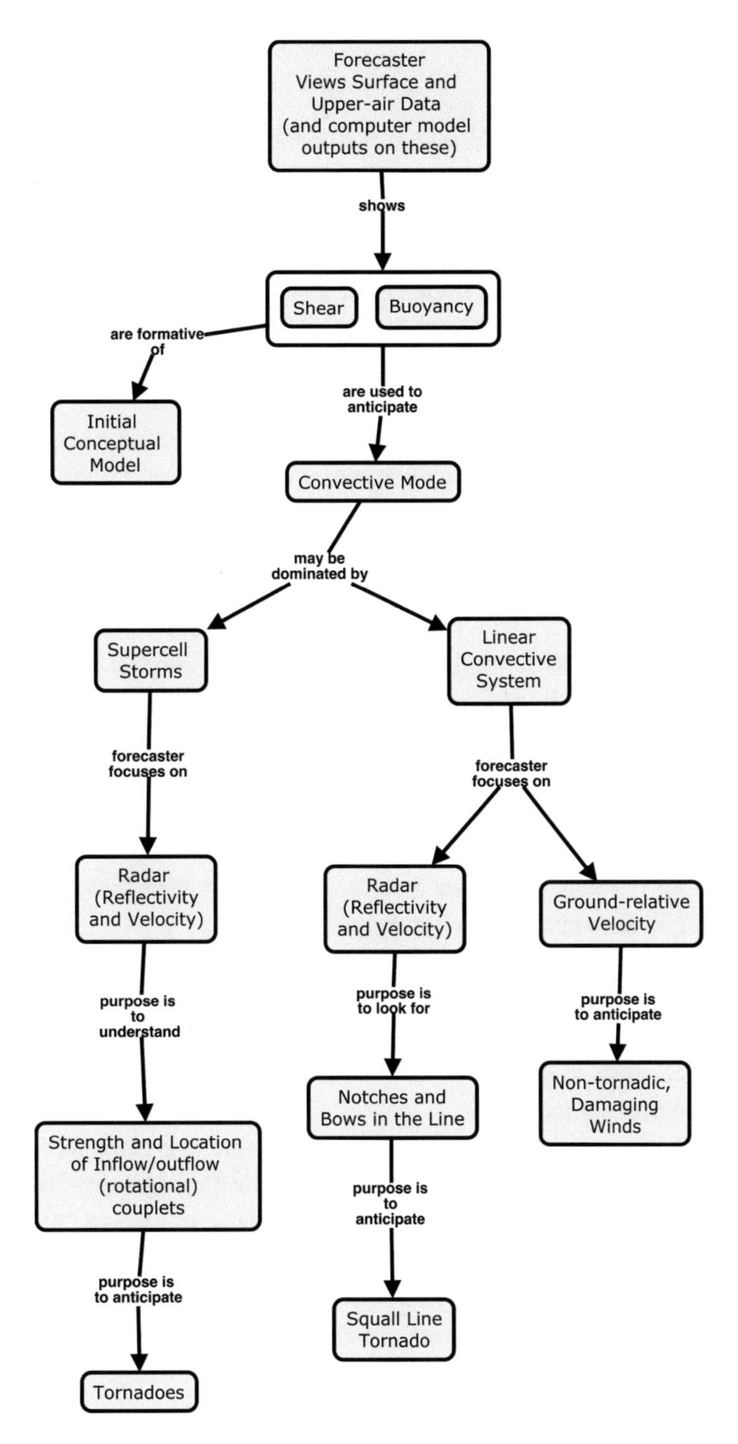

Figure 10.13

A tornado anticipation strategy for the hour to minutes prior to tornado touchdown.

the possible evolution of severe storms that day; they knew the sorts of warnings that might have to be issued, and they had workload management plans in place. They would employ their usual procedures, which included developing a conceptual model of the supercell (see chapter 4), referencing ground-truth data and observations, referencing the radar, and relying on human expertise. The storm cells were large and well defined, making them ideal for detection by the radar algorithm. In addition, there were many NWS-trained spotters in the area, who provided ground-truth observations. Despite these advantages, the forecasters would have to adapt.

The first storm was detected with indications of hail. Tornadogenesis was seen about 30 minutes later, and the first of many tornado warnings was issued. After this first event, additional supercell thunderstorms quickly developed, spanning the distance from north Texas to southern Kansas, with a powerful F-5 tornado striking Bridge Creek, Moore, and Oklahoma City, Oklahoma. The warning for that tornado used strong phrasing, including "tornado emergency" and "extremely dangerous and life-threatening." Such phrasing had never been used before. By the end of the event, eight different supercell thunderstorms had produced 72 tornadoes, with 66 of those tornadoes occurring in the Norman National Weather Service Forecast Office's area of responsibility. At one time that evening, four tornadoes were occurring at the same time in different storms in the Forecast Office's area. Some of these storms tracked over similar areas as well, with the area between Crescent and Mulhall Oklahoma being hit by two different tornadoes about an hour apart. Over an eight-hour period, a warning was issued on average every four minutes. Nearly every supercell within the storm system was prolific in spawning tornadoes, and these persisted for most of the lifetime of their parent supercells. Tracking individual storm cells, maintaining an awareness of their evolution, and actual forecasting became difficult and stressful. Family and friends of the forecasters—and the forecasting facility—were in or near the path of the largest tornado that night.

The individual forecasters had to adapt, as did the entire work system at the forecasting facility. Four warning sectors allowed each warning forecaster to focus on a smaller area, and a fifth meteorologist was designated as warning coordinator. He oversaw operations and used a then-new warning detection system with advanced mesoscyclone and tornado detection algorithms to help ensure that warnings and forecast statements were issued. He monitored the flow of forecasts and compared these to ground-truth.

In addition to adaptation of their work methods "on the fly," during the May 3, 1999, tornado outbreak, forecasters had to take subjective information into account when interpreting the computer model outputs. Prior to the tornado outbreak, there was some evidence that severe storms would break out, but there was also evidence that they might not. The computer model outputs were based on data sets formed from inadequate sampling of temperature and moisture over nearby water areas at various heights in the atmosphere during conditions in which the jet stream was showing an anomalous pattern. The result was uncertainty (in the forecaster's mind) about where convection would most likely occur, if it were to occur. The mesoscale model supported the notion of severe convection in the form of multiple supercells. Sites of convective initiation would be in areas where there was ample potential energy (i.e., higher temperatures due to the ground's absorption of the sun's rays and advection of moisture into the area).

> It is important to recognize, however, that without forecaster confidence in the model, such a substantial revision to the [initial] hypothesis could not be made. ... Forecasters require information about the performance characteristics of the model, such as an understanding of the model "climatology" and false-alarm rates for particular phenomena. How often does the model produce long-lived supercells? How often do such forecasts verify? (Roebber et al., 2002, p. 425)

Forecasters made inferences about convection based on satellite images (suggesting whether the high-level cirrus clouds represented the advection of moisture from the subtropics) and radar data suggestive of wind shear. "Forecasters have become adept at using indirect diagnosis strategies" (Roebber, et al., 2002, p. 427).

In addition to flexecuting with regard to the weather, forecasters also have to flexecute (and sensemake) with regard to their technology. One reason that flexecuting the technology is crucial to forecasting is not just that unexpected circumstances arise, but simply that there is so much technology available. The SAFA expert system (see chapter 11) was intended to help forecasters make decisions about which computer model to select for forecasting hurricane tracks. It was created on the basis of expert knowledge of how the various models work, their biases, and so on.

When new technologies come along, there is always a learning curve, but there is also relearning. For example, new phased array radars generate more frequent and higher resolution scans of storms (Zrnić et al., 2007), presenting details of storm structure and dynamics (e.g., tornadogenesis). This technology is empowering in terms of validating the forecaster's mental model and predicting tornado formation (Heinselman et al., 2015). But the best use of the new scans involves careful comparison of the data with what the forecaster would expect to see in the standard NEXRAD radar scan data (Heinselman et al., 2012). After using the new radar in an experimental evaluation of tornado forecasting (using archived data), one forecaster responded to the question, "Did you do anything that would be atypical of normal work habits?" with this:

I loaded up the four-panel [display] with the lowest four tilts [scan angles] because I was able to see almost real time, updating of each elevation. That was a big help in overall monitoring of the strength of circulation. [I] would not necessarily use this before a warning. The four-panel allowed more hands-off. All tilts is more hands-on, as you go up–down, through time. Felt like I could hold off on warning. Because I had more data to back up decisions. And more data to see the evolution.

Most forecasters can tell stories that involved flexecution with the technology. One U.S. Navy forecaster reported an event in which there was an outbreak of severe storms in the region of the airfield (Hoffman, Coffey, and Ford, 2000). Unexpectedly, the link to the airfield's radar was lost. To compensate, the forecaster made telephone calls to other regional airfields and airports to grab whatever reports and information he could. Most NWS forecasters will explain that they frequently adjust the scan parameters of their NEXRAD system. One NWS forecaster said to us:

Yes, there are frequent needs to change the settings of local 88Ds. Most common are changing the volume coverage pattern to fit the expected weather (always a challenge), or adjusting the pulse repetition frequency to get a storm out of the "purple haze." There are a myriad others too like adjusting the rain rate algorithm or a first guess on storm motion.

Perhaps the best way to understand the processes of sensemaking and flexecuting the weather to create a forecast is to review narratives of actual forecasting situations in more detail. Table 10.3 presents an example, written from the forecasters' perspective, of the May 1999 tornadic event in the Norman, Oklahoma, area.

We will return to this weather event in our discussion of human–computer interdependence in chapter 12. The final outcome on May 3, 1999, was a remarkable achievement in the interdependence of forecasters and their computers.

Appendix B presents the extended narratives for two cases. Case 1 is about forecasting severe storm, and case 2 is about hurricane track forecasting. These narratives are a window on the actual process of forecasting. They exemplify how the issuing of a forecast is not an end point. Because weather events develop, mature, dissipate, and move, a forecast will often be amended, sometimes many times.

Conclusions

Results from studies of forecasting dovetail with the findings in the literature on expertise. The formation of mental models and the processes of sensemaking and flexecution are key ingredients in the forecasting activity. Forecasters ask sensemaking questions (What is happening? Why is it happening?) and mentally project to the future (What will happen? Why will it happen?) (Bosart, 2003). Upon forming a

Table 10.3
A timeline of forecaster reasoning for the May 3, 1999, tornado outbreak

1. *Sensemaking: The Big Picture*
The anomalous southerly jet stream was poorly sampled owing to its offshore location, yet it had considerable influence on the evolution of the computer model forecasts. Given higher resolution forecast information, it might have been possible to reject some hypotheses as inconsistent as the observations came in, and continue this interrogation in an iterative fashion. But in the case of the 1999 outbreak, this was not possible.
2. *Sensemaking: Initial Hypothesis Formation*
It was apparent that a favorable environment for severe convection would exist in the region. Diagnosis using observations and radar wind profiles indicated that upper level winds were likely to be stronger than forecast by the computer models and suggested the possibility of supercell organization.
3. *Sensemaking: Mental Model Refinement*
However, the situation did not conform to the existing conceptual model of severe weather outbreak because there were weak signs of convective initiation. Hence, the initial hypothesis that was formed at the Storm Prediction Center was for late-afternoon convective development, with a transition from isolated supercells to a line of storms.
4. *Flexecution: Hypothesis Testing*
In the stage of hypothesis testing, several lines of evidence were examined. Computer model forecasts showed little run-to-run consistency in the outbreak area, which, combined with weak (observed and forecast) convergence, contributed to much uncertainty regarding convective initiation.
5. *Flexecution: Uncertainty*
It was conceivable that an intense or a prolific outbreak would not form and supercells would not develop at all had the convergence been stronger. Hence, although there was evidence to support severe convection, the prospects for convective initiation were mixed, the information supporting supercell organization was ambiguous until late, and no observational or computer model evidence existed to support an outbreak scenario. These uncertainties delayed the upgrade of the categorical convective outlook from "slight"' to "moderate" risk.
As data came in, these ideas were reexamined:
Observational evidence from the wind profiler network later in the afternoon reinforced the notion that sufficient shear would exist to support supercell organization.
Tornadic supercells ultimately developed earlier, were more numerous, and produced more significant tornadoes than anticipated. An outbreak scenario was always possible, but along with the location of the resulting convection, it was highly sensitive to the analysis details of the southern jet stream anomaly.

Source: Adapted from Roebber et al. (2002).

hypothesis, forecasters determine whether the right ingredients are in play to cause the types of weather that are anticipated (Doswell, 1996). Doing this requires flexible exploration of the data and flexible interaction with the technology (Joslyn and Jones, 2008). Forecasters have to be flexible, if only because they are often interrupted for various reasons. More important, forecasters engage in flexecution especially during unusual weather events. They see technology as a tool having both strengths and limitations. They want to know how various algorithms and automated techniques work so they will be better able to interpret the information coming from them.

Although reasoning can be described generically in terms of such mental process notions as sensemaking and flexecution, there is no one single forecasting process. Forecasting is neither a single process nor a fixed process, and, therefore, it is not an activity that can be decontextualized or strictly prescribed. In addition to skill level, experience, and other human factors, the forecasting process depends on:

- the weather at hand (e.g., severe storms vs. fog),
- the weather as it develops (e.g., tornado development vs. hurricane development),
- the season and locale (e.g., winter in a city in California vs. summer in the forests of Nova Scotia),
- the available data,
- human factors (e.g., individual differences in proficiency, styles, and strategies), and
- the available technological capabilities, including logistical and situational factors (e.g., changes in the technology, loss of a radar uplink).

Forecasters have outlined general approaches to creating a forecast, such as: (1) the forecast funnel of Leonard Snellman (1982), (2) the core questions approach of Lance Bosart, (3) the ingredients-based approach of Doswell (1996), (4) the Base Model of expertise from Hoffman et al. (2006), and (5) the goal sequence model from Josslyn and Jones (2008). These generic descriptions can be thought of as slightly different perspectives on the same thing. Each of them is valid. Each can be regarded as descriptive but also as somewhat prescriptive in spirit. All of them are useful in understanding how forecasters reason. Although they can be used to craft specific models that describe specific forecasting processes after the fact (as in figures 10.4 through 10.11), the act of forecasting is a high-level cognitive domain where *fixed* steps in creating a forecast cannot be universally prespecified and many core activities are continuous and parallel cycles.

This is not the same, however, as saying that none of the processes, strategies procedures that forecasters follow, can be specified or formalized. The next two chapters consider two questions: Can computers be made to think like forecasters? Will computers replace forecasters? These may seem like two versions of the same question. After all, if a computer could mimic the forecaster, then it could replace the forecaster. But these two questions are not the same. Each charts a different path in the evolution of modern forecasting technology and the cognitive work it supports, and each leads to a different resolution of the matter.

Box 10.2
Show Me the Data

As we have shown in this chapter, forecasters need to be able to "drill down" into a great deal of data, some of which is highly technical. Although many citizens prefer succinct and actionable guidance (e.g., Will it rain here today?), many are also genuine weather enthusiasts. They want to see the data sets, maps, animated imagery, and more. These same people often find most forecasts insufficient, including those provided in televised news broadcasts. We suggest that readers consider:

1. Undertaking a targeted search to locate websites (governmental, private sector, and university) that provide free real-time and/or archived satellite, radar (other) data that meet their needs. This applies to both computer and portable appliance applications.
2. Finding sites that provide detailed forecast discussions. Reading these is an effective way to quickly learn which computer models are currently preferred by professional forecasters, why they are preferred, and some information about the current biases of these models. Other aspects of forecaster reasoning are also conveyed, including insight into how forecasters are thinking about the challenging aspects of that week's forecast, and how certain they are about the forecast they have issued. NWS websites routinely offer forecast discussions. So, too, do many university and private-sector sites. If such background information is important, people should find sites that offer this content.
3. Many local television stations have apps and alerting features for severe weather.
4. Subscribing to reliable and credible social media and blog sites will ensure that real-time information (forecasts and warnings) is delivered, rather than having to search for it.
5. Many websites offer smartphone apps and laptop "weather walls" designed specifically for weather enthusiasts.
6. The NWS website immediately presents a U.S. map showing all weather alerts and advisories [http://www.weather.gov]. The site links to the local NWS Weather Forecast Offices and to radar and GOES satellite images [http://www.srh.noaa.gov].
7. The website of the National Center for Atmospheric Prediction presents visualizations of current weather and computer model outputs and model output statistics [http://www.wpc.ncep.noaa.gov].
8. Many of the "experimental" pages at the Storm Prediction Center's website are geared toward severe storms/tornado forecasting [http://www.spc.noaa.gov/exper/].
9. Some university websites are quite good, in fact ideal, for those who might like to try their hand at forecasting. The web pages link to many weather data maps and high-resolution radar products. We are hesitant to list specific sites because they are in a constant state of flux, there are so many good ones, and any list we might provide could be misinterpreted as favoritism or arbitrary. A list of URLs can be provided on request by emailing to rhoffman@ihmc.us.

11 Can a Machine Imitate the Human?

The question "Can computers be made to think like forecasters?" takes us into the literature of expert systems, in which attempts were made to create computer programs that would reason in ways that human forecasters reason. That literature gives us some insight into how forecasters reason, and to that extent, it is valuable to our purpose of understanding how forecasters think about the weather.

Can Computers Be Made to Think Like Forecasters?

In the 1980s and continuing to this day, there have been attempts to generate computer models that predict the weather not by modeling the atmosphere, but by modeling the reasoning of human forecasters. These are called "expert systems." In this chapter, we mine the literature on expert systems to see what it tells us about how human forecasters reason. After all, to build an expert system for weather forecasting based on how humans do it, you have to find out how humans do it. We forego analysis of the attempts to use "neural networks" in weather forecasting because they are not based on explicit conceptual models (for examples, see Frankel, Schiller, Draper, and Barnes, 1990; Kyle, 1985; Smotroff, 1991). Neural networks are mathematical systems that represent causal relations as a network, with probabilities or strengths associated with the relations. As the network is exposed to instances, the probabilities or strengths automatically adjust, and over time the network becomes a model of the category or concept being analyzed. Such models do not really have much of anything to do with actual neurons or neural networks and even less to do with the modeling of human cognition (although for some tasks in some applications, their performance might be quite good). It is the literature on knowledge-based systems (KBSs) for forecasting which offers useful evidence concerning forecaster reasoning.

The field of expert systems emerged in the 1970s and burgeoned in the 1980s. To build an expert system, you need a knowledge base of domain concepts and causal models and an inference engine of "if-then" procedural rules to enable a computer to act as an expert (within limits) (see Brule and Blount, 1989; Clancey, 1989, 1992; Hayes-Roth, Waterman, and Lenat, 1983; Hoffman, 1992). The computer systems often relied on a type of representation called a "frame," which describes conceptual entities in terms of their properties. The systems also could rely on causal networks. All of these components were derived from interviews with domain experts, a method of cognitive task analysis (Hart, 1989; McGraw and Harbison-Briggs, 1989).

The technology of expert systems was based on a new premise in the field of Artificial Intelligence (AI). For some of the pioneers of the field, the goal of AI was to mimic general human intelligence (see, e.g., Dreyfus, 1979; Winston, 1984). To others in the field, the goal of AI was to replicate human performance, without necessarily mimicking humans, that is, without reasoning the same way that humans reason. But in the 1970s, another idea emerged: to capture human knowledge and skill in highly specific domains rather than attempt to create some form of general intelligence that could solve problems of any sort. Attempts at building a "general intelligence" that would apply general logic in problem solving without reliance on domain-specific knowledge were not as successful as had been hoped in the early days of AI. Thus, the more successful expert systems have been ones that are task specific (e.g., diagnosis tasks, planning tasks, financial decision making, industrial process control) (Buchanan et al., 2017).

Expert systems were developed to assist in tasks such as medical diagnosis, electronics debugging, computer programming, diagnosis of bacterial infections, decision making in the search for mineral deposits, airline piloting, manufacturing engineering, and industrial process control, to name but a few. The elicitation, preservation, analysis, and dissemination of the knowledge and skills of experts for the purpose of creating expert systems came to represent one of the most important uses of computer technology (Hoffman, 1992; Hoffman et al., 1995) and remains a topic area within the field of AI (see Giarratano and Riley, 2004).

The era of expert systems can be dated from about 1971, when Edward Feigenbaum and his colleagues (Feigenbaum, Buchanan, and Lederberg, 1971; see also Lederberg and Feigenbaum, 1968) created a software system to capture the reasoning of expert chemists in the interpretation of mass spectrograms. Other seminal expert systems were MYCIN (Buchanan and Shortliffe, 1984; Shortliffe, 1976) for diagnosing bacterial infections and PROSPECTOR (Duda, Gaschnig, and Hart, 1979) for determining site potential for geological exploration.

The expert systems approach was applied early to problems in weather forecasting. For discussions of the earliest expert systems in weather forecasting, see Collopy,

Adya, and Armstrong (2001), Dyer (1987), Moninger (1988) or Roberts et al. (1990). A number of attempts at building meteorology KBSs met with some success. We focus on these systems, developed in the period spanning the 1970s—1990s, because they were premised on the need to understand how expert forecasters reason.

Scientific meteorology and forecasting rely heavily on computational methods and algorithms. An algorithm is a description of a series of steps that if followed results in a definite and correct answer. An example would be the basic algorithm in many computational weather forecasting models. This is to start with data on parameters such as the temperature, air pressure, winds, and moisture content of parcels of air, for parcels of so many square kilometers, in a grid of some number of parcels. Equations of dynamics are applied to determine how the adjacent parcels interact at their boundaries, and this process is iterated across all the parcels and across time, to generate a representation of what the atmosphere would look like at some point in the future. Depending on the size of the grid, this process is computationally intensive. Such computational models are extremely sensitive to initial conditions, which explains why models provide non-definitive predictions (each model provides a slightly different solution). There are vast areas of the globe that are undersampled (e.g., the Pacific Ocean is an extreme example, but smaller areas exist even in the relatively well sampled United States). Progress on massive computability is one of the reasons that there have been great advances recently in computational modeling of the atmosphere (see chapter 2; also Sumner, 2015).

But KBSs are a different sort of beast from the computer models used in atmospheric modeling (discussed in more detail in chapter 12). KBSs rely on heuristics, also called "rules-of-thumb." A heuristic is a decision rule that is usually qualitative rather than quantitative, and it is not guaranteed to result in a correct solution or inference but often does work satisfactorily. Box 11.1 presents an example.

Some expert systems for weather forecasting were hybrid systems, having elements of human knowledge and reasoning combined with algorithms. Gaffney and Racer's (1983) system for predicting the likelihood of severe storms expressed storm probability in terms of qualitative labels (i.e., "weak," "moderate," "strong"), but performed computations based on a mathematical equation that combined the weighted values of a number of atmospheric parameters (e.g., wind speeds and vorticity). Gaffney and Racer (1983) regarded their algorithm as an expert system for two reasons. First, the parameters were selected on the basis of experienced forecasters' judgments. Second, the algorithm could learn by adjusting the weightings of the parameters. However, Gaffney and Racer's (1983) system did not engage in logical reasoning based on an expert's weather concepts and principles.

Box 11.1
A Heuristic for the Lifting of Fog

At an air base, the pilots and trainees need to log practice hours, and on days when there is morning fog, they are keen to learn from the forecasters about when the fog might lift. One forecaster had learned from experience how to quickly assess lifting. From the observation platform on the roof of the air terminal building, it was possible to see a tall hotel miles in the distance. Over years of observation, a simple rule was determined: If the fog lifted to where one could see a certain floor of the building by a certain time in the morning, then it would be possible for planes to take off before noon. This informal rule could not be used to make official forecasts, certainly, although one could determine the specific physics in play. But the simple observation allowed the forecasters to quickly size up the day's flying conditions.

This heuristic addresses how the rate of solar heating is affecting the fog (although it is based on a casual observation). A similar heuristic can be applied to fog formation as well. If a distant building and other lights twinkle, fog is less likely to form overnight; if lights don't twinkle, it means fog is starting/more likely to form.

> In the weather forecasting expert system, rules are primarily used to express relationships that have some basis in physical laws ... [but] rules are not used to express purely mathematical relationships ... an experienced forecaster keys the forecast to the primary factors in a rather intuitive way, rather than attempting to account for every factor in a mathematically rigorous process. (Jasperson and Venne, 1990 pp. 16–17)

Our focus is on KBSs derived by interviewing experts and hence capturing (at least in part), the terminology, causal associations, and steps by which they solve problems. To develop an expert system, one must engage experts in one or another knowledge elicitation procedure to reveal their knowledge and reasoning (see Cooke and Rowe, 1994; Crandall, Klein, and Hoffman; 2006; Hoffman and Lintern, 2006; Hoffman et al., 1995; Rowe and Wright, 2001; Shadbolt and Burton, 1990). This made expert system development projects a potential source of knowledge concerning forecaster reasoning. In KBSs, expert explanations are explicitly represented as a qualitative model of domain concepts and their definitions and production rules (including hierarchical and causal relations). Thus KBSs replicate at least in part the language in which experts describe and articulate arguments about the weather.

The Structure of Forecasting KBSs

Most of the forecasting expert systems combined a good many rules (100 to 200) composed into decision trees (Clancey, 1985). Here is an example of the sorts of procedural rules used to capture domain reasoning in weather forecasting expert systems (Jasperson and Venne, 1990). Such rules can be cryptic and sometimes complicated, but a simple example is:

IF sky cover is less than or equal to 3/10
and
IF air mass is modified or tropical
THEN conclude a normal temperature variation amplitude.

In various expert systems, one finds combinations of forward-chaining and backward-chaining in the control processes.

- Forward chaining is when a set of current conditions leads to a conclusion: *IF there are conditions x,y, and z, THEN thunderstorms will happen.*
- Backward chaining is when the reasoning starts off assuming a final condition and then works backward through the rules to find a path that leads to the known conditions: *El Niño will develop this winter, and IF El Niño is developing, there will be a mild winter in Canada.*

Table 11.1 is a sample of the kinds of forecasting expert systems that were created (see also Armstrong, 2001c; Bullas, McLeod, and de Lorenzis, 1990; Campbell, 1988; Campbell and Olson, 1987; Diak et al., 1998; Kyle, 1985; Moninger, 1990; Orgill, Kincheloe, and Sutherland, 1992; Roberts et al., 1990). For some of these, the names are obviously not acronyms, for some they are, and for some of those, the reports did not explain the names.

Box 11.2
What Is an "Air Mass"?

An air mass is a synoptic-scale body of air that is relatively homogenous in terms of its surface temperature and moisture content. Frequently used air-mass categories are topical and arctic (temperature) and maritime and continental (moisture). When coupled, these categories describe four air mass types (e.g., maritime tropical), each of which has strong implications for weather dynamics. A modified air mass is one that has not lost its mass characteristics but has nonetheless been affected by interactions with modifying influences (e.g., passing over cold water).

Table 11.1
Some weather forecasting knowledge-based systems

System Name (or Acronym)	Description	Reference
Willard	Hybrid system for forecasting severe thunderstorms, tornadoes, hail, and wind gusts. The system developer was a meteorologist and so did "self-knowledge elicitation" along with interviews of other forecasters. The input to Willard was data from a computational model.	Zubrick, 1984, 1988; Zubrick and Riese, 1985
ARCHER	Identified severe weather events (fronts, microbursts, etc.) from Doppler radar signatures. In addition to if-then rules, it relied on mathematical models of atmospheric dynamics.	Moninger, 1986
METEOR	Predicted storms based on observational data (e.g., cloud cover). Relied on heuristic rules to interpret the outputs of a computer model based on climate statistics.	Elio and de Haan, 1986; Elio, de Haan, and Strong, 1987
Itaska	Hybrid system to conduct short-range single-station forecasting using only surface and upper air data. The system used heuristic rules organized around a standard conceptual model of forecasting and also a "neural network" module. It was one of the first expert systems to use a graphical user interface.	Jasperson and Venne, 1990
ZEUS	Expert system for predicting fog and visibility at U.S. Air Force bases. It relied on heuristic rules derived in a group discussion of the causes of fog, as well as computer models of atmospheric dynamics.	Dyer, 1989; Dyer and Freeman, 1989; Stunder, Dyer, and Koch, 1987a; Stunder, Koch, Sletten, and Lee, 1987b
HAIL	Expert system for predicting hail, tornadoes, and strong winds. Procedural rules operated on radar data, input by the forecaster, including tilt to the storm core, reflectivity of storm core at a low and mid-level, a gradient in reflectivity situated on the advancing flank of the storm, strong convergence of shear, and differential reflectivity indicative of hail.	Merrem and Brady, 1988; see also Stewart et al., 1989
AESOP	Expert system for shipboard prediction of maritime fog and haze. It relied on a physical model of advection, a statistical model, and observations made onship.	Peak, 1988; Peak and Tag, 1989

Table 11.1 (continued)

System Name (or Acronym)	Description	Reference
WIND	Expert system for predicting downslope winds in northern Colorado. It relied on physical and statistical models and took its inputs directly from the AWIPS.	Weaver and Philips, 1990; Rockwood et al., 1992
TIPS (Thunderstorm Intelligence Prediction System)	Expert system for single-station forecasting of microburst occurrence and type (i.e., dry, wet, severe) based on upper atmosphere observations (temperatures, winds, stability, etc.). The rule base was derived from meteorological principles and also the subjective judgments of the forecaster.	Lee and Passner, 1993
SIAMES (Satellite Image Analysis Meteorological Expert System)	Expert system for training in satellite image interpretation. A CD-ROM was used to depict images and accompanying explanatory text. The system prompted visual search and analysis going from major features (i.e., fronts, lows, etc.) to subtler features (i.e., effects of sun glint, internal waves, reef effects); trainees could also navigate through the image library. The source of the expertise was a documentation analysis of Navy Tactical Applications Guides (containing thousands of high-resolution images).	Fett et al., 1997

The systems listed in table 11.1 were all aimed at the forecasting task. But it is also possible to create a KBS to help forecasters deal with their technology. The Systematic Approach Forecast Aid (SAFA) was developed at the U.S. Naval Postgraduate School by Lester Carr and his associates for forecasting the tracks of tropical cyclones in the Pacific Ocean (Carr, Elsberry, and Peak, 2001). A number of computer models are used for predicting the tracks of hurricanes (in the Atlantic Ocean) and cyclones (in the Pacific Ocean). They differ in a number of ways: some are statistical, some take into account ocean surface temperatures, some are based on models of atmospheric dynamics in a region, some are based on models of atmospheric dynamics in an entire hemisphere, and some are global in scope.

Each model has certain systematic tendencies or "biases." For example, a model may be overly sensitive to the effects of one cyclone on another in the same region of the ocean. A model may tend to predict tracks for westward moving cyclones that show too much poleward movement at longer forecast intervals. A model might be particularly sensitive to the effects of vertical wind shear on cyclone evolution. As a result of the differences in how the models work, and their various biases, for any given cyclone, the predictions of tracks from different models might show strong agreement, moderate agreement, moderate disagreement, or strong disagreement. The agreement (or disagreement) might be modest in shorter-term forecasts (e.g., 6 hours) and might remain that way over longer forecast intervals (e.g., 24 to 72 hours) or the agreement (or disagreement) might decrease or increase. With experience, forecasters learn to take model differences and tendencies into account as they create their forecasts. "The meteorological reasoning of the forecaster is highly important, albeit qualitative, component of the official forecast" (Carr et al., 2001, p. 355). But under certain circumstances, it can still be hard for the forecaster to tell which model (or models) is best.

Carr and his colleagues (2001) interviewed a number of experienced forecasters about the various model biases and developed a tool to help forecasters select those particular models that would converge on a best prediction, based on the knowledge of the systematic biases of each individual model. Next, a team of six forecasters used the expert system to aid them in "simulated real-time forecasting" of tracks for a set of cyclones (based on archived data). The SAFA system identified those models that generated track predictions suspected to have an error (at 72 hours) of 300 nautical miles and then recalculated a consensus track for the models that did not "bust." The SAFA system determined the track, winds, and sea surface pressure predictions of the various models and displayed its results in animations over time of the model outputs, thus guiding the forecasters in their deliberations:

> Perhaps the most satisfying success during the Beta test was during Typhoon Gloria [1965] when the three global models were predicting a left turn toward the Philippines and the regional models were predicting a right turn. Such widely diverse track guidance is a difficult situation for the forecaster ... in this case the forecaster noticed the turn [in one of the models] and an examination of the wind and sea level pressure indicated a direct cyclone interaction of Gloria with a cyclonic circulation over the Philippines ... [a] regional high-resolution model predicted a more compact cyclone such that a direct cyclonic interaction was not predicted to occur. ... Because satellite imagery suggested that the storm at this time was small, the decision was to accept the two regional model tracks and reject the three global models tracks. That was the correct decision. (Carr et al., 2001, p. 359)

Note that the SAFA system did not make decisions, which had been the vision for earlier generations of expert systems. In this newer incarnation, the expert system is "expert" at helping human forecasters by conducting analyses that would otherwise consume time and resources, enabling best use of the computer model guidance, that is, selection of those models that are likely to provide the best consensus track for a given cyclone and circumstance. The various computer models tend to cluster. For a given cyclone, two or three of the models might show a track going in one general direction with one particular curvature pattern. Such agreement might suggest that that one has identified the best forecast, but it might be misleading because the clustering models might be subject to the same error tendencies. "When one model track within [a] cluster is found likely to be erroneous, it would take only a little time to animate the other model predictions from the cluster to see if the same error mechanism is present" (Carr et al., 2001, p. 364). As another example of decision aiding, adding more models to generate a combined track forecast does not necessarily result in a more accurate forecast because if an error is introduced, then it would take valuable forecaster time to identify and correct the error "and not provide information worth the effort" (Carr et al., 2001, p. 364).

As an indication that there is yet much more research and development activity to be done in this area, for the set of cyclones studied, there were more than a dozen track forecast errors that could not be attributed to any of the bias mechanisms that had been identified by experts and therefore included in the SAFA system. Nevertheless,

> This is a step toward a new paradigm in which it is acknowledged that the primary guidance for tropical cyclone track forecasting is the dynamical models. ... Then, the forecaster can add value by applying the conceptual models as guided by the SAFA ... a more informative prognostic reasoning message can be issued (p. 367).

But doing today what Carr et al. (2001) did would be more difficult. Models are being updated far more frequently, sometimes more than once in a year, challenging forecasters to develop understanding of the models' biases.

A main take-away is that these expert systems were not merely sets of if-then procedural rules derived from experts' knowledge. Nearly all of them relied on computer models of atmospheric dynamics or statistical/climate models. They also relied on computer science technologies in addition to procedural rules, such as neural nets, frames, goal hierarchies, decision trees, uncertainty representations, and other techniques.

So How Well Did the Expert Systems Perform?

The researchers who developed expert systems for weather forecasting did not advocate for the notion that the expert systems would replace the human forecaster, by virtue of their success at imitating forecaster reasoning. Rather, the goal was to replace unaided judgment (see Collpoy, Adya, and Armstrong, 2001). That said, some expert systems were directly subject to evaluation by comparison of their forecasts to those of human forecasters. All such evaluation processes resulted in ideas about how to improve the technology and the expert systems approach (Bullas, McLeod, and de Lorenzis, 1990; Diak et al., 1998). In many cases, improvements were needed in terms of support for visualization; those improvements pertain to all forecasting technology and not just expert systems (see chapter 9). Expert systems that focused on particular aspects of a forecast, such as cloud cover (e.g., Stunder, Koch, Sletten, and Lee, 1987b), tended to reveal their own weaknesses—typically, they needed to have enriched knowledge bases, and they needed to incorporate better models of atmospheric dynamics.

As we explained in previous chapters, forecast accuracy is assessed in terms of a skill score. Using the skill score, the performance of some expert systems compared favorably to that of NWS-issued forecasts (e.g., "Willard"; Zubrick, 1988). A USAF fog forecasting system reported by Rosemary Dyer and her colleagues (Dyer and Freeman, 1989; Stunder et al., 1987a, 1987b) (see table 11.1) showed skill scores in the range of +0.16 to +0.46, with any positive value reflecting the gain from the forecast over and above the forecast climatological norm. Weaver and Philips (1990; see their table 2) reported detection hit rates of greater than 90% on their downslope wind system, and skill scores in the range of +0.8 to +0.9. Lee and Passner (1993 see their table 2) reported hit rates of about +0.7 to +0.9 on their single-point microburst system, with skill scores in the range of +0.16 to +0.62.

Stewart, Moninger, Grassia, Brady, and Merrem (1989) compared the HAIL system for diagnosing severe weather from Doppler radar data (Merrem and Brady, 1988) to forecasts made by seven NOAA research meteorologists. The forecasters made probability forecasts for hail and severe hail for each of 75 sets of radar data, with each data set consisting of the values for seven radar parameters (including reflectivity at low, mid, and upper levels of the storm; the tilt of the storm; and rotation or convergence of the storm core). Hail was verified by observations made by "storm chasers" who were participating in the PROFS Program (see chapter 2). Hail had actually occurred in about 15% of the cases and severe hail in about 5% of the cases. As one might expect, assuming that the meteorologist participants were proficient forecasters, they tended to agree—intercorrelations of 0.75 to 0.91 for hail and 0.78 to 0.95 for severe hail. Although the forecasts showed some only a slight improvement compared to the climatological norm (i.e., low skill scores), the forecasters were able to distinguish between hail-producing and nonhail-producing storms to some extent.

The regression models computed across the 75 forecasts showed that a single radar parameter (reflectivity at low and mid levels of the storm) could account for 80% to 92% of the variance in the individual forecasts. The various cues were positively correlated. In other words, they provided partially redundant information, with only one of the cues standing out slightly in terms of its relative importance. In this sort of situation, differing underlying reasoning strategies could result in similar forecasts and correlated skill scores; different meteorologists attached different importance to different cues. As a result,

> In the relatively infrequent [radar] scans when cues diverge i.e., when some cues indicate hail and other cues indicate no hail, disagreements among meteorologists will emerge. Thus, meteorologists can be expected to disagree most when forecasting is most difficult. (Stewart et al., 1989, p. 31)

Given the level of agreement and consistency in the meteorologist's forecasts, and the consequent reasonably good fit of the statistical models to those forecasts, it is not surprising that human forecasts fared slightly better than those of the HAIL expert system, which is based on the reasoning of an individual forecaster. The correlations between the HAIL forecasts and those of the meteorologists were all positive, ranging from 0.70 to 0.85 for hail and from 0.63 to 0.79 for severe hail. In predicting hail, the expert system performed only slightly better than the worst-performing meteorologist. However, in diagnosing severe hail, the expert system performed at about the level of the best-performing meteorologist. The range of correlations between the actual weather and the HAIL forecasts fell at about the range of the correlations between the actual weather and the meteorologist's forecasts, and the correlations between the actual weather and the forecasts were based on the linear models of the individual forecasters. HAIL may have come in second place in this competition, but it was a close race in which all the competitors were dealing with limited information.

"Willard" had problems in forecasting severe weather for a day in which there were numerous severe weather reports. The input to Willard consisted of the outputs of a mathematical model of atmospheric dynamics, and the large grid scale of that model did not represent fine-scale features (short-wave low-pressure troughs) (see Zubrick, 1988). Some evaluations of expert systems reported resounding success, in the sense that the expert advisory systems were received favorably and actually used. More objective criteria for success were also applied. For his expert system for visibility forecasting, Peak (1988) reported a hit rate of about 50% to 70% depending on what was being forecast—haze, fog, and so on.

We have been discussing expert systems as a thing of the past, which they mostly are. As the technology of expert systems "took off" in the 1980s, problems emerged. Forecaster users had lots to say about how the first-generation systems needed to be changed, basically, by making them more user-friendly and less obviously designed at the convenience of the programmers (see Rockwood, Weaver, Brown, Jamison, and Holmes, 1992). This was an era in which there were still some forecasters who had had little experience with computers, which set the stage for skepticism. The forecasting methods and procedures that had been instantiated in expert systems often did not match those utilized by forecasters, in part because forecasters do not all use the same process (e.g., they look at different data and do so at different stages in their sensemaking). Thus, different forecasters did not get the same results from the use of an expert system. In addition, it took upward of 30 minutes for the forecaster to input the data and answers to questions that the expert system needed to be able to make inferences about the weather (although the entry claim for the expert systems was often that their use would take only minutes). Another important element that contributed to the decline of expert systems—in all the application domains and not just weather forecasting—was the incapacity of the systems to provide rich (let alone satisfying) explanations of how the system came to its conclusions. For most reports on expert system development, there was no indication that there was any emphasis on the requirement for an expert system to be able to explain the how and why of its outputs (an exception is AESOP). The expert systems were opaque, and a print-out of the rule sequences followed for particular cases was too long and cryptic to be of much explanatory value. An example of one piece of a cryptic rule appears in box 11.3.

The upshot was that less experienced forecasters—presumably the ones for whom an expert system would be designed—were limited in their ability to evaluate the outputs of the expert system. Success in the use of expert systems was related to the expertise of the forecasters: Those who got the best results from using the expert system were those who needed it the least! Thus, the expert systems came to be used in training, helping less experienced forecasters learn what to look for in the weather data. That certainly is a useful application, though not the one initially intended.

Box 11.3
Example of a Cryptic Rule

```
if (group_str_find ("F GF IF," press_weather)){
    if (vis < 4:
   else
   return (3);
}
else if ((time>==5X 6X 7X 8X," press _weather))
else
    return (1)
```

Peering into the Black Box

Most of the reports on KBSs did not provide many details on how the expert knowledge was acquired or, for that matter, concrete evidence that the forecasters whose knowledge was tapped would actually qualify as experts (as opposed to "experienced" or "highly experienced" forecasters; see chapter 7). But it seems clear that most of the knowledge acquisition was based on documentation analysis and unstructured interviews with only one or a few forecasters, who were often the system developers. Most system developers relied on unstructured interviews, which accounts for some of the difficulties that were noted—knowledge elicitation via unstructured interviews is exhausting and time-consuming (see Hoffman, 1987a). A case in point is the HAIL system by Merrem and Brady (1988). The system was based on unstructured interviews with a single research meteorologist, and "since development of an expert system is extremely time-consuming, it was not possible to develop one for the other meteorologists in this experiment" (Merrem and Brady, 1988, p. 28).

This said, the knowledge elicitation procedures highlighted the richness and depth of expert knowledge in forecasting. The process of developing an expert system can support the expert in explicating and refining their knowledge. A number of expert system developers commented on the value of the knowledge elicitation procedure for actually helping the forecasters to refine their expertise. That is, the act of attempting to express knowledge and reasoning strategies in the form of implementable rules encourages the expert to think about their knowledge and reasoning from a fresh perspective and enables the expert to compare their reasoning to that provided by other experts. Kyle (1985) noted this benefit to his mathematical model-based expert system, commenting that:

> Meteorologists seem to be able to converse about the topics being input into the expert system with almost no mathematics. The difference is that each of the people in the conversation already has a mental image of the phenomena involved ... graphical information is especially helpful to the user in transforming his mental image into the system result and vice versa, and that is important in achieving system credibility. (pp. 246–247)

Going beyond this, Elio and de Haan (1986) speculated that, "the 'side-effect' of designing expert systems—compelling experts to think about domain and knowledge in new ways—may be the more significant benefit" (p. 544; see also Elio, de Haan, and Strong, 1987). Developers also commented about the ways in which the expert forecaster understands the limitations of the mathematical/statistical algorithms that are incorporated into expert systems:

the [statistical model] implicitly embodies expert knowledge about how measurable meteorological conditions can be related to infer large-scale weather dynamics ... [but] the meteorologist does not use the model as a "black box." He understands the factors that can fool the model's predictor variables, and he compensates for them. This kind of knowledge distinguishes the use of the model from that of less experienced forecasters. (Elio and de Haan, 1986, p. 527)

The expert systems literature also highlights the qualitative nature of forecasters' mental models. Specifically, forecasters rely on qualitative reasoning in addition to some mathematical reckoning. This emphasis on qualitative reasoning is most evident in forecasters' description of what the weather is or will be when they participated in knowledge elicitation sessions (Elio and de Haan, 1986; Jasperson and Venne, 1990). Nearly every report includes an acknowledgment of the role of the forecaster's conceptual model of the atmosphere. Some expert systems, such as that of Merrem and Brady (1988), were said to have been designed "to represent as closely as possible the thinking process used by the chosen expert meteorologist" (Stewart et al., 1989, p. 28). A statement by Jasperson and Venne (1990) is the most explicit about what the meteorologist's thinking process is supposed to be:

The meteorological forecasting process that is imitated begins with the assimilation of all the pertinent data into a diagnostic model. At the conclusion of this assimilation, the meteorologist has a mental model of the current synoptic situation. This is perhaps the most important part of the forecasting process. ... The meteorologist then uses this model to project a consistent forecast of future meteorological events. (p. 4)

This description entirely duplicates the generic models of forecaster reasoning described in chapter 10. The expert system that Jasperson and Venne (1990) developed had an architecture in which the domain concepts and rules were organized according to this analysis of forecaster decision making. In the first phase, the expert system used input data to call up the appropriate frames and then used slots and slot inheritance to infer additional parameter values. After a representation of the weather situation was refined, a second set of rules and frames were invoked to conduct the forecast in a decision-tree fashion.

Looking back at expert systems applications broadly, the more successful expert systems were task-specific (e.g., diagnosis tasks, planning tasks, industrial process control) (Buchanan et al., 2017). The attempts to develop expert systems for weather forecasting certainly bear out this generalization: The superior performance of experts is made possible by the experts' extensive and specialized knowledge of the domain concepts and causal relationships, and explicit knowledge of their own problem-solving strategies.

Developers of the weather forecasting expert systems emphasized the need to incorporate local knowledge (Elio and de Haan, 1986; Elio, de Haan, and Strong, 1987; Stunder, Dyer, and Koch, 1987a). To be useful, a weather forecasting expert system cannot be generic; rather it must be tightly coupled to particular contexts (e.g., by including frames for "objects" that specify regions or geographical features) (Jasperson and Venne, 1990; Orgill, Kincheloe, and Sutherland, 1992; Sumner, 2015). "The local expertise and idiosyncratic factors (e.g., topographical influences) are often diluted, or ignored, in general weather models; this is precisely the kind of information that is best represented in a knowledge-based system" (Elio and de Haan, 1986, p. 544).

A main contributing factor to the success of the Zeus system for predicting visibility at airbases (see table 11.1; Stunder et al., 1987b) was precisely the incorporation of location-specific information and trends (e.g., air trajectories in the Fort Bragg region, the utility for fog forecasting of surface layer information at Seymour Johnson Airbase). Given this, many expert systems are not transportable, not just because they need local knowledge but also because the process of incorporating local knowledge into a generic system entails that the entire rule base must be reconfigured (Dyer, 1989; Dyer and Freeman, 1989). Experienced forecasters insist that the consideration of local factors (geographically based) is quite often key to the success (or failure) of a forecast. Thus, some expert systems had knowledge bases that were layered or modularized, separating domain-general principles from "regional" and "local" layers (Dyer, 1989; Lee and Passner, 1993; Orgill, Kincheloe, and Sutherland, 1992).

Dyer (1987, 1990) and Peak and Tag (1989) noted that another side benefit of expert system development projects, possibly a major benefit of the approach, is that by incorporating local expertise, one is helping to preserve and disseminate hard-gained knowledge and experience (which was a focus in chapter 8).

Metamorphosis from Expert Systems to Intelligent Systems

Expert systems ran into a number of problems, which we highlighted earlier: Those who benefited the most from using an expert system were those who needed it the least; the expert systems were limited in their ability to explain their reasoning, and the best success was achieved only if the expert system was highly context-specific. Over time, AI researchers and application software engineers came to use the term "intelligent systems" instead of "expert systems." The IEEE journal *Expert Systems* was retitled *Intelligent Systems* in 2001. This is slightly but importantly different from Artificial Intelligence (AI). The traditional vision for AI is to replicate human intelligence, based on the seminal ideas of Alan Turing and others. Intelligent systems are computer software systems that perceive, reason, learn, and rely on mechanisms of inference, but they do not necessarily mimic the human's problem-solving and reasoning methods. In addition, intelligent systems technologies are often oriented toward applications as decision support systems. In other words, they assist in cognitive work, but they do not necessarily do so by attempting to mimic the human or substitute for the human (see Turban, 1992).

> An expert system to process weather radar data ... highlighted those regions of the radar scope where the expert system dictates that microbursts or gust fronts might occur ... [the system did not] perform the analysis the way the human forecaster would. There had been an examination of the physical phenomena and of the precursors to the phenomena. The resulting computer program cannot, by any stretch of the imagination, be said to mimic the human expert. The only justification for this approach is simply that it works. (Dyer, 1987, p. 23)

This statement by senior forecaster Rosemary Dyer foreshadowed the evolution of expert systems into a notion that the intelligence—the expertise—emerges when the machine helps the human be better at being human.

An example of the development of an intelligent system for weather decision making is a project conducted by leading cognitive systems engineer Emilie Roth and her colleagues (Roth et al., 2006; Scott et al., 2002, 2005). In the paradigm of cognitive systems engineering, a process of cognitive task analysis is conducted to generate models of the cognitive work. This can involve observing the work as well as conducting structured interviews to understand the cognitive work requirements and factors that contribute to cognitive complexity (see chapter 8). Roth et al. (2006) conducted such analyses in the context of U.S. Air Force airlift operations, which naturally has a concern with aviation weather.

The cognitive task analysis focused on the work of weather forecasters who are responsible for predicting weather patterns and their likely impact on scheduled flight missions in different portions of the world. It also focused on the work of flight managers who are the consumers of the weather forecast products. Flight managers monitor and manage airlift flights during execution (takeoff to landing). They are responsible for rerouting flights if weather conditions necessitate it.

One of the striking results of the cognitive task analysis was that managers had no visualizations to support the specific collaborative work in deciding whether rerouting due to weather conditions would be needed, and in coming up with appropriate reroutes. They had weather maps, and flight managers had ways to visualize flight mission routes, but there were no ways to visualize the mission routes superimposed on the weather and no way to be alerted to weather patterns that could endanger scheduled flights.

The cognitive task analyses revealed a number of "work-arounds," where personnel had developed methods to compensate for the deficiencies in their software and other support systems. The cognitive task analysis also revealed a number of "leverage points" where intelligent systems would be helpful. Three such important leverage points were: (1) searching, collecting, and integrating information from multiple sources to generate more accurate weather forecasts; (2) developing an overall situational awareness of the aviation-pertinent weather as it related to planned routing of scheduled airlift missions; and (3) alerting to dynamic changes in weather in geographic areas of interest (as defined by the presence of scheduled missions).

The intelligent systems element to their Work-Centered Support System for Global Weather Management (WCSS-GWM) involved the creation of "software agents." These chunks of software that acquire data for outside sources, find information that supports or matches a given forecast, and, given some mission plan, identify weather that would impact a mission. The forecaster could identify watch areas and the software agents could alert forecaster attention regarding those areas. The intelligence here is that the software agents can "go out into the world" and conduct certain fairly simple search and data analysis tasks, bring important information back to the decision maker, and thereby make the human–machine work system more capable and efficient. Especially important were forecasts for airfields and upper air forecasts and pilot reports on weather that impacts aviation. The WCSS-GWM software system is, in effect, a tool to help nonprogrammers create software agents that can assist them.

With regard to the need to maintain situational awareness of the aviation-pertinent weather, WCSS-GWM makes the activities of the agents visible to the forecasters and flight planners on a display that is geo-referenced (i.e., it relies on map overlays). Such a display had not previously been available. By implementing software agents and displays that take advantage of the leverage points, the WCSS-GWM system not only helps forecasting but also helps monitor weather with respect to individual missions and helps in other aspects of the cognitive work, such as the creation of forecast products and the collaboration of forecasters with flight mission planners.

The WCSS-GWM was installed at an operations center and was refined iteratively over a period of three years, based on observations, interviews, and feedback from the Air Force personnel. One of the most striking findings over this period involved the rapid pace of change in the nature of the cognitive work: changes in goals, the scale and scope of missions, team and organizational structure, introduction of new information sources and data types, cohort-related ebbs and flows in the overall the proficiency level of the operational workforce, and complexity as more situations and problems emerged. These changes required changes to roles and responsibilities, the creation of new roles (e.g., a forecaster dedicated to high-risk missions), and other operational changes.

> A majority of the system change requests arose from changes in how the system was used, changes in work processes, organizational changes, changes in systems [WCSS-GWM] communicated with, and other environmental changes. One of the most common reasons ... was expansion of the role of the WCSS-GWM within the organization—either its use by a new category of user or by expanding its use into a new area of work. The WCSS-GWM was originally conceived as a tool to aid collaboration between weather forecaster and flight manager in identifying mission-endangering weather en route. As it came into daily use ... a number of system change requests were made in order to expand the utility of the system. Most of these changes involved bringing new data into the system and overlaying new information on the map. (Roth et al., 2006, p. 697)

The success at using WCSS-GWM to support these adaptations counts as one of the many success stories of cognitive systems engineering. The concept of "evolvable systems" adduced by Roth et al. (2006) represents a major contribution to the field of intelligent systems.

The wave of excitement and effort at making forecasting expert systems dissipated as intelligent systems technologies became more advanced, as the field of cognitive systems engineering emerged, and as cognitive task analysis methodologies came to be more widely utilized in technology research and development (see Crandall and Hoffman, 2013; Crandall, Klein, and Hoffman, 2006; Militello and Hutton, 2000; Schraagen, Chipoman and Shalin, 2000). There are still some efforts at making forecasting expert systems, but the technological tide has turned to computer systems that model the weather, rather than systems that model the forecaster. This latter methodology leads to the second question: "Will computers replace forecasters?"

12 Can a Machine Replace the Human?

It always escapes me why [some meteorologists] feel that they must work toward replacing the forecaster rather than couching their [computer model outputs] in positive terms of assisting the forecaster, especially in difficult situations. I believe that part of the answer to this rhetorical question is that [computer model outputs are] not especially helpful and more often than not misleading in economically significant situations. (Snellman, 1978, p. 4)

The creation of forecasts by expert systems that imitate or duplicate the reasoning of human forecasters is just one approach to computational modeling. Another approach, the primary one, is to create computer systems that model the physics and dynamics of the atmosphere. As the early computational models became more powerful and capable, some meteorologists claimed that the computers would eventually replace the human forecaster (see Snellman's [1978] lament on this stance above). Claims are sometimes made today that computational models outperform human forecasters, leaving some speculating about how soon such models will replace human forecasters. Before addressing this question, we need to say more about how the computer models work.

Computational Weather Prediction Models

Although attempts to computationally model the atmosphere date to the 1950s, it was in the 1980s when computer models started to become useful in forecasting by virtue of their finer grain of analysis and their increased processing capacity. Computational weather prediction models have weather data as inputs (called the "initialization"), and then they run equations that are simulations of various atmospheric-oceanic dynamics (World Meteorological Organization, 2016). Computer models that are intended to support research simulate a variety of things: horizontal transport of temperature and humidity, scattering of solar radiation (by water vapor, ozone, clouds), cloud cover, cloud dynamics (buoyancy, downdrafts, evaporation, etc.), drag caused by the interaction of winds with the terrain, and many other physical and thermodynamic processes. Computer models intended for use in operational forecasting produce upper level maps, surface charts, humidity fields, and other projections that are valuable to forecasters.

The equations driving the simulations are nonlinear partial differential equations. The equations work across a grid representing model-specific areas of the Earth's surface and upward through layers of the atmosphere above. Each sector in the three-dimensional grid is described in terms of a set of parameters (i.e., temperature, pressure, winds). The physical equations are thermodynamic expressions of the ways in which the air at a sector in the grid interacts with the air in each of the sectors that are adjacent to it. The layers near the Earth's surface can also be subject to equations that take into account terrain and land–water interactions. The sectors may be spaced within a coarse or fine grid. A model can cover a region (say, at the continental scale), a hemisphere, or the entire globe. Some models take ocean surface temperatures or snow cover into account.

When a model is run, all the physical interactions of all the sectors in all the layers in the grid are iterated over many time steps (which can also be large or small). For this reason, it can take many minutes to hours for a computer model to output a description of the weather, say, two weeks into the future, even using the most advanced and powerful computer arrays and numerical processing techniques. In addition, adjustments are applied to the outputs to accomplish a number of things: correct for biases of the models, accommodate the outputs to climatological trends and norms, and generate specific forecast products (e.g., predictions of changes in tropospheric jet stream winds looking 6, 12, 24, 36, or 42 hours into the future and animated forecast maps). Table 12.1 lists some of the main computer models. Not listed are models that are primarily intended for research rather than forecasting (for reference, see Lynch, 2008; Shuman, 1989; [http://celebrating200years.noaa.gov/foundations/numerical_wx_pred/welcome.html]).

The computer models have a number of constraints (Ramage, 1993). The input data come from fairly widely-spaced observation points, that is, wider than the horizontal scale at which weather can vary dramatically. Data are also not uniformly spaced; for example, land areas are data rich, but ocean areas are data poor. Acquiring even more data (especially over land areas) would not overcome a second type of constraint—even for relatively large grid sizes, the computations are intensive and time-consuming. It is only recently that high-resolution numerical models have been created to make more precise predictions. The ability to make precise predictions depends on using a

Table 12.1
Some of the major computer models for making weather predictions

Model	Source	Description
Coupled Ocean/ Atmosphere Mesoscale Prediction System (COAMPS)	Naval Research Laboratory, Marine Meteorology Division, U.S. Navy Meteorology and Oceanography Command [http://www.nrlmry.navy.mil/ coamps-web/web/home]	Short-term and Nowcast capabilities for any given region of the Earth in both the atmosphere and ocean. Began running on workstations onship in 1997 and is not part of the Navy's Distributed Atmospheric Mesoscale Prediction System. Useful in hurricane and cyclone track forecasting because it takes ocean surface temperatures into account.
Global Forecast System (GFS)	NOAA National Centers for Environmental Prediction [https://www.ncdc.noaa.gov/ data-access/model-data/model-datasets/global-forcast-system-gfs]	Medium range (up to 16 days) predictions for the entire Earth.
North American Mesoscale Model (NAM), formerly the Eta Model	NOAA National Centers for Environmental Prediction [https://www.ncdc.noaa.gov/ data-access/model-data/model -datasets/north-american -mesoscale-forecast-system-nam]	Predictions for the continental United States with higher resolution predictions possible for specified regions.
European Centre for Medium-Range Weather Forecasts Model (ECMWF)	European Centre for Medium-Range Weather Forecasts [http://www.ecmwf.int]	A model of the Earth's atmosphere, for predictions up to medium range (two weeks). Takes into account both terrain and ocean surface circulation.
Navy Global Environmental Model (NAVGEM) formerly the Navy Operational Global Atmospheric Prediction System (NOGAPS)	U.S. Navy Meteorology and Oceanography Command [http://www.nrlmry.navy.mil/ metoc/nogaps/]	Global model

Table 12.1 (continued)

Model	Source	Description
Weather Research and Forecasting (WRF)	Collaboration of the National Center for Atmospheric Research (NCAR), the National Centers for Environmental Prediction (NOAA), the Forecast Systems Laboratory (NOAA), the Air Force Weather Agency (AFWA), the Naval Research Laboratory (NRL), the University of Oklahoma (OU), and the Federal Aviation Administration (FAA) [http://www.wrf-model.org/index.php]	Designed as a research as well as a forecasting tool, WRF can generate atmospheric simulations based on data, analyses, or hypotheticals, from microscale to mesoscale. It is being used operationally but is continually being developed, and is used in research.
Fifth Generation Mesoscale Model (MM5)	Collaboration of The National Center for Atmospheric Research and Pennsylvania State University [http://www2.mmm.ucar.edu/mm5/]	This is actually a research model, not a forecasting model, unlike WRF. MM5 makes predictions for weather and climate change at the regional (subcontinental) scale, made possible by taking local terrain into account.

small grid size. This also lowers the uncertainty in the description of the initial conditions and allows for modeling at regional scales (Brooks, Doswell, and Maddox, 1992; Roebber, 1999a). An example is a microscale model called the Regional Atmospheric Modeling System (RAMS) used at the 1996 Olympics for forecasting weather down to the times and locations of the individual athletic venues (Treinish, 1997; Treinish and Rogowitz, 1997).

One path for the improvement of computer models is to increase the resolution of some individual and more powerful models, and another is to combine the results from multiple models (Brooks and Doswell, 1993). In addition to the many individual models that differ in terms of such features as the area covered, the physical processes that are modeled, terrain versus ocean factors, how far into the future they look, and so forth, there are a number of meta-model software systems called ensembles (see chapter 5). The ensembles use statistical methods to generate multiple and slightly different initial conditions, run the individual models, and compare the simulation results to see whether they agree or "converge" (e.g., for hurricane tracks). The individual models can be adjusted to take into account their particular biases. The ensemble outputs portray the outputs of multiple models on the same map to help forecasters identify ways in which the different models agree or differ (cf. Fauerbach, Edsall, Barnes, and

MacEachren, 1996; Roebber, 2015). An added value of the ensemble method is that the differences among individual models can be used to calculate the uncertainties associated with the various predicted weather parameters. This capability has been quite useful for forecasters (see Novak, Bright, and Brennan, 2008).

An ensemble that has proved useful is the Interactive Grand Global Ensemble (IGGE) created by the Observing System Research and Predictability Experiment. This international program was established by the World Meteorological Organization. The goal was to support research and also to improve medium-range forecasts, covering the entire Earth. IGGE takes as its inputs the outputs from global models that are run by the U.S. ECMWF, United Kingdom, China, Japan, the United States, Canada, Australia, Brazil, Korea, and France [see https://software.ecmwf.int/wiki/display/IGGE/Home].

In using the various models in forecasting, the outputs of the models and ensembles are received at the WFOs, military forecasting centers, and commercial forecasting service companies through a NOAA-sponsored communication network. Ensembles can be used to generate a form of consensus forecast. The forecasts by highly proficient forecasters can sometimes be just as good as a consensus forecast, and an averaged forecast can sometimes be less good than that of the more proficient individual forecasters (Roebber, 1996b), which we take as indications that genuine expertise in forecasting is achievable (see chapter 5). But another interpretation is that, at least in some cases, this is because the consensus acts as a filter, reducing the impact of the more extreme predictions (Roebber, 2010). Consensus has been used as a standard of forecast excellence, as in the comparison of scores from different forecast periods and varying sites in the National Collegiate Weather Forecasting Contest (e.g., Roebber et al. 1996b).

Particular observations can significantly alter a computer model's forecast, and consensus forecasts created by the ensemble process can sometimes fail spectacularly (Zhang, Snyder, and Rotunno, 2002). Hence, even a consensus based on the outputs of multiple computer models can lead forecasters astray (Roebber, 2010). What is currently emerging is the idea that the commonest approach—simple averaging—is not the best method for generating a model-based consensus. This is one of the many focus areas for research on computerized weather forecasting.

How Well Do the Computer Models Perform?

As we suggested earlier and will show further, computer models perform well. But better questions are: "When do they perform well, and why?", and "When do they not perform so well, and why?" A computer model can only add utility to a forecast when the current weather cannot be linearly extrapolated from the current conditions. At that point, the utility depends on the weather dynamics: A computer model can only begin to help the forecaster depending on how fast the weather is changing or weather situations are evolving (Brooks, Doswell, and Maddox, 1992). Specifically, there is some degree of scale dependence, with computer models doing well at synoptic scales but having difficulty modeling weather at the mesoscale down to the temporal scale of regional nowcasting. Given the linkage of spatial and temporal scales, what this means is that when the changes in the weather are slow, it can be a long time before a model's outputs become useful, in the sense that a good forecast can be composed based largely, if not entirely, on persistence. It is also true that the models don't pick up regime changes well (e.g., the weather in a region is dominated for days by a high-pressure system and then a front moves in rapidly). The models do better after the "new" regime kicks in and before it changes again. When the weather changes are fast, as they often are at the mesoscale, the models tend not to do well, and nowcast is in a sense the "Goldilocks zone," where forecasters have a lot of opportunity to add skill/utility.

In chapter 3 on forecaster training, we mentioned a study by Roebber and Bosart (1996b), who examined data from a forecasting contest involving students and meteorology faculty. The results showed that beginning in the late 1960s, when the first computer models came out, and across the years during which the numerical models were developed, improved, and operationalized, as the numerical models got better, there was less value added by the humans' adjustments of the computer guidance. Individual computer models sometimes produce good predictions, especially for precipitation and temperature (see Roebber and Bosart, 1996a) and especially when the various models converge on the same predictions. This trend continues today, although it should not be interpreted to mean that forecasters provide no value added.

Computer models have performed well in forecasting competitions dating to the 1970s (Sanders, 1973). Vislosky and Fritsch (1995, 1997), and subsequently Baars and Mass (2005), compared forecasts with a consensus prediction created by combining the outputs of a number of computer models. One way of combining the model outputs was by calculating a simple average (e.g., of the predicted high temperatures), and a second way involved multiplying each individual prediction by a factor reflecting the variability (average error) of the predictions from each of the models. Baars and Mass (2005) examined a year's data from 29 different locations/climates.

For a considerable number of the predictions (365 days times 29 stations times three forecast variables), all of the predictions—those by the models, the consensus predictions derived from them, and the human-generated forecasts—were about equally accurate. The same held for the degree of difference in the least accurate predictions; they did not differ from each other by more than about two degrees in the case of the temperature predictions. There were many days when the human-generated forecasts were more accurate than the model predictions. One of the individual models did best for precipitation, and the human forecasters did better for temperature predictions than they did for precipitation. The consensus models were more often the more accurate, whereas the humans were least often the least accurate.

The unadjusted predictions of the individual models showed the worst performance (i.e., more days when the predictions were least accurate). The weighted consensus model did better overall in terms of its average error. A consensus prediction can be as good and sometimes better than a human-generated forecast for certain weather variables, and better still is a consensus prediction based on a weighted average of the models' outputs. But the models perform less well for conditions where the temperatures depart considerably from the climatological norm, as one would expect. In such circumstances, the human forecasters perform the best. A good consensus prediction, one that takes model biases into account, can be consistent, but when it fails, it fails spectacularly.

The accuracy of the predictions of computer models has been steadily increasing since the advance of the ensemble forecasts that integrate the predictions of multiple computer models. For some weather parameters, skill scores have improved steadily. The 500-mb geopotential height is the height in the atmosphere that roughly divides half of the mass of the atmosphere above and below (roughly about 18,000 feet). Ensemble predictions of that weather parameter have shown improvement of about 2% per decade. Skill scores for predictions of precipitation have also increased fairly steadily (see, e.g., [http://www.wpc.ncep.noaa.gov/images/hpcvrf/wpc05yr.gif] and [http://www.wpc.ncep.noaa.gov/images/hpcvrf/WPCmdlsd110yrly.gif]) (Roebber, Schultz, Colle, and Stensrud, 2004; Toth et al., 1997; Vitart, Monteni, and Buiza, 2014; see Leutbecher and Palmer, 2008 for a review.) Figure 12.1 (plate 18) shows the "threat scores" for predictions generated by computer models (NAM, see table 12.1) at NOAA's Weather Prediction Center for the years 1961 through 2015. The predictions are for rain amounts of 0.5 inches. The threat score can be thought of as a type of skill score. It is calculated as square kilometers where precipitation was observed divided by square kilometers over which precipitation was forecast minus the square kilometers over which precipitation was not observed. The data show increasing skill for predictions out over 1, 2, and 3 days. The slopes are roughly the same, showing an annual increase of about 2% per decade.

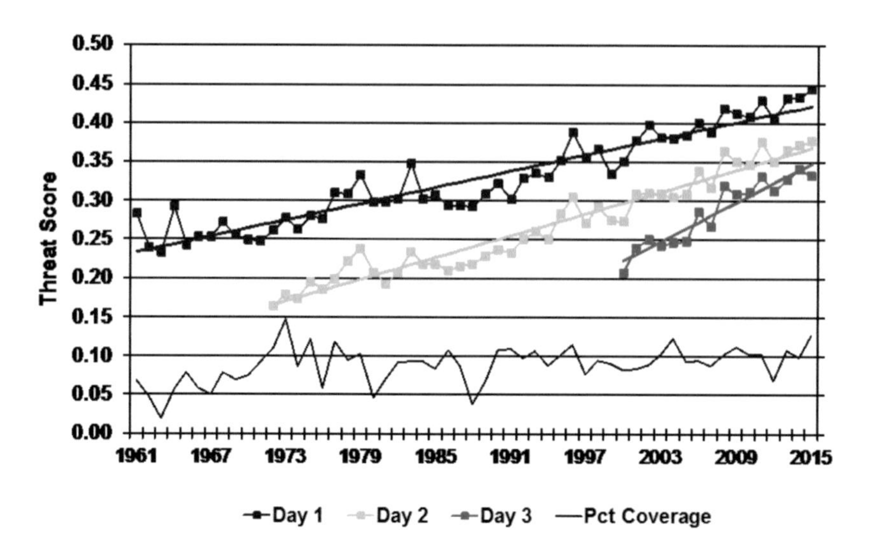

Figure 12.1
(plate 18) "Threat" (performance) scores for computer models across the years 1961 through 2015 [downloaded March 28, 2016, from http://www.wpc.ncep.noaa.gov/images/hpcvrf/wpc05yr.gif].

Among the more powerful models is the one operated by the European Center for Medium-Range Weather Forecasting (ECMWF). This model has always done well in predicting the location of high- and low-pressure systems, out to about one week in the future, after which its skill drops. More difficult to forecast is heavy rain or snow, but forecasting those has benefited considerably from computational models, that is, the model outputs are used by human forecasters to develop their conceptual models. Also difficult to forecast are the tracks of hurricanes. In the 1970s, 48-hour error in the forecasts of hurricane tracks was on the order of 500+ nautical miles, when looking two days out. Currently, errors for the same forecast period are less than 100 nautical miles. Kerr (2012) credits this entirely to the computer models, claiming that, "by the 1990s models forecasting hurricane tracks surpassed human performance" (p. 735). The success of the models can be attributed to the dynamics that the models consider, specifically ocean surface temperatures. One can think of a hurricane as being "anchored" to a warm pool of water at the ocean surface. As that pool moves, so will the hurricane be steered. In addition, the hurricane is driven in the upper atmosphere by the upper level winds. These too will steer the hurricane. The newest models take those into account. Hence, track forecasts can do well, especially when the steering component is significant.

Another computer model showing considerable skill is the High Resolution Rapid Refresh Model (HRRM), being run at the National Center for Atmospheric Research in Boulder, Colorado. It has performed well at predicting large storms and hurricane tracks. Its skill specifically at predicting winds has received much attention, especially its prediction of the winds accompanying Hurricane Sandy, 15 hours before the hurricane struck the New Jersey coast.

Overall, it is clear that the predictions of computer models have shown continuous and considerable improvement in their accuracy (skill scores) over the years spanning 1955 to the present, when one looks at specific predictions (e.g., the height in the atmosphere where the air pressure is 500 millibars) (see figures 2.1 and 2.2 in Gordon et al., 2003). While examples of model improvements and successes can be understood in terms of why they do well, one should also consider why they sometimes do not perform well. After all, one purpose of in the evaluation of model performance is to improve the modeling and its theoretical foundations. Again consider hurricane track forecasting. The models are still poor when the steering is weak because then the motion of the storm depends more on the storm's internal dynamics. This is also the problem with hurricane intensity forecasts, which have improved more slowly and are still quite difficult to get right. Eye-wall replacement cycles happen quickly and can lead to substantial variations in hurricane intensity over short time periods (less than one day). These cycles are not predicted well.

Computer Models Do Not Generate Forecasts

"Numerical prediction models do not produce a weather forecast. They produce a form of guidance that can help a human being decide upon a forecast of the weather" (Brooks, Doswell, and Maddox, 1992, p. 121; see also Doswell, 1986a). Computer models make particular predictions, such as the expected rainfall in a specified region over some specified period of time or the minimum surface pressure in a particular area at some specified time. Forecast verification studies (see chapter 5) refer to the skill of models versus human forecasts, when all the models actually predict is such specifics as surface temperatures and probability of precipitation. A forecast is more than a set of particular predictions; it is an integrated and explanatory projection of what will be happening in the weather. Box 12.1 shows a simple example of a prediction versus a forecast. Figure 1.6 is an example of MOS guidance.

Box 12.1
Prediction versus Forecast

Prediction

For Regis County Kansas on 25 August 2005

High temperature today 97 degrees Fahrenheit.

Low temperature 81 degrees Fahrenheit.

Winds 2 knots or less and out of the west-southwest.

Probability of precipitation 20%.

Forecast

Today: Mostly sunny, with a high temperature of 97 degrees. Light southwest winds. 20% chance of a late afternoon shower or thunderstorm.

Tonight: Lingering showers, followed by clear skies. Low temperature near 81.

Wednesday: Sunny skies followed by a chance of thunderstorms. High near 98. Chance of rain 30%.

The outlook for the rest of the week: Mostly sunny skies with only a slight chance of a late-day thunderstorm. Daytime highs will reach well up into the 90s, and overnight lows will only drop to near 80 degrees.

The reason that a forecast adds value is that, although it is generalized and not addressed to specific uses (weather decision support), it is formulated such that it can be used by recipients in that way that depends on their needs.

Not only is a prediction different from a forecast, there are different kinds of forecasts. Much of the difference has to do with the primary intended beneficiary of the forecast. Each spring, NOAA and Colorado State University issue hurricane season outlooks, including a discussion of the reasoning behind the forecast. The forecasts useful to the aviation community need not refer to anticipated ocean wave heights, and the forecasts of value to the maritime community need not refer to upper level winds. Special forecasts are issued regarding drought conditions, forest fires, flooding, hurricanes, and other significant events. In addition, published discussions focus on short- and long-term weather trends (monthly and seasonal outlooks). In other words, the NWS produces many tailored products. All of them include anticipated weather parameters, and they often explain the parameters by referencing the pertinent atmospheric dynamics.

The best *technical* forecasts are the Area Forecast Discussions (AFDs) posted by the Weather Forecast Offices of the National Weather Service. AFDs are posted online for sharing/access among all the forecast offices, TV station staff, emergency managers, the public, and others. AFDs result from internal NWS discussions and interoffice discussions. In high-impact situations, AFDs may reference other discussions, such as those issued by the National Hurricane Center, the Storm Prediction Center, and/or the National Centers for Environmental Prediction.

AFDs do present predictions of weather variables, but they lay out the forecasters' analyses of what is happening, why it is happening, and what might (or might not) be expected over the foreseeable future. The AFDs also address uncertainties and other factors that impact the forecast. Forecast discussions are valuable to forecasters as a way of "comparing notes." In her studies of forecaster career tracks, LaDue (2011) found that young forecasters read the discussions from across the country to help them learn how to do their jobs (see chapter 3). Educators recommend forecast discussions to meteorology students in forecasting classes, such as those offered by INNOVATIVE WEATHER (see chapter 3). An example forecast discussion is presented in table 12.2. This table references the severe winter storm event described in chapter 5. This AFD also illustrates how a forecast is more than just the listing of projected weather parameters. It explains the projection in terms of the atmospheric dynamics.

Public forecasts are also more than just the listing of projected weather parameters. They often focus on explaining what the citizen can anticipate in terms of the consequences of the weather. Referencing the January 2015 winter storm that was discussed in chapter 5 and referenced in table 12.2, a televised public forecast or a forecast included in a regional newspaper might say that citizens can expect heavy snowfalls of so many inches per hour, strong winds resulting in low visibility, snow drifts topping so many feet in some locations, and possible coastal flooding. The forecast might say that the local emergency management protocols have been initiated, trucks are beginning to salt the roads and highways, citizens should consider stocking up on essentials, and travel should be avoided after 6:00 PM. This information is all useful and actionable, but note that, unlike the AFD, there is no reference to technical details of the atmospheric dynamics. Although public forecasts sometimes include quotations from an AFD, or statements made in interviews with NWS forecasters, the forecasts often do not refer to the key meteorological concepts: Is there a cyclonic circulation? When and where did it form? Where is it now? Is it deepening, or when might it deepen? Where might the snowfall gradient be? How uncertain is the projection of that gradient?

Table 12.2
An example Area Forecast Discussion

AREA FORECAST DISCUSSION
National Weather Service New York NY
1:52 PM Eastern Standard Time, Saturday, January 24, 2015
SYNOPSIS
Rapidly deepening low pressure will track to Cape Cod this evening and then up into the Canadian maritimes on Sunday.
A cold front will move across the area on Sunday.
High pressure builds north of the area for the start of the new week.
Meanwhile, low pressure passes south of the area and intensifies over the western Atlantic.
Weak low pressure impacts the region late in the week.
Near Term /until 6 PM This Evening
Low pressure southeast of Montauk Point will continue to rapidly intensify as it tracks to the east of Cape Cod this evening.
In response to approaching northern and then southern stream energy, operational and high resolution models [are] still signaling a deformation band of precipitation developing over the region this afternoon and slowly translating east into early this evening.
An additional 1 to 3 tenths of [precipitation] possible with this band, mainly east of the Hudson and highest amounts across southeast Connecticut/eastern Long Island.
Thermal profiles and dynamic cooling will be enough to transition any light wintry mix to snow across interior tri-state with a transition from liquid to snow across city/coast...
Snow could be briefly moderate to heavy with this band across Long Island/Connecticut.
So an additional coating to 1 inch of snow possible for New York City metro and eastern lower Hudson valley.
Coating to 2 inch potential across Long Island/Connecticut.
A worst case of an additional 2 to 4 inches across southern Connecticut.
Low probability for this worst case threat to extend down to Long Island.
Any snow tapers off from west to east late this afternoon into early evening.
Advisories have been dropped west of the Hudson based on above and may be able to drop New York City/eastern Hudson at 4 PM based on banding development.
Advisories farther east will likely be extended. ...

Source: Downloaded October 29, 2015 from http://www.srh.noaa.gov.

Our main point (that the computer models do not actually generate forecasts) needs particular emphasis because lack of clarity on this matter, and the attendant misuse of the word "forecast," clouds the arguments about man versus machine. Consider, for example, a blog posted by Greg Postel of the "Capital Weather Gang" on April 24, 2012 [http://www.washingtonpost.com/blogs/capital-weather-gang]. Postel's point was to explain the scale dependence of the predictability of weather parameters. At the synoptic and hemispheric scale of large and slow-moving events (such as air masses), weather dynamics are predictable many days into the future, whereas for more local and fast-evolving events (such as severe storms or tornadoes), forecasts can be good but only on short time frames. Postel referred to data on a particular kind of skill score for computer model outputs, a type of skill score that measures the deviation of a predicted pattern from the climatology (see Roebber and Bosart, 1998). The computer model outputs being referenced were for the jet stream (vectors and speeds). Throughout his blog, Postel referred to computer model outputs as forecasts, but they are *not* forecasts, they are predictions.

Perhaps the clearest example of how model predictions are not forecasts involves the context of use (see chapter 5). For weather parameters that have critical thresholds (e.g., freezing mark for temperature) and public forecasts in situations in which public safety is at issue, the models don't really have anything to say directly. What they do is provide the forecaster some context for understanding what the risk is. So the computer model is a tool (one of many) that forecasters can use to do their job, especially in those relatively few instances during the year when forecasts are critical (Roebber and Bosart, 1996a).

In addition to this cautionary tale that model outputs are actually not forecasts, there is also the consideration that the computer models outputs can be misused.

Computer Models can Be Misused

"Just as with any information source, numerical models can help or hinder a forecaster, depending on his or her experience, understanding of the model and its shortcomings, and the weather situation" (Brooks, Doswell, and Maddox, 1992, p. 121). Although there is evidence that operational forecasters are aware of the problems and limitations of the models (e.g., Fawcett, 1969), there is also evidence of the misuse or improper use of models.

In our chapter 5 discussion of forecaster performance, we began with the story of the 2000 surprise snowstorm that hit the Washington, DC, area. We can revisit that event, with the advantage of hindsight, and discern the impact of overreliance on the computer models. According to leading meteorologist Lance Bosart (2003), the forecasters focused too much on the question "What is going to happen?" and not enough on the question "What is happening and why?" Up through January 24, the computer models were all forecasting that the storm would be far enough offshore to spare the major cities from North Carolina to New England. The result was a consistent forecast of "no snow" in Washington, DC.

The agonizing reappraisal began as ... snowfall rates broke out in the Raleigh–Durham North Carolina region, and then became a raging torrent after the first [computer model] forecasts were received early on 25 January as the Carolina snows accumulated rapidly and spread into southern Virginia. Although the forecast reappraisal came in time to ... have sanders, salters, and plows operational before morning and thus to avert a regional transportation disaster, it was too late to warn many members of the general public in the DC area who had already gone to bed blissfully content that the next morning would dawn cloudy (at worst) and dry. Still, the weather analysis and forecast process failed ... as the now-famous boldface headlines and highly critical articles of the "snow job" in the *Washington Post* (and other newspapers) the next day made abundantly clear. (Bosart, 2003, p. 522)

Bosart (2003) argued that the "damage to forecaster credibility was self-inflicted." Satellite imagery showed a clear sign of cyclogenesis (S-shape to the inflow of moist air into a spiral-shaped low-pressure system). Observational evidence pointed to a strong coastal storm, but forecasters remained focused on what the computer models were saying. This begs the question of why the models had it wrong. Perhaps certain data about the winds at upper levels of the atmosphere had been miscoded or missampled, causing the computer models to misappraise the forces that were influencing the developing storm.

This situation illustrates the potential hazard of running automated quality-control systems on autopilot. ... The massive forecast failure in the face of compelling observational evidence that the model and guidance forecasts were "going off the rails" raises the possibility that forecaster big-picture satellite and radar analysis and interpretation skills have deteriorated from disuse. (Bosart, 2003, p. 523).

In chapters 7 and 10, we mentioned the connection between the forecaster's proficiency level and the limited or improper use of computer models. Over the years since the introduction of computer models, some forecasters have come to be "anchored" by the computer model outputs. Less experienced forecasters, generally, have been observed to begin their process by examining the computer model outputs and relying on that as their process continues (see chapter 7). Apprentices and even some journeymen rely uncritically on the computer models. They tend to treat computer model outputs as a whole and not look to details or treat the details as guidance. What the computer models print out, the forecasters put out as their forecasts. Apprentices and even journeymen sometimes find themselves in situations where they would be unable to make forecasts at all without using the models.

Apprentices and some journeymen forecasters look at guidance from more than one computer model but are less likely than experts to engage in comparison or integration across the computer model outputs. They chose which computer model outputs to focus on at depending on things they are told about which model works well (the "model of the day") and which models do not. They look at a small set of the model output statistics, and are more focused on producing forecasts than on developing their mental model. One result is that when the computer models are in conflict or the outputs seem deficient, they are in trouble. Another result is that their confidence in their forecasts is primarily a function of their confidence in their preferred model (Joslyn and Jones, 2008).

In addition to the consideration that the computer model outputs can be misused, reliance on computer models can lead to de-skilling of the human forecaster.

The Unintended Negative Impact of Computer Models

Snellman (1977) found that "Forecasters are operating more as communicators and less as meteorologists. The forecaster can ... accept guidance, put it into words, and go home. Not once does he have to use his knowledge and experience" (p. 1036). A number of forecasters have expressed concern about this overreliance or misplaced reliance on computer models, arguing that the models encourage the forecasters to "make less than full use of their knowledge and diagnostic abilities" (Murphy and Brown, 1984, p. 387). Brooks, Doswell, and Maddox (1992) echoed the concern:

> Numerical models, at any scale, should not be designed to produce weather forecasts or warnings directly, but rather to provide guidance for meteorologists.... Intelligent interpretation of that guidance is an essential link in the forecast process. (pp. 130–131)

In the Canadian study discussed above (McCarthy et al., 2007), interviews with the forecasters made a strong case about the design limitations to the forecasting tools:

> One of the common comments ... was that their operational tools were inappropriate for the job. Forecasters had become increasingly reliant on model information over the years ... [most of the software] was designed to display and interrogate model data, while tools to aid in their analysis and diagnosis of observed data remained neglected. (p. 9)

In other words, the increasing reliance on the models over recent years had actually handicapped the forecasters in certain ways. This led to recommendations for a new workstation system. "The fears of Snellman and others (e.g., Doswell, 1986) were well founded" (McCarthy et al., 2007, p. 10).

The Bosart–Doswell Lament

Leading forecaster Charles Doswell of NOAA (1986a, 1986b, 2004; Doswell, Lemon, and Maddox, 1981) bemoaned this downside to new computer technology—that the reliance on automated quantitative analyses had become separated from the process of understanding. Doswell et al. (1981) raised the question of whether forecasting had improved since the advent on computer models, suggesting that it had not. They quoted Leonard Snellman (1977), saying:

Today's forecaster can, if he chooses, and many do, come to work, accept [computer model outputs], put this into words, and go home. Not once does he have to use his meteorological knowledge and experience. This type of practice is taking place more and more across the United States, and [AFOS] will make it easier to do. (Doswell et al., 1981)

Following the lead of Leonard Snellman (1977, 1978), Doswell et al. (1981) called this "meteorological cancer." To Doswell, the clarion call for technology—that the human could be freed from having to conduct routine, laborious, and mathematical analyses (such as map preparation)—reflected a superficial understanding of what it means to achieve and exercise forecasting expertise: "Somehow, the act of drawing lines on a chart can greatly improve one's comprehension of the data" (Doswell, 1986c, p. 694). "If dependence on [computer models] discourages forecasters from working with the meteorological data, then they likely will not develop the proper intuition about atmospheric structure and behavior" (Doswell, 2004, p. 1123; see also Doswell and Maddox, 1986).

This view was echoed by another senior meteorologist Lance Bosart (1989, 2003), who lamented the automation by the NWS of surface analysis (i.e., the automated mapping of isobars, surface fronts, and low- and high-pressure systems):

Over the next couple of years we are going to be flooded with data from new observational systems. The process of converting all this data to usable information is going to require considerable human talent. Credible mesoscale analyses are not going to be made by meteorologists who have lost their synoptic-scale analysis skills from disuse ... this is the time ... to implement automated procedures where human beings are always in control. ... The analyst [could] chose to accept, discard, or modify the automated analysis. (p. 271)

The Bosart–Doswell Lament was also expressed by Australian meteorologist P. S. Targett (1994), who said,

The only forecasters required will need long experience and be highly skilled thus leaving fewer lesser skilled positions available as a training ground for future forecasters to gain experience. The result may be a loss of forecasting skills by the next generation of forecasters which may be exaggerated by increasing isolation from synoptic forecasting and from the observational data due to the trend to computer based analysis ... the most likely impact of computer developments is the near future is that there will be less forecasters. (p. 52)

The Bosart–Doswell Lament was echoed by Maja-Lisa Perby (1989), whose study of aviation forecasting and the impacts of computer models (see chapter 10) suggested that:

Meteorologists have not accepted some types of changes in work which would lead to a weakening or deterioration of the inner picture. ... When the meteorologist gradually forms his understanding of the weather situation ... some traits in the weather pattern emerge as a natural part of the process ... the introduction of many new kinds of information and materials leads to a *weakening* of the traditional parts of the work process. (pp. 49–50; italics original)

This weakening happens in two ways: Computer models provide the less experienced forecaster with a convenient crutch and make the experienced forecaster less adaptive and more passive (Stuart, Schultz, and Klein, 2007). "There is a risk that human forecaster skills will atrophy from disuse in an environment in which the quality of the numerical models and forecast guidance is steadily improving unless steps are taken to ensure that forecasters remain actively engaged in the complete weather analysis and forecasting process" (Bosart, 2003, p. 525).

The Bastardi Lament

In 2000, Joe Bastardi, a senior forecaster with AccuWeather, distributed a white paper associated with a talk he gave at the New York Academy of Sciences, a paper now famously known as "The Bastardi Manifesto." In it, he discussed this question of machine replacing man and what it would take to create the "perfect forecast." He acknowledged that the skill scores of the computer models had gotten to the point where if the forecaster simply churned out what the models said, the forecaster would be right more often than not. He then discussed a number of cases of severe weather forecasting indicating how one model would do a good job at predicting pressures but another would do a good job at predicting storm position. One model would do well for a tropical system, another would do well when the system reached higher latitudes, still another would do well in the Southern Hemisphere but not so well in the Northern Hemisphere, and so forth. He discussed a number of cases where the computer model predictions were busts, which in turn led to human-generated forecasters that were also busts.

In Bastardi's view, the models did the best on the "no-brainer" forecasts. Actually, all forecasts are best in an absolute sense for the simple reason that the weather outcome can be self-evident. It is easy to tell when there is zero chance of rain. But if we restrict ourselves to cases when the forecast is not self-evident, how much of a difference can a forecaster make in those cases? This might be expressed in terms of the number of events per year when the model forecast might bust. When talking about temperature forecasts, for example, there are not many such chances (Brooks, Doswell, and Maddox, 1992). The implication for Bastardi was the notion that the skill scores for the models need to be weighted: The more extreme the weather event and the further out in time the event is expected, the more weight should be given in evaluating the models' performance: "It shows what skill the model really has in the more important and bigger events that have bigger effects on the public" (handout page). A stiffer standard might be used to assess model performance (rather than just comparing to climatology or persistence).

This is actually an open area for research. It is not entirely clear that a model or ensemble forecasts perform best for the more routine forecast situations. Well-calibrated ensembles do add the most value for weather events at the extremes. But the same is true for the human-generated forecasts. As we pointed out in chapter 5, the question of how to measure the goodness of forecasts (whether human or machine generated) involves many measurement issues. Also as we said there, the problem from a verification standpoint is translating the many, varying considerations, even for one forecast parameter at one time range, into some sort of catch-all metric. That really can't be done.

Setting aside this measurement issue, Bastardi's point was that the "verification mentality that exists among some forecasters" was used to justify the models (all quotations are from a handout). With regard to the ensembles, "Now one may argue that if we get all the models together, understand their biases we can come up with the perfect forecast. Fine, if the weather is not doing anything ... consensus can destroy the chance to hit the extreme events."

One of the main purposes of any scientific model is to help us recognize when we are wrong and then investigate why and how we are wrong in our understanding. If one regards this as a main function of computer models for weather analysis, and not assume that the only purpose is to make predictions, then the outlook changes. In his discussion of the tropical weather in the summer of 2000, Bastardi noted, "it took several busts on the part of the model, and some detective work" to come up with an understanding of why the main model in use at that time had overforecast tropical storm development: There was colder air at higher altitudes (in the tropopause), and this inhibited the development of large storms.

> It has always been a great suspicion of mine, and one that may haunt me till the day I die, that without computer models the forecaster of the 50s who knew his stuff could beat me today. ... I have seen [a senior forecaster] walk into his office, look at two maps, make a statement, and turn out right ... a lot of forecasters of the past not up on today's models would clean the clocks of the hotshots today given an even playing field with no models.

Through the 1980s to about 2000, there were these laments about the improper use, overreliance on, and misuse of models. Joe Bastardi was especially set against using ensembles to generate consensus. But things have changed since 2000.

Since 2000 ...

Today's senior forecasters, who learned how to forecast when the models had less verisimilitude, are familiar with the problem Bastardi addressed. In the early days, one could take one look at a model's output and know right away whether it should be dismissed. Today, if the models were tossed out completely, skill would decrease. Decades ago, the five-day forecast trended to climatology because the forecaster could not beat climatology that far out. Climatology is not competitive at five days anymore, and that is thanks to the computer models. Perhaps senior forecasters would do better than younger forecasters because the former did know how to use the limited available information, but in absolute terms all the forecasts would not be as good as they are today.

The computer models (both individual runs and ensembles) provide useful information, and of course this can be superseded by other data and informed by forecaster experience. Leading meteorologists advocate for looking at models late in the process, not early in the process: The forecaster should form hypotheses initially based on observations and then modify that view as new information comes in, which is any combination of new observations and models as the situation demands and may also involve scale refinements (e.g., if you are doing storm scale forecasting and need to incorporate more information about the local scale).

Some of the problems with the models (and model verification methods) have been resolved, including how the model consensus is derived. But we still see reports that younger, less experienced people are misusing and over relying on models. One can certainly get seduced by the models because the realism is there, which is why many forecasters advocate for forming initial ideas from observations and then using the models as well as other information to interrogate the data. Experience with using models helps. Younger people need to learn, some of it trial by fire. We wonder whether the "improper" use of models is less rampant in WFOs than one might suppose. We suspect model misuse/overuse will remain rampant especially in military forecasting because of the job pressures and limited training time to achieve expertise. Misuse of the models is how aerographers adapt to the demands of their job. It is not because of something fundamentally wrong with their process. Their process is imposed. Training and expectations (job pressures) are both critical, especially training to understand the strengths and limitations of the models.

Despite these cautionary tales about computer models—that they do not actually produce forecasts, that their outputs can be misused, and that over reliance on them leads to de-skilling—there is another current in the debates about weather forecasting using computer models. Will the models eventually replace the human altogether?

Pitting Man against Machine

In the experience of leading forecasters such as Charles Doswell, individual forecasters develop heuristics that enable them to make good forecasts for their particular region. Because each forecaster's experiences will be unique, it should perhaps not be surprising that, given some set of data, the computer models would be consistent, whereas different forecasters might come to different conclusions (e.g., Doswell, 2004; Uccellini et al., 1992). Comparisons have been made of human- versus machine-generated forecasts. For example, Patrick McCarthy and his colleagues (2007) at the Environment Canada Storm Prediction Center conducted an experiment in which a team of forecasters generated short-term forecasts of winds and precipitation. They relied on the familiar data types, including radar, satellite imagery, surface observations, and so on. Their forecasts' errors were then compared to the prediction errors of various weather parameters forecast by a system called SCRIBE, which took its inputs from the primary computer model in use by Environment Canada at that time. Interestingly, the forecasting teams included a mix of more- and less-experienced forecasters. They had no access to any computer model or forecast products other than the raw SCRIBE predictions. Over a two-week period:

> The meteorologists did handily top the accuracy of the SCRIBE forecasts in the shorter term, with the gap closing rapidly beyond 24 hours. The most intriguing development was that the [forecasting] teams managed a significantly better performance than their [Storm Prediction Center] counterparts in the shorter term, suggesting the importance of a greater reliance on data and short-term meteorological techniques. (McCarthy et al., 2007, p. 4)

The researchers next successfully replicated the experiment. The superior performance of the forecasting teams was attributed to three things:

1. The teams were motivated to "beat the model,"
2. During the exercise there were daily opportunities for the teams to get feedback and engage in discussion of forecasts and forecast errors, and
3. The forecasts by the Storm Prediction Center were conservative, in that it was less likely for forecasters to modify a forecast that had been issued by other forecasters on the previous shift; thus, the average error was greater.

Because the SCRIBE forecast is an automated text product derived from a computer model, it is not strictly true that the forecasters had no model information, but it was quite limited. So the method McCarthy et al. used can be viewed as a quite effective way of leveraging human talent, as this experiment shows. In a sense, it is a way of putting the forecasters on a "computer model diet": They still got some useful information from the model, but they were not allowed to rely heavily on the model.

As a result of this project, the McCarthy et al. method was subsequently used as a case study-based training method to train new forecasters, and the method for short-term forecasting was adopted throughout the Canada Meteorological Service.

Another way of "pitting man against machine" is to estimate the improvement made to a forecast when the forecaster relies on a computer model but produces a forecast that differs from what the model says. Novak et al. (2011) looked at the precipitation forecasts of forecasters at the Hydrometeorological Prediction Center in comparison with the predictions from a number of computer models (NAM, GFS, and ECMWF; see table 12.1). By their estimation, over two decades, human forecasters contributed a consistent 20% to 40% improvement over the model predictions (relative to an error threshold of one inch per 24 hours). For high-impact events (error threshold of three inches per 24 hours), again there was human value added, in some years as much as 10%. Although for longer term forecasts (three days out) the models do very well, the finding for shorter term and high-impact events confirmed what a number of meteorologists had argued, that human forecasts are particularly valuable for improving on computer model predictions for high-impact events (e.g., Bosart, 2003; Mass, 2003a; McCarthy et al., 2007; Stuart et al., 2006). Novak et al. (2011) also looked at temperature forecasts at the medium range, that is, looking from three to seven days out. Although the human value added has lessened over the past 30 years, there was again a human value added of about 10%. According to Novak et al. (2011), the human forecaster is skilled at recognizing when the model outputs are making larger errors. For small adjustments made to the model forecasts, the human adds relatively less value (about 5%), but for cases when the human makes larger adjustments to what the models say, the value added is greater (i.e., more than about 8%). "As the history of [computer modeling] and the human forecaster role continues to be written, the overall evidence ... suggests that active and engaged forecasters can continue to make incremental improvement to [model] predictions despite radical improvements in [the models]" (Novak et al., 2011, p. 4).

A study by David Reynolds (2003; see also Olson, Junker, and Korty, 1995) analyzed precipitation forecasts for the continental United States, for which there was improvement in the model predictions by the forecaster, comparing this to the number of years it took to develop a model that reached that same level of accuracy. Over the 37-year period under study, there was an increase of 56% in the threat scores for precipitation forecasting, about a 1.6% increase per year, and this roughly matched the improvement in the accuracy of the models themselves. The value added by the human forecaster was about 30% in the early 1990s, but it decreased about 10% per decade thereafter. Although the models improved, so did the human forecasts, but the humans continued to add some value to the forecasts. Reynolds (2003) argued that there was a trade-off occurring. As the cycle of improving the models and implementing refined models shortened from about five years to less than one year, the forecasters had less time to become adept at understanding model tendencies and biases, and thereby knowing when and how to adjust their forecasts. In their final calculation, Reynolds (2003) determined that the added human value equated to about 14 years of improvement in the computer models.

Despite these differences (sometimes subtle and sometimes particular) in things that the models do well and things that humans do well, there is also kind of parity: "The most accurate of the model outputs, temperature, is one of the human forecaster's most accurate forecasts and one of the least accurate model outputs, rainfall amount, is one of the human forecaster's least accurate" (Targett, 1994, p. 47). Nevertheless, there is still a current of thought based on the view that man and machine are in a competition. Despite strong evidence—not just of the human's value added but of the *necessity* of the human—some alleged visionaries and some pundits still advocate (or even take for granted) the view that forecasts could be better and humans can be eliminated simply by throwing more raw computing power at the problem.

All Hail the Computer!

Looking back to the late 1980s and early 1990s, as more computer models were introduced into operational forecasting context, more predictions were made that computer models would outperform the human within the next ten years, including the ability to automatically generate forecasts expressed in natural language (e.g., McPherson, 1991). "[The] human's advantage over the computer may eventually be swamped by the vastly increased number crunching ability of the computer ... as the computer driven models will simple get bigger and better" (Targett, 1994, p. 50). At the same time, the optimism of the visionaries was countered by credible skeptics: "The serious difficulties that exist at all levels of the numerical forecasting process are not going to be solved simply by faster computers and increasingly powerful technology" (Brooks, Doswell, and Maddox, 1992, p. 130).

As the computer models increased in resolution and expanded the set of variables that entered into the computations, the models' predictions did improve. The successes of models such as ECMWF in hurricane track forecasting can certainly be attributed to their computing power. In commenting on the success of the ECMWF model, Tim Hewson of the European Center remarked, "You need a really good supercomputer" (quoted in Kerr, 2012). But the impression given is that it is raw computing power alone that leads to model success.

Articles in the popular and scientific press often present a stance of human versus machine, as when Kerr (2012) uses the phrase "All Hail the Computer!" or blog posts proclaim such things as "supercomputer powered models are about to make weather forecasts more accurate" [http://www/motherboard.vice.com; February 3, 2015]. Articles in both the popular press and scientific outlets also ask whether "machines are taking over" (Kerr, 2012, p. 734).

> It is getting increasingly difficult for human forecasters to improve upon [computer model outputs] ... [they] cannot consistently beat [computer model] precipitation forecasts for virtually all of the locations. (Baars and Mass, 2005, p. 1045)

This conveys two attitudes. One is that this is indeed a competition, and one in which the humans cannot "beat" the computers. The other tacit attitude is not just that the humans are losing, but that some people, apparently, want the humans to lose. Such presentations are mostly premised on the stance that machines do it better, and humans are fallible and disposable. The primary evidence that is usually emphasized is the drastic improvement in the forecasts made by computers over the past few years, compared with what is said to be the relatively stable performance of humans charted over a span of decades.

A recent breakthrough in the field of artificial intelligence (AI) was in the game Go (Chouard, 2016). The IBM machine that beat the Go champion worked through a combination of raw computing power (by itself not enough because the number of possible moves in Go is a huge number raised to a huge power) and computer algorithm training regarding strategy based on pattern recognition and feedback (identifying what strategies work in which situations). There is a lesson here for weather forecasting. The primary human contributions are pattern recognition and strategizing outcomes based on those patterns, informed by both formal training and experience. In contrast, AI systems are narrowly defined to solve particular problems, not the multiple problems and changing goals that can arise in a weather forecast setting (e.g., the uplink to the radar is lost). AI systems do not yet flexecute. The analogy would be if one were playing a game of Go and suddenly the rules of the game were changed.

The valuation of computational capabilities over human reasoning is typified by Nate Silver's (2012) *The Signal and the Noise: Why So Many Predictions Fail*. This popularization often places the word expert within scare quotes, as if experts are all phoneys. Silver presents a paradoxical view on man versus machine. On the one hand, Silver (2012) says, "The forecasts you usually see reflect a combination of computer and human judgment," and on the other hand, he says, "Humans can make the computer forecasts better or they can make them worse" (Silver, 2012, p. 122). Both statements are true, but the second statement needs considerable qualification. On the one hand, Silver (2012) refers to the human–computer "partnership," and on the other hand, he seems to value computer modeling more than human expertise. He begins with the assertion that there had been no progress in weather forecasting until the introduction of computer models. This is arguably a huge overstatement, and is certainly misleading. Over the time period in which computer modeling was first developed, meteorology advanced via other technologies, especially as more observing stations provided more data, Doppler radar technology advanced, and capabilities of weather satellites expanded. Furthermore, the computer models of the early decades (1950s–1960s) were of such limited power and resolution as to make their value minimal, their forecasts little better than guesses. This situation persisted for many years.

Silver (2012) attributes the slow progress in forecast quality and accuracy to three things. The first reason given is the inherent unpredictability and uncertainty of weather events. Because the computer model outputs are highly dependent on the initial conditions, the further into the future the computer model predicts, the lower its skill: "after seven or eight days, the models ... actually displayed negative skill: they are worse than what you or I could do sitting around at home and looking up a table of long-term weather averages" (Silver, 2012, p. 133). The second reason for slow progress is the fact that ground truth data are limited and sparse. The third reason is the fact that even the most powerful supercomputers are not super enough.

Actually there is a fourth reason, which Silver does not discuss. We are not yet optimally leveraging the data and tools we already have. Additional leverage is gained by improved scientific understanding and applications of that understanding. Additional leverage is also gained by making the best use of the available information, whether that is by humans, computers, or some combination.

If these were the only reasons, and if the inherent unpredictability of certain weather events is beyond our control, then the conclusion from Silver's argument is that the primary path to better forecasts is more data, more computer power, and finer resolution in the computational models.

Silver (2012) rightly acknowledges that humans can do things computers cannot, noting that for much of its history, forecasting relied heavily on hand chart work, the perception of patterns in the data, and the development of conceptual models for visualizing atmospheric dynamics (see figures 1.1 and 4.1).

According to [NWS] statistics, humans improve the accuracy of precipitation forecasts by about 25 percent over the computer guidance alone, and temperature forecasts by about 10 percent ... these ratios have been relatively constant over time ... as much progress as the computers have made, forecasters continue to add value on top of that. Vision accounts for a lot. (Silver, 2012, p. 125)

But Silver's (2012) overall stance is man versus machine, shown when he asks, "What is it, exactly, that humans can do better than computers that can crunch numbers at seventy-seven teraflops?" (p.123), implying that the machine's capacity for number crunching is more highly valued than the human's capacity for sensemaking.

The fight over man versus machine is not restricted to the rhetoric in the popular press. It has been a matter of debate within meteorology for some time. In a review of a report that computer models outperform forecasters, the Chief of the NWS' Western Region Scientific Services Division, Leonard Snellman (1978), said:

[The researcher] notes that Western Region forecasters make significant improvements on temperature and probability of precipitation forecasts, but little or no improvement on other parameters. What [he] failed to include was that the only local forecasts that he used which were actually issued to the public were the temperature and POP forecasts. The other forecast elements he compared were ... completed on the midnight and day shifts after all the public forecasts are issued. These latter forecasts are never seen by the public. At the time this is done, the forecaster is often mentally and physically drained and eating his lunch ... conclusions based on a comparison of these data with [computer model outputs] are questionable. Further, [he] failed to comment on comparisons of [computer model outputs] and local aviation forecasts, where our data show that Western Region forecasters as a whole are 25% better than [computer model outputs] with some stations over 40% better. (p. 4)

Snellman was writing in 1978. Although computer models have certainly improved significantly since then, there is a lesson here: The details that are left "under the hood" in the man-versus-machines comparisons are often crucial if one wants to achieve a clear understanding of the matter. Snellman (1978) went on to say:

My concern is that articles [such as this] are having a degrading effect on operational forecasters. The message from high-ranking management people is "follow [the computer models] don't use [them]," But intelligent use of [model outputs] should improve our public forecasts ... My forecast is that we in meteorology shall have to learn over again. ... that you can automate only so far before getting diminishing returns in quality of the final product. ... [We need to] motivate forecasters to using, not just following [model outputs] so the considerable contribution that they can bring to the quality of the final product is realized. (p. 4)

Just Imagine ...

It is widely and almost unanimously predicted that improvement in weather forecasts will continue as more and more computer power becomes available. Pointers are made to such innovations as massively parallel computing and petaflop speeds. The computational models have gotten more sophisticated by calculating over smaller grids (e.g., five square kilometers) and wider areas (i.e., an entire hemisphere) and by incorporating more variables (e.g., ocean surface temperatures). Computers have gotten more powerful in memory and processing speed, and thus the model runs take less time and can be iterated more frequently. Thus, one reads claims to the effect that, because even our rudimentary computer models do so well now, *just imagine* what will happen when more powerful methods such as neural nets are employed (Baars and Maas, 2005).

Many meteorologists have discussed hypothetical futures for forecasting in light of projections about increasing computer capabilities in terms of resolution and processing speed (e.g., Brooks and Doswell, 1993; McIntyre, 1988, 1994, 1999). At forums on the future of forecasting held at annual meetings of the American Meteorological Society (Stuart et al., 2006, 2007), some forecasters have expressed the view that the future forecaster will have only these two primary responsibilities:

1. supervisor or manager of the forecasts generated by computers, only resorting to traditional methods (e.g., Snellman's forecasting funnel) when the computers break, and
2. communicator of the forecasts to the public and other customers.

An expansion of this vision is presented in a "roadmap" for the National Weather Service prepared by the National Academy of Sciences (Gordon et al., 2003). The Academy conducted a major review of an NWS continuing modernization and restructuring effort and presented a visionary scenario. The primary focus was on integrating new technologies for weather research and weather forecasting: more automated weather observation and data collection systems and improved computational capabilities for data assimilation and weather modeling. "Rapid development, testing and implementation of new algorithms" (Gordon et al., 2003, p. 2) was said to be a primary means by which the NWS would evolve continuously, rather than by episodic overhauls.

In the visionary scenario, a forecaster is working alongside an emergency manager in a "weather planetarium" at the start of a winter storm event. They scan the dome above them to see a virtual representation of regional weather as it unfolds and as it is expected to unfold over time, with overlaid windows providing weather specifics in various regions. The emergency manager asks to see a worst possible case, and the computers show what that would be like, with an assigned probability of occurrence. Next, the forecaster selects "Impacts" and the manager asks the computer to "Delete fog," and they are able to view what the city would look like following the worst part of the blizzard. The computer assigns the event a "catastrophic" value on a weather impact scale.

The weather computers and automated scene generators that are envisioned in this scenario would certainly be wonders, and none of it lies beyond the pale of what is conceivable, even with today's technology. But where in the scenario is the forecaster doing any forecasting? It would seem that this is a visionary statement in which the forecaster is a mere button-pusher. Looking across the Panel's recommendations, the focus basically is on creating more and more powerful computers. The Panel imagines a day when more than a thousand different computer models churn out predictions every hour, which are assembled and then assigned numbers that let the end-user know how reliable the model outputs are. The computers of the future can guarantee their own forecasts and assign probabilities with mechanical certainty. Everyone's car will tell them whether to pack an umbrella (and then drive them to work?). No one will ever die in a weather-related aviation accident because airplanes will steer themselves around dangerous weather. Better and more timely warnings will mean that no Little League baseball player will ever be struck by lightning.

The Panel report mentions the need for "increasing the value of weather information" and generating "user-oriented services and products," emphasizing the importance of weather forecasts to consumers and the overall economy. But in addition to turning the forecaster into a button-pusher, the vision for more and better computers turns the end-user into a juggler of probabilities. The quest for numerosity in public forecasts would result in a system in which weather events are assigned probabilities, another probability-like number is tacked on to tell the end-user how reliable the probability estimate is, and then more numbers are tacked on to express such things as likely degree of impact (see Gordon et al., 2003, p. 41).

The National Academies of Science report noted that the increasing accuracy of computer model predictions is not the same as the goodness of forecasts: "There is a long way to go before forecasts of rain, snow, severe weather and other phenomena will be as accurate as is theoretically possible" (p. 31). The tacit goal here seems obvious: Better computers will eliminate the human. The Panel's review of developments in computational modeling notes some "remarkable" forecasts:

On March 12–14 1993 the eastern third of the United States was hit by a major winter storm ... [it] produced extensive heavy snow, severe coastal flooding, spawned tornados and damaging winds ... The formation of the storm was forecast five days in advance. The unusual intensity of the storm was forecast three days in advance. (p. 34)

The computers definitely helped the forecasters see the superstorm start to develop well in advance. But the impression given in the Panel report is that this success was entirely attributable to the computer models. While the forecasters relied on computer models, enabling them to issue warnings well in advance of the storm, some things were underestimated: the storm's rapid development and the snowfall in the Midwest. It was the forecasters who recognized the mesoscale aspects of the storm as it began to develop. The computer model outputs differed considerably, and the forecasters had to exercise considerable judgment and evaluation to make difficult decisions when the computer models disagreed (Kocin and Uccellini, 2004). Forecasters generated and issued an extensive number of warnings, and NWS forecasters successfully coordinated with private-sector forecasters, media meteorologists, and government officials in time to prepare the public, marine, and aviation interests to take precautions (Uccellini et al., 1995). The 1993 Superstorm still ranks as one of the top storms causing insurance losses (see Risk Management Solutions, 2008).

In reaching for its conclusions, the National Academies Panel seems to say that human expertise remains crucial to weather forecasting, including the use of computer models:

Automation will change but not eliminate the role of people in weather observation and forecasting. ... Most important, of course, continued refinements and improvements of the models will depend on human expertise. (p. 49)

But then a different underlying stance is revealed and prioritized: When circumstances are typical or nominal, there is really no role at all for the human forecaster, and when the weather is unusual, the human is relegated to the role of supervisor:

In the future, highly skilled experts will be needed for interpreting ... data or ensuring that unusual circumstances (those not covered by the processing rules. Human quality control will continue to be necessary to check the basic realism of the output, especially in situations involving unusual or catastrophic weather ... the forecast/warning process is now trending toward ... greater dependence on explicit science, captured in forecasting tools and routines, rather than relying on tacit knowledge of skills of individuals ... more effort will go into refining, extending and validating the models and other tools of automated prediction. (pp. 49, 58)

When projected forward, elements of the Panel's technocentric vision are at odds with the human-centric goals (better forecasting and better forecast products) that motivate the Panel's recommendations for bringing about evolutionary improvement in weather forecasting. The disconnect between the vision and what is known in the field of cognitive systems engineering is manifest, and it is disturbing. The insertion of increasing levels of automation into the workplace leads to the degradation of expertise. It turns the operator into a supervisor. This has been shown repeatedly in many modern domains such as industrial process control and aviation (cf. Hollnagel, Woods, and Leveson, 2006) and military automated weapon systems (Hoffman, Cullen, and Hawley, 2016; Hoffman, Hawley, and Bradshaw, 2014). Automation always triggers "automation surprises" when the operator has to reengage and figure out what is going on after having spent some period of time not staying on top of the situation (Bainbridge, 1983). Roebber (2010, p. 4413) stated:

> Forecasters run the risk of losing situational awareness by being relegated to the role of supervisor of a largely automated forecast system, that is, a system that is driven primarily by [computer] model guidance. The challenge is akin to that of a modern airline pilot, who must maintain alertness and skill over long inactive periods in preparation for those few instances where intervention is truly needed. In such a situation, it is not hard to imagine that a usual lack of necessary observations will increase the likelihood of forecasters failing to recognize key observations when and where they do exist. This may particularly be the case in the context of forecast outliers.

It would seem that lessons learned have become lessons forgotten.

Conclusions

The quotation of Leonard Snellman (1978) that began this chapter can be put into context: The business of computer modeling the weather has changed quite drastically in the past 35 to 40 years. In the early 1980s, the models were easily dismissed because they were often irrelevant or barely relevant. Not so today. That said, it is important to understand that the computer model outputs are specific predictions of certain weather parameters; they are not forecasts. Model outputs are useful and, some would say, necessary input for the forecaster's process of forming a mental model and generating a sensible forecast. As expert forecasters such as Joe Bastardi have said, "You can't predict the weather well by using the computer models unless you can predict the weather well without using the computer models." Although the computer model predictions will no doubt continue to improve, and although much of the improvement will come from advances in computational systems, *forecasts* do not get better merely because one throws more raw computing power at the problem. Advances in computing techniques and the science of meteorology have greatly increased the skill of weather prediction models. At times these models outperform humans with regard to predicting the values of certain specific weather parameters in specific kinds of situations. Performance measurements have shown a fairly consistent human edge because as the *predictions* made by the models improve, so do the *forecasts* made by the human forecasters who rely on the model outputs. It is true that the human edge has been diminishing, but it is also true that there remains a huge value to that difference (e.g., a small edge in energy demand forecasting has a huge monetary value).

The claims that computer models would eventually replace the human forecaster led senior forecaster Charles Doswell to begin integrating the literature from cognitive psychology on human judgment and decision making into the meteorologists' conception of forecasting. But in that literature, he found the substitution myth in ascendance:

Generally speaking, it is accepted the judgment and decision making literature that analysis replaces intuition wherever possible (Hammond, 1996). Nebeker (1995, p. 40) describes this as a drive to replace "art" with "science." Of course, the continuing incompleteness of meteorological theory leaves room for human intuition, but the path to the future might be interpreted to suggest the inevitable dominance of analysis over intuition. This certainly is the vision that pervaded the early history of [computer modeling] and apparently persists today. ... However, barring some presently unforeseen breakthrough, the weather is going to remain resistant to becoming deterministically predictable even as [computer models] and observational technology continue to improve. ... Judgment and decision making studies have shown that analytic reasoning tends to produce highly accurate results, but occasionally produces very large errors. ... On the other hand, intuitive reasoning produces results that may have small average error but [errors that are] more widely dispersed ... [and] in under some circumstances analysis can fail to give any answer at all. (2004, p. 1123)

Humans play a crucial role in the forecasting process (e.g., Bosart, 2003; Targett, 1994). "human forecasters fill an irreplaceable role—to explain and interpret the probabilistic and deterministic forecasts generated by automated numerical weather prediction systems and to provide critical advisories, watches and warnings" (Mass, 2003a, p. 79). This chapter opened with a quotation from Leonard Snellman, writing in the 1970s. We close the chapter with another quotation, from the 1950s:

The development of high-speed computing machines has vastly increased the possibility of obtaining timely forecasts by objective techniques ... a real bridge between theory and applications has been provided. In spite of progress in the development of quantitative techniques, the conventional forecaster will have an important part to play. His wide experience of local and regional conditions ... will be invaluable on supplementing machine-made forecasts. While machines provide the answers that can be computed routinely, the forecaster will have the opportunity to concentrate on the problems which can only be solved only by resort to scientific insight and experience. Furthermore, since the machine-made forecasts are derived, at least in part, from idealized models, there will always be an unexplained residual which invites study. (Petterssen, 1956, pp. vi–vii)

This quote was written by senior Norwegian meteorologist Sverre Petterssen in 1956. To the question, "Will computers replace human forecasters?" the short answer is *No*, but a better answer is, *This is the wrong question*. The question is not about man *versus* machine but about man *plus* machine. Rather than valuing computational modeling over human expertise, or valuing human expertise over computers, there is a third path. This is the focus of the next chapter. What future do we envision in which the human forecaster has a critical role? Can we support this vision with what we have discussed in this and other chapters?

13 Forecaster–Computer Interdependence

The evolution of the forecast process toward a forecaster-machine mix has been based on improvement in numerical models. ... It has also depended on the ability of forecasters to interpret the model outputs ... recognizing the strengths and weaknesses of each, and then to make a judgment on the most probable solutions. (Uccellini et al., 1995, p. 184)

The idea that there should be some appropriate "man-machine mix" in the forecasting process actually dates to when computational modeling was being envisioned (Sutton, 1954). In the intervening years, such a man-plus-machine notion fell by the wayside as the models got better and people adopted the man-versus-machine stance. In a 1977 paper on the modernization of weather forecasting, leading forecaster Leonard Snellman (of the NWS Western Region Scientific Services Division) referred to what he and others called "meteorological cancer"—the increasing tendency for forecasters to abdicate practicing meteorological science and become more and more just a conduit of information generated by computers (see Snellman, 1977, 1991). Snellman (1991) saw it as becoming too easy for forecasters to get "scientifically lazy in this computer age" (p. 3). "Certainly the forecaster's current preparation routines will be changing considerably as new data and technology become available. However ... the new technology [must support] the 'man' part of the 'man-machine' mix ... even with new tools available there is a need for in-depth diagnosis of [computer-generated] data" (Snellman, 1991, pp. 3, 4). At the same time, Snellman was optimistic about the new computer models and computerized workstation systems, such as AFOS (see chapter 2) and the Automated Surface Observation System (ASOS):

The ability to issue forecasts with details of time and intensity of severe weather as well as ordinary day-to-day weather changes will be great fun as well as more helpful to users. This increased job satisfaction, and being the best you can be, will take place only if forecasters are given time to be a scientist (thermostat) and not just a communicator (thermometer) ... [management needs to lead] in showing how important it is to give forecasters enough time to operate as scientists. (Snellman, 1991, p. 3)

Snellman (1991) contrasted the computer models, as forecasting tools, with the workstations and NEXRAD radar that were providing forecasters with more and richer observational data, saying, "my misgivings are fading … once forecasters are informed and taught the merits and use of new data, meteorological cancer will be diminished" (p. 4).

Such prescient statements, made 20 to 40 years ago, are still in force. In chapters 7 and 12 we pointed out how some forecasters, especially the less experienced ones, can rely heavily, even too heavily, on the outputs of computer models. However, it is equally clear that forecasters do not relate to the computers as if they are in a competition. In her study of National Weather Service forecasters, Daipha (2007) noted forecasters' deep desire to "produce the most accurate NWS forecast possible" (p. 79), and for that, forecasters exploited what models offered while attempting to improve on them.

This discussion thread on the man-versus-machine issue has also taken place in the United Kingdom. On the one hand, Tennekes (1988), writing in the Royal Meteorological Society's journal *Weather*, warned that computer models were marginalizing the human because the forecaster provided little value. Eventually, the human forecaster would be bypassed. He regarded this as an issue of responsibility—that it was the forecaster who was ultimately responsible for determining the quality of forecasts. At the same time, he argued that automation had not gone far enough: "As national weather services attempt to conquer a share of the market, they will have to learn to work at lightning speed: automated observations, automated updating of local short-range forecasts, automated delivery of forecast products, automatic transmission of digitized radar and satellite data, and so on" (Tennekes, 1988, p. 167).

McIntyre (1988, 1994, 1999) was less concerned that the human would be marginalized and instead asserted that there would be an enhanced, not a diminished, role for the forecaster—although forecasting tasks can be daunting and should benefit as much as possible from automation. Supporting his argument were three core ideas: (1) advances in our theoretical understanding are the fundamental driver for advances in computer modeling, (2) human perception and sensemaking are unmatchable by computers, and (3) human perception and sensemaking of data allow forecasters to adjust the initial conditions for computer model runs, resulting in better computer model outputs.

Best Use of Computer Models

Experts and senior journeymen do not make wholesale judgments that computer model output is either "right" or "wrong." When looking at the outputs of each particular computer model, they examine subelements for particular data types (e.g., wind speeds). They adjust their reliance on particular pieces of information as the weather unfolds (Roebber, 1998). They look for convergence of computer model outputs—features or developments on which various models agree. Experts are almost always skeptical of computer model guidance and typically know the biases and weaknesses of each of the various computer model products. They know, based on their experience and understanding of the weather situation, which models to view for a confirming match—which models give good versus poor information depending on the nature of the weather situation. (There is information in the raw data, for example, spatial and temporal patterns in physical fields and climate data.)

Although the data used to initialize a computer model can comprise an adequate description of the current weather conditions, as the equations run out over longer intervals, the initial uncertainties can grow rapidly and change the forecast. The computer models are usually "supervised" (see Ballas, 2007; Kirschenbaum, 2004; McIntyre, 1988). The data used for initialization may not be as timely or reliable as the forecaster might like. The different computer models (based on different subsets of physics) often make different weather predictions, and all have certain tendencies or biases (e.g., a model may tend to overpredict the depth of low-pressure centers that form over the eastern U.S. coastline after "skipping over" the Appalachian mountains). When utilizing a computer model's outputs, it is always the job of the human forecaster to choose the most probable future state based on his or her own understanding of the current conditions (Roebber, 1999b). The expert forecaster's "intuitive assimilation" and "general picture of the situation" begin with the examination of observations and then additional inputs, including satellite images and computer models. All are used to explore and modify the initial conceptualization (Murphy and Winkler, 1974b). Correction of model bias, as well as intimate knowledge of local effects, are believed to enable forecasts that outperform the consensus of the unsupervised computer models (Stuart, Schultz, and Klein, 2007).

This can be understood as an interesting turn of the wheel. The attempts during the 1970s and 1980s to create expert systems revealed that such systems worked well if they were specific and local (e.g., for forecasting such things as lightning at the Kennedy Space Flight Center; see chapter 11). The vision for general expert systems for weather forecasting was not easily achieved. However, the idea of local "procedural rules" is now manifested in the ways in which forecasters interpret the outputs of the ensemble forecasts, compare the outputs of various models, and "tweak" the computer model outputs to account for model biases and to take local effects into account (Joslyn and Jones, 2008; Roebber, 2015).

Forecasters rely on the computer model outputs at least to some extent in generating their forecasts, so one cannot draw a clear line between predictions made by computers and forecasts made by humans. There is no clear-cut comparison of computer-generated versus human-generated forecasts. Humans rely on the computer, and as the computers get better, so too do humans. And, lest we forget, it is human experts who developed the theoretical and mathematical understandings that made computational modeling possible in the first place.

"Today, National Hurricane Center forecasters consult a half-dozen different models before predicting a hurricane position" (Kerr, 2012, p. 735). The key word is "consult." It is humans who issue the forecasts. Between the computer model output (predictions) and the human-generated forecast is human judgment. So if the forecast is a good one, do we credit the human or the machine? Well, we have to credit both. Some forecasters may just send out the results from the computer models. But as we have seen (chapter 12), expert forecasters do not do that. They exercise their judgment as they refine their conceptual models. The computers do not generate their forecasts. Using the computer models, forecasters have doubled their skill (Kerr, 2012), but note that it is still humans who generate the forecasts; the computer models help.

In saying that the humans "consult" the models, Kerr (2012) recognizes that the computer models all have limitations, especially in forecasting rapidly developing storms and tornadoes. But the point of computational modeling is to allow us to learn about what we do not understand, and thereby make improvements to both the computer models and the forecasters' conceptual models. Computer models are tools, a service provided to forecasters to help them make forecasts, but they are also tools to help meteorologists learn and to help both forecasters and meteorologists know when to be surprised. Ultimately, or in the long run, these latter functions are the more important ones for the science of meteorology and its application to forecasting.

In this regard, a shortfall of the computer models involves scale dependence. It is interesting to note that the word "mesoscale" appears in nearly every row in table 12.1, which lays out the various computer models. It would seem that the computer modeling has focused on that scale of analysis. Bosart (2003) and others we have cited have argued, based on evidence, that the human forecaster adds significant value to forecasts at the mesoscale. Yet skill at the mesoscale cannot be achieved without a robust understanding of the atmosphere at the synoptic scale.

> The absence of real-time high quality mesoscale surface analysis is a significant roadblock to forecaster ability to detect, track, diagnose, and predict important mesoscale circulation features associated with a rich variety of weather of interest to the general public. ... Although [models] are capable of simulating all kinds of mesoscale detail, it is by no means obvious how to sort out

the limited wheat from the abundant chaff. ... The absence of independent, high-resolution, real-time mesoscale analyses makes it very difficult for forecasters to evaluate critically the abundance of detailed output from these models. (Bosart, 2003; pp.520, 527–528)

Defining Forecaster–Computer Interdependence

The human forecaster is heavily dependent on technology. For example, the computer displays of systems such as the Advanced Weather Interactive Processing System (AWIPS; see chapter 2) enable forecasters to develop their conceptual models by visualizing all the available data fields, such as winds, temperatures, satellite image loops, and computer model outputs. The key to its functionality is integrated geolocation: showing all the available data fields projected onto a map (see [http://www.goes-r.gov/downloads/ScienceWeek/2013/presentations/03-18/02-Jordan-Gerth.pdf]).

As another example, the NEXRAD radar enables the forecaster to develop a mental model of storm structures (e.g., downdrafts, updrafts, vorticity) and anticipate storm evolution. "The tornado detection algorithm, when used intelligently, helps to focus attention on threatening features of convective storms that might go unnoticed in a busy situation" (Andra et al., 2002, p. 559). Forecasters also benefit greatly from a considerable number of additional products created with the Doppler radar data: "The meteorologist concerned with flash flooding saves enormous time and resources by using radar-derived rainfall estimates that rely on algorithms versus manually tracking and integrating [data]" (Andra et al., 2002, p. 559).

Although forecasters' dependence on technology is apparent, the best use of the technology depends on the forecaster's skill and flexecution. For example, forecasters must occasionally change the default rainfall rate calculation, such as for tropical systems, which have a smaller raindrop size distribution. It takes expertise at using the technology, at understanding the displays and visualizations, to perceptually discriminate the important features.

With regard to computer models, forecasters have to understand and take into account their constraints and limitations. Although the data used to initialize a numerical model can permit an adequate description of the current weather conditions, as the equations project out, the initial uncertainties can grow rapidly and change the longer term predictions. Thus, when utilizing a computer model's outputs, it is always the job of human forecasters to choose the most probable future state based on their own understanding of the current conditions (Roebber, 1999b).

Also with regard to radar, the forecaster has to understand the limitations of the technology; although the radar has "skill" (as meteorologists define that word; see chapter 5), it is by no means perfect. The radar cannot sample an entire volume of the atmosphere all at once. Spurious feature detections can arise in high-turbulence regions. The experienced forecaster can recognize such data artifacts. The radar algorithms serve the forecaster in a support role rather than as a primary tool for feature recognition (Andra et al., 2002, p. 561). What the forecaster does is "interrogate the storm." This process involves tasking the radar to make particular sweeps and particular elevations, looking through the radar data across a number of scans of volumes of the atmosphere at various heights taken every few minutes, and looking at the direction and velocity of the winds and trying to match the data to a conceptual model of storm structure. In addition, the radar from several adjacent radar locations can be used to look at storms from different distances and at different angles, and this is a significant help in developing conceptual models of storm structure. "Expert subjective analysis of basic reflectivity and velocity is *critical* both to early warnings and to minimization of false alarms" (Andra et al., 2002, p. 561).

A third way in which the forecaster–computer interdependence is manifest is that the forecaster uses computer technology to issue warnings. To issue a severe storm or tornado warning, the forecaster uses the AWIPS graphical workstation (see chapter 2) to define a region and time period for which the warning is valid. The warning regions are seen in the radar visualizations provided by the NWS and such private forecasting services as the Weather Channel. AWIPS enables forecasters to create "macros," which are routines for displaying certain radars, data fields, and map backgrounds as needed depending on circumstances, geography, and other factors. Thus, what the technology shows and how it shows it depends on the human forecaster's understanding.

In chapter 5, we discussed the winter 1993 Superstorm that hit the northeastern United States as an example of forecasting performance and success. We can also see that case as an example of human–machine interdependence (see table 5.1).

> The improved spatial and temporal resolution of the [NEXRAD] radar and a local cooperative observer network enabled the local forecasting office to provide more detailed and accurate products with more timely updates than ever before … the performance of the forecasting community during the March 1993 Superstorm can be shown to represent an important milestone in an effort of over 40 years in which weather forecasting has been transformed to a science-based and user-oriented service. … The general success of the forecasts can be related, in part, to the performance of the operational global numerical models that produced forecasts of a major cyclone event 5 to 6 days in advance. Nevertheless, the role of the forecasters was a crucial

element in (1) rectifying sometimes conflicting numerical guidance and (2) refining the forecasts to pinpoint the areas of heavy snow, the positions of the rain-snow boundary, and the timing for the development of blizzard conditions over a large area of the eastern United States. (Uccellini et al., 1995, pp. 198–199)

Andra et al. (2002) described the forecaster–computer interdependence process in this way:

Strategy combines conceptual models, key datasets, and technology in a way that maximizes the meteorologist's potential ... based on this conceptual model, the forecaster configures the workstation displays with datasets relevant to detecting convective initiation and evolution ... this proactive strategy allows the warning decision maker to anticipate events—the hallmark of a situationally aware expert ... the meteorologist quickly reviews [radar] algorithm results and mentally compares the algorithm-detected features with those found in the subjective analysis. The meteorologist then reconciles any discrepancies [and adjusts the] conceptual model as required ... [the algorithms] benefit meteorologists by detecting features overlooked in the subjective analysis. ... In other instances the algorithm detections are erroneous. (p. 562)

Interdependence holds for technologies in addition to the radar and the main AWIPS workstation. In chapters 4 and 10, we pointed out that the best use of the outputs of computer models involves the active engagement of human expertise.

- The computer models—from their highest level thermodynamic algorithms down to the individual lines of program code—are essentially the result of human scientific reasoning about the dynamics of the atmosphere.
- Having developed their own conceptual model of the weather situation, the forecaster will inspect the outputs of a number of computer models, and the outputs of the ensemble forecast programs that integrate the outputs of multiple computer models, and select the preferred "model of the day" to use as further guidance in forecasting.
- In ensemble forecasting, the data used to determine the initial conditions that are input to models are often uncertain; as the computer models calculate out over time, uncertainties increase. Thus, the model outputs depend crucially on the initial conditions. Forecasters will adjust the initial conditions that computer models use, resulting in better model outputs (McIntyre, 1988; Swets, Dawes, and Monahan, 2000).
- Sometimes the adjustments are based on the forecaster's mental model, and sometimes the adjustments compensate for what is perceived to be a bias in the model outputs.
- The forecaster interprets model outputs based on an awareness of the particular biases of the individual models—their tendency to over- or underpredict certain kinds of trends or weather events in certain circumstances. Forecasters take into account the effects of topography as well as other tendencies that they know a given model does

not adequately capture (Morss and Ralph, 2007). Forecasters will differentially weigh the computer predictions depending on the specific forecast context (Bastardi, 2000; Roebber, 1998).

• Finally, "the forecaster must remain involved with the data through the forecast shift. Computer resources will need to be used to compare the model forecasts and the observations rapidly. Flexible methods of interacting with both the data and the models are necessary in this process" (Brooks and Doswell, 1993, pp. 375–376; Doswell, 1992).

In other words, the computer depends on the human and the human depends on the computer. Human–machine interdependence can be understood as described in table 13.1.

The notion of interdependence can perhaps be made most concrete by reference to the first row under "forecasters." Forecasters need computers to keep them informed of ongoing events. One might suppose that forecasters keep track of ongoing events by looking at data, that is, observations from barometers and other devices. Forecasting does indeed rely heavily on data, but the "data" include many variables that can be either determined or mapped only by computers. Consider, for instance, the GOES

Table 13.1
The central notion of human–computer interdependence, as the relation between forecasters and computational weather forecasting models

Computer Models Are Constrained in That...	And They Need the Forecaster to...
The variables they "understand" are limited in scope.	Extend and adjust the parameters in their equations.
They can be blind to anomalies.	Keep them stable given the variability and dynamics in the atmosphere.
They are not "aware" of the fact that their model of the atmosphere is not the atmosphere.	Keep the model aligned with the world.
Forecasters Are Not Limited in That...	**Yet They Need Computers to...**
Sensitivity to context is high and is driven by knowledge and experience.	Keep them informed of ongoing events.
Sensitivity to change is high and is driven by the recognition of anomaly.	Help them align and repair their perceptions and conceptions by bringing in new data.
Adaptability to change is high and is goal-driven	Implement changes following changes in situations.
They are aware of the fact that the computer model of the atmosphere is not the atmosphere.	Computationally instantiate their mental models of the world.

satellite images (see figures 1.2 and 1.3 [plates 1 and 2]). The sensors on the satellites detect the radiant energies. The images derived from the radiometric data are just that—*derived*. Complex equations are needed to take the pixel-by-pixel radiometric values and calculate the temperature associated with each pixel. In short, without computers, "there are no data" (Somerville, 2011).

Forecasters have to be adaptive in many respects, including how they use and depend on their technologies. Even on a day without any severe weather, they have to adapt their activities and understanding of the weather as requests for information come in (Joslyn and Jones, 2008; Pliske et al., 1997). Forecasters especially have to adapt to severe weather. In chapter 10, we mentioned forecaster sensemaking and flexecution activity during the outbreak on May 3, 1999, in Oklahoma. This is also a good example of human–machine interdependence.

The Norman Oklahoma Area Severe Weather Outbreak of 3 May 1999

National Weather Service forecasters at the Norman Oklahoma regional office anticipated the evolution of severe storms on May 3, 1999; they knew the sorts of warnings that might have to be issued, and they had workload management plans in place. They would employ their usual procedures, which included using the conceptual model of the supercell (see chapter 4), referencing ground-truth data and observations, referencing the radar, and relying on "human expertise" (Roebber et al., 2002). The storm cells were large and well defined, making them ideal for detection by the radar algorithm. In addition, there were many NWS-trained spotters in the area, who provided ground-truth observations. Despite these advantages, the forecasters would still have to adapt.

The first storm was detected early, with indications of hail. However, the first tornado spawned by a supercell can be difficult to anticipate because the radar evidence is not always clear, and there had been no ground-truth verification that tornadoes were being spawned. Tornadogenesis was seen about 30 minutes later after the first severe storm warning, and a tornado warning was then issued. After this first event, additional storms developed quickly, from north Texas to southern Kansas, with a powerful EF-5 tornado in Moore, Oklahoma. Warning statements used such phrasing as "tornado emergency" and "extremely dangerous and life-threatening." Such phrasing had never been used before.

Over an eight-hour period, a warning was issued, on average, every four minutes. Sixty-six tornadoes were created by the eight supercells on this day, with as many as many as four tornadoes detectable at a given time. These persisted for the lifetime of their supercell and were often close to population centers.

Although the radar tornado detection algorithm did well overall, it was not useful for detecting the rotational (shear) signatures for the initial tornado spawned by a supercell. For the biggest difference, for the second of the eight supercells, the radar detection algorithm detected tornadogenesis one minute before it verified, whereas the forecasters had anticipated the tornado 48 minutes before it verified. For another of the eight supercells, the radar algorithm detected tornadogenesis with a ten-minute lead time, but the forecasters did not determine tornadogenesis until eight minutes after the tornado verified (Andra, Quoetone, and Bunting, 2002; Quoetone, Andra, Bunting, and Jones, 2001; Roebber, Schultz, and Romero, 2002).

Tracking individual storm cells and maintaining an awareness of their evolution and the emergence of tornadoes became difficult and particularly stressful, because the families and friends of the forecasters lived in the path of the tornadoes. The individual forecasters had to adapt, as did the entire work system at the NWS forecasting facility. A meteorologist was designated as a warning coordinator. He used a then-new warning detection system that had additional algorithms to verify the warnings and forecast statements that were issued, and he monitored the flow of forecasts and compared to ground-truth.

In addition to adaptation of their work methods, flexecution during the 3 May 1999, tornado outbreak involved forecasters' ability to take subjective information and their mental models into account when interpreting the computer model outputs. The computer model outputs were based on data sets formed from inadequate sampling of temperature and moisture at heights in the atmosphere during conditions in which the jet stream was showing an anomalous pattern. The result was uncertainty (in the forecaster's mind) about where convection would be most likely to occur, if it were to occur at all. The mesoscale computer model supported the notion of severe convection in the form of multiple supercells. Sites of convective initiation would be in areas where there was ample potential energy (i.e., due to the daytime absorption of the sun's rays) and advection (movement of air differing in temperature or moisture).

> It is important to recognize, however, that without forecaster confidence in the model, such a substantial revision to the [initial] hypothesis could not be made. ... Forecasters require information about the performance characteristics of the model, such as an understanding of the model climatology and false-alarm rates for particular phenomena. How often does the model produce long-lived supercells? How often do such forecasts verify? (Roebber et al., 2002, p, 425)

Forecasters made inferences based on satellite images (suggesting whether the high-level cirrus clouds represented the advection of moisture northward from the subtropics) and radar data suggestive of wind shear. "Forecasters have become adept at using indirect diagnosis strategies" (Roebber et al., 2002, p. 427).

Across the entire event, the forecasting office issued 48 tornado warnings, and about 90% of the predicted tornados actually happened with a median lead time of 23 minutes. This was a huge achievement. It took human experts and advanced technology ... relying on each other. Forecasters and computer models are necessarily interdependent, and all the presentations on the "replacement" of humans by machines question actually adduce this interdependence.

The forecaster–computer model relation is not competitive, but rather interdependent and collaborative. Studies of forecast verification, which compare the forecasts made by weather forecasters to those made by computer models, have resulted in some specific ideas about best use of computer models and best use of human forecasters. For example, Baars and Mass (2005) concluded that the models can generate tables of predictions looking many hours out into the future and thereby save the forecaster from having to do what had traditionally been a laborious semi-manual process. In this way, it is said that the forecaster can "retrieve the skill increase" attributable to the computer model. This frees forecasters to "spend most of their time on the short-term (1–12 hours) problem, where the combination of superior graphical interpretation and physical understanding ... will allow profound improvements" (Baars and Mass, 2005, p. 1047).

Appendix D presents details on a second case study in human–computer interdependence, expressed in a detailed timeline. This was the case of the March 1993 winter Superstorm that hit the eastern United States. In summarizing this forecasting success, Louis Uccellini and his colleagues (1995) said:

> The general success of the forecasts can be related, in part, to the performance of the operational global numerical models that produced forecasts of a major cyclone event 5 to 6 days in advance. Nevertheless, the role of the forecasters was a crucial element in (1) rectifying sometimes conflicting numerical guidance and (2) refining the forecasts to pinpoint the areas of heavy snow, the positions of the rain-snow boundary, and the timing for the development of blizzard conditions over a large area of the eastern United States. (pp. 198–199)

The users of weather forecasts are typically interested in how weather will influence their activities. They are concerned about or sensitive to the weather in the context of decisions they need to make. Sensitivity to this relates to the key role for human forecasters into the future—to integrate all of the meteorological information, including the computer model predictions, in the weather-decision context.

> The human forecaster will continue to play a central role in the preparation of forecasts. This will include interpretation of numerical forecasts and translation of the numerical guidance into practical forecasts and warnings. It will include tracking and interpretation of current weather information from a variety of observational systems. (Serafin, MacDonald, and Gall, 2002, p. 382)

A View of the Future

In their case study of severe weather forecasting, Andra et al. (2002) concluded by saying, "We must take care to ensure that automation does not *interfere* with or *replace* human expertise, but serves to *enhance* it" (p. 563). Forecasts that include a forecaster's adjustments to computer model outputs do not always improve on the models' predictions (Baars and Mass, 2005), but it misses the point to take this as the conclusion for the comparison of man and machine. McIntyre's (1999) conclusion seems rather more correct: "There is extraordinary potential for effective, efficient person-machine interaction in future weather prediction operations" (p. 338). The domain is difficult, the models are limited, and there are limitations to human expertise.

Although current human–machine interdependence in weather forecasting is an empirical fact, as we have shown in this chapter, only on the human side of the equation do we see the capacity to adapt to requirements that could not have been foreseen when the technology was initially designed. In chapter 11, we culminated the discussion of expert systems by highlighting the work of cognitive systems engineers who have been taking a new approach to the question of how to make intelligent technologies to aid in cognitive work. This field emphasizes the need for computers to be "team players" in a human–computer interdependence relationship (Christoffersen and Woods, 2002; Johnson et al., 2014; Klein et al., 2004).

For example, Roth et al. (2006) developed a workstation system to aid U.S. Air Force operators in mission planning and execution in a way that could take mission-endangering weather into account. Their system development effort included three years of workplace observations *after* the workstation system was installed. Over that period, numerous changes were made iteratively to the system. The researchers were able to list the design principles necessary to create "decision aids" that would adapt to changing demands and the changing nature of the cognitive work. These include designed-in capabilities for bringing new data types into the work process, adding new data types and formats into existing displays, seamlessly acquiring data from a new source, altering the ways that data are presented in some existing display, and allowing integration of locally created work-arounds and strategies. This concept of evolvable support systems raises the notion of human–machine interdependence to a new level with respect to the procurement of human- or work-centered technologies (Potter et al., 2000; Woods and Dekker, 2000). "Systems need to explicitly incorporate mechanisms to enable users to adapt the system to evolving requirements" (Greenbaum and Kyng, 1991; Roth et al., 2006, p. 703).

14 Conclusions and Prospects

It is becoming something of a legend that a select few in the forecasting profession have begun to develop an almost uncanny level of skill ... and such individuals are coming to be known as ace forecasters. A few of these individuals have, in news media interviews, described their subjective experience as something like merging with, or becoming part of, the weather, or not just being in the eye of the storm but becoming the eye of the storm, as one of them put it. And the urge to aspire to such skill and join the ranks of internationally famous ace forecasters is intense. ... It is just one small but significant part of the drama. (McIntyre, 1999, p. 349)

Major Conclusion

This book has presented conclusive evidence that genuine expertise has been achieved in the domain of weather forecasting. Expertise in forecasting develops in the ways it does in other professional domains and reveals the same developmental factors and phenomena as in other domains (chapters 3 and 7). Highly experienced and proficient forecasters tell the same sorts of life stories. They demonstrate that they are not only expert at forecasting, but expert at their organization and its procedures (chapter 3). Highly proficient and experienced forecasters possess extensive and highly organized knowledge, much as that shown by studies of experts in other professional domains (chapters 6, 8, and 11). The reasoning of highly proficient and experienced forecasters follows the same patterns of reasoning (sensemaking and flexecuting) and perception (perceptual learning) that we have seen in other professional domains (chapter 8, 9, and 10). Highly proficient and experienced forecasters achieve the critical level of performance that is widely used to benchmark expertise in other professional domains, including the extent of their experience and their objectively measurable performance (chapters 5 and 7 and appendixes B and C). Finally, the actual cognitive work of forecasters, involving the reliance on a great deal of highly sophisticated technology, shows the same patterns of human–computer interdependence that we see in other professional sociotechnical domains (chapters 12, 13 and appendix D). We summarized these general features of expertise in table 14.1.

Table 14.1
The defining features of expertise that also characterize highly experienced and proficient weather forecasters

Judgment
• Experts' judgments are highly accurate and reliable.
• Experts can rapidly size up situations and know immediately what actions to take (recognition-primed decision making).
• Experts rarely say, "This is what I believe." They do not support their actions or judgments by citing their own authority. They know that is neither sufficient nor helpful to colleagues.
• When the expert generates a judgment, he or she anticipates the value and impact of the judgments for their clients, customers, or users.
• When the expert generates a judgment, he or she anticipates the consequences throughout the work system or organization, in addition to the collateral consequences to other related systems or subsystems.

Skill
• Experts' performance shows consummate skill (i.e., more effective and/or qualitatively different strategies) and economy of effort (i.e., more efficient).
• For routine activities, experts display signs of "automaticity" where the expert seems to be carrying out a task without significant cognitive load; conscious processing is reserved for strategic control and/or more complex activities.
• Experts have high metacognitive ability: awareness of their own strategies and limitations.

Perception
• Experts can perceive patterns in data and conceive meaning in the patterns that nonexperts cannot.
• The patterns, meanings, and relations define the functions and dynamics of the domain problems, and the patterns sometimes exist only across multiple data types.
• Experts form rich mental models of cases or situations to support sensemaking and anticipatory thinking.

Knowledge
• Experts' domain knowledge is extensive, detailed, and highly organized according to domain principles.
• Experts usually have experience, knowledge, and skills in particular subdomains.

Adaptivity and Motivation
• Experts are intrinsically motivated to work hard on hard problems.
• Experts can adapt "on the fly" to rare, tough, or unusual cases.
• Experts have vivid memories of their past mistakes and are driven to avoid ever making them again.
• Experts are intrinsically curious and motivated to stretch their skills, knowledge, and capabilities; they are not comfortable with their own ignorance; typical compensation packages are not the only, or even most important, reward they seek.

Table 14.1 (continued)

Franchise Abilities
• Experts are not only highly proficient with regard to their job, tasks, and problem domain, but they are also highly proficient in terms of their knowledge of their organization and its history, culture, and operations.
• Experts are highly regarded by peers. People depend on them for mission-critical, complex technical guidance or high-stakes decision making. Colleagues use the expert's phone number as a hotline. Their absence from the workplace can be difficult for others.
• Experts appreciate and consider the perspectives of others. They know that others do not think like they do, and they show patience in dealing with others, helping them to gain an understanding of problems or situations.
• Experts know which other experts to go to for consultation on particular types of problems.
• Experts think in detail about training to high proficiency; what it would take for someone else to achieve their level of proficiency.

Source: Ericsson et al. (2006); Feltovich et al. (2004); Hoffman (1992, 1998, 2007); Hoffman and Fiore (2007); Hoffman et al. (2011)

It is important to recognize that the work noted in this book is almost exclusively focused on governmental (NWS and military) meteorologists. Broadcast and private-sector meteorologists have not been studied and analyzed to anywhere near the same degree. However, from interactions with these communities, it is clear to the authors that the scenarios and findings would be comparable. That is because expertise is highly marketable. If the expertise is lacking, job offers and a viable client base would quickly vanish. Still, the achievement of high proficiency in these other weather forecasting professions would be a prime subject for further study.

Prospects

Many significant theoretical and applied questions remain, and a number of avenues of applied psychological research hold promise for contributions to the cognitive science of expertise, meteorological education and training, creation of intelligent, human-centered computing systems, and human factors involved in work system design.

Human Factors of Workstation Design

The original workstation for NEXRAD had all the look of the slickest science fiction movie (see chapter 2). But it relied on a graphics pad that lacked usability from a human- or task-centered perspective (Hoffman, 1997). The workstation for the radar operator relied on an obsolete command code interface, and to make things worse, the command codes were confusing and, in some cases, ambiguous, which human factors psychologists recognize as constituting poor design (see Sanger et al., 1995). NEXRAD actually had distinct workstations (system operations, forecasting, and product generation), and operations of the system relied on dozens of technical manuals. Complexity notwithstanding, the fundamental problem with the initial NEXRAD workstation design was that decisions about system specification did not embrace any significant human factors or human-centered computing considerations and, hence, did not benefit as much as they might have from empirical human factors engineering analysis. WFOs had to engage in a considerable effort to "correct" problems that never should have been created in the first place: usability flaws that, alas, were discovered after the systems were becoming operational (see Bias and Mahew, 1994; Hoffman, 1997). The NEXRAD radar is a marvel of technological and computational capability, much of which is still being explored. But the history of computer interface design teaches that it would be unwise to assume that current workstation systems interfaces and displays are far better in terms of human factors than previous systems merely because they are "current."

Perceptual Learning

This topic brings us full circle to the "data overload" problem that introduced this book. Given the ever-increasing volume of remotely sensed data awaiting analysis, and given the continuing improvements in sensors and information processing systems, the emphasis in meteorology, and the broader field of remote sensing, has been widening the data acquisition bottleneck through the use of automated data analysis approaches called "data fusion," using such techniques as merging different data types into single visualizations and algorithmic pattern recognition and spectral analysis (for an introduction, see Campbell, 1996; Liggins, Hall, and Llinas, 2008). However, this cannot be to the neglect of the human factor. Automated pattern recognition and image processing techniques remain limited (see Friedl, Estes, and Starr, 1988).They are not a magic bullet to reveal hidden truths and thereby make decisions for the human. In fact, the human must be in the loop, on the loop and all over the loop, because the human is capable of perceptual learning and can form and reform concepts: "Although [computer analysis of pixel values] offers the benefits of simplicity and economy, it is not capable of exploiting information contained in relationships between each pixel and those that neighbor it" (Campbell, 1996, p. 314).

The challenge for computational analysis is just not to move ever closer toward humanity's dream of artificial intelligence that can substitute for the human, but to yield analytical results that do not inadvertently hide meaning. This is the potential trap of the data fusion approach to the data overload problem. Experts always need to be able to see, and drill down on, the raw data. They do not always need someone else's or some machine's interpretations (Klein and Hoffman, 1992).

Even if bulk image processing is conducted by machine, humans will continue to make the important interpretations and, ultimately, decisions. They will continue to be the perceiving and sensemaking agents who create the algorithms used to process the data, and they will continue to do so by fiddling with their displays and color-coding schemes and by reasoning around the interpretation anomalies that can arise as display schemes are applied to actual data sets (see, e.g., Artis and Carnahan, 1982; Campbell, 1996; Carnahan and Larson, 1990; Childs, 1988; Hoffman, 1990; Hoffman and Conway, 1990; Hoffman and Pike, 1995.) According to the "Moving Target Rule" of human-centered computing, the sociotechnical workplace is constantly changing (Ballas, 2007; Hoffman and Woods, 2005). Constant change in the display technologies in the workplace will entail constant change in the cognitive work to be accomplished. Change in cognitive work happens because of changes in goals (i.e., new tasks, new challenges) but especially because of changes in technology, including changes in data and display types. For instance, the NEXRAD radar revolutionized radar meteorology and forecasting, and new radar algorithms are being introduced all the time, resulting in new data products (hence, new displays) and new combinations of data types (e.g., Sat-Rad displays). In the modern sociotechnical work context, the expert must engage in frequent, if not nearly continuous, perceptual *relearning*. Patterns previously learned and perceived in one way, come to be perceived frequently and conceived in one or more new ways.

Perceptual relearning of dynamic information defined over sets of integral cues that are transmodal (they exist over different data types) becomes the "holy grail" for expertise studies and applications to training. This is especially true in domains such as weather forecasting, where the Moving Target Rule can be seen in operation all the time, ranging from software upgrades to entirely new intelligent systems and computer models.

Accelerated Expertise

Throughout the chapters of this book, we have shown from a variety of sources—forecasters' intuitions, tools that have been made by and for forecasters, psychologists' studies, and various forms of latent knowledge—that forecasters create and rely on mental models. Their mental models have a functional use, namely, to integrate information into a coherent whole that allows them to understand the dynamics of the weather and make quality, actionable forecasts. What is noteworthy is that the evidence comes from so many places and is so consistent across the different sources. Given the impacts of climate change and even routine storminess on a growing population, combined with the "grey tsunami," it is more critical than ever that efforts be undertaken to capture the knowledge and reasoning strategies of expert forecasters as well as ensure that these are passed on to future generations of forecasters.

Technology developers and trainers need to know more about how the differing reasoning styles manifested by less proficient forecasters evolve into the expert style. Research is needed to advance our understanding of the reasoning of forecasters who do not seem to follow the expert pattern (i.e., the cognitive styles of journeymen revealed by Pliske et al., 1997; Pliske, Crandall, and Klein, 2004). Training inadequately prepares forecasters for what they actually experience and have to do in the operational context. The contexts in which forecasters work can make it difficult for them to develop and refine their expertise (i.e., lack of formal procedures to provide timely feedback). Research recounted in this book involved initial attempts at performance measurement for proficiency scaling. Much more work is needed to assess forecaster proficiency in ways that are organizationally appropriate and that use criteria that are knowledge-based rather than the traditional skill score assessment (i.e., a final judgment hit rate). Programs should be created for training in the operational context and should allow new forecasters to practice under the guidance of an experienced mentor.

It takes years to achieve expertise—thousands of hours of deliberate practice to reach world-class caliber in chess (or sports or weather forecasting; see chapter 3; Ericsson et al., 2006). Any method for accelerating the achievement of expertise will hinge on the ability to support the processes of learning and perceptual relearning of dynamic cue configurations, including those that exist across multiple data types. The value of on-the-job training is clear, and it merits increased management and organizational support (Stanard et al., 2002). A key challenge is whether there can be any "shortcuts" to mastery or ways to accelerate the perceptual learning and perceptual relearning processes. Accelerating the progression from senior apprentice to journeyman and from junior journeyman to junior expert, while perhaps shaving off only a few years, would represent a huge cost savings and a significantly enhanced workforce (Hoffman et al., 2014).

In domains of expertise where perceptual skill is paramount, such as weather forecasting, it seems reasonable to speculate that providing critical exemplars of targets makes the perceptual learning process possible. But it may not accelerate it. Klein and Hoffman (1992) argued that it may take upward of ten years to achieve expertise because experts (by definition) are capable of dealing with tough discriminations and challenging cases that are (by definition) rare. If the opportunity to work on such cases were somehow time compressed, perceptual learning and relearning might be accelerated. This is called "tough case time compression."

Time-compressed, case-based practice will depend on having at hand a large and explorable set of cases, in all their rich detail. Furthermore, the full case library, and therefore both the training and testing sets of cases, will need to include cases that are routine and frequent, cases that are non-routine and rare, cases that are easy and simple, cases that are complex and tough. Generating sets of scenarios is a job for cognitive task analysis, a well-understood methodology (Crandall, Klein, and Hoffman, 2006). Training regimes are needed, through which the learner is able to experience the full range of the meaningful event patterns. The case information will have to be packaged so that cases can be relived, explored as they are relived, and compared as they are explored in "compressed" time. Compression may be accomplished by splicing to remove chunks of time or shortening delays (e.g., letting thunderstorms develop over a span of minutes rather than hours, etc.), but such an approach is likely not to be the only, or even the best, way to compress time. Cases find their meaning in how sequences or parallelisms of events hang together across time, thus making time one of the cues within a configuration. So "dynamic" is not just a qualifier, but rather a variable that the learner might need to manipulate. In other words, multiple methods for compressing time will likely be needed.

We hypothesize that an additional benefit from the research and development activity that we propose would be intelligent systems that have been called "Janus Machines" (Hoffman, Lintern, and Eitelman, 2004). This notion dissolves the traditional distinction between training support systems and performance support systems. A software-based decision aid that is good for training should be good for actually doing the job, and vice versa. Hence, there is a logical link to the two-headed God, Janus. One of Janus's faces was the mentor standing at the gates to Knowledge; the other was the apprentice, who held the keys to the gates.

The acquisition of skill at making perceptual discriminations will correlate with increasing knowledge about the underlying meanings, dynamics, and causal relations that are formative of the perceptible patterns. Similarly, increasing skill at perceptual discrimination is related to increasing sophistication in understanding and integrating the relevant perceptual dimensions. Strategies that are embodied in intelligent systems might help learners (initiates, apprentices, journeymen) ramp up their knowledge and skill more quickly when they move into an unfamiliar region or when the nature of the work changes abruptly. Janus machines for perceptual relearning would be of considerable benefit in many work domains, especially ones linked to significant workforce issues.

Climate Change

This book would not be complete without some consideration of the implications that climate change might have for forecasting. The question of whether severe weather can be linked to the climatology of the past 100 years has been a focus for research (see for example, Johns, 1982, 1984). Although much of the scientific literature on climate change focuses on change in terms of the past 100 to 300 years (Pachauri and Meyer, 2014; [http://ipcc.ch/report/ar5/]), climate, in its more general context, spans eons. Climate has always been an integral part of discussions of various geologic time periods, which typically span many millions of years. Still, there is no doubt that carbon dioxide levels have been rising unabated since records began at Mauna Loa Observatory in the 1950s [http://www.esrl.noaa.gov/gmd/ccgg/trends/full.html]. There has been a corresponding period of rising air temperatures, although some question whether urbanization is driving all or part of this rise. It is certainly true that forecasters, like the general public, have been "primed" by the occurrence of recent severe storms and other forms of extreme weather such as droughts (Jones and Gray, 2016). Such events have great impacts on society [http://www.ncdc.noaa.gov/billions/mapping], compounded by such things as population growth in high-risk weather areas (e.g., coastal zones).

Although there is considerable ongoing effort on the part of meteorologists and other environmental scientists to model climate and climate change (Risbey and Kandlikar, 2002; see appendix E), those models are distinct from the models used in forecasting. One might think that the computer models used to support forecasting (see chapter 11) have something like a "climatological drift" term in the dynamical equations, but they don't. One of the problems of numerical weather processing, in fact, is that the models change sufficiently rapidly that we don't actually know what their climatology is and how that compares to the actual climatology.

However, aside from governmental administrative and policy issues, and research funding levels, local NWS forecasters and even managers mostly stand clear of the climate change issues. This is due in part to politicization, but additionally, forecasters have to focus on the here-and-now of the weather. Their primary involvement with climate is to link various events (e.g., record-breaking events) to a climatological database. Climate, and the comparison of current events to historical events and trends, is addressed primarily by National Centers: The Storm Prediction Center, the National

Center for Environmental Prediction, the National Hurricane Center, the Drought Information Center, and various regional hydrological or snow centers. The Storm Prediction Center maintains long-term radar and radiosonde data. A new national effort named the Community Collaborative Rain, Hail, and Snow Network [http://www.cocorahs.org] has more than 10,000 weather observers and is developing a long-term precipitation database. The National Climatic Data Center, recently renamed the National Centers for Environmental Information (NCEI [https://www.ncdc.noaa.gov]), keeps long-term weather and climate records for official weather stations, cooperative observers, radar and satellite data, and more. The NCEI addresses climate trends (at least in the short term). The Climate Prediction Center (CPC), part of NCEP [http://www.cpc.ncep.noaa.gov], addresses relatively short-term trends involving the El Niño–La Niña cycle, ocean temperatures, and persistent weather patterns as it generates forecasts from eight days to a year ahead (including seasonal hurricane outlooks).

This being said, we are not concerned here with the understanding, misunderstanding, and/or beliefs about weather data and climate change on the part of laypersons or politicians. This has become and will continue to be a topic for extensive discussion elsewhere (e.g., Drury, 2014; Hering, 2016; National Research Council, 2006, 2010; Nickerson, 2014; Tak, Toet, and van Erp, 2015), and is an emerging topic in the fields of psychology and sociology (see Pearson, Schuldt, and Romero-Canyas, 2016). Our concern here is with the implications of climate change for forecasting (see Warner, 2011).

Although climate change may be affecting the weather in ways that raise crucial issues and challenges (e.g., periods of extreme drought followed by periods of excessive rainfall, increasing frequency of severe weather events; see Yoon, et al., 2015), climate change per se is not part of the day-to-day weather forecasting equation. Experienced forecasters have certainly noticed the changes and trends that seem to be occurring in recent years and which may accompany climate change. Forecasters certainly see value whenever a trend (such as increasing variability in weather or increases in extreme weather events) can be attributed. But the forecasting process will continue to involve the activities and reasoning strategies described in the models presented in chapter 10. Forecasting looks primarily at the short-term and regional evidence: surface observations, data on winds at various heights, satellite images showing moisture, radar, and so on. At most, a forecaster considers climate trends only on the 5- to 10-day forecast, but such medium-range forecasts are generally understood more in terms of seasonal tendencies than climate. One study of forecast verification suggests that much of the added skill in medium-range forecasts comes from consideration of strong El Niño or La Niña (ENSO) oscillation conditions and temperature trends

(Livezey and Timofeyeva, 2008). Nevertheless, the skill attributable to the consideration of seasonal trends is small, compared with the value added by consideration of the regional and short-term evidence (snow cover observations, persistency of upper level wind patterns, etc.).

Broadcast meteorologists may or may not get involved in climate change issues, especially in reporting about scientific studies and news releases (see, e.g., [http://www.cbsnews.com/videos/weathermans-take-on-climate-change/]). Some television stations allow their "station scientist" the freedom to produce special reports and/or segments that address the science behind climate change. However, based on station management, there may be more or less climate reporting than at other stations in a local market. TV meteorologists, much like their governmental counterparts, generally do not include climate as a factor within local short- and long-term forecasts.

Weather forecasters forecast weather, not climate. As one NOAA forecaster put it when asked about the implications of climate change for forecasting:

> I'm not sure how much climate change is directly influencing forecaster reasoning—as we tend to focus on such short time scales compared to the longer trend of climate change. But we do notice when something extraordinary occurs (i.e. Hurricane Sandy) and realize that perhaps the background climate change is affecting outcomes even on a shorter time scale.

Expertise at forecasting will still be characterized as it is now, and forecasters will still have to continually expand their knowledge, adapt to changing technology, and learn how to perceive meanings in new displays and data types. So, much will remain the same.

Box 14.1
Space Weather

In addition to the possible future implications of climate change for forecasting, weather forecasting in the near future will also have to consider "space weather." Space weather focuses on solar activity and its impact on the Earth-Ocean-Atmosphere system. The Space Weather Prediction Center (SWPC [http://www.swpc.noaa.gov]) located in Boulder, Colorado, focuses on geomagnetic storms, solar flares, the solar wind, and sunspot activity. The impacts are less weather-based and more electromagnetic (power grids, satellite systems, aircraft navigation). Still, many of the same scenarios described in this book for NWS and military forecasters play out at the SWPC—display systems, operational forecasts and data integration, team interaction, and more. Public and even targeted client group dissemination of atypical information (such as "proton fluxes" and "coronal mass ejections") is not an easy task.

Codicil: Funding for the Future of Forecasting

Working together, research meteorologists and operational forecasters will continue to advance our understanding of weather and climate, and with that understanding will come advances in our ability to forecast the weather, provided that the funding remains sufficient. The future of forecasting depends primarily on the general public and its willingness to advocate for funding of weather research and the development of our forecasting capability. Writing in *The Kansas City Star* in 1978, Allen Pearson, then-Director of the National Severe Storms Forecast Center, said: "The average per capita cost for the National Weather Service comes to about what you would pay for a large hamburger, fries, and soft drink" (p. 3). Correcting for the increase in the U.S. population (to about $320 million today), the current NWS budget of about $972 million translates to a per capita cost that is now about that of a hamburger alone. (Satellite systems are part of the NOAA budget, not the NWS budget.)

In chapter 1, we motivated this book by mentioning some rather startling statistics: billions of dollars in weather-related property damage and thousands of lives lost. Despite remarkable forecasting success stories, some of which we have recounted in this book, meteorologists and forecasters will never rest as long as lives and property are at risk of severe weather hazards.

The way forward must also involve the integration of meteorology and cognitive science.

> A potentially fruitful collaboration could develop between weather forecasters and those who study cognition as well as decision making ... learning more about how human forecasters [improve on the guidance they receive from computer models] perhaps with special attention paid to cognitive styles between those forecasters who consistently perform best and those who perform poorly, might result in an improved human forecast product overall. We do not know how to combine analytical methods and products with human intuitive approaches to produce the most accurate human-produced forecasts. Thus, we do not know how to go about raising the overall performance level of human forecasts in comparison to objective methods, nor does there seem to be any evidence for a commitment to learn about such things. If the operative assumption is that analysis will drive out intuition entirely, then the absence of research aimed specifically at human weather forecasting is a moot point ... for the management of forecasting organizations to be *demonstrably* committed to a future for humans in the process, the dedication of resources to this critically important task is essential. A consistent collaboration between meteorologists, cognitive psychologists, and others involved in judgment and decision making research will be necessary if the goal of improving human weather forecasting is to be achieved. Such interdisciplinary work is often underfunded and, consequently, usually has more lip service than results. The failure to commit significant resources to this collaboration is tantamount to conceding the forecasting role to purely objective methods. (Doswell, 2000, p. 1125)

Appendix A: List of Acronyms

This list does not include the names of the various expert systems; see chapter 11.

AFD	Area Forecast Discussion
AFGL	Air Force Geophysics Laboratory
AFOS	Automation of Field Operations and Services
AFWA	Air Force Weather Agency
AI	Artificial Intelligence
AMS	American Meteorological Society
ASOS	Automated Surface Observation System
AWIPS	Advanced Weather Interactive Processing System
CCM	Certified Consulting Meteorologist
CDM	Critical Decision Method
COAMPS	Coupled Ocean/Atmosphere Mesoscale Prediction System
COMET	Cooperative Program for Operational Meteorological Education and Training
CONUS	Continental United States
COR	Condition of Readiness
COSMIC	Constellation Observing System for Meteorology, Ionosphere, and Climate
CPC	Climate Prediction Center
DIC	Drought Information Center
ECMWF	European Centre for Medium-Range Weather Forecasts
FDO	Forecast Duty Officer
GEFS	Global Ensemble Forecast System
GFS	Global Forecast System
GOES	Geostationary Operational Environmental Satellite

HRRM	High Resolution Rapid Refresh Model
IFPS	Interactive Forecast Preparation System
IGGE	Interactive Grand Global Ensemble
ITCZ	Inter-Tropical Convergence Zone
KBS	Knowledge-Based System
LPATS	Lightning Position and Tracking System
McIDAS	Man–Computer Interactive Data Access System
METOC	NAVal TRAining METeorology and OCeanography FACility
MM5	Fifth-Generation Mesoscale Model
MMP	Macrocognitive Modeling Procedure
MOS	Model Output Statistics
NAM	North American Mesoscale Model (formerly the Eta Model)
NASA	National Aeronautics and Space Administration
NASP	Naval Air Station Pensacola
NAVGEM	Navy Global Environmental Model (formerly NOGAPS)
NCAR	National Center for Atmospheric Research
NCDC	National Climatic Data Center (now NCEI)
NCEI	National Centers for Environmental Information
NCEP	National Centers for Environmental Prediction
NEXRAD	Next Generation Weather Radar
NGM	Nested Grid Method
NHC	National Hurricane Center
NMC	National Meteorological Center (now NCEP)
NOAA	National Oceanic and Atmospheric Administration
NOGAPS	Navy Operational Global Atmospheric Prediction System
NWA	National Weather Association
NWS	National Weather Service
POP	Probability of Precipitation
PRAVDA	Perceptual Rule-Based Architecture for Visualizing Data Accurately
PROFS	Program for Regional Observing and Forecasting Services
RAMS	Regional Atmospheric Modeling System
RPD	Recognition-Primed Decision Making
SAFA	Systematic Approach forecast Aid
SAND	Satellite Alpha Numeric display workstation (U.S. Navy)
SIGMET	Significant Meteorological information
SOP	Standard Operating Procedure

SPC	Storm Prediction Center
STORM-LK	System to Organize Representations in Meteorology-Local Knowledge
SWPC	Space Weather Prediction Center
TAF	Terminal Aerodrome Forecast
WCSS-GWM	Work-Centered Support System for Global Weather Management
WFO	Weather Forecast Office
WFO-A	Weather Forecast Office-A workstation
WRF	Weather Research and Forecasting (computer model)

Appendix B: Extended Narratives of Two Cases of Forecasting Severe Weather

This appendix describes the results from two Critical Decision Method (CDM) procedures. In the first case, an expert discusses a case of severe storm forecasting. In the second case, an expert forecaster describes a case of hurricane track forecasting. Case 1 includes the CDM probe questions and the participants' responses. The narrative for case 2 has the responses to probes integrated into the narrative, for ease of exposition. The task analysis coding categories were: Observation or Situation Assessment (OSA), Decision (D), and Action (A). METOC is an abbreviation for NAVal TRaining METeorology and OCeanography FACility. LPATS is the Lightning Position and Tracking System. NASP is Naval Air Station Pensacola.

Case Number 1: Storms Associated with a Stalled Front in the Gulf Coast Region (5 PM – 5 AM shift)

OSA = Observation / Situation Assessment
D = Decision
A = Action

Event	OSA, D, A
March 1999 Before arrival at METOC I was skywatching. I saw cirrus to the southwest—anvil cirrus blowing off the tops. You can see this even though the main clouds might be 100–200 miles away. It was not a blue or gold sunset.	OSA
5:00 PM There were clear skies and high pressure over the region. There were two Lows between here and Corpus Christi, TX. It was like a stalled front. The Lows were waves on a stationary front. They were out over the Gulf. One was SSW of New Orleans, the other was NNE of Corpus Christi, also over the Gulf. The Lows were losing 1 millibar per hour.	OSA

PROBE Did this case remind you of a previously encountered case?	RESPONSE This was a textbook scenario—a stalled front with waves and energy approaching it.
This implies deepening.	D
PROBE What were your alternative courses of action?	RESPONSE Verify the strengthening and approach toward the Charlie Areas of Responsibility and inform the customers.

The participant drew a diagram, re-created in figure B.1.

At sunset I saw cirrus to the southwest, with lightning. I could see anvil cirrus blowing off the tops. This confirmed that there was energy out there.	OSA
PROBE What were your goals? What would you say to a relief officer?	RESPONSE I knew I had to cover "Charlie 1" and alert to possible gale-force winds. I'd say that there were not enough data yet. We had to query the buoys.
This implied that something in the atmosphere was turning over. Southwest is the magic direction. There must be some creature out there generating it. Something strong was out in the "Charlie 1" area.	D

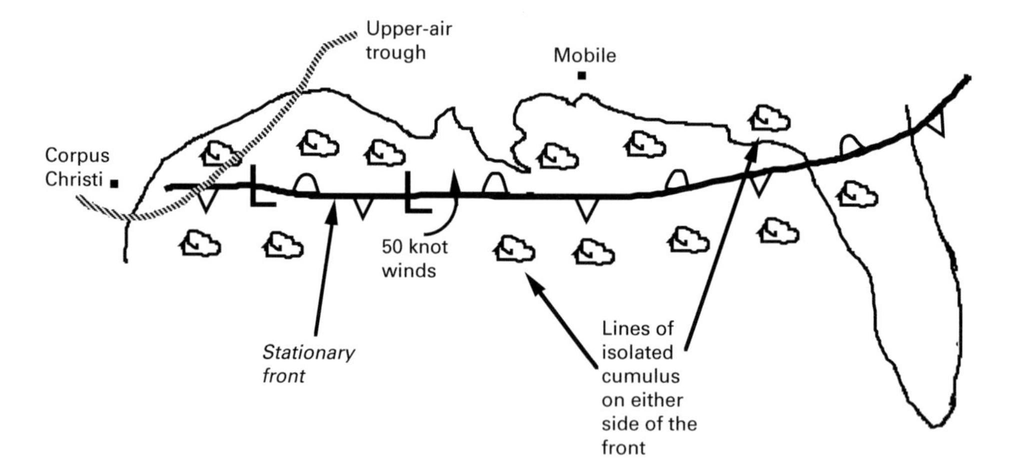

Figure B.1

6:00 PM I started analyzing the buoy data.	A
PROBE What were you seeing?	RESPONSE Winds were associated with the Lows. I could see the pressure falling and knew I could put out a warning for strong winds. LPATS showed a ring of lightning around the Low. This was unusually symmetrical, but showed that the Low was well organized. From a hand plot of buoy data I could plot the front, the Lows (position, movement, rate of movement). I did about one plot per hour, about 6 or 8 in all. Enough to know that the warning had to go out and then 2 or 3 more plots to show that it really was out there. After that I just checked the buoy data and added notes onto the plots I had done.
PROBE: Alternatives?	The analysis process is cut-and-dried. You can see the trends in the pressure and wave heights. If no buoy data were available, then this would be a tough call. I'd look at LPATS and NEXRAD for the Doppler effect. GOES for the cirrus. But you can't get surface winds from NEXRAD because of Earth's curvature below the horizon. But you could read the winds a few thousand feet above the surface. You might be able to get some ship reports—scan for them. Also, oil rigs put out data. You'd have to look for alternative sources of the surface data.

Figure B.2 shows the buoy data as received at METOC, and figure B.3 is the chart prepared by the participant.

I knew what the situation was.	D
PROBE What were your goals at this time? What training or experience was necessary in handling this case?	RESPONSE To get ready to begin alerting our customers. This was a textbook case. A stalled front off the Texas coast. You look out to the southwest and if you see any approaching trough, vorticity, or a vorticity maximum, any Low or wave on the front will develop one or two storm systems. It is taught in the school and is discussed in the Local Handbook. But you still need to experience it first-hand a few times. Experience makes all the difference. If you get burned once, then you learn. Trainees are given training with sets of cases so they get exposed to it.

23L

Offshore Data at 05Z Dec 21

DAY/ HOUR	ID	Latit (-degrees-)	Longit (-degrees-)	Temp (---C---)	Dewp (---C---)	Wind (---degr/knots---)	Gust	MaxGst	Press (millibars)	PTend	SeaT (C)	Wvht (m)	WvPd (s)
21/05	42001	25.9	-89.7	25.0	22.1	150 19	G 25	150 27	1012.2	1.1	25.0	2.0	7
21/05	42002	25.9	-93.6	22.2	20.4	290 8	G 8	290 12	1011.4	1.9	23.8	2.0	7
21/05	42003	25.9	-85.9	24.2		130 19	G 23	130 23	1014.4	-0.3	26.4	1.5	7
→ 21/05	42007	30.1	-88.8	12.3	11.3	020 29	G 35	020 41	1015.5	0.0	16.4	2.0	5
21/05	42020	26.9	-96.7	15.4	13.2	350 23	G 29		1015.4	2.0		3.0	9
21/05	42035	29.2	-94.4	11.4	9.9	360 19	G 23		1016.5	2.6	15.9	2.0	7
21/05	42036	28.5	-84.5	19.9	13.2	090 17	G 21		1016.3	-0.3	21.1		
21/05	42039	28.8	-86.0	20.7	15.7	110 23	G 27	090 27	1014.8	-1.3	22.4	1.5	5
21/05	42040	29.2	-88.2	21.0	17.9	120 31	G 39	110 43	1009.0	-3.7	21.5	3.0	8
21/05	42041	27.2	-90.4	19.4		320 19	G 29	330 29	1012.1	3.5	24.5	2.5	8
21/05	9VBK	26.4	-88.7	23.5	22.1	140 23			1012.5	-0.7	23.1		
→ 21/05	BURL1	28.9	-89.3	13.5	13.4	010 41	G 46	360 49	1010.7	2.7			
21/05	CDRF1	29.1	-82.9	16.2		110 10	G 12	110 12	1018.2	-0.3			
21/05	CSBF1	29.7	-85.2	17.3	14.2	110 4	G 8	090 13	1016.0	-1.1			
→ 21/05	DPIA1	30.3	-88.0	11.6		010 33	G 38	350 41	1016.2	0.0			
21/05	DRYF1	24.6	-82.7	24.2		120 12	G 14	110 18	1016.2	-0.3	24.6		
21/05	GDIL1	29.3	-89.9	11.9	11.2	020 23	G 29	010 35	1014.5	0.5	15.0		
21/05	KTNF1	29.8	-83.5	12.1	10.1	070 7	G 8	070 8	1017.6	-0.6			
21/05	PTAT2	27.8	-97.0	8.4	6.9	340 18	G 22	350 23	1018.7	1.0	16.6		
21/05	SANF1	24.1	-82.0	24.6		120 15	G 16	110 18	1015.8	-0.4	24.5		
21/05	SRST2	29.7	-94.0	9.4	8.7	350 11	G 13	010 20	1016.9	1.3			
21/05	VENF1	27.1	-82.4	19.0	14.8	090 8	G 10	090 14	1017.4	-0.5	19.5		

Reports: 22

Figure B.2

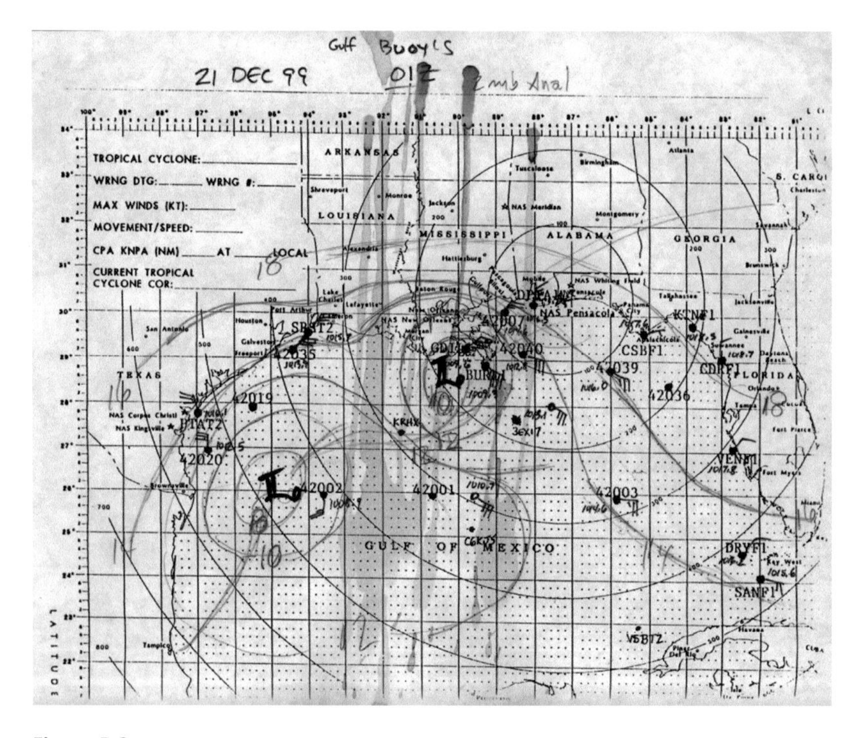

Figure B.3

Figure B.4 shows the numerical model output indicating two regions of vorticity in the Gulf, indicated by "Xs" in enclosed dashed-line ellipses, one near the coast of Texas and the other south of New Orleans.

<table>
<tr><td colspan="2">Pressures continued to drop in the Lows.
The forecast out of Norfolk had nothing on wave heights or winds.
I knew what the situation was.
There were no alternative interpretations.
My goal was to get ready to begin alerting our customers.
The forecast out of Norfolk fit the scenario, actually.
They mostly focus on the Atlantic; they look east.
It was not totally unexpected that they would not be forecasting what I saw.</td><td>OSA</td></tr>
<tr><td>PROBE
Alternative courses of action?</td><td colspan="2">RESPONSE
The Lows developing along the front could have blind-sighted Norfolk.
I wanted to alert them.
And they just might have been looking at it and working on it.
I needed to see what stage of the game they were in, so that we could avoid putting out conflicting warnings.
Norfolk should have put out guidance about something developing, but little fronts in the Gulf are not their main concern.
If you are the little guy in the Gulf and a sleeper jumps up and bites you...
And they just might have been looking at it and working on it.
I needed to see what stage of the game they were in, so that we could avoid putting out conflicting warnings.
The pressures might have started going up again of the system, causing the deepening went out of phase and the trough moved quickly.
Upper air data covers only 12 hours.
Satellite fixes can be used to determine trough speeds, and you can look to see whether the trough speed is greater than the speed of the Low at the surface.
If I lost the buoy data and the pressure started to decrease, then I would have contacted other stations and have them give me the buoy data.
You're never really blinded.</td></tr>
</table>

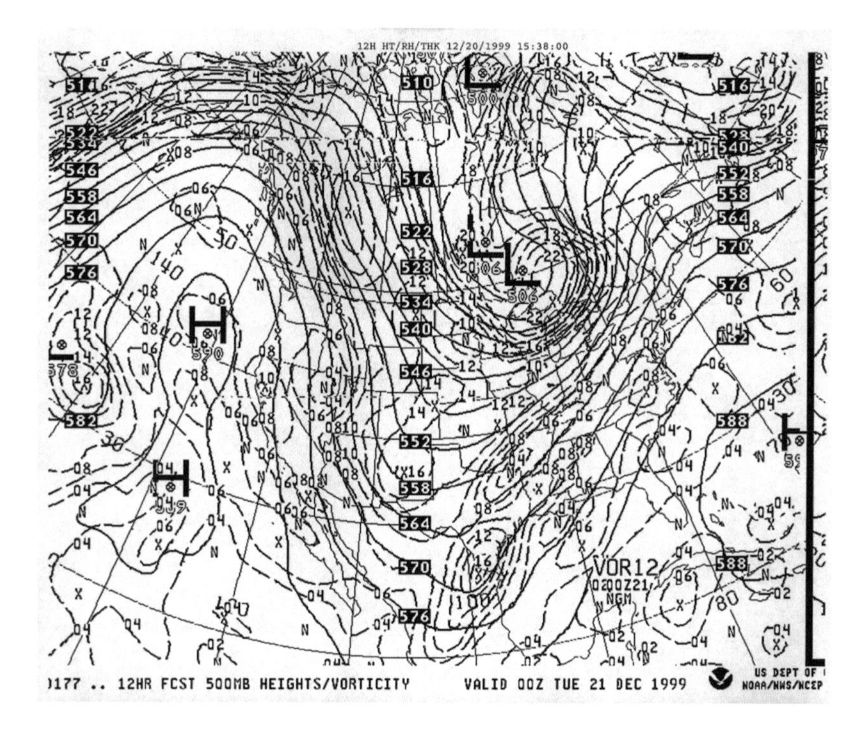

Figure B.4

7:00 PM I decided to call the Duty Officer at Norfolk and asked: "Are you forecasting storms for here in Pensacola? Do you want to modify your forecast?" The Officer replied that he had just gotten on duty. I went to them for confirmation since they are responsible for Gulf warnings.	A
He told me to cover my own area. He looked at the data and agreed with me. This confirmed my assessment. And I knew that the customers would not get conflicting data.	D
8:00 PM I put out a non-tropical gale warning (calling for 34–47 knot winds), for Charlie 1 and Charlie 2, valid through the following day (18 hours out).	A

PROBE	RESPONSE
What were your options at this point?	The warning could have been more specific (e.g., speeds). It is better to over-warn than under-warn because you can always stop a warning. But extending or changing a warning implies that they do not really know the situation. And when a warning goes out, people go to work doing things like buttoning stuff down. They may not be there to see a modification or an extension. The NWS puts out bulletins at standard times for 12–24 hours or long-range warnings. But I had no time to look at those. For fast-firing systems like this one (12 hours or less), the NWS bulletins do not help.
PROBE: What might a novice have done?	RESPONSE A novice would have done the same.

Figure B.5 shows the warning that was issued.

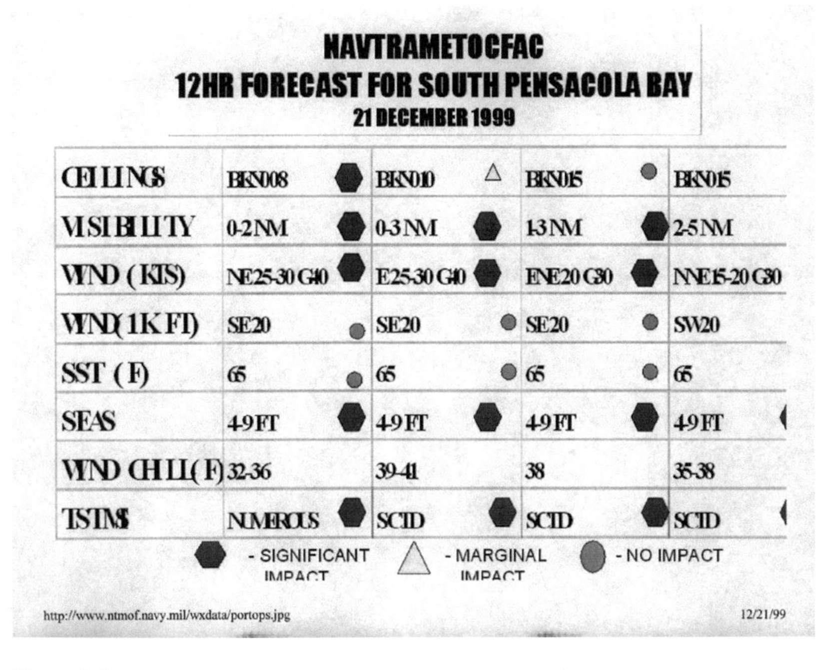

NAVTRAMETOCFAC
12HR FORECAST FOR SOUTH PENSACOLA BAY
21 DECEMBER 1999

CEILINGS	BKN008	BKN010	BKN015	BKN015
VISIBILITY	0-2NM	0-3NM	1-3NM	2-5NM
WIND (KTS)	NE25-30G40	E25-30G40	ENE20G30	NNE15-20G30
WIND (1K FT)	SE20	SE20	SE20	SW20
SST (F)	65	65	65	65
SEAS	4-9FT	4-9FT	4-9FT	4-9FT
WIND CHILL (F)	32-36	39-41	38	35-38
TSTMS	NUMEROUS	SCTD	SCTD	SCTD

- SIGNIFICANT IMPACT - MARGINAL IMPACT - NO IMPACT

http://www.ntmof.navy.mil/wxdata/portops.jpg 12/21/99

Figure B.5

The following two visible GOES images (figures B.6 and B.7) show what appeared over the Gulf at 10:41 PM Central time and 2:12 AM Central time.

<table>
<tr><td colspan="2">I called the Operations Officer in case there were Navy operations going on.
People might have had special needs.</td><td>A</td></tr>
<tr><td>PROBE
Were there any options at this point?</td><td colspan="2">RESPONSE
No, this is checklist stuff you have to do.</td></tr>
<tr><td colspan="2">Midnight
The trough moved across the Low and moved ahead of it and out of phase.
Subsidence caught the Low and weakened.
But at sea they had lightning (seen on LPATS) and gale-force winds (from buoy data).</td><td>OSA</td></tr>
<tr><td>PROBE:
Might things have gone differently?</td><td colspan="2">RESPONSE
This all fit the standard scenario.
It was unlikely that things would have unfolded any differently.
Maintained gale-force winds require major storm systems.
This is rare.
Storm of March 1993 hit western Florida with 112 mph winds.
That situation was similar to this one—everything lines up perfectly.
But major storms out of this scenario are rare.
These were minor storms.</td></tr>
</table>

Figure B.6

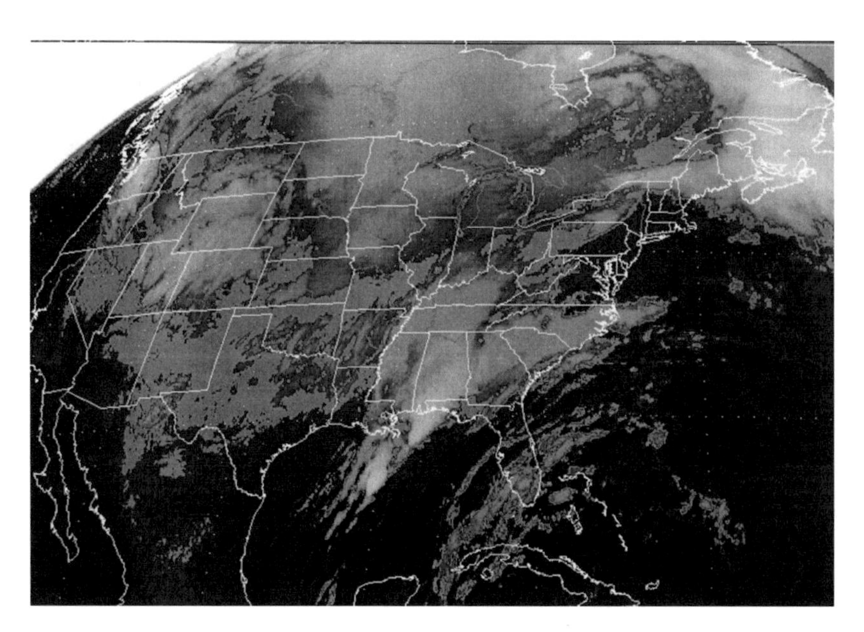

Figure B.7

<table>
<tr><td colspan="2">I kept monitoring the data—radar, LPATS, but the main data were from the buoys because they told winds and sea heights.</td><td>OSA</td></tr>
<tr><td>PROBE
What were you thinking at this point?</td><td colspan="2">RESPONSE
Were the Lows intensifying and moving eastward?
Intensification would imply a need to upgrade the warning.
Would people need to do preparations at the Base?</td></tr>
<tr><td colspan="2">The Lows carried on for a while out at sea.</td><td>OSA</td></tr>
<tr><td>PROBE
Mental modeling?</td><td colspan="2">RESPONSE
This was seen in the buoy data.
The winds flattened out and then dissipated.
The seas flattened.
The energy causing it had gone away.
Re-intensification was not possible.
There was only one energy source—the one trough.</td></tr>
<tr><td colspan="2">Through the night.
Once the trough moves out of phase with the Lows, the pressure rose, the thunderstorms dissipated, and the sea heights dropped (seen in the buoy data).</td><td>OSA</td></tr>
</table>

<table>
<tr><td>PROBE
Mental modeling?
What were your goals at this point?</td><td colspan="2">RESPONSE
Dissipation also fit the standard scenario.
My goal at this point had to do with the warning that had been issued.
The warning was for too long a valid interval.
Relief would have to change the warning to free people up.
There was no need for them to react to the warning.
They could go back to normal operations.</td></tr>
<tr><td colspan="2">I decided that the intensity was lessening.</td><td>D</td></tr>
<tr><td>PROBE:
What mistakes might a novice make at this point?</td><td colspan="2">RESPONSE:
Error was unlikely at this point.
A novice might make an error if they did not keep looking at the buoy data.</td></tr>
<tr><td colspan="2">5:00 AM
Watch change.
I explained the situation to the relief personnel at watch turnover.
They would have to decide whether to let the warning ride or cut it short.</td><td>A</td></tr>
<tr><td>PROBE
What were your goals at this point?</td><td colspan="2">RESPONSE
I knew the warning would have to be cut short.</td></tr>
<tr><td>PROBE:
What might have happened differently?</td><td colspan="2">RESPONSE:
Under a gale warning, the watch change briefing would never be minimal.
The Operations Officer gives it a high level of attention because it has implications for the day's training activities.
If there were a need to extend the warning, then they would have to decide that at the beginning of the watch, during the turnover.
If the warning remained in effect, then that would imply limits to what could be done in the Charlie areas.</td></tr>
<tr><td colspan="2">8:00 AM
They canceled the warning.</td><td>A</td></tr>
<tr><td colspan="3">Participant's Analysis of This Case</td></tr>
<tr><td colspan="3">Non-tropical weather can kick off quickly from the warm water.
You need to look at where the stationary fronts are with weak Lows rippling along them and where upper air troughs come across them.
The upper air troughs move into phase with surface systems, and for about 6 hours they intensify the Lows and make them pump up in their lower levels.
The trough causes difluence aloft, and the air gets sucked up through the Lows.
This kicks off storms around the Lows.
Nor' Easters start the same way—a weak Low influenced by the upper air.
Then the troughs overrun the Lows and fill them in.</td></tr>
</table>

Decision Requirements	
Cues and Variables	• Skywatching provides critical information before a watch period, as well as during it.
Needed information	• A thorough knowledge of standard scenarios needs reinforcement by lived experience. • Causes of lifting. • Sometimes there is no effective substitute for surface data to inform about low-pressure systems, but helpful information can be gotten from LPATS (i.e., storm organization and intensification) and NEXRAD (mid- to upper level winds). • Understanding and forecasting of Gulf weather is critically reliant on data from buoys.
Hypotheticals	• Rules-of-thumb for hypothetical reasoning in standard scenarios can often be stated succinctly (e.g., stationary front over the Gulf with weak lows can be energized by upper level troughs overrunning them from the southwest). "You look out to the southwest, and if you see any approaching trough, vorticity, or a vorticity maximum, any Low or wave on the front will develop one or two storm systems." • Forecasters need to be able to reason speculatively about what might cause intensification or dissipation of developing storm systems.
Options	• Mutual reliance among forecasters at various stations to coordinate warnings and share information.
Goals	• Reasoning about determining the valid interval for warnings depends on thorough understanding of client needs and the activities in which clients engage as a result of issued warnings. • Coordination of warnings among responsible forecasting offices.
Rationale	• Reasoning about determining the valid interval for warnings depends on thorough understanding of client needs and the activities in which clients engage as a result of issued warnings. • It is better to extend a warning out for a longer rather than a shorter valid interval. Any need to amend a warning implies a lack of understanding of the weather situation. Warnings can always be cut short at the watch change.
Situation Assessment	• What might cause intensification or dissipation in a developing storm system? • It is important to have thorough knowledge of typical scenarios, but also to have had enough lived experience so as to have had the chance to learn from errors during the typical scenarios.
Time/Effort	• In severe weather situations, hand chart-work and hand charting skills can be of critical importance to both understanding and forecasting. • Monitoring of data for long periods (many hours) is sometimes necessary.

Case Number 2: Hurricane Track Forecasting (5 PM – 5 AM shift)

Can you remember a situation where you did not feel you could trust or believe certain data, such as a computer model or some other product—a situation where the guidance gave a different answer than the one you came up with?
All the time! Hurricane Georges. NHC said the eye would go one way, but it hit Biloxi. They were wrong on where the eye was. We could see it on the radar. We had to go with the official forecast. I did my own track. It was a Sunday that it made landfall. It was midwatch. We could see the eye coming up on the radar. I was the forecaster. There wasn't much to do except a few advisories. So we watched the radar. As it came to landfall, the eye was south of NASP. The NHC had it shifting northwest to Louisiana, more of a westward track. But we could see it heading due north toward Biloxi. It made landfall in the AM Monday morning. We kept monitoring it on radar. We were here with blankets, books, food, and flashlights, and we camped out all weekend.

OSA = Observation / Situation Assessment D = Decision A = Action		
Event	**OSA, D, A**	**Time**
Hurricane Georges. September 26–29.		
Georges crossed the Florida Keys on 25–26 September. See the enhanced infrared image of Georges entering the Gulf of Mexico.	OSA	

Figure B.8 (plate 19) shows a GOES satellite image of Hurricane George just after it crossed the Florida Keys.

I came on midwatch duty Saturday evening, 27 September. I was the forecaster. COR-4 had been set four days earlier. METOC was in COR-3 and was going to COR-2. We were here with blankets, books, food, and flashlights, and we camped out all weekend. The National Hurricane Center had Georges tracking west-northwest. See NHC advisory #48 (below).	OSA, D	5:00 PM

Figure B.8

ZCZC MIATCPAT2 ALL
TTAA00 KNHC DDHHMM
BULLETIN
HURRICANE GEORGES ADVISORY NUMBER 48
NATIONAL WEATHER SERVICE MIAMI FL
4 AM CDT SUN SEP 27 1998
...DANGEROUS HURRICANE GEORGES APPROACHING THE WARNING AREA...BE PREPARED...
A HURRICANE WARNING IS IN EFFECT FROM MORGAN CITY LOUISIANA TO PANAMA CITY FLORIDA. A HURRICANE WARNING MEANS THAT HURRICANE CONDITIONS ARE EXPECTED IN THE WARNED AREA WITHIN 24 HOURS. PREPARATIONS TO PROTECT LIFE AND PROPERTY SHOULD BE RUSHED TO COMPLETION.
A TROPICAL STORM WARNING AND A HURRICANE WATCH ARE IN EFFECT FROM EAST OF PANAMA CITY FLORIDA TO ST. MARKS FLORIDA. A HURRICANE WATCH IS IN EFFECT FROM WEST OF MORGAN CITY TO INTRACOASTAL CITY LOUISIANA.
AT 4 AM CDT...0900Z...THE CENTER OF HURRICANE GEORGES WAS LOCATED NEAR LATITUDE 28.1 NORTH...LONGITUDE 87.6 WEST. THIS POSITION IS ABOUT 110 MILES SOUTHEAST OF THE MOUTH OF THE MISSISSIPPI RIVER AND ABOUT 200 MILES SOUTHEAST OF NEW ORLEANS.

GEORGES IS MOVING TOWARD THE NORTHWEST NEAR 10 MPH AND THIS MOTION IS EXPECTED TO CONTINUE TODAY WITH SOME DECREASE IN FORWARD SPEED. OUTER BANDS SHOULD GRADUALLY BEGIN TO SPREAD ACROSS THE COASTAL SECTIONS WITHIN THE WARNING AREA SOON AND HURRICANE FORCE WINDS SHOULD BEGIN TO AFFECT THE AREA LATER TODAY.
MAXIMUM SUSTAINED WINDS ARE NEAR 110 MPH WITH HIGHER GUSTS. GEORGES IS A STRONG CATEGORY TWO HURRICANE ON THE SAFFIR/SIMPSON HURRICANE SCALE. SOME FLUCTUATIONS IN INTENSITY ARE EXPECTED BEFORE LANDFALL.
HURRICANE FORCE WINDS EXTEND OUTWARD UP TO 115 MILES FROM THE CENTER...AND TROPICAL STORM FORCE WINDS EXTEND OUTWARD UP TO 175 MILES MAINLY TO THE EAST.
ESTIMATED MINIMUM CENTRAL PRESSURE IS 970 MB... 28.64 INCHES.
STORM SURGE FLOODING OF 10 TO 15 FEET... UP TO 17 FEET AT THE HEADS OF BAYS... ABOVE NORMAL TIDE LEVELS IS POSSIBLE IN THE WARNED AREA AND WILL BE ACCOMPANIED BY LARGE AND DANGEROUS BATTERING WAVES.
FLOODING RAINS ARE LIKELY IN ASSOCIATION WITH GEORGES AND WILL BECOME PARTICULARLY SEVERE IF GEORGES FORWARD MOTION DECREASES NEAR LANDFALL AS IS NOW FORECAST.
SMALL CRAFT FROM INTRACOASTAL CITY TO HIGH ISLAND TEXAS SHOULD REMAIN IN PORT. SMALL CRAFT ALONG THE WEST COAST OF THE FLORIDA PENINSULA SHOULD REMAIN IN PORT UNTIL WINDS AND SEAS SUBSIDE.
REPEATING THE 4 AM CDT POSITION... 28.1 N... 87.6 W. MOVEMENT TOWARD...NORTHWEST NEAR 10 MPH. MAXIMUM SUSTAINED WINDS...110 MPH. MINIMUM CENTRAL PRESSURE... 970 MB.
AN INTERMEDIATE ADVISORY WILL BE ISSUED BY THE NATIONAL HURRICANE CENTER AT 7 AM CDT FOLLOWED BY THE NEXT COMPLETE ADVISORY AT 10 AM CDT.

Figure B.9 shows the track predicted by the NHC.

METOC administration had decided to continue the regular watch sections over the weekend.	D	
Planes had sortied out by Saturday and Sunday, and remaining planes had been stored by Sunday afternoon. Ships had left by Friday to head toward Yucatan. See NHC Advisory #50 (below).	OSA;	4:00 PM

ZCZC MIATCPAT2 ALL
TTAA00 KNHC DDHHMM
BULLETIN
HURRICANE GEORGES ADVISORY NUMBER 50
NATIONAL WEATHER SERVICE MIAMI FL
4 PM CDT SUN SEP 27 1998
...DANGEROUS HURRICANE GEORGES CLOSING IN ON THE CENTRAL GULF COAST...
A HURRICANE WARNING IS IN EFFECT FROM MORGAN CITY LOUISIANA TO PANAMA CITY FLORIDA. A HURRICANE WARNING MEANS THAT HURRICANE CONDITIONS ARE EXPECTED IN THE WARNED AREA WITHIN 24 HOURS. PREPARATIONS TO PROTECT LIFE AND PROPERTY SHOULD BE RUSHED TO COMPLETION.

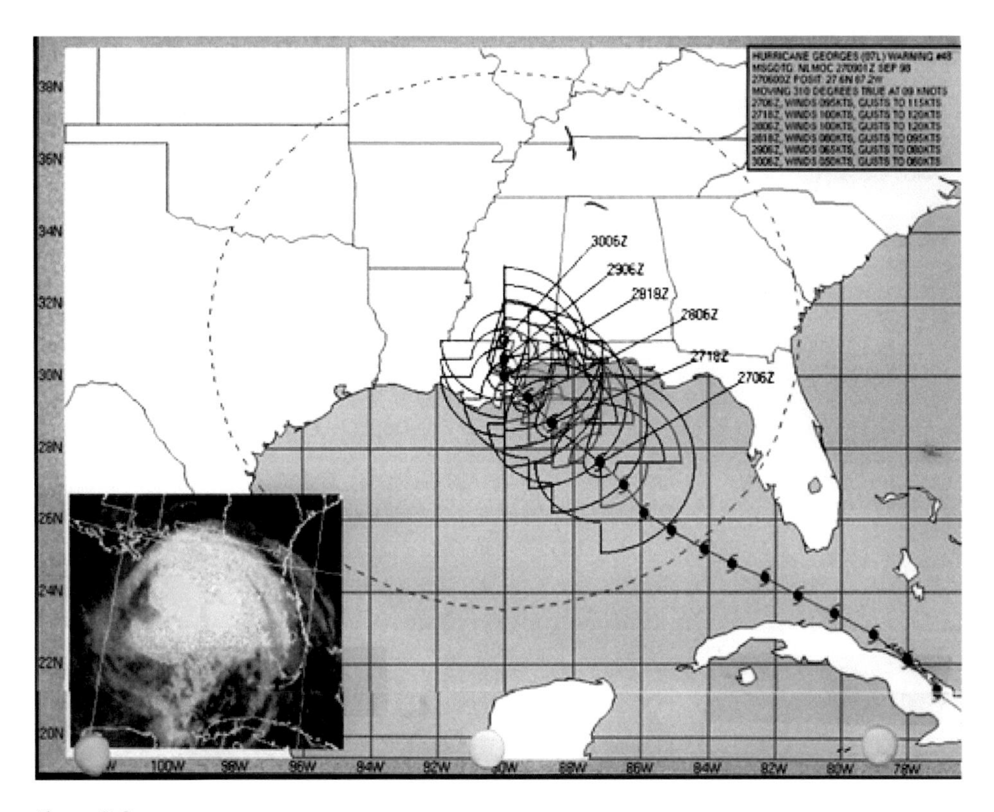

Figure B.9

A TROPICAL STORM WARNING IS IN EFFECT FROM EAST OF PANAMA CITY FLORIDA TO ST. MARKS FLORIDA. A HURRICANE WATCH IS IN EFFECT FROM WEST OF MORGAN CITY TO INTRACOASTAL CITY LOUISIANA.
AT 4 PM CDT... 2100Z... THE HURRICANE WATCH FROM EAST OF PANAMA CITY FLORIDA TO ST. MARKS FLORIDA IS DISCONTINUED.
AT 4 PM CDT THE CENTER OF HURRICANE GEORGES WAS LOCATED NEAR LATITUDE 29.0 NORTH... LONGITUDE 88.5 WEST. THIS POSITION IS ABOUT 40 MILES EAST OF THE MOUTH OF THE MISSISSIPPI RIVER AND ABOUT 125 MILES EAST-SOUTHEAST OF NEW ORLEANS LOUISIANA.
THE TRACK OF GEORGES IS WOBBLING A LITTLE ALONG A NORTHWEST HEADING NEAR 8 MPH AND THIS MOTION IS EXPECTED TO CONTINUE THROUGH TONIGHT BRINGING THE CORE OF THE HURRICANE NEARER TO THE COAST. DO NOT FOCUS ON THE PRECISE LOCATION AND TRACK OF THE CENTER. THE HURRICANES DESTRUCTIVE WINDS... RAIN... AND STORM SURGE COVER A WIDE SWATH.
MAXIMUM SUSTAINED WINDS REMAIN NEAR 110 MPH WITH HIGHER GUSTS. SOME FLUCTUATIONS IN STRENGTH ARE LIKELY BEFORE LANDFALL AND GEORGES COULD STILL BECOME A CATEGORY 3 HURRICANE.

HURRICANE FORCE WINDS EXTEND OUTWARD UP TO 70 MILES EAST OF THE CENTER... AND TROPICAL STORM FORCE WINDS EXTEND OUTWARD UP TO 185 MILES. RAINBANDS OF GEORGES ARE SPREADING ACROSS PORTIONS OF THE WARNING AREA. THE AUTOMATED OBSERVING SITE AT DAUPHIN ISLAND ALABAMA RECENTLY REPORTED A 58 MPH WIND SUSTAINED OVER TWO MINUTES WITH A PEAK GUST TO HURRICANE FORCE... 74 MPH. WINDS WILL CONTINUE TO INCREASE IN THE WARNING AREA THIS EVENING AND TONIGHT. AN AIR FORCE RESERVE UNIT HURRICANE HUNTER PLANE REPORTED A MINIMUM CENTRAL PRESSURE OF 961 MB... 28.38 INCHES. STORM SURGE FLOODING OF 10 TO 15 FEET ABOVE NORMAL TIDE LEVELS... AND UP TO 17 FEET AT THE HEADS OF BAYS... IS POSSIBLE IN THE WARNED AREA AND WILL BE ACCOMPANIED BY LARGE AND DANGEROUS BATTERING WAVES. FLOODING RAINS OF 15 TO 25 INCHES... WITH LOCALLY HIGHER AMOUNTS ARE LIKELY IN ASSOCIATION WITH THIS SLOW-MOVING HURRICANE. ISOLATED TORNADOES ARE POSSIBLE EAST AND NORTHEAST OF THE TRACK OF GEORGES. SMALL CRAFT FROM INTRACOASTAL CITY LOUISIANA WESTWARD AND SOUTHWARD ALONG THE COAST OF TEXAS SHOULD REMAIN IN PORT. SMALL CRAFT ALONG THE WEST COAST OF THE FLORIDA PENINSULA SHOULD REMAIN IN PORT UNTIL WINDS AND SEAS SUBSIDE. REPEATING THE 4 PM CDT POSITION... 29.0 N... 88.5 W. MOVEMENT TOWARD... NORTHWEST NEAR 8 MPH. MAXIMUM SUSTAINED WINDS... 110 MPH. MINIMUM CENTRAL PRESSURE... 961 MB. INTERMEDIATE ADVISORIES WILL BE ISSUED BY THE NATIONAL HURRICANE CENTER AT 6 PM CDT AND 8 PM CDT FOLLOWED BY THE NEXT COMPLETE ADVISORY AT 10 PM CDT.

Figure B.10 shows the GOES image of Georges at this time.

Figure B.11 shows the hurricane track forecast at this time.

Even if the hurricane had shifted more to the east, there would have been little for us to do. We had to provide information to local people (e.g., Disaster Preparedness). METOC was boarded up by Thursday.	D, A	
Georges was off the southeast shore of Louisiana. It was well defined on radar. We could see the eye coming up.	OSA	5:00 PM–8:00 PM
The National Hurricane Center (NHC) had the wrong track. They said the eye would go one way, but it ended up hitting Biloxi. The NHC ended up being off by about four hours on predicted landfall.	D	
The models had it going every which way after landfall. (See the model forecasts below.)	OSA	

Figure B.10

Figure B.12 shows the divergence of various forecast tracks from different computer models.

They were wrong on where the eye was. We could see it on radar.	D, OSA	
The center was erratic. You could see the eye wobble on the GOES loop and the radar loop. The eye was running in and out and sometimes was defined and sometimes was not. It was moving slowly and shifting from side to side.	OSA	
Sunday night there was an NHC phone discussion.	OSA	9:00 PM
They were still wrong on where the eye was, but they did say it was erratic. (See the NHC Bulletin #51 below.)	D	

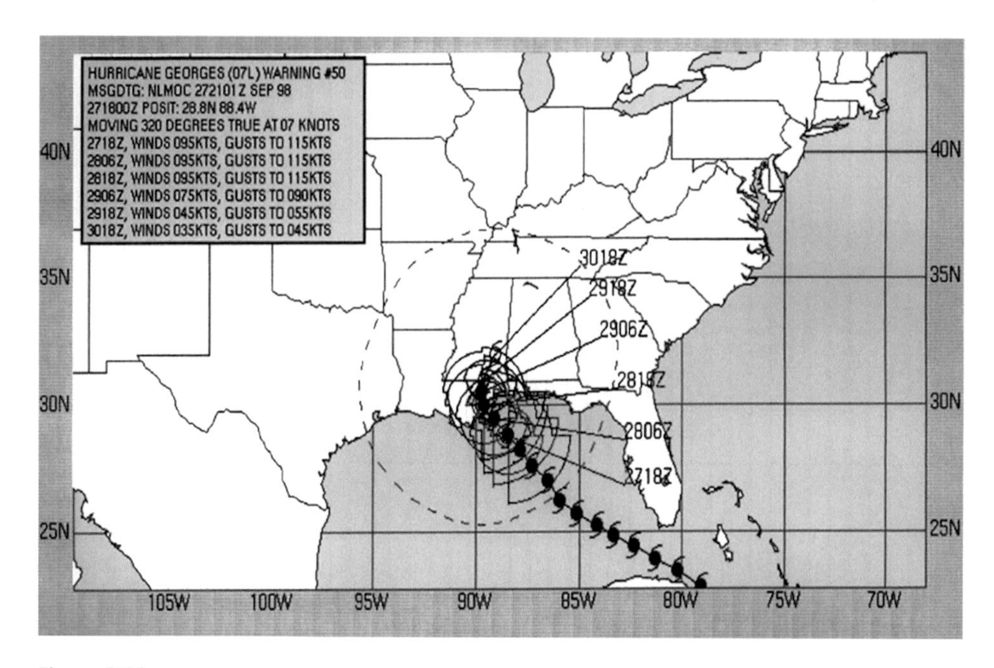

Figure B.11

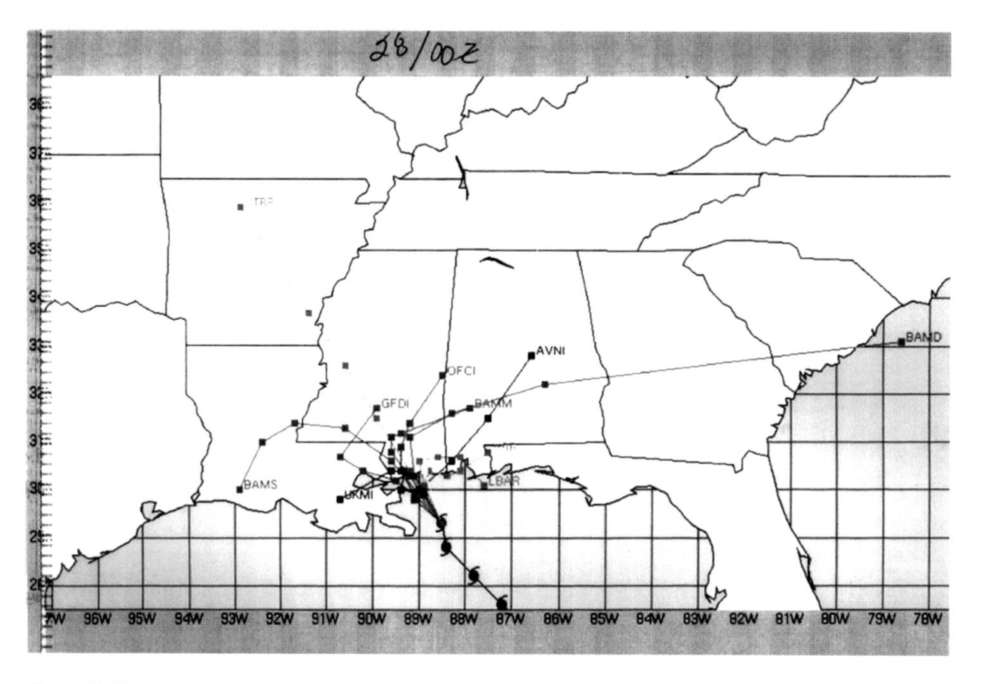

Figure B.12

BULLETIN HURRICANE GEORGES ADVISORY NUMBER 51 NATIONAL WEATHER SERVICE MIAMI FL 10 PM CDT SUN SEP 27 1998 GEORGES SLOWLY APPROACHING THE CENTRAL GULF COAST. A HURRICANE WARNING REMAINS IN EFFECT FROM MORGAN CITY LOUISIANA TO PANAMA CITY FLORIDA. A HURRICANE WARNING MEANS THAT HURRICANE CONDITIONS ARE EXPECTED IN THE WARNED AREA WITHIN 24 HOURS. PREPARATIONS TO PROTECT LIFE AND PROPERTY SHOULD BE RUSHED TO COMPLETION. A TROPICAL STORM WARNING AND A HURRICANE WATCH REMAIN IN EFFECT FROM EAST OF PANAMA CITY FLORIDA TO ST. MARKS FLORIDA. AT 10 PM CDT. THE HURRICANE WATCH IS DISCONTINUED FROM WEST OF MORGAN CITY TO INTRACOASTAL CITY LOUISIANA. AT 10 PM CDT 0300Z. THE CENTER OF HURRICANE GEORGES WAS LOCATED NEAR LATITUDE 29.5 NORTH, LONGITUDE 88.6 WEST. THIS POSITION IS ABOUT 70 MILES SOUTH-SOUTHEAST OF BILOXI MISSISSIPPI AND ABOUT 105 MILES EAST-SOUTHEAST OF NEW ORLEANS LOUISIANA. GEORGES IS MOVING ERRATICALLY BUT GENERALLY TOWARD THE NORTH NORTHWEST NEAR 6 MPH. ON THIS TRACK THE CORE OF THE HURRICANE SHOULD REACH THE COAST LATE MONDAY. DO NOT FOCUS ON THE PRECISE LOCATION AND TRACK OF THE CENTER. THE HURRICANE'S DESTRUCTIVE WINDS, RAIN, AND STORM SURGE COVER A WIDE SWATH. MAXIMUM SUSTAINED WINDS ARE NEAR 105 MPH 165 KM/HR WITH HIGHER GUSTS. SOME FLUCTUATIONS IN WIND SPEED ARE LIKELY UNTIL THE CENTER MOVES OVER LAND. HURRICANE FORCE WINDS EXTEND OUTWARD UP TO 45 MILES (75 KM.) FROM THE CENTER, AND TROPICAL STORM FORCE WINDS EXTEND OUTWARD UP TO 175 MILES (280 KM.). THE NEW ORLEANS WSR-88D RADAR INDICATES THAT HURRICANE FORCE WINDS MAY BE OCCURRING NEAR THE GROUND OVER EXTREME SOUTHERN MISSISSIPPI. KEESLER AIR FORCE BASE RECENTLY REPORTED A GUST TO 79 MPH. THE LATEST ESTIMATED MINIMUM CENTRAL PRESSURE IS 961 MB. (28.38 INCHES). STORM SURGE FLOODING OF 10 TO 15 FEET ABOVE NORMAL TIDE LEVELS AND UP TO 17 FEET AT THE HEADS OF BAYS IS POSSIBLE IN THE WARNED AREA AND WILL BE ACCOMPANIED BY LARGE AND DANGEROUS BATTERING WAVES. FLOODING RAINS OF 15 TO 25 INCHES WITH LOCALLY HIGHER AMOUNTS ARE LIKELY IN ASSOCIATION WITH THIS SLOW-MOVING HURRICANE. ISOLATED TORNADOES ARE POSSIBLE EAST AND NORTHEAST OF THE TRACK OF GEORGES.	But another factor was when we saw the eye come up.

SMALL CRAFT FROM WEST OF MORGAN CITY LOUISIANA WESTWARD AND SOUTHWARD ALONG THE COAST OF TEXAS SHOULD REMAIN IN PORT. SMALL CRAFT ALONG THE WEST COAST OF THE FLORIDA PENINSULA SHOULD REMAIN IN PORT UNTIL WINDS AND SEAS SUBSIDE. REPEATING THE 10 PM CDT POSITION 29.5 N., 88.6 W. MOVEMENT TOWARD NORTH NORTHWEST NEAR 6 MPH. MAXIMUM SUSTAINED WINDS 105 MPH. MINIMUM CENTRAL PRESSURE 961 MB. INTERMEDIATE ADVISORIES WILL BE ISSUED BY THE NATIONAL HURRICANE CENTER AT MIDNIGHT CDT AND 2 AM CDT FOLLOWED BY THE NEXT COMPLETE ADVISORY AT 4 AM CDT MONDAY.		
But another factor was when we saw the eye come up.	OSA	
We looked at buoy data every few hours and did our own charts. See the charts I did at 00Z (below).	OSA, A	Midnight

Figure B.13 shows the participants' hand chart work.

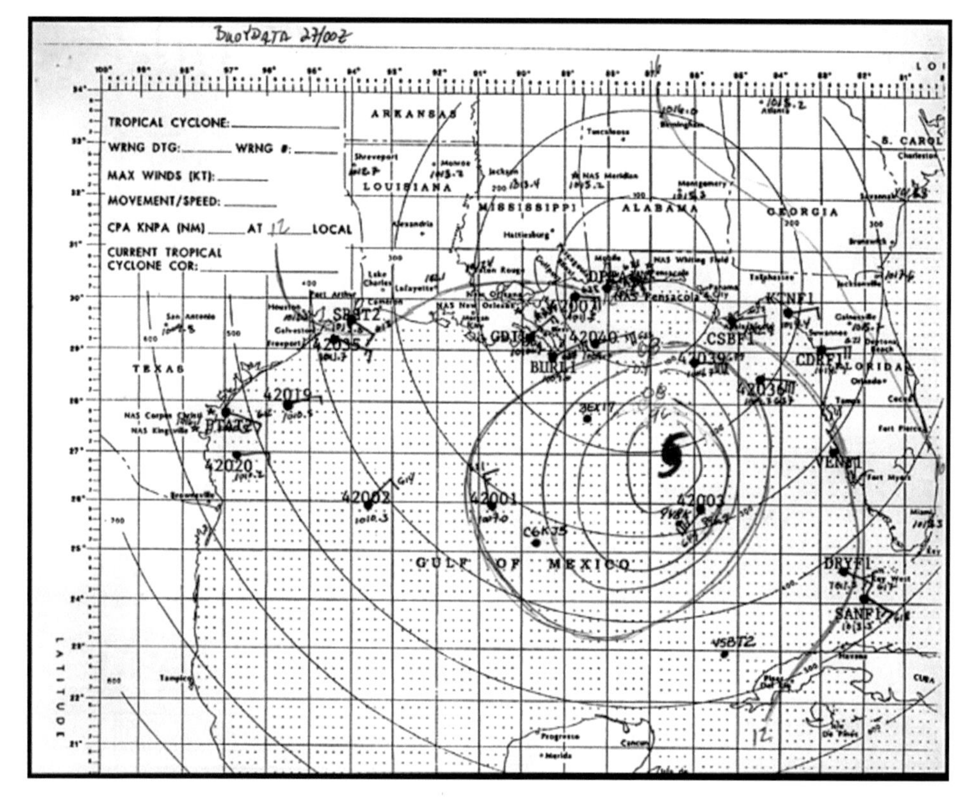

Figure B.13

The NHC had it making landfall later on Monday.	OSA	
You can't blame the NHC. They already had their forecast out, and they had to follow it. So did we. The NHC could always update every three hours, changing where they put the storm surge watches. But usually they stick with their hourly forecast.	D	
At 2:00 AM the NHC had it between New Orleans and Biloxi.	OSA	2:00 AM
The NHC shifted the track a little to the east out to Gulfport, but we were leery about that track. They were still off.	OSA, D	
It picked up speed right after the NHC conference call, so there was not much they could do. The hurricane sped up and headed straight north. Based on buoy data, we could tell that it was heading north.	OSA, D	
The NHC had it shifting northwest to Louisiana, more of a westward track. But we could see it heading due north toward Biloxi. Still, it hit within the area of their forecast.	OSA, D	
There wasn't much to do except put out a few advisories for heavy weather—every six hours. So we watched the radar.	A, OSA	
It made landfall Monday morning 28 September between Biloxi and Ocean Springs. (See the GOES visible image, the GOES colorized image, the GOES enhanced IR image, the NEXRAD reflectivity image, and the NEXRAD velocity image below.)	OSA	approx. 7:30 AM
As it came to landfall, the eye was south of NASP.	OSA	

Figures B.14 and B.15 (plate 20) show the GOES visible and infrared images at the time of landfall of the center of Hurricane George, the morning of 28 September 1998.

Figure B.16 (plate 21) shows the radar image for Hurricane George at the time the center made landfall.

Figure B.14

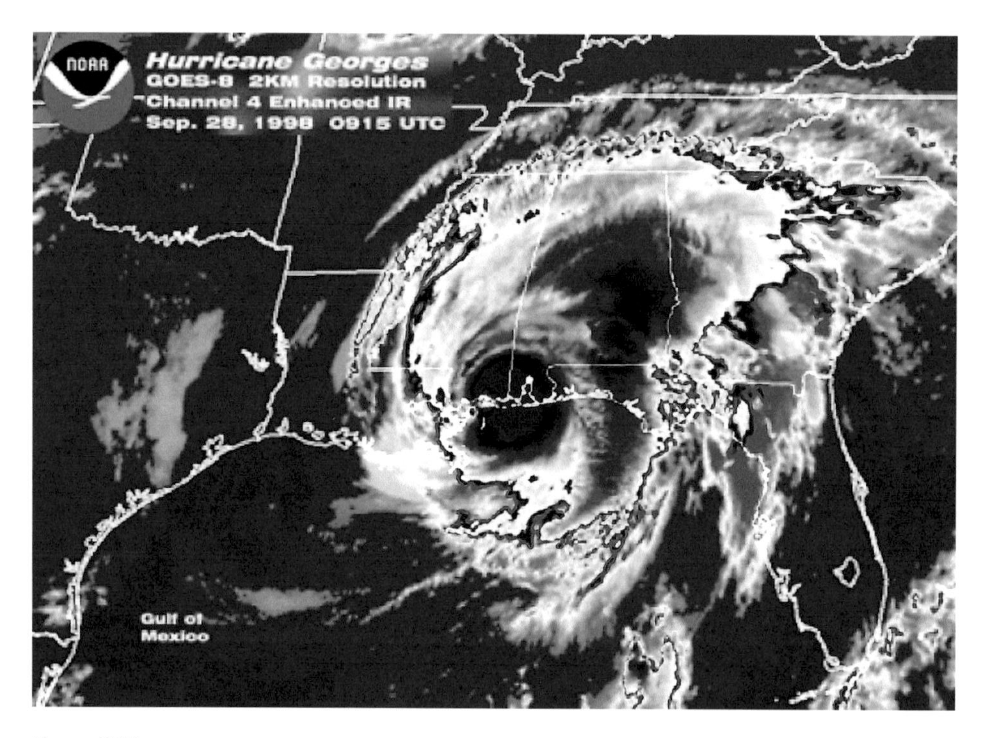

Figure B.15

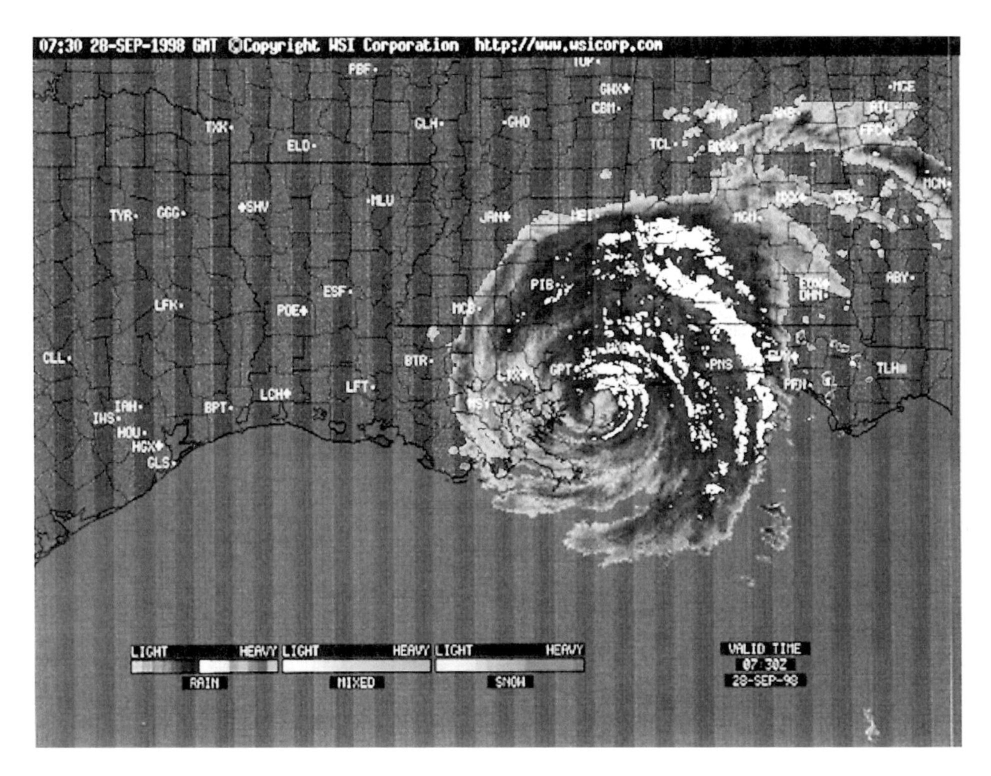
07:30 28-SEP-1998 GMT ©Copyright WSI Corporation http://www.wsicorp.com
LIGHT
HEAVY
RAIN
MIXED
SNOW
VALID TIME
07:30Z
28-SEP-98

Figure B.16

Appendix C: Example Synoptic Analyses of GOES Visible Images

The reader is invited to inspect figures C.1 through C.5 and then review the following analyses provided by a NWS forecaster.

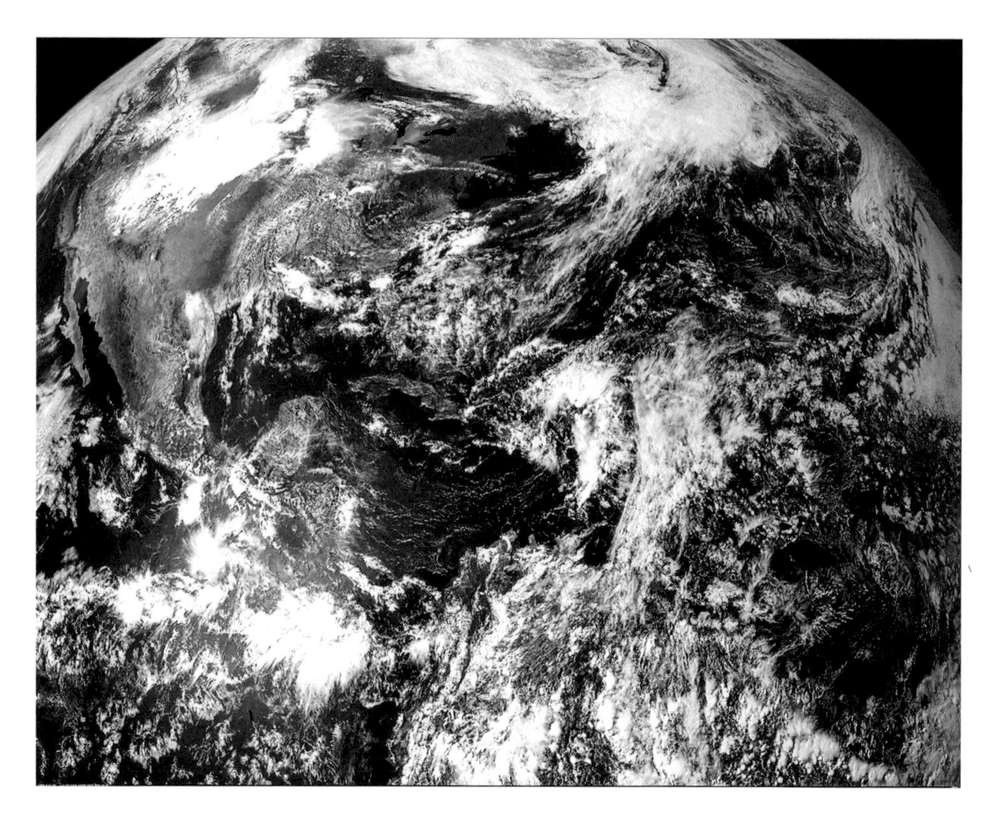

Figure C.1

Continental U.S./Western Atlantic Ocean view, James Bay south to South America. Notable features include convection as part of the Inter-tropical convergence zone (ITCZ) south of Mexico in eastern Pacific Ocean. Also, a subtropical Low was located just north of Puerto Rico with associated convection. In addition, a frontal system was noted extending from the Canadian Maritimes southwest to the Carolina coast and the Gulf Coast. There was also a low- pressure system noted at the northern end of this frontal system in the North Atlantic Ocean. There was also a low-pressure system located across south central Canada, with a frontal system extending southwest into the central Rockies. Diurnal cumulus clouds were also found across portions of the Mississippi Valley and interior southeast states.

Figure C.2

Eastern continental U.S./Western Atlantic Ocean, James Bay south to Cuba. A large subtropical Low can be seen just east of the Bahamas. Also, a frontal system extended from Quebec into the Great Lakes region, extending south/southwest to Texas. Along this front, a possible wave of low pressure can be seen in the vicinity of Iowa/northern Illinois. To the south of this frontal system, a large and broad upper level ridge of high pressure was encompassing much of the eastern CONUS, with some patchy high cirrus clouds embedded within this ridge. Some stratocumulus was noted north of the aforementioned subtropical low. In addition, low clouds were noted on the north side of the frontal boundary, likely indicative of low-level cold air advection across portions of northwest Ontario and the northern Great Lakes.

Figure C.3

West Coast continental United States and eastern Pacific Ocean, from U.S./Canada border south to southern Baja California. Low stratus/stratocumulus clouds were noted along and off the California coast into the eastern Pacific Ocean. A frontal system extended from the northern Rockies/southwest Canada region west-southwest to off the Washington/Oregon coasts. Convection was also noted across portions of the Rockies and western Plains, possibly associated with a frontal system and/or dry line boundary, with additional scattered convection across portions of Texas and Mexico. There also was a small upper level vorticity maximum (pocket of mid/upper tropospheric cyclonic circulation) noted across southwest Canada extending into north central Montana.

Figure C.4
Western continental United States into central Pacific Ocean, southern Canada south to equator. Most notable were two distinct tropical cyclones—one south of Baja California and another one further west. There was additional, less organized convection along the Inter-tropical convergence zone (ITCZ) just north of the Equator. Also, a distinct mid-latitude cyclone was noted approaching the Pacific northwest/northern California coast. A possible upper level jet stream maximum was located from Southern California northeast across the north central Rockies and into the northern Plains. Low stratus/stratocumulus cloud patches were also evident throughout the eastern Pacific Ocean.

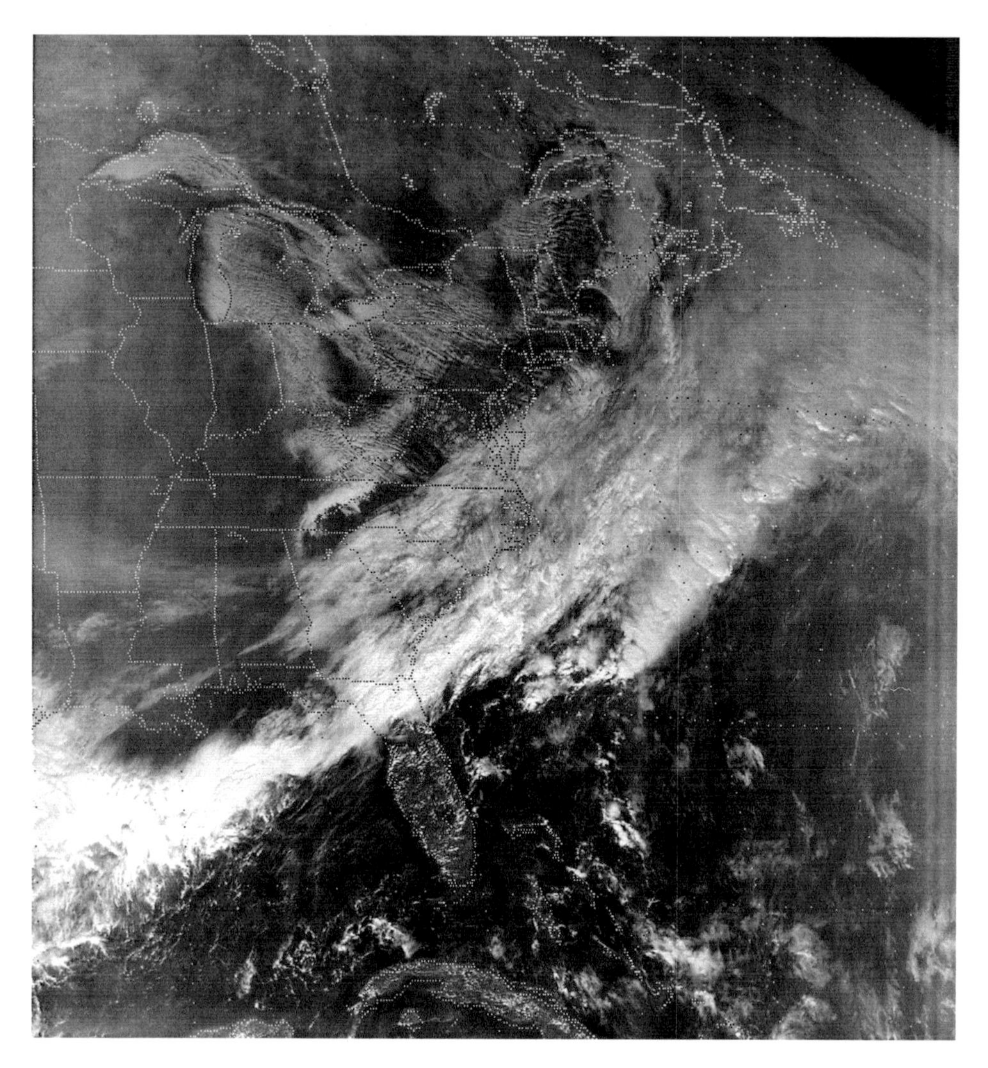

Figure C.5

Eastern continental United States/Western Atlantic Ocean, Canadian Maritimes south to Cuba. A frontal system extended from the Canadian Maritimes south/southwest to just off the eastern seaboard to northern Florida and into the Gulf of Mexico. Deep convection was evident to the south and east of this front from off the southeast U.S. coast to just west of Bermuda. Also, scattered cumulus clouds were noted ahead of this front over Florida. Lake-effect cloud streamers (likely producing bands of lake-effect precipitation) were located across the interior northeast states/northern Appalachians into northern New England and southeast Quebec. Some of this was probably associated with low-level instability induced by strong cold air advection in the wake of the aforementioned frontal system.

Appendix D: A Case Study in Human–Machine Interdependence

The March 1993 Winter Superstorm

Timeline based on Uccellini et al. (1995) and Kocin, Schumacher, Morales, and Uccellini (1995).

Hindsight Analysis

"Cyclogenesis along the east coast was predicted up to five days in advance. The unusual intensity of the storm was highlighted three days in advance, with snowfall amounts exceeding 12 inches predicted over a large area with unprecedented lead times. Numerous blizzard watches and warnings were also issued with unprecedented lead times, allowing the media and government officials to prepare the public, aviation and marine interests to take necessary precautions ... the increasing confidence of forecasters to predict major storm events, although hard to quantify, was perhaps the key ingredient for the unprecedented lead times ... that led people to believe the forecasters and take appropriate action. ... The improved spatial and temporal resolution of the [NEXRAD] radar and a local cooperative observer network enabled the local forecasting office to provide more detailed and accurate products with more timely updates than ever before ... the performance of the forecasting community during the March 1993 superstorm can be shown to represent an important milestone in an effort of over 40 years in which weather forecasting has been transformed to a science-based and user-oriented service. ... The forecasts for heavy snow and rate of snowfall were consistent across the entire event, although the snow in eastern Kentucky was underforecast. The winter storm watches issued by WFOs in the regions expected to receive the most snowfall were issued with 25 to 40 hours lead-time. The winter storm warnings and special weather statements issued by the WFOs on 11 March were issued with 10-20 hours lead-time, before a single snowflake had fallen. The long lead times allowed emergency response to coordinate with utilities, implement shelter plans, advise health centers to stock additional supplies, activate emergency broadcast systems, etc." (pp. 197–199)

Table D.1

Date	What the Computer Models Said	What the Forecasters Thought	What the Forecasters Did
March 7	Storm would develop along the East Coast of the United States. The new "ensemble" forecasts and forecasts based on statistical analysis predicted an 85% chance of precipitation (2–3 inches of snow) in West Virginia.	Forecasters at the National Meteorological Center (NMC) felt that cyclogenesis along the east coast United States would be unlikely because the models had overforecasted previous events of this type. Also, the weak cyclones that had developed tracked further inland and stronger ones that had developed tracked out over the Atlantic.	Continued analysis of model outputs and observational data.
March 7–11	Models consistently predicted a major cyclone along the East Coast.	NMC forecasters' skepticism of the model outputs diminished.	Continued analysis of model outputs and observational data.
March 10–11	One model predicted cyclogenesis in the Gulf of Mexico, whereas other models predicted cyclogenesis off the East Coast.	Consternation over the inconsistency of the outputs of the various computer models.	Local Weather Forecast Offices (WFOs) along the east coast United States began issuing discussions of the potential for a severe storm, with blizzard conditions. They commenced frequent briefings with local emergency response managers. NWS Eastern Region Headquarters advised the Federal Emergency Management Agency of the possibility of blizzard conditions.

Table D.1 (continued)

Date	What the Computer Models Said	What the Forecasters Thought	What the Forecasters Did
March 11	The storm event began to unfold. A jet stream pattern in the western United States that the models had predicted developed as predicted and would influence the East Coast cyclogenesis.		National Meteorological Center began to issue storm summary statements, predicting "unusually severe" and "perhaps record-breaking" snowfall of "historic proportions." Hurricane hotline was activated. WFOs along the East Coast began to exchange information and develop a consensus on which models they could rely. WFOs issued the first winter weather watches.
March 11–12	Details of the developing storm differed from what computer models were predicting.	Difficulty in predicting the location, intensity, and track of the developing cyclone.	
March 12–13	The model that was predicting storm development along the East Coast underestimated the rapid cyclogenesis occurring in the Gulf of Mexico.		NMC relied more on the outputs of the computer models that model the entire globe, rather than those that model just North America, because regional models had tended to over-predict the central low pressure.
March 13	Models began to converge on cyclogenesis in the Gulf, off the Louisiana coast.	The model differences were significant. Forecasters continued to compromise between their own analyses and the predictions of the computer models about cyclone position and the central pressure of the Low.	NMC forecasted a developing low-pressure center in the southeastern United States. NMC adjusted the predicted position of the rain–snow line further south, into central Alabama. With each successive model run, the forecasters predicted a lower and lower central pressure. Blizzard warnings were issued by all eastern region WFOs.

Table D.1 (continued)

Date	What the Computer Models Said	What the Forecasters Thought	What the Forecasters Did
March 13	The Sterling, VA WFO had one of the first NEXRAD radar installations. Individual bands of snowfall could be tracked.	Although the Sterling, VA forecasters were busy because of the weather event and the need to work using a new technology, NEXRAD enabled forecasting that was previously impossible.	The Sterling, VA WFO issued frequent location-specific half-hourly nowcasts. They confirmed radar scan data with surface observations made by a network of cooperative observers.
March 14	Models began to converge in forecasting a major storm along the East Coast, but differed in predicting its track. Earlier that winter, two of the models correctly predicted a track west of the Appalachian mountains when a third model had kept the storms along the East Coast.	Forecasters had to make judgments concerning snowfall amounts and the timing of when snow would change to ice or rain and then back to snow. Snow amounts were of special concern because a record-breaking storm was expected. Forecasters felt that the models were showing underdevelopment of the cyclone. Different models were predicting different scenarios. Forecasters relied on different models to predict different things (low pressure, storm track, etc.) based on past experience with the models' successes.	NMC forecasters compromised among the models and their own judgments about the central pressure and location of the cyclone. Forecasters began to shift the predicted rain–snow line further to the north.
March 14	Models began to converge on the storm track, placing it along the New England coastline.		

Appendix E: References on Visualization Design and Display Technology

The design of new visualizations and displays is an active area of research and development in meteorology, information visualization sciences, and technology more broadly. Although reference is made in this literature to "visualization," the research is about display design more than the psychology of perception, visual comprehension, or sensemaking. Hence, we provide these references as an appendix and do not review or integrate this material into the body of the chapters of this book. A full bibliography and review of the area of weather data visualization and display design would be a major compendium in and of itself. This appendix lists many pertinent references, by category. This listing is suggestive and by no means exhaustive. Although this list focuses on recent work, it includes a few classic publications. The authors thank Marc Rautenhaus of the Computer Graphics and Visualization Group, Technische Universität München, Garching, Germany, for providing these references.

Visualizations for Climate Research

Böttinger, M., H. Pohlmann, N. Röber, K. Meier-Fleischer, and D. Spickermann. 2015. Visualization of 2D uncertainty in decadal climate predictions. In *Proceedings of the Workshop on Visualization in Environmental Sciences* (EnviroVis 2015). Sponsored by the Eurographics Working Group on Data Visualization and the IEEE Visualization and Graphics Technical Committee. doi:10.2312/envirvis.20151083.

Dasgupta, A., J. Poco, E. Bertini, and C. T. Silva. 2016. Reducing the analytical bottleneck for domain scientists: Lessons from a climate data visualization case study. *Computing in Science & Engineering* 18:92–100.

Dasgupta, A., J. Poco, Y. Wei, R. Cook, E. Bertini, and C. T. Silva. 2015. Bridging theory with practice: An exploratory study of visualization use and design for climate model comparison. *IEEE Transactions on Visualization and Computer Graphics* 21:996–1014.

Glaas, E. (with eight others). 2015. Facilitating climate change adaptation through communication: Insights from the development of a visualization tool. *Energy Research & Social Science* 10:57–61.

Jänicke, H., M. Böttinger, U. Mikolajewicz, and G. Scheuermann. 2009. Visual exploration of climate variability changes using wavelet analysis. *IEEE Transactions on Visualization and Computer Graphics* 15:1375–1382.

Jin, H., and D. Guo. 2009. Understanding climate change patterns with multivariate geovisualization. In *Proceedings of the IEEE International Conference on Data Mining* (pp. 217–222). New York: IEEE.

Johansson, J., T.-S. S. Neset, and B.-O. Linnër. 2011. Evaluating climate visualization: An information visualization approach. In *Proceedings of the 14th International Conference on Information Visualization* (pp. 156–161). New York: IEEE.

Ladstädter F. (with eight others). 2010. Exploration of climate data using interactive visualization. *Journal of Atmospheric and Oceanic Technology* 27:667–679.

Matuschek, O., and A. Matzarakis. 2011. A mapping tool for climatological applications. *Meteorological Applications* 18:230–237.

Max, N., R. Crawfis, and D. Williams. 1993. Visualization for climate modeling. *IEEE Computer Graphics and Applications* 13:34–40.

Nocke, T., S. Buschmann, J. F. Donges, N. Marwan, H. J. Schulz, and C. Tominski. 2015. Review: Visual analytics of climate networks. *Nonlinear Processes in Geophysics* 22:545–570.

Nocke, T., M. Flechsig, and U. Böhm. 2007. Visual exploration and evaluation of climate-related simulation data. In *Proceedings of the IEEE Conference on Simulation* (pp. 703–711). New York: IEEE.

Poco J. (with eight others). 2014. Visual reconciliation of alternative similarity spaces in climate modeling. *IEEE Transactions on Visualization and Computer Graphics* 20:1923–1932.

Qu, H., W.-Y. Chan, A. Xu, K.-L. Chung, K.-H. Lau, and P. Guo. 2007. Visual analysis of the air pollution problem in Hong Kong. *IEEE Transactions on Visualization and Computer Graphics* 13:1408–1415.

Santos, E., J. Poco, Y. Wei, S. Liu, B. Cook, D. N. Williams, and C. T. Silva. 2013. UV-CDAT: Analyzing climate datasets from a user's perspective. *Computing in Science & Engineering* 15:94–103.

Schneider, B. 2012. Climate model simulation visualization from a visual studies perspective. *WIREs Climate Change* 3:185–193.

Steed, C. A. (with seven others). 2014. Web-based visual analytics for extreme scale climate science. *In Proceedings of the IEEE International Conference on Big Data* (pp. 383–392). New York: IEEE.

Stephens, E. M., T. L. Edwards, and D. Demeritt. 2012. Communicating probabilistic information from climate model ensembles: Lessons from numerical weather prediction. *WIREs Climate Change* 3:409–426.

Tominski, C., J. F. Donges, and T. Nocke. 2011. Information visualization in climate research. In *Proceedings of the 15th International Conference on Information Visualization* (pp. 298–305). New York: IEEE.

Wong, P. C., H.-W. Shen, R. Leung, S. Hagos, T.-Y. Lee, X. Tong, and K. Lu. 2014. Visual analytics of large-scale climate model data. In *Proceedings of the 4th Symposium on Large Data Analysis and Visualization* (pp. 85–92). New York: IEEE.

Visualization of Atmospheric Dynamics and Severe Weather

Abram, G., and L. Treinish. 1995. An extended data-flow architecture for data analysis and visualization. In *Proceedings of the 6th Conference on Visualization 95*. New York.

Bally, J. 2004. The thunderstorm interactive forecast system: Turning automated thunderstorm tracks into severe weather warnings. *Weather and Forecasting* 19:64–72.

Dobashi, Y., K. Kaneda, H. Yamashita, T. Okita, and T. Nishita. 2000. A simple, efficient method for realistic animation of clouds. In *Proceedings of the 27th Annual Conference on Computer Graphics and Interactive Techniques* SIGGRAPH 2000 (pp. 19–28.). New York: Association for Computing Machinery Press/Addison-Wesley Publishing Co.

Doraiswamy, H., V. Natarajan, and R. S. Nanjundiah. 2013. An exploration framework to identify and track movement of cloud systems. *IEEE Transactions on Visualization and Computer Graphics* 19:2896–2905.

Fjukstad, B., J. M. Bjorndalen, and O. Anshus. 2014. Uncertainty estimation and visualization of wind in weather forecasts. In *IVAPP 2004: The International Conference on Information Visualization Theory and Applications* (pp. 321–328). New York: IEEE.

Gallus, W. A., C. Cervato, C. Cruz-Neira, G. Faidley, and R. Heer. 2005. Learning storm dynamics with a virtual thunderstorm. *Bulletin of the American Meteorological Society* 86:162–163.

Graedel, T. E. 1977. The wind boxplot: An improved wind rose. *Journal of Applied Meteorology* 16:448–450.

Günther, T., M. Schulze, A. Friederici, and H. Theisel. 2015. Visualizing volcanic clouds in the atmosphere and their impact on air traffic. *IEEE Computer Graphics and Applications*. doi:10.1109/MCG.2015.121.

Harris, M. J., and A. Lastra. 2001. Real-time cloud rendering. *Computer Graphics Forum* 20:76–85.

Harned, S., S. Businger, and M. Stephenson. 1997. The application of three and four dimensional visualization of meteorological data fields in weather forecasting. In *Symposium on Weather Forecasting* (pp. 50–53). Boston, MA: American Meteorological Society.

Hufnagel, R., and M. Held. 2012. A survey of cloud lighting and rendering techniques. In *Proceedings of the 20th International Conference on Computer Graphics, Visualization, and Computer Vision* (WSCG'12) (pp. 53–63). New York: Association for Computing Machinery.

Martin, J. P., J. E. Swan, R. J. Moorhead, Z. Liu, and S. Cai. 2008. Results of a user study on 2D hurricane visualization. *Computer Graphics Forum* 27:991–998.

Max, N., R. Crawfis, and D. Williams. 1992. Visualizing wind velocities by advecting cloud textures. In *Proceedings of the IEEE Conference on Visualization* (pp. 179–184). New York: IEEE.

Paterson, M. P., and S. F. Benjamin. 1975. Better than a wind rose. *Atmospheric Environment* 9:537–542.

Shen, B.-W., B. Nelson, W.-K. Tao, and Y.-L. Lin. 2013. Advanced visualizations of scale interactions of tropical cyclone formation and tropical waves. *Computing in Science & Engineering* 15:47–59.

Ware, C., J. G. W. Kelley, and D. Pilar. 2014. Improving the display of wind patterns and ocean currents. *Bulletin of the American Meteorological Society* 95:1573–1581.

Visualization of Radar, Satellite, and Remote Sensing Data

Gerstner, T., D. Meetschen, S. Crewel, M. Griebel, and C. Simmer. 2002. A case study on multiresolution visualization of local rainfall from weather radar measurements. In *Proceedings of the IEEE Conference on Visualization* (pp. 533–536). New York: IEEE.

Hasler, A. F. 1981. Stereographic observations from geosynchronous satellites: An important new tool for the atmospheric sciences. *Bulletin of the American Meteorological Society* 62:194–212.

Kristof, P., B. Benes, C. X. Song, and L. Zhao. 2013. A system for large-scale visualization of streaming Doppler data. In *Proceedings of the 2013 IEEE International Conference on Big Data* (pp. 33–40). New York: IEEE.

Murray, D., B. Hibbard, T. Whittaker, and J. Kelly. 2001. Using VisAD to build tools for visualizing and analyzing remotely sensed data. In *Proceedings of the International Geoscience and Remote Sensing Symposium* (Vol. 1, pp. 204–206). New York: IEEE.

Phipps, M., and S. Rowe. 2010. Seeing satellite data. *Public Understanding of Science (Bristol, England)* 19:311–321.

Visualizations of the Outputs of Computer Models and Ensembles

Atger, F. 1999. Tubing: An alternative to clustering for the classification of ensemble forecasts. *Weather and Forecasting* 14:741–757.

Bensema, K., L. Gosink, H. Obermaier, and K. Joy. 2015. Modality-driven classification and visualization of ensemble variance. *IEEE Transactions on Visualization and Computer Graphics* 47:218–224.

Demir, I., C. Dick, and R. Westermann. 2014. Multi-charts for comparative 3D ensemble visualization. *IEEE Transactions on Visualization and Computer Graphics* 20:2694–2703.

Ferstl, F., K. Bürger, and R. Westermann. 2016. Streamline variability plots for characterizing the uncertainty in vector field ensembles. *IEEE Transactions on Visualization and Computer Graphics* 22:767–776.

Froude, L. S. R., L. Bengtsson, and K. Hodges. 2007. The prediction of extratropical storm tracks by the ECMWF and NCEP ensemble prediction systems. *Monthly Weather Review* 135:2545–2567.

Gneiting, T., and A. E. Raftery. 2005. Weather forecasting with ensemble methods. *Science* 310:248–249.

Hamill, T. M., M. J. Brennan, B. Brown, M. DeMaria, E. N. Rappaport, and Z. Toth. 2012. NOAA's future ensemble-based hurricane forecast products. *Bulletin of the American Meteorological Society* 93:209–220.

Höllt T. (and seven others). 2014. Ovis: A framework for visual analysis of ocean forecast ensembles. *IEEE Transactions on Visualization and Computer Graphics* 20:1114–1126.

Köthur, P., C. Witt, M. Sips, N. Marwan, S. Schinkel, and D. Dransch. 2015. Visual analytics for correlation-based comparison of time series ensembles. *Computer Graphics Forum* 34:411–420.

Marzban, C., and S. Sandgathe. 2009. Verification with variograms. *Weather and Forecasting* 24:1102–1120.

Mirzargar, M., R. T. Whitaker, and R. M. Kirby. 2014. Curve boxplot: Generalization of boxplot for ensembles of curves. *IEEE Transactions on Visualization and Computer Graphics* 20:2654–2663.

Sanyal, J., S. Zhang, J. Dyer, A. Mercer, P. Amburn, and R. Moorhead. 2010. Noodles: A tool for visualization of numerical weather model ensemble uncertainty. *IEEE Transactions on Visualization and Computer Graphics* 16:1421–1430.

Schumacher, R. S., and C. A. Davis. 2010. Ensemble-based forecast uncertainty analysis of diverse heavy rainfall events. *Weather and Forecasting* 25:1103–1122.

Treinish, L. A. 2002. Case study on the adaptation of interactive visualization applications to web-based production for operational mesoscale weather models. In *Proceedings of the IEEE Conference on Visualization* (pp. 549–552). New York: IEEE.

Whitaker, R. T., M. Mirzargar, and R. M. Kirby. 2013. Contour boxplots: A method for characterizing uncertainty in feature sets from simulation ensembles. *IEEE Transactions on Visualization and Computer Graphics* 19:2713–2722.

Visualizations of Uncertainty. Probability, and Risk

Bonneau, G.-P., H.-C. Hege, C. Johnson, M. Oliveira, K. Potter, P. Rheingans, and T. Schultz. 2014. Overview and state-of-the-art of uncertainty visualization. In *Scientific Visualization*, eds. C. D. Hansen et al. (pp. 3–27). London: Springer.

Cox, J., D. House, and M. Lindell. 2013. Visualizing uncertainty in predicted hurricane tracks. *International Journal for Uncertainty Quantification* 3:143–156.

Demuth, J. L., B. H. Morrow, and J. K. Lazo. 2009. Weather forecast uncertainty information: An exploratory study with broadcast meteorologists. *Bulletin of the American Meteorological Society* 90:1614–1618.

Demuth, J. L., R. E. Morss, J. K. Lazo, and D. C. Hilderbrand. 2013. Improving effectiveness of weather risk communication on the NWS Point-and-click web page. *Weather and Forecasting* 28:711–726.

Demuth, J. L., R. E. Morss, B. H. Morrow, and J. K. Lazo. 2012. Creation and communication of hurricane risk information. *Bulletin of the American Meteorological Society* 93:1133–1145.

Johnson, C. R., and A. R. Sanderson. 2003. A next step: Visualizing errors and uncertainty. *IEEE Computer Graphics and Applications* 23:6–10.

Jones, D. W., and S. Josslyn. 2004. The MURI uncertainty monitor (MUM). In *Proceedings of the 84th AMS Conference*. Boston, MA: American Meteorological Society.

Josslyn, S., K. Pak, D. Jones, J. Pyles, and E. Hunt. 2007. The effect of probabilistic information on threshold forecasts. *Weather and Forecasting* 22:804–812.

MacEachren, A. M., A. Robinson, S. Hopper, S. Gardner, R. Murray, M. Gahegan, and E. Hetzler. 2005. Visualizing geospatial information uncertainty: What we know and what we need to know. *Cartography and Geographic Information Science* 32:139–160.

Mass, C. (with nine others). 2009. PROBCAST: A web-based portal to mesoscale probabilistic forecasts. *Bulletin of the American Meteorological Society* 90:1009–1014.

Nadav-Greenberg, L., S. L. Joslyn, and M. U. Taing. 2008. The effect of uncertainty visualizations on decision making in weather forecasting. *Journal of Cognitive Engineering and Decision Making* 2:24–47.

Pang, A. T., C. M. Wittenbrink, and S. K. Lodha. 1997. Approaches to uncertainty visualization. *Visual Computer* 13:370–390.

Roebber, P. J. 2009. Visualizing multiple measures of forecast quality. *Weather and Forecasting* 24:601–608.

Wittenbrink, C. M., A. T. Pang, and S. K. Lodha. 1996. Glyphs for visualizing uncertainty in vector fields. *IEEE Transactions on Visualization and Computer Graphics* 2:266–279.

Zehner, B., N. Watanabe, and O. Kolditz. 2010. Visualization of gridded scalar data with uncertainty in geosciences. *Computers & Geosciences* 36:1268–1275.

3-D and 4-D and Perspectival Visualizations

Alpert, J. C. 2003. 3-dimensional animated displays for sifting out medium range weather events. In *Proceedings of the 19th Conference on Interactive Information and Processing Systems (IIPS) for Meteorology, Oceanography, and Hydrology*. Paper 15.2. Boston, MA: American Meteorological Society.

Barjenbruch, D. B., E. Thaler, and E. J. Szoke. 2002. Operational applications of three dimensional air parcel trajectories using AWIPS D3D. *In Proceedings of the 18th Conference on Interactive Information Processing Systems (IIPS) for Meteorology, Oceanography, and Hydrology*, Paper J5.1. Boston, MA: American Meteorological Society.

Carlin, A. V. 1954. Meteorological charts in three dimensions. *Monthly Weather Review* 82:97–100.

Diehl, A., L. Pelorosso, C. Delrieux, C. Saulo, J. Ruiz, M. E. Gröller, and S. Bruckner. 2015. Visual analysis of spatio-temporal data: Applications in weather forecasting. *Computer Graphics Forum* 34:381–390.

Fernández, A., A. M. González, J. Díaz, and J. R. Dorronsoro. 2015. Diffusion maps for dimensionality reduction and visualization of meteorological data. *Neurocomputing* 163:25–37.

Helbig, C., L. Bilke, H.-S. Bauer, M. Böttinger, and O. Kolditz. 2015. MEVA: An interactive visualization application for validation of multifaceted meteorological data with multiple 3D devices. *PLoS One* 10. doi:10.1371/journal.pone.0123811.

Hibbard, W. L. 1986. Computer-generated imagery for 4-D meteorological data. *Bulletin of the American Meteorological Society* 67:1362–1369.

Hasler, A. F., H. Pierce, K. R. Morris, and J. Dodge. 1985. Meteorological data fields "in perspective." *Bulletin of the American Meteorological Society* 66:795–801.

Hibbard, W. L. 1997. Future directions in 3-D meteorological display. In *Proceedings of the 6th ECMWF Workshop on Meteorological Operational Systems* (pp. 186–190). European Centre for Medium-range Weather Forecasts (ECMWF) Reprint #2465.

McCaslin, P. T., P. A. McDonald, and E. J. Szoke. 2003. 3D visualization development at NOAA forecast systems laboratory. *Computer Graphics* 34:41–44.

Von der Haar, T. H., A. C. Meade, R. J. Craig, and D. L. Reinke. 1988. Four-dimensional imaging for meteorological applications. *Journal of Atmospheric and Oceanic Technology* 5:136–143.

Xie, J., H. Yu, and K.-L. Ma. 2014. Visualizing large 3D geodesic grid data with massively distributed GPUs. In *Proceedings of the 4th Symposium on Large Data Analysis and Visualization* (pp. 3–10). New York: IEEE.

Visualizations Using Open Source and GIS

Bergholt, L. 2008. Diana: An opens source production and visualization package. In *Proceedings of the 11th ECMWF Workshop on Meteorological Operational Systems* (pp. 150–152). Reading, England: ECMWF.

Blower, J. D., A. L. Gemmell, G. H. Griffiths, K. Haines, A. Santokhee, and X. Yang. 2013. A web map service implementation for the visualization of multidimensional gridded environmental data. *Environmental Modelling & Software* 47:218–224.

Cox A., R. Lucksinger, and D. Eck. 2010. Supporting operational weather forecasting with Google Earth. In *Proceedings of the 26th Conference on Interactive Information and Processing Systems (IIPS) for Meteorology, Oceanography, and Hydrology* (P. J8.2). Boston, MA: American Meteorological Society. [https://ams.confex.com/ams/90annual/techprogram/programexpanded_583.htm]

Sun, X., S. Shen, G. G. Leptoukh, P. Wang, L. Di, and M. Lu. 2012. Development of a web-based visualization platform for climate research using Google Earth. *Computers & Geosciences* 47:160–168.

Wang, Y. Q. 2014. MeteoInfo: GIS software for meteorological data visualization and analysis. *Meteorological Applications* 21:360–368.

Wang, Y., G. Huynh, and C. Williamson. 2013. Integration of Google Maps/Earth with microscale meteorology models and data visualization. *Computers & Geosciences* 61:23–31.

Visualization Using Stereographic and Virtual Reality Techniques

Gallus, W. A., D. N. Yarger, C. Cruz-Neira, and R. Heer. 2003. An example of a virtual reality learning environment. *Bulletin of the American Meteorological Society* 84:18–20.

Hasler, A. F., M. des Jardins, and A. J. Negri. 1981. Artificial stereo presentation of meteorological data fields. *Bulletin of the American Meteorological Society* 62:970–973.

Helbig, C., H.-S. Bauer, K. Rink, V. Wulfmeyer, M. Frank, and O. Kolditz. 2014. Concept and workflow for 3D visualization of atmospheric data in a virtual reality environment for analytical approaches. *Environmental Earth Sciences* 72:3767–3780.

Liu, P., J. Gong, and M. Yu. 2015. Graphics processing unit-based dynamic volume rendering for typhoons on a virtual globe. *International Journal of Digital Earth* 8:431–450.

Papathomas, T. V., J. Schiavone, and B. Julesz. 1987. Stereo animation for very large data bases: Case study in meteorology. *IEEE Computer Graphics and Applications* 7:18–27.

Smith, T. M., and V. Lakshmanan. 2011. Real-time, rapidly updating severe weather products for virtual globes. *Computers & Geosciences* 37:3–12.

Turk, F. J., J. Hawkins, K. Richardson, and M. Surratt. 2011. A tropical cyclone application for virtual globes. *Computers & Geosciences* 37:13–24.

Webley, P. W. 2011. Virtual globe visualization of ash: Aviation encounters, with the special case of the 1989 Redoubt-KLM incident. *Computers & Geosciences* 37:25–37.

Yalda, S., G. Zoppetti, R. Clark, and K. Mackin. 2012. Interactive immersion learning: Flying through weather data onboard the GEOpod. *Bulletin of the American Meteorological Society* 93:1811–1813.

Ziegler, S., R. J. Moorhead, P. J. Croft, and D. Lu. 2001. The MetVR case study: Meteorological visualization in an immersive virtual environment. In *Proceedings of the 2001 IEEE Visualization Conference* (pp. 489–596). New York: IEEE.

Workstation Designs

Achtor, T., T. Rink, T. Whittaker, D. Parker, and D. Santek. 2008. McIDAS-V: A powerful data analysis and visualization tool for multi and hyperspectral environmental satellite data. In *Proceedings SPIE* (Society of Photographic Instrumentation Engineers), *7085*. Bellingham, WA: International Society for Optical Engineering.

Daabeck, J. 2005. Overview of meteorological workstation development in Europe. In *Proceedings of the 21st International Conference on Interactive Information Processing Systems (IIPS) for Meteorology, Oceanography, and Hydrology,* Paper J8.1. Boston, MA: American Meteorological Society.

Limbach, S., M. Sprenger, E. Schömer, and H. Wernli. 2014. IWAL: An interactive weather analysis laboratory. *Bulletin of the American Meteorological Society* 96:903–909.

Reviews

Dow, A. K., E. M. Dow, T. D. Fitzsimmons, and M. M. Materise. 2014. Harnessing the environmental data flood: A comparative analysis of hydrologic, oceanographic, and meteorological informatics platforms. *Bulletin of the American Meteorological Society* 96:725–736.

Fox, P., and J. Hendler. 2011. Changing the equation on scientific data visualization. *Science* 331:705–708.

Gregg, W. R., and I. R. Tannehill. 1937. International standard projections for meteorological charts. *Monthly Weather Review* 65:411–415.

Grotjahn, R., and R. M. Chervin. 1984. Animated graphics in meteorological research and presentations. *Bulletin of the American Meteorological Society* 65:1201–1208.

Haase, H., M. Bock, E. Hergenröther, C. Knöpfte, H. J. Koppert, F. Schröder, A. Trembinski, and J. Weidenhausen. 1999. Where weather meets the eye: A case study on a wide range of meteorological visualizations for diverse audiences. In *Proceedings of Data Visualization'99* (pp. 261–266). New York: Springer.

Hibbard, W. L. 2005. Vis5D, Cave5D, and VisAD. In *The visualization handbook,* ed. C. D. Hansen and C. Johnson (pp. 673–688). New York: Academic Press.

Hibbard, W., and D. Santek. 1989. Visualizing large data sets in the earth sciences. *Computer* 22:53–57.

Keeling, S. J. 2010. Visualization of the weather: Past and present. *Meteorological Applications* 17:126–133.

Lundblad, P., H. Löfving, A. Elovsson, and J. Johansson. 2011. Exploratory visualization for weather data verification. In *Proceedings of the 15th International Conference on Information Visualization* (pp. 306–313). New York: IEEE.

Middleton, D., T. Scheitlin, and B. Wilhelmson. 2005. Visualization in weather and climate research. In *The visualization handbook*, ed. C. D. Hansen and C. Johnson (pp. 845–871). New York: Academic Press.

Papathomas, T. V., J. A. Schiavone, and B. Julesz. 1988. Applications of computer graphics to the visualization of meteorological data. *Computer Graphics* 22:327–334.

Quinan, P. S., and M. Meyer. 2016. Visually comparing weather features in forecasts. *IEEE Transactions on Visualization and Computer Graphics* 22:389–398.

Reinders, F., F. H. Post, and H. J. W. Spoelder. 2001. Visualization of time-dependent data with feature tracking and event detection. *Visual Computer* 17:55–71.

Rhyne, T. (with 11 others). 1992. Visualization requirements in the atmospheric and environmental sciences (five case study reports). In *Proceedings of the IEEE Conference on Visualization 1992* (pp. 428–435). New York: IEEE.

Schiavone, J. A., and T. V. Papathomas. 1990. Visualizing meteorological data. *Bulletin of the American Meteorological Society* 71:1012–1020.

Song, Y., J. Ye, N. Svakhine, S. Lasher-Trapp, M. Baldwin, and D. S. Ebert. 2006. An atmospheric visual analysis and exploration system. *IEEE Transactions on Visualization and Computer Graphics* 12:1157–1164.

Ware, C., and M. D. Plumlee. 2013. Designing a better weather display. *Information Visualization* 12:221–239.

References

Adams, R. L. A. 1973. Uncertainty in nature, cognitive dissonance, and the perceptual distortion of environmental information: Weather forecasts and New England beach trip decisions. *Economic Geography* 49:287–297.

Adelson, B. 1981. Problem solving and the development of abstract categories in programming languages. *Memory & Cognition* 9:422–433.

Ahlstrom, U. 2003, July/September. Current trends in the display of aviation weather. *Journal of Air Traffic Control*, 14–21.

Alexander, P. A. 2003. The development of expertise: The journey from acclimation to proficiency. *Educational Researcher* 32:10–14.

Allen, D. R. 2001. The genesis of meteorology at the University of Chicago. *Bulletin of the American Meteorological Society* 82:1905–1909.

Allen, G. 1982. Probability judgment in weather forecasting. In *Proceedings of the 9th Conference on Weather Forecasting and Analysis* (pp. 1–6). Boston, MA: American Meteorological Society.

Altmann, E. M., and J. G. Trafton. 2002. Memory for goals: An activation-based model. *Cognitive Science* 26:39–83.

American Meteorological Society. 1996. *Proceedings of the Eighth Conference on Satellite Meteorology and Oceanography*. Boston, MA: American Meteorological Society.

American Meteorological Society. 2003. *Curricula in the Atmospheric*. Oceanic, Hydrologic, and Related Sciences. [Downloaded 13 December, 2016, from http://www.ametsoc.org/AMSUCAR_CURRICULA/index.cfm]

Amirault, R. J., and R. K. Branson. 2006. Educators and expertise: A brief history of theories and models. In *The Cambridge Handbook of Expertise and Expert Performance*, ed. K. A. Ericsson, N. Charness, P. J. Feltovich, and R. R. Hoffman (pp. 69–86). New York: Cambridge University Press.

Anderson, J. R. 1982. Acquisition of a cognitive skill. *Psychological Review* 89:369–406.

Anderson, J. R. 2005. *Cognitive Psychology and Its Implications*. 6th ed. New York: Worth Publishers.

Anderson, J. R., F. G. Conrad, and A. T. Corbett. 1989. Skill acquisition and the LISP tutor. *Cognitive Science* 13:467–505.

Anderson, M., and B. Meyer. 2013. *Diagrammatic Representation and Reasoning*. New York: Springer.

Andra, D. L., E. M. Quoetone, and W. F. Bunting. 2002. Warning decision making: The relative roles of conceptual models, technology, strategy, and forecaster expertise on 3 May 1999. *Weather and Forecasting* 17:559–566.

Ariley, D. 2008. *Predictably Irrational: The Hidden Forces That Shape Our Decisions*. New York: HarperCollins.

Armstrong, E., and K. Thompson. 2003. Strategies for increasing minorities in the sciences: A University of Maryland, College Park, model. *Journal of Women and Minorities in Science and Engineering* 9:159–167.

Armstrong, J. S., ed. 2001a. *Principles of Forecasting: A Handbook for Researchers and Practitioners*. Norwell, MA: Kluwer.

Armstrong, J. S. 2001b. Judgmental bootstrapping: Inferring experts' rules for forecasting. In *Principles of Forecasting: A Handbook for Researchers and Practitioners*, ed. J. S. Armstrong (pp. 169–192). Norwell, MA: Kluwer Academic Publishers.

Artis, D. A., and W. H. Carnahan. 1982. Survey of emissivity variability in thermography of urban areas. *Remote Sensing of Environment* 12:313–329.

Ausubel, D. P., J. D. Novak, and H. Hanesian. 1978. *Educational Psychology: A Cognitive View*. 2nd ed. New York: Holt, Rinehart and Winston.

Baars, J. A., and C. F. Mass. 2005. Performance of National Weather Service forecasts compared to operational, consensus, and weighted model output statistics. *Weather and Forecasting* 20:1034–1047.

Bader, M., and T. Waters, eds. 1987. *Satellite and Radar Imagery Interpretation*. Darmstadt, Germany: European Organization for the Exploitation of Meteorological Satellites (EUMETSAT).

Bainbridge, L. 1983. Ironies of automation. *Automatica* 19:775–779.

Ballas, J. A. 2007. Human centered computing for tactical weather forecasting: An example of the "Moving Target Rule." In *Expertise Out of Context*, ed. R. R. Hoffman (pp. 316–326). Mahwah, NJ: Erlbaum.

Ballas, J. A., T. Tsui, J. Cook, D. Jones, K. Kerr, E. Kirby, et al. 2004. Improved workflow, environmental effects analysis, and user control for tactical weather forecasting. In *Proceedings of the Human Factors and Ergonomics Society 48th Annual Meeting* (pp. 320–324). Santa Monica, CA: Human Factors and Ergonomics Society.

Barker-Plummer, D., R. Cox, and N. Swoboda, eds. 2006. *Diagrammatic Representation and Inference. Proceedings of Diagrams 2006: The 4th International Conference*. New York: Springer.

Barnston, A. G., M. H. Glantz, and Y. He. 1999. Predictive skill of statistical and dynamical climate models in SST Forecasts during the 1997–98 El Niño episode and the 1998 La Niña onset. *Bulletin of the American Meteorological Society* 80:217–243.

Bartlett, F. 1932. *Remembering: A Study in Experimental and Social Psychology*. London: Cambridge University Press.

Bastardi, J. 2000, October. *The perfect weather forecast... Can computer models beat the best meteorologists?* Presentation at the New York Academy of Sciences, New York, NY.

Bauer, M. I., and P. N. Johnson-Laird. 1993. How diagrams can improve reasoning. *Psychological Science* 4:372–378.

Baum, W. A. 1975. The roles of universities and weather services in the education of meteorological personnel. *Bulletin of the American Meteorological Society* 56:226–228.

Baumgart, L. A., E. J. Bass, B. Philips, and K. Kloesel. 2008. Emergency management decision making during severe weather. *Weather and Forecasting* 23:1268–1279.

Becerra-Fernandez, I., A. Gonzalez, and R. Sabherwal. 2004. *Knowledge Management: Challenges, Solutions, and Technologies*. Upper Saddle River, NJ: Prentice-Hall.

Belville, J. D., and G. A. Johnson. 1982. The role of decision trees in weather forecasting. In *Preprints of the Ninth Conference on Weather Forecasting and Analysis* (pp. 7–11). Boston, MA: American Meteorological Society.

Bennett, K. B., and J. M. Flach. 2011. *Display and Interface Design: Subtle Science and Exact Art*. Boca Raton, FL: CRC Press.

Bereiter, C., and M. Scardamalia. 1993. *Surpassing Ourselves: An Inquiry into the Nature and Implications of Expertise*. Chicago: Open Court.

Berry, D. C. 1987. The problem of implicit knowledge. *Expert Systems: International Journal of Knowledge Engineering and Neural Networks* 4:144–151.

Berry, D. C., and D. E. Broadbent. 1984. On the relationship between task performance and associated verbalizable knowledge. *Quarterly Journal of Experimental Psychology* 36:209–231.

Bertin, J. 1967/1983. *Sémiologie Graphique: Les diagrammes—les réseaux—les cartes (Semiology of Graphics)*. Madison, WI: University of Wisconsin Press.

Bias, A. D., and D. J. Mahew. 1994. *Cost-Justifying Useability*. Boston: Academic Press.

Bias, R. G., and R. R. Hoffman. 2013, November/December. Where is the rigor in the field of usability analysis? *IEEE Intelligent Systems* 66–72.

Biederman, I., and M. M. Shiffrar. 1987. Sexing day-old chicks: A case study and expert systems analysis of a difficult perceptual-learning task. *Journal of Experimental Psychology: Learning, Memory, and Cognition* 13:640–645.

Bjerkness, V. 1919. On the structure of moving cyclones. *Geofysiske Publikasjoner* 1:1–8.

Bogacz, S., and J. G. Trafton. 2002. Understanding static and dynamic visualizations. In *Proceedings of the Second International Conference on Diagrammatic Representation and Inference.* [DOI:10.1007/3-540-46037-3_35]

Bogacz , S., and J. G. Trafton. 2005. Understanding dynamic and static displays: Using images to reason dynamically. *Cognitive Systems Research* 6:312–319.

Bolton, M. J., and W. G. Blumberg. 2015, October. *Learning disorders in the meteorological community: Implications for communication and education.* Paper presented at the annual meeting of the National Weather Association, Oklahoma City, OK. [Downloaded May 26, 2016, from http://nwas.org/meetings/nwas15/abstracts-html/2481.html]

Bond, N. A., and C. F. Mass. 2009. Development of skill by students enrolled in a weather forecasting laboratory. *Weather and Forecasting* 24:1141–1148.

Borland, D., and R. M. Taylor. 2007. Rainbow color map (still) considered harmful. *IEEE Computer Graphics and Applications* 27:14–17). New York: IEEE.

Bosart, L. F. 1983. An update on trends of skill in daily forecasts of temperature and precipitation at the State University of New York. *Bulletin of the American Meteorological Society* 64:346–354.

Bosart, L. F. 1989. Automation: Has its time really come? *Weather and Forecasting* 4:271.

Bosart, L. F. 2003. Whither the weather analysis and forecasting process? *Weather and Forecasting* 18:520–529.

Bowden, K. A., P. L. Heinselman, and Z. Kang. 2016. Exploring applications of eye-tracking in operational meteorology research. *Bulletin of the American Meteorological Society.* [Available online at http://journals.ametsoc.org/doi/abs/10.1175/BAMS-D-15-00148.1]

Brafman, O., and R. Brafman. 2009. *Sway: The Irresistible Pull of Irrational Behavior.* New York: Broadway Books.

Bransford, J. D. 1979. *Human cognition: Learning, Understanding and Remembering.* Belmont, CA: Wadsworth.

Bransford, J. D., and J. J. Franks. 1971. The abstraction of linguistic ideas. *Cognitive Psychology* 2:331–350.

Breslow, L. A., R. M. Ratwani, and J. G. Trafton. 2009. Cognitive models of the influence of color scale on data visualization tasks. *Human Factors* 51:321–338.

Breslow, L. A., J. G. Trafton, and R. M. Ratwani. 2009. A perceptual process approach to selecting color scales for complex visualizations. *Journal of Experimental Psychology: Applied* 15:25–34.

Brier, G. W. 1950. Verification of forecasts expressed in terms of probability. *Monthly Weather Review* 78:1–3.

Broadbent, D. E., and B. Aston. 1978. Human control of a simulated economic system. *Ergonomics* 21:1035–1043.

Broadbent, D. E., P. FitzGerald, and M. H. P. Broadbent. 1986. Implicit and explicit knowledge in the control of complex systems. *British Journal of Psychology* 77:33–50.

Brooking, A. 1999. *Corporate Memory: Strategies for Knowledge Management.* London: International Thomson Business Press.

Brooks, H. E., and C. A. Doswell. 1993. New technology and numerical weather prediction—A wasted opportunity? *Weather* 48:173–177.

Brooks, H. E., C. A. Doswell, and R. A. Maddox. 1992. On the use of mesoscale and cloud-scale models in operational forecasting. *Weather and Forecasting* 7:120–132.

Brotzge, J., K. Hondl, B. Philips, L. Lemon, E. J. Bass, D. Rude, and D. L. Andra. 2010. Evaluation of distributed collaborative adaptive sensing for detection of low-level circulations and implications for severe weather warning operations. *Weather and Forecasting* 25:173–189.

Brown, J. S., A. Collins, and P. Duguid. 1989. Situated cognition and the culture of learning. *Educational Researcher* 18:32–42.

Brown, T. J., M. Berliner, D. S. Wilks, M. B. Richman, and C. K. Wilke. 1999. Statistics education in the atmospheric sciences. *Bulletin of the American Meteorological Society* 80:2087–2098.

Brundage, J. M. 1986. The evolution of the user interface menus for the PROFS operational workstation. In *Proceedings of the 1st Workshop on Operational Meteorology* (pp. 4–6). Ottawa: Canadian Meteorological and Oceanographic Society.

Brule, J. F., and A. Blount. 1989. *Knowledge Acquisition.* New York: McGraw-Hill.

Buchanan, B. G., R. Davis, and E. A. Feigenbaum. 2006. Expert systems: A perspective from computer science. In *Cambridge Handbook of Expertise and Expert Performance,* ed. K. A. Ericsson, N. Charness, P. Feltovich, and R. Hoffman (pp. 87–104). New York: Cambridge University Press.

Buchanan, B. G., and E. H. Shortliffe. 1984. Explanation as a topic of AI research. In *Rule-Based Expert Systems: The MYCIN Experiments on the Stanford Heuristic Programming Project,* ed. B. G. Buchanan and E. H. Shortliffe (pp. 331–337). Reading, MA: Addison-Wesley.

Buchanan, B., R. G. Smith, R. Davis, and E. A. Feigenmaum. 2017. Expert systems: A perspective from computer science. In *Cambridge Handbook on Expertise and Expert Performance,* 2nd ed., ed. K. A. Ericsson, A. Kozbelt, M. Williams, and R. R. Hoffman. New York: Cambridge University Press.

Bullas, J. M., J. C. McLeod, and B. de Lorenzis. 1990. Knowledge augmented severe storms predictor (KASSPr)-An operational test. In *Preprints of the 16th Conference on Severe Local Storms* (pp. 106–111). Boston, MA: American Meteorological Society.

Bullock, C. S. 1985. *A proposed forecast methodology.* Report No. AC-20104, Program for Regional Observing and Forecasting Services, NOAA, Boulder, CO.

Bullock, C. S., J. S. Wakefield, J. M. Brundage, D. S. Walts, T. J. LeFebvre, and P. A. Amstein. 1988. The DARE workstation and some lessons learned from its operational use. In *Proceedings of the 4th*

International Conference on Interactive and Information Processing Systems for Meteorology, Oceanography, and Hydrology (pp. 70–76). Boston, MA: American Meteorological Society.

Byrne, M. D., R. Catrambone, and J. T. Stasko. 1999. Evaluating animations as student aids in learning computer algorithms. *Computers & Education* 33:253–278.

Byrne, R. M. J. 2002. Mental models and counterfactual thinking. *Trends in Cognitive Sciences* 6:405–445.

Calderwood, R., B. Crandall, and G. Klein. 1987. Expert and novice fireground command decisions. Report MDA903-85-C-0327. U.S. Army Research Institute, Alexandria VA.

Campbell, J. B. 1996. *Introduction to Remote Sensing*. New York: The Guilford Press.

Campbell, S. E. 1988. Microburst precursor recognition using an expert system approach. In *Proceedings of the Fourth International Conference on Interactive Information Processing Systems for Meteorology* (pp. 300–307). Boston, MA: American Meteorological Society.

Campbell, S. E., and S. Olson. 1987. Recognizing low-altitude wind shear hazards from Doppler weather radar: An artificial intelligence approach. *Journal of Atmospheric and Oceanographic Technology* 4:5–18.

Cañas, A. J., G. Hill, R. Carff, N. Suri, J. Lott, T. Eskridge, G. Gómez, M. Arroyo, and R. Carvajal. 2004. CmapTools: A Knowledge Modeling and Sharing Environment. In A. J. Cañas, J. D. Novak & F. M. González (Eds.), *Concept Maps: Theory, Methodology, Technology, Proceedings of the 1st International Conference on Concept Mapping* (pp. 125–133). Pamplona, Spain: Universidad Pública de Navarra.

Cañas, A. J., J.W. Coffey, M.J. Carnot, P. Feltovich, R. R. Hoffman, J. Feltovich, and J. D. Novak. 2003. A Summary of Literature Pertaining to the Use Of Concept Mapping Techniques and Technologies for Education and Performance Support. Report to The Chief of Naval Education and Training, prepared by the Institute for Human and Machine Cognition, Pensacola FL. [http://www.ihmc.us/users/acanas/Publications/ConceptMapLitReview/ IHMC%20Literature%20 Review%20on%20Concept%20Mapping.pdf]

Canham, M., and M. Hegarty. 2010. Effects of knowledge and display design on comprehension of complex graphics. *Learning and Instruction* 20:155–166.

Carlson, T., P. Knight, and C. Wyckoff. 2014. *An Observer's Guide to Clouds and Weather*. Boston: American Meteorological Society.

Carnahan, W. H., and R. C. Larson. 1990. An analysis of an urban heat sink. *Remote Sensing of Environment* 33:65–71.

Carr, L. E., R. L. Elsberry, and J. E. Peak. 2001. Beta test of the systematic approach expert system prototype as a tropical cyclone track forecasting aid. *Weather and Forecasting* 16:355–368.

Carter, G., and P. Polger. 1986. *A 20-year summary of National Weather Service verification results for temperature and precipitation*. Technical Memorandum No. NWS FCST 31. Washington, DC: National Oceanic and Atmospheric Administration.

Ceci, S. J., and J. K. Liker. 1986. A day at the races: A study of IQ, expertise, and cognitive complexity. *Journal of Experimental Psychology. General* 115:255–266.

Charba, J. P., and W. H. Klein. 1980. Skill in precipitation forecasting in the National Weather Service. *Bulletin of the American Meteorological Society* 61:1546–1555.

Charmaz, K. 2006. *Constructing Grounded Theory: A Practical Guide Through Qualitative Analysis*. Thousand Oaks, CA: SAGE.

Charness, N., E. M. Reingold, M. Pomplun, and D. M. Stampe. 2001. The perceptual aspect of skilled performance in chess: Evidence from eye movements. *Memory & Cognition* 29:1146–1152.

Chase, W. G. 1983. Spatial representations of taxi drivers. In *Acquisition of Symbolic Skills*, ed. D. R. Rogers and J. H. Sloboda (pp. 391–405). New York: Plenum.

Chase, W. G., and H. A. Simon. 1973. Perception in chess. *Cognitive Psychology* 5:55–81.

Chatters, G. C., and V. E. Suomi. 1975. The application of McIDAS. *IEEE Transactions on Geoscience Electronics* GE-13:137–139.

Chi, M. T. H., P. J. Feltovich, and R. Glaser. 1981. Categorization and representations of physics problems by experts and novices. *Cognitive Science* 5:121–152.

Chi, M. T. H., R. Glaser, and M. J. Farr, eds. 1988. *The Nature of Expertise*. Mahwah, NJ: Erlbaum.

Chi, M. T. H., R. Glaser, and E. Rees. 1982. Expertise in problem solving. In *Advances in the Psychology of Human Intelligence*, vol. 1, ed. R. J. Sternberg (pp. 7–75). Hillsdale, NJ: Erlbaum.

Chi, M. T. H., and R. D. Koeske. 1983. Network representation of a child's dinosaur knowledge. *Developmental Psychology* 19:29–39.

Chiesi, H., G. J. Spilich, and J. F. Voss. 1979. Acquisition of domain-related information in relation to high and low domain knowledge. *Journal of Verbal Learning and Verbal Behavior* 18:257–283.

Childs, C.M. 1988. Interpreting composite imagery requires training. *Tech-Tran* (Newsletter of the Engineer Topographic Laboratories, U.S. Army Corps of Engineers), *13*, 3. Fort Belvoir, VA: U.S. Army Corps of Engineers.

Chisholm, D. A., A. J. Jackson, M. E. Niedzeilski, R. Schechter, and C. F. Ivaldi. 1983. *The use of interactive graphics processing in short-range terminal weather forecasting: An initial assessment*. Report AFGL-83–0093, Air Force Geophysics Laboratory, Hanscom AFB, MA.

Choo, C. W. 1998. *The Knowing Organization: How Organizations Use Information to Construct Meaning, Create Knowledge and Make Decisions*. New York: Oxford University Press.

Chouard, T. 2016. The Go files: AI computer clinches victory against Go champion. *Nature*. [Blog downloaded June 6, 2016, from http://www.nature.com/news]

Christ, R. E. 1975. Review and analysis of color coding research for visual displays. *Human Factors* 17:542–570.

Christ, R. E., and G. M. Corso. 1983. The effects of extended practice on the evaluation of visual display codes. *Human Factors* 25:71–84.

Christensen-Szalanski, J. J., and L. R. Beach. 1984. The citation bias: Fad and fashion in the judgment and decision literature. *American Psychologist* 39:75–78.

Christensen-Szalanski, J. J., P. H. Diehr, J. B. Bushyhead, and R. W. Wood. 1982. Two studies of good clinical judgment. *Medical Decision Making* 3:275–284.

Christoffersen, K., and D. D. Woods. 2002. How to make automated systems team players. In *Advances in Human Performance and Cognitive Engineering Research*, vol. 2, ed. E. Salas (pp. 1–12). St, Louis, MO: Elsevier.

Clancey, W. J. 1985. Heuristic classification. *Artificial Intelligence* 27:289–350.

Clancey, W. J. 1989. Viewing knowledge bases as qualitative models. *IEEE Intelligent Systems* 4:18–23.

Clancey, W. J. 1992. Model construction operators. *Artificial Intelligence* 53:1–115.

Clemen, R. T. 1985. Extraneous expert information. *Journal of Forecasting* 4:329–348.

Clemen, R. T. 1989. Combining forecasts: A review and annotated bibliography. *International Journal of Forecasting* 5:559–583.

Clemen, R. T., and A. H. Murphy. 1989. Objective and subjective precipitation probability forecasts: Some methods for improving forecast quality. *Weather and Forecasting* 1:213–218.

Coffey, J. W., and R. R. Hoffman. 2003. The PreSERVe Method of Knowledge Modeling Applied to the Preservation of Institutional Memory. *Journal of Knowledge Management* 7:38–52.

Cohen, M. S. 1993a. The naturalistic basis of decision biases. In *Decision Making in Action: Models and Methods*, ed. G. Klein, J. Orasanu, R. Calderwood, and C. E. Zsambok (pp. 51–102). Norwood, NJ: Ablex.

Cohen, M. S. 1993b. Three paradigms for viewing decision making processes. In *Decision Making in Action: Models and Methods*, ed. G. Klein, J. Orasanu, R. Calderwood, and C. E. Zsambok (pp. 36–50). Norwood, NJ: Ablex.

Collins, H. M. 1992. *Changing Order: Replication and Induction in Scientific Practice*. 2nd ed. Beverly Hills, CA: Sage.

Collins, A., and R. Evans. 2007. *Rethinking Expertise*. Chicago: University of Chicago Press.

Collpoy, F., M. Adya, and J. S. S. Armstrong. 2001. Expert systems for forecasting. In *Principles of Forecasting: A Handbook for Researchers and Practitioners*, ed. J. S. Armstrong (pp. 285–300). Norwell, MA: Kluwer.

Colucci, S. J., P. C. Knappenberger, and T. K. Cepa. 1992. Evaluation of a nonprobabilistic weather forecasting experiment. *Weather and Forecasting* 7:507–514.

Conway, E. D. 1997. *An Introduction to Satellite Image Interpretation*. Baltimore, MD: Johns Hopkins University Press.

Cooke, E. W. 1906. Forecasts and verifications in Western Australia. *Monthly Weather Review* 34:23–24, 274–275.

Cooke, N. J. 1992. Modeling human expertise in expert systems. In *The Psychology of Expertise: Cognitive Research and Empirical AI*, ed. R. Hoffman (pp. 29–60). Mahwah, NJ: Erlbaum.

Cooke, N. M., and F. Durso. 2008. *Stories of Modern Technology Failures and Cognitive Engineering Successes*. Boca Raton, FL: CRC Press.

Cooke, N. M., and J. E. McDonald. 1986. A formal methodology for acquiring and representing expert knowledge. *Proceedings of the IEEE* 74:1422–1430.

Cooke, N. M., and A. L. Rowe. 1994. Evaluating mental model elicitation methods. In *Proceedings of the Human Factors and Ergonomics Society 38th Annual Meeting* (pp. 261–265). Santa Monica, CA: Human Factors and Ergonomics Society.

Corbett, J., C. Mueller, C. Burghart, K. Gould, and G. Granger. 1994. ZEB: Software for integration, display, and management of diverse environmental data sets. *Bulletin of the American Meteorological Society* 75:783–792.

Corbin, J., and A. L. Strauss. 2008. *Basics of Qualitative Research*. 3rd ed. Los Angeles: SAGE.

Coulson, R. L., P. J. Feltovich, and R. J. Spiro. 1997. Cognitive flexibility in medicine: An application to the recognition and understanding of hypertension. *Advances in Health Sciences Education: Theory and Practice* 2:141–161.

Crandall, B., and R. R. Hoffman. 2013. *Cognitive Task Analysis. Oxford Handbook of Cognitive Engineering*. Oxford: Oxford university Press.

Crandall, B., G. Klein, and R. R. Hoffman. 2006. *Working Minds: A Practitioner's Guide to Cognitive Task Analysis*. Cambridge, MA: MIT Press.

Croft, P. J. 1999. Assessing the "Excitement of Meteorology!" for young scholars. *Bulletin of the American Meteorological Society* 80:879–891.

Curran, P. J. 1987. Remote sensing methodologies and geography. *International Journal of Remote Sensing* 8:1255–1275.

Curtis, J. C. 1992, September. *The meteorological data flood: An overview*. Paper presented at the Fourth Workshop on Operational Meteorology, sponsored by the Canadian Meteorological and Oceanographic Society.

Curtis, J. C. 1998. The forecast process—One forecaster's perspective. Presentation at the First Conference on Artificial Intelligence, Phoenix, AZ. Sponsored by the American Meteorological Society, Boston, MA.

Daan, H., and A. H. Murphy. 1982. Subjective probability forecasting in the Netherlands: Some operational and experimental results. *Meteorologische Rundschau* 35:92–112.

Daipha, P. 2007. *Masters of Uncertainty: Weather Forecasters and the Quest for Ground Truth.* Doctoral Dissertation, The University of Chicago.

Daipha, P. 2012. Weathering risk: Uncertainty, weather forecasting, and expertise. *Sociology Compass* 6:15–25.

Daipha, P. 2015. From bricolage to collage: The making of decisions at a weather forecast office. *Sociological Forum* 30:787–808.

Daipha, P. 2016. *Masters of Uncertainty: Weather Forecasting and the Quest for Ground Truth.* Chicago: University of Chicago Press.

Davenport, C. E., C. S. Wohlwend, and T. L. Koehler. 2015. Motivation for and development of a standardized introductory meteorology assessment exam. *Bulletin of the American Meteorological Society* 96:305–312.

Davenport, T. H., and L. Prusak. 1998. *Working Knowledge: How Organizations Manage What They Know*. Cambridge, MA: Harvard Business School Press.

Davidoff, J. 1987. The role of color in visual displays. *International Review of Ergonomics* 1:21–42.

Davies, C., W. Tompkinson, N. Donnelly, L. Gordon, and K. Cave. 2006. Visual saliency as an aid to updating digital maps. *Computers in Human Behavior* 22:672–684.

Davies, D., D. Bathurst, and R. Bathurst. 1990. *The Telling Image: The Changing Balance Between Pictures and Words in a Technological Age*. Oxford: Oxford University Press.

Dawes, R. 2001. *Everyday Irrationality: How Pseudo-scientists, Lunatics, and the Rest of Us Systematically Fail to Think Rationally*. Boulder, CO: Westview Press.

Dawes, R., P. Meehl, and D. Faust. 1989. Clinical versus statistical prediction of human outcomes. *Science* 243:1668–1674.

Dawes, R. M., and B. Corrigan. 1974. Linear models in decision making. *Psychological Bulletin* 81:95–106.

Day, E. A., W. Arthur, and D. Gettman. 2001. Knowledge structures and the acquisition of a complex skill. *Journal of Applied Psychology* 86:1022–1033.

Day, S. B., and R. L. Goldstone. 2012. The import of knowledge export: Connecting findings and theories of transfer of learning. *Educational Psychologist* 47:153–176.

Deakin, J. M., and S. Cobley. 2003. An examination of the practice environments in figure skating and volleyball: A search for deliberate practice. In *Expert Performance in Sports: Advances in Research on Sport Expertise*, ed. J. Starkes and K. A. Ericsson, 90–113. Champaign, IL: Human Kinetics.

De Elìa, R., and R. Laprise. 2005, September. The unbearable lightness of probabilities. *Bulletin of the American Meteorological Society* 86:1224–1225.

de Groot, A. D. 1948/1978. *Thought and Choice in Chess*. 2nd ed. The Hague: Mouton.

Demuth, J. L., B. H. Morrow, and J. K. Lazo. 2009. Weather forecast uncertainty information: An exploratory study with broadcast meteorologists. *Bulletin of the American Meteorological Society* 10:1614–1618.

Demuth, J. L., R. E. Morss, B. H. Morrow, and J. K. Lazo. 2012. Creation and communication of hurricane risk information. *Bulletin of the American Meteorological Society* 93:1133–1145.

Derouin, R. E., T. J. Parrish, and E. Salas. 2005, Spring. On-the-job training: Tips for ensuring success. *Ergonomics in Design* 13:23–26.

Diak, G. R., M. C. Anderson, W. L. Bland, J. M. Norman, J. M. Mecikalski, and R. M. Aune. 1998. Agricultural management decision aids driven by real-time satellite data. *Bulletin of the American Meteorological Society* 79:1345–1355.

Dodson, D. C. 1989. Interaction with knowledge systems through connection diagrams: Please adjust your diagrams. In *Research and Development in Expert Systems V*, ed. B. Kelly and A. L. Rector, 35–46. Cambridge: Cambridge University Press.

Dorsey, D. W., G. E. Campbell, L. L. Foster, and D. E. Miles. 1999. Assessing knowledge structures: Relations with experience and post-training performance. *Human Performance* 12:31–57.

Doswell, C. A. 1986a. The human element in weather forecasting. *National Weather Digest* 11:6–18.

Doswell, C. A. 1986b. The role of diagnosis in weather forecasting. In *Proceedings of the 11th Conference on Forecasting and Weather Analysis* (pp. 177–182). Boston: American Meteorological Society.

Doswell, C. A. 1986c. Short-range forecasting. In *Mesoscale Meteorology and Forecasting*, ed. P. S. Ray (pp. 689–719). Boston, MA: American Meteorological Society.

Doswell, C. A. 1990. Personal communication. National Severe Storms Laboratory, NOAA, Norman, OK.

Doswell, C. A. 1992. Forecaster workstation design: Concepts and issues. *Weather and Forecasting* 7:398–407.

Doswell, C. A. 2003. What does it take to be a good forecaster?" Personal Blog. [http://www.flame.org/~cdoswell/forecasting/Forecaster_Qualities.html]

Doswell, C. A. 2004. Weather forecasting by humans—Heuristics and decision making. *Weather and Forecasting* 19:1115–1126.

Doswell, C. A., H. Brooks, and R. A. Maddox. 1996. Flash flood forecasting: An ingredients-based methodology. *Weather and Forecasting* 11:560–581.

Doswell, C. A., L. R. Lemon, and R. A. Maddox. 1981. Forecaster training: A review and analysis. *Bulletin of the American Meteorological Society* 62:983–988.

Doswell, C. A., and R. A. Maddox. 1986. The role of diagnosis in weather forecasting. In *Proceedings of the 11th Conference on Weather Forecasting and Analysis* (pp. 177–182). Boston, MA: American Meteorological Society.

Dreyfus, H. 1979. *What Computers Can't Do*. Cambridge, MA: MIT Press.

Dreyfus, H., and S. E. Dreyfus. 1986. *Mind Over Machine*. New York: Free Press.

Drury, C. G. 2014. Can human factors professionals contribute to global climate change solutions? *Ergonomics in Design* 22:30–33.

Duda, J., J. Gaschnig, and P. Hart. 1979. Model design in the PROSPECTOR consultant system for mineral exploration. In *Expert Systems in the Micro-Electronic Age*, ed. D. Michie (pp. 153–167). Edinburgh: Edinburgh University Press.

Duncker, K. 1945. On problem solving. *Psychological Monographs* 58:1–113.

Durrett, J. H., ed. 1987. *Color and the Computer*. New York: Academic Press.

Dutton, J. A. 2002. Opportunities and priorities in a new era for weather and climate services. *Bulletin of the American Meteorological Society* 83:1303–1311.

Dvorak, V. F. 1975. Tropical cyclone intensity analysis forecasting from imagery. *Monthly Weather Review* 103:420–430.

Dyer, R. M. 1987. Expert systems in weather forecasting and other meteorological applications. *AI Applications in Environmental Science* 1:19–24.

Dyer, R. M. 1989. Adapting expert systems to multiple locations. *AI Applications in Environmental Science* 3:11–16.

Dyer, R. 1990. Variations amongst users of weather prediction systems. In *Proceedings of Resource Technology 90: The 2nd International Symposium on Advanced Technology in Natural Resource Management*, Session on Human Factors (pp. 534–538). Sponsored by the American Society of Photogrammetry and Remote Sensing, Bethesda, MD.

Dyer, R. M., and G. L. Freeman. 1989. *Rule-based systems for visibility forecasts*. Report No. GL-TR-89–0116, Air Force Geophysics Laboratory, Hanscom AFB, MA.

Ebbesen, E. B., and V. K. Konecni. 1980. On the external validity of decision-making research: What do we know about decisions in the real world? In *Cognitive processes in choice and decision behavior*, ed. T. S. Wallsten, 21–45. Hillsdale, NJ: Erlbaum.

Economist. 1995, June. Garbage in, garbage out. *The Economist [US]:70*. Academic OneFile.

Eggleston, R. G., E. M. Roth, and R. A. Scott. 2003. A framework for work-centered product evaluation. In *Proceedings of the Human Factors and Ergonomics Society 47th Annual Meeting* (pp. 503-507). Santa Monica, CA: Human Factors and Ergonomics Society.

Ehrendorfer, M. 1997. Predicting the uncertainty of numerical weather forecasts: A review. *Meteorologische Zeitschrift* 6:147–183.

Einhorn, H. J., and R. M. Hogarth. 1981. Behavioral decision theory: Processes of judgment and choice. *Annual Review of Psychology* 32:53–88.

Eley, M. G. 1988. Determining the shapes of land surfaces from topographical maps. *Ergonomics* 31:355–376.

Elias, N. 1982. *Power and Civility. Vol. II. The Civilizing Process*. New York: Pantheon Books.

Eliassen, A. 1995. *Jacob Bjerknes 1897–1975: A Biographical Memoir*. Washington, DC: National Academies Press.

Elio, R., and J. de Haan. 1986. Representing quantitative and qualitative knowledge in a knowledge-based storm forecasting system. *International Journal of Man-Machine Studies* 25:523–547.

Elio, R., J. de Haan, J., and G. S. Strong. 1987. METEOR: An artificial intelligence system for convective storm forecasting. *Journal of Atmospheric and Oceanographic Technology* 4:19–28.

Ellsaesser, H. W. 1982. Comments on "Forecaster Training—A Review and Analysis." *Bulletin of the American Meteorological Society* 63:782–783.

Ellrod, G. P. 1989. An index for clear air turbulence based on horizontal deformation and vertical wind shear. In *Proceedings of the Third International Conference on the Aviation Weather System* (pp. 339–344). Boston, MA: American Meteorological Society.

Ericsson, K. A., ed. 2009. *Development of Processional Expertise*. Cambridge: Cambridge University Press.

Ericsson, K. A., N. Charness, P. J. Feltovich, and R. R. Hoffman, eds. 2006. *Cambridge Handbook of Expertise and Expert Performance*. New York: Cambridge University Press.

Ericsson, K. A., R. T. Krampe, and C. Tesch-Römer. 1993. The role of deliberate practice in the acquisition of expert performance. *Psychological Review* 100:363–406.

Ericsson, K. A., and A. C. Lehman. 1996. Expert and exceptional performance: Evidence on maximal adaptations on task constraints. *Annual Review of Psychology* 47:273–305.

Ericsson, K. A., and H. A. Simon. 1993. *Protocol Analysis: Verbal Reports as Data*. 2nd ed. Cambridge, MA: MIT Press.

Ericsson, K. A., and J. Smith, eds. 1991. *Toward a General Theory of Expertise*. Cambridge: Cambridge University Press.

Errico, R. M. 1999. Workshop on assimilation of satellite data. *Bulletin of the American Meteorological Society* 80:463–471.

Ettenson, R. T., J. Shanteau, and J. Krogstad. 1987. Expert judgment: Is more information better? *Psychological Reports* 60:227–238.

Evans, J. St. B. T. 1989. *Bias in Human Reasoning*. Hillsdale, NJ: Erlbaum.

Evetts, J., H. Mieg, and U. Felt. 2006. Professionalization, scientific expertise, and elitism: A sociological perspective. In *Cambridge Handbook of Expertise and Expert Performance*, ed. K. A. Ericsson, N. Charness, P. J. Feltovich, and R. R. Hoffman (pp. 105–123). New York: Cambridge University Press.

Fabrikant, S. I., S. R. Hespanha, and M. Hegarty. 2010. Cognitively inspired and perceptually salient graphic displays for efficient spatial inference making. *Annals of the Association of American Geographers* 100:1–17.

Fahle, M., and T. Poggio, eds. 2002. *Perceptual Learning*. Cambridge, MA: The MIT Press.

Fauerbach, E., R. Edsall, D. Barnes, and A. MacEachren. 1996. Visualization of uncertainty in meteorological forecast models. In *Proceedings of the 7th International Spatial Data Handling Conference* (pp. 17-28). Delft, The Netherlands: Taylor and Francis.

Fawcett, E. B. 1969. Systematic errors in baroclinic prognoses at the National Meteorological Center. *Monthly Weather Review* 97:670–682.

Feigenbaum, E. A., B. G. Buchanan, and J. Lederberg. 1971. On generality and problem solving: A case study using the DENDRAL program. In *Machine Intelligence 6*, ed. B. Meltzer and D. Michie (pp. 165–190). Edinburgh: Edinburgh University Press.

Feltovich, P. J., K. M. Ford, and R. R. Hoffman, eds. 1997. *Expertise in Context*. Cambridge, MA: The MIT Press.

Feltovich, P. J., R. R. Hoffman, and D. D. Woods. 2004, May/June. Keeping it too simple: How the reductive tendency affects cognitive engineering. *IEEE Intelligent Systems* 19:90–95.

Feltovich, P. J., M. J. Prietula, and K. A. Ericsson. 2006. Studies of expertise from psychological perspectives. In *Cambridge Handbook of Expertise and Expert Performance*, ed. K. A. Ericsson, N. Charness, P. J. Feltovich, and R. R. Hoffman (pp. 41–68). New York: Cambridge University Press.

Feltovich, P. J., R. J. Spiro, and R. L. Coulson. 1989. The nature of conceptual understanding in biomedicine: The deep structure of complex ideas and the development of misconceptions. In *Cognitive Science in Medicine*, ed. D. A. Evans and V. L. Patel (pp. 115–171). Cambridge, MA: The MIT Press.

Feltovich, P. J., R. J. Spiro, and R. L. Coulson. 1993. Learning, teaching and testing for complex conceptual understanding. In *Test Theory for a New Generation of Tests*, ed. N. Frederiksen, R. J. Mislevy, and I. I. Bejar (pp. 181–217). Hillsdale, NJ: Erlbaum.

Feltovich, P. J., R. J. Spiro, and R. L. Coulson. 1997. Issues of expert flexibility in contexts characterized by complexity and change. In *Expertise in Context*, ed. P. J. Feltovich, K. Ford, and R. R. Hoffman (pp. 125–146). Menlo, CA: AAAI Press/The MIT Press.

Fett, R. W., M. E. White, J. E. Peak, S. Brand, and P. M. Tag. 1997. Application of hypermedia and expert system technology to Navy environmental satellite image analysis. *Bulletin of the American Meteorological Society* 78:1905–1915.

Fields, C. A., J. E. Newberry, H. D. Pfeiffer, C. A. Soderlund, S. F. Kirby, and G. B. McWilliams. 1992. MERCURY: A heterogeneous system for spatial extrapolation of mesoscale meteorological data. *International Journal of Man-Machine Studies* 36:309–326.

Fine, G. A. 2007. *Authors of the Storm: Meteorologists and the Culture of Prediction.* Chicago: The University of Chicago Press.

Fiore, S. M., F. Jentsch, R. Oser, and J. A. Cannon-Bowers. 2000. Perceptual and conceptual processing in expert/novice cue pattern recognition. *International Journal of Cognitive Technology* 5:17–26.

Fiore, S. M., S. Scielzo, and F. Jentsch. 2004. Stimulus competition during perceptual learning: Training and aptitude considerations in the X-ray security screening process. *International Journal of Cognitive Technology* 9:34–39.

Fischhoff, B. 1982. For those condemned to study the past: Heuristics and biases in hindsight. In *Judgment Under Uncertainty: Heuristics and Biases*, ed. D. Kahneman, P. Slovic, and A. Tversky (pp. 335–351). New York: Cambridge University Press.

Fischoff, B., and R. Beyth. 1975. "I knew it would happen": Remembered probabilities of once and future things. *Organizational Behavior and Human Performance* 13:1–16.

Fisher, A. 2005, March 21. How to battle the coming brain drain. *Forbes Magazine*, pp. 121–131.

Flanagan, J. C. 1954. The critical incident technique. *Psychological Bulletin* 51:327–358.

Florida, R. 2002. *The Rise of the Creative Class: How It's Transforming Work, Leisure and Everyday Life*. New York: Basic Books.

Florida, R. 2005. *The Flight of the Creative Class: The New Global Competition for Talent.* New York: HarperCollins.

Ford, K. M., A. J. Cañas, J. W. Coffey, J. Andrews, E. J. Schad, and H. Stahl. 1992. Interpreting functional images with NUCES: Nuclear Cardiology Expert System. In *Proceedings of the Fifth Annual Florida Artificial Intelligence Research Symposium,* ed. M. B. Fishman (pp. 85–90). Ft. Lauderdale, FL: FLAIRS.

Ford, K. M., J. W. Coffey, A. Cañas, E. J. Andrews, and C. W. Turner. 1996. Diagnosis and explanation by a nuclear cardiology expert system. *International Journal of Expert Systems* 9:499–506.

Forsyth, D. E., M. J. Istok, T. D. O'Bannon, and K. M. Glover. 1985. *The Boston area NEXRAD demonstration (BAND).* Report AFGL-85–0098, Air Force Geophysics Laboratory, Hanscom AFB, MA.

Frankel, D., I. Schiller, J. S. Draper, and A. A. Barnes, Jr. 1990. Investigation of the prediction of lightning strikes using neural networks. In *Preprints of the 16th Conference on Severe Local Storms* (pp. 7–12). Boston, MA: American Meteorological Society.

Fraser, J. M., P. J. Smith, and J. W. Smith. 1992. A catalog of errors. *International Journal of Man-Machine Studies* 37:265–307.

Friedhoff, B. W. 1991. *Visualization: The Second Computer Revolution*. San Francisco: W. H. Freeman.

Friedl, M. A., J. E. Estes, and J. L. Star. 1988. Advanced information extraction tools in remote sensing for earth science applications: AI and GIS. *AI Applications in Natural Resource Management* 2:17–30.

Friedman, R. M. 1989. *Appropriating the Weather: Vilhelm Bjerknes and the Construction of a Modern Meteorology*. Ithaca, NY: Cornell University Press.

Fritsch, J. M. 1992. Operational meteorological education and training: Some considerations for the future. *Bulletin of the American Meteorological Society* 73:1843–1846.

Gaffney, J. E., and R. I. Racer. 1983. A learning interpretive decision algorithm for severe storm forecasting support. In *Preprints of the 13th Conference on Severe Local Storms* (pp. 274–276). Boston, MA: American Meteorological Society.

Gagné, R. M. 1968. Learning hierarchies. *Educational Psychologist* 6:1–9.

Gaines, B., and M. Shaw. 1995. Collaboration through concept maps. Paper presented at CSCL95: The First International Conference on Computer Support for Collaborative Learning, New York: Association for Computing Machinery. [http://dl.acm.org/citation.cfm?id=222020]

Gallus, W. A., D. N. Yarger, and D. E. Herzmann. 2000. An interactive severe weather activity to motivate student learning. *Bulletin of the American Meteorological Society* 81:2205–2212.

Galton, F. 1869. *Hereditary Genius: An Inquiry into Its Laws and Its Consequences*. London: Macmillan.

Galton, F. 1874. *English Men of Science: Their Nature and Nurture*. London: Macmillan.

Gedzelman, S. D. 1978. Forecasting skill of beginners. *Bulletin of the American Meteorological Society* 59:1305–1309.

Gentner, D., and D. R. Gentner. 1983. Flowing waters or teeming crowds: Mental models of electricity. In *Mental Models*, ed. D. Gentner and A. Stevens (pp. 99–130). Hillsdale, NJ: Erlbaum.

Gentner, D., and A. Stevens, eds. 1983. *Mental Models*. Hillsdale, NJ: Erlbaum.

Ghirardelli, J. E., and B. Glahn. 2010. The Meteorological Development Laboratory's aviation weather prediction system. *Weather and Forecasting* 25:1027–1051.

Giarratano, J. C., and G. D. Riley. 2004. *Expert Systems: Principles and Programming*. 4th ed. Independence, KY: Course Technology.

Gibson, J. J., and E. J. Gibson. 1955. Perceptual learning: Differentiation or enrichment? *Psychological Review* 62:32–41.

Gigerenzer, G., R. Hertwig, E. van den Broek, B. Fasolo, and K. V. Katsikopoulos. 2005. "A 30% chance of rain tomorrow": How does the public understand probabilistic weather forecasts? *Risk Analysis* 25:623–629.

Gilbreth, F. B. 1911. *Motion Study*. New York: Van Nostrand.

Gilbreth, L. M. 1934. Time and motion study techniques. *Occupations* 12:35–38.

Gilbreth, L. M., and F. B. Gilbreth. 1917. *Applied Motion Study*. New York: Sturgis and Walton.

Gilmore, M. S., and L. Wicker. 1998. The influence of midtropospheric dryness on supercell morphology and evolution. *Monthly Weather Review* 126:943–958.

Gilovich, T., D. Griffin, and D. Kahneman, eds. 2002. *Heuristics and Biases: The Psychology of Intuitive Judgment*. Cambridge: Cambridge University Press.

Gimeno, L., and R. Garcia. 1998. Project EUROMET: Meteorological computer-assisted learning in Europe. *Bulletin of the American Meteorological Society* 79:1079–1081.

Giraytys, J. 1975. AFOS Experimental program. *IEEE Transactions on Geoscience Electronics* GE-13:111–115.

Gladwell, M. 2004, December 13. The picture problem. *The New Yorker*, pp. 74–81.

Glahn, B. 2003. Comments on "IFPS and the Future of the National Weather Service." *Weather and Forecasting* 18:1299–1304.

Glaser, R. 1987. Thoughts on expertise. In *Cognitive Functioning and Social Structure over the Life Course*, ed. C. Schooler and W. Schaie (pp. 81–94). Norwood, NJ: Ablex.

Glasgow, J., N. H. Narayanan, and B. Chandrasekaran, eds. 1995. *Diagrammatic Reasoning: Cognitive and Computational Perspectives*. Cambridge, MA: MIT Press.

Glenberg, A. M., and D. A. Robertson. 2000. Symbol grounding and meaning: A comparison of high-dimensional and embodied theories of meaning. *Journal of Memory and Language* 43:379–401.

Glickman, T. S., ed. 2000. *Hook Echo: Glossary of Meteorology*. 2nd ed. Boston, MA: American Meteorological Society.

Gobet, F., and G. Campitelli. 2007. The role of domain-specific practice, handedness and starting age in chess. *Developmental Psychology* 43:159–172.

Godske, C. L., T. Bergeron, J. Bjerknes, and R. C. Bundgaard. 1957. *Dynamic Meteorology and Weather Forecasting*. Boston, MA: American Meteorological Society.

Golden, J. H., C. F. Chappell, C. G. Little, A. H. Murphy, E. B. Burton, and E. W. Pearl. 1978. What should the NWS be doing to improve short-range weather forecasting: A panel discussion with audience participation. *Bulletin of the American Meteorological Society* 59:1334–1342.

Goldstone, R. L. 1998. Perceptual learning. *Annual Review of Psychology* 49:585–612.

Goldstone, R. L. 2000. Unitization during category learning. *Journal of Experimental Psychology: Human Perception and Performance* 26:86–112.

Goodman, S. J., J. Gurka, M. DeMaria, T. J. Schmit, A. Mostek, G Jedlovec, C. Siewert, W. Feltz, J. Gerth, R. Brummer, S. Miller, B. Reed, and R. R. Reynolds. 2012. The GOES-R Proving Ground:

Accelerating user readiness for the next-generation geostationary environmental satellite system. *Bulletin of the American Meteorological Society* 93:1029–1040.

Gordon, S. E., K. A. Schmierer, and R. T. Gill. 1993. Conceptual graph analysis: Knowledge acquisition for instructional system design. *Human Factors* 35:459–481.

Gordon, W. E. (Chair) 2003. A Vision for the National Weather Service: Road Map for the Future. Report of the Panel on the Road Map for the Future of the National Weather Service (W. E. Gordon, Chair), National Weather Service Modernization Committee, National Academy of Sciences. Washington, DC: National Academies Press.

Gray, W. D., and M. C. Salzman. 1998. Damaged merchandise? A review of experiments that compare usability evaluation methods. *Human-Computer Interaction* 13:203–261.

Greenbaum, J., and M. Kyng. 1991. *Design at Work: Cooperative Design of Computer Systems.* Hillsdale, NJ: Erlbaum.

Greeno, J. G. 1978. The nature of problem solving abilities. In *Handbook of Learning and Cognitive Processes,* vol. 5, ed. W. K. Estes (pp. 239–270). Hillsdale, NJ: Erlbaum.

Gregg, W. R., and I. R. Tannehill. 1937. International standard projections for meteorological charts. *Monthly Weather Review* 65:411–415.

Grenci, L. 2001, July/August. Crossing the line. *Weatherwise* 54:50–51.

Grice, G. K. 1983. Comments on "Forecaster training—A review and analysis." *Bulletin of the American Meteorological Society* 64:517.

Groen, G. J., and V. L. Patel. 1988. The relationship between comprehension and reasoning in medical expertise. In *The Nature of Expertise,* ed. M. T. H. Chi, R. Glaser, and M. J. Farr (pp. 287–310). Hillsdale, NJ: Erlbaum.

Grote, U. H., and C. S. Bullock. 1997. User interface design and the WFO-Advanced Workstation. In *Preprints of the 13th Conference on Interactive Information and Processing Systems* (pp. 320–323). Boston, MA: American Meteorological Society.

Hahn, B. B., E. Rall, and D. W. Klinger. 2002. Cognitive task analysis of the warning forecaster task. Final Report for National Weather Service (Office of Climate, Water, and Weather Service, Order No. RA1330-02-SE-0280). Fairborn, OH: Klein Associates Inc.

Hallett, J., M. Wetzel, and S. Rutledge. 1993. Field training in radar meteorology. *Bulletin of the American Meteorological Society* 74:17–22.

Hamblyn, R. 2001. *The Invention of Clouds: How an Amateur Meteorologist Forged the Language of the Skies*. New York: Farrar, Straus, and Giroux.

Hambrick, D.Z., F. Ferreira, and J. M. Henderson. 2014a. The 10,000 hour rule is wrong and perpetuates a cruel myth. [Downloaded January 25, 2015, from slate.com]

Hambrick, D. Z., F. L. Oswald, E. M. Altmann, E. J. Meinz, F. Gobet, and G. Campitelli. 2014b. Deliberate practice: Is that all it takes to become an expert? *Intelligence* 45:34–45.

Hamill, T. M. 1999. Hypothesis tests for evaluating numerical precipitation forecasts. *Weather and Forecasting* 14:155–167.

Hammond, K. R. 1980. *The integration of research in judgment and decision theory*. Technical Report No. 226, Center for Research on Judgment and Policy, University of Colorado, Boulder, CO.

Hammond, M., and R. Collins. 1991. *Self-Directed Learning: Critical Practice*. New York: Nichols/GP Publishing.

Hanes, L., and M. Gross. 2002. Capturing valuable undocumented knowledge: Lessons learned at electric utility sites. Presentation at the IEEE 7th Conference on Human Factors and Power Plants, Scottsdale, AZ.

Harned, S., S. Businger, and M. Stephenson. 1997. The application of three and four dimensional visualization of meteorological data fields in weather forecasting. In *Symposium on Weather Forecasting* (pp. 50–53). Boston, MA: American Meteorological Society.

Harrington, J. A., R. S. Cerveny, and J. S. Hobgood. 1991. Competitive learning experiences: The role of weather forecasting contests in geography programs. *Journal of Geography* 90:27–31.

Hart, A. 1989. *Knowledge Acquisition for Expert Systems*. New York: McGraw Hill.

Hartley, R. T. 1981. How expert should an expert system be? In *Proceedings of the Seventh International Joint Conference On Artificial Intelligence* (pp. 862–867). Vancouver, BC: International Joint Conferences in Artificial Intelligence.

Harvey, N. 2001. Improving judgment in forecasting. In *Principles of Forecasting: A Handbook for Researchers and Practitioners*, ed. J. S. Armstrong, 59–80. Norwell, MA: Kluwer.

Hayes, J. R. 1985. Three problems in teaching general skills. In *Thinking and Learning Skills: Vol. 2. Research and Open Questions*, ed. S. F. Chipman, J. W. Segal, and R. Glaser (pp. 391–405). Hillsdale, NJ: Erlbaum.

Hayes-Roth, F., D. A. Waterman, and D. B. Lenat, eds. 1983. *Building Expert Systems*. Reading, MA: Addison-Wesley.

Hegarty, M. 1992. Mental animation: Inferring motion from static displays of mechanical systems. *Journal of Experimental Psychology: Learning, Memory, and Cognition* 18:1084–1102.

Hegarty, M. 2004. Mechanical reasoning as mental simulation. *Trends in Cognitive Sciences* 8:280–285.

Hegarty, M. 2011. The cognitive science of visual-spatial displays: Implications for design. *Topics in Cognitive Science* 3:446–474.

Hegarty, M. 2013. Cognition, metacognition and the design of maps. *Current Directions in Psychological Science* 22:3–9.

Hegarty, M., M. S. Canham, and S. I. Fabrikant. 2010. Thinking about the weather: How display salience and knowledge affect performance in a graphic interference task. *Journal of Experimental Psychology: Learning, Memory, and Cognition* 36:37–53.

Hegarty, M., H. S. Smallman, and A. T. Stull. 2012. Choosing and using geospatial displays: Effects of design on performance and metacognition. *Journal of Experimental Psychology. Applied* 18:1–17.

Hegarty, M., H. S. Smallman, A. T. Stull, and M. Canham. 2009. Naïve Cartography: How intuitions about display configuration can hurt performance. *Cartographica* 44:171–187.

Hegarty, M., M. Stieff, and B. Dixon. 2014. Reasoning with diagrams: Towards a broad ontology of spatial thinking strategies. In *Space in Mind: Concepts for Spatial Learning and Education*, ed. D. R. Montello, K. Grossner, and D. G. Janelle (pp. 75–98). Cambridge, MA: MIT Press.

Heideman, K. F., T. R. Stewart, W. R. Moninger, and P. Regan-Cirincione. 1993. The weather information and skill experiment (WISE): The effect of varying levels of information on forecast skill. *Weather and Forecasting* 8:25–36.

Heinselman, P., D. LaDue, D. M. Kingfield, and R. R. Hoffman. 2015. Tornado warning decisions using phased array radar data. *Weather and Forecasting* 30:57–78.

Heinselman, P., D. LaDue, and H. Lazrus. 2012. Exploring impacts of rapid-scan radar data on NWS warning decisions. *Weather and Forecasting* 27:1031–1044.

Heisterman, M. (with 10 others). 2015. The emergence of open-source software for the weather radar community. *Bulletin of the American Meteorological Society* 96:117–128.[Downloaded from https://www.arm.gov/science/highlights/RNTU2/view]

Henson, R. 2010, July/August. Blue skies and green screens: The history of weathercasting graphics. *Weatherwise* 63:32–37.

Hering, L. 2016. *The climate requires us to change*. Presentation at the World Resources Simulation Center. [Downloaded 1 December 2016 from http://www.wrsc.org/presentation/climate-requires-us-change]

Heuer, R. J. 1999. *The Psychology of Intelligence Analysis*. Washington, DC: Center for the Study of Intelligence.

Hoch, S. J. 1988. Who do we know: Predicting the interests and opinions of the American consumer. *Journal of Consumer Research* 15:315–324.

Hodur, R. M. 1997. The Naval Research Laboratory's Coupled Ocean/Atmosphere Mesoscale Prediction System (COAMPS). *Monthly Weather Review* 125:1414–1430.

Hoekstra, A., K. Klockow, R. Riley, J. Brotzge, H. Brooks, and S. Erickson. 2011. A preliminary look at the social perspective of Warn-on Forecast: Preferred tornado warning lead time and the general public's perceptions of weather risks. *Weather, Climate, and Society* 3:128–140.

Hoffman, R. R. 1987a. *General guidance for the development of the advanced meteorological processing system*. Final Report, Contract No. F49620–85–00113. Air Force Geophysics Laboratory, Hanscom AFB, MA.

Hoffman, R. R. 1987b, Summer. The problem of extracting the knowledge of experts from the perspective of experimental psychology. *AI Magazine* 8:53–67.

Hoffman, R. R. 1990. Remote perceiving: A step toward a unified science of remote sensing. *Geocarto International* 5:3–13.

Hoffman, R. R. 1991. Human factors psychology in the support of forecasting: The design of advanced meteorological workstations. *Weather and Forecasting* 6:98–110.

Hoffman, R. R., ed. 1992. *The Psychology of Expertise: Cognitive Research and Empirical AI*. Hillsdale, NJ: Erlbaum.

Hoffman, R. R. 1997, February. *Human factors in meteorology*. Paper presentation at the Short Course on Human Factors Applied to Graphical User Interface Design, held at the 77th annual meeting of the American Meteorological Society, Los Angeles, CA.

Hoffman, R. R. 1998. How can expertise be defined? Implications of research from cognitive psychology. In *Exploring Expertise*, ed. R. Williams, W. Faulkner, and J. Fleck (pp. 81–100). New York: Macmillan.

Hoffman, R. R., J. W. Coffey, and K. M. Ford. 2000. *A Case Study in the Research Paradigm of Human-Centered Computing: Local Expertise in Weather Forecasting. Report on the Contract, "Human-Centered System Prototype*. Washington, DC: National Technology Alliance.

Hoffman, R. R., J. W. Coffey, K. M. Ford, and M. J. Carnot. 2001, October. STORM-LK: A human-centered knowledge model for weather forecasting. In *Proceedings of the 45th Annual Meeting of the Human Factors and Ergonomics Society* (p. 752). Santa Monica, CA: Human Factors and Ergonomics Society.

Hoffman, R. R., J. W. Coffey, K. M. Ford, and J. D. Novak. 2006. A method for eliciting, preserving, and sharing the knowledge of forecasters. *Weather and Forecasting* 21:416–428.

Hoffman, R. R., and J. Conway. 1990. Psychological factors in remote sensing: A review of recent research. *Geocarto International* 4:3–22.

Hoffman, R. R., B. Crandall, and G. Klein. 2008. Protocols for cognitive task analysis. Report, Florida Institute for Human and Machine Cognition, Pensacola FL. [Downloaded from www.dtic.mil/cgi-bin/GetTRDoc?AD=ADA475456]

Hoffman, R. R., B. Crandall, and N. R. Shadbolt. 1998. Use of the Critical Decision Method to elicit expert knowledge: A case study in the methodology of cognitive task analysis. *Human Factors* 40:254–276.

Hoffman, R. R., T. M. Cullen, and J. K. Hawley. 2016. Rhetoric and reality of autonomous weapons: Getting a grip on the myths and costs of automation. *Bulletin of the Atomic Scientists*, 72. [DO I:10.1080/00963402.2016.1194619]

Hoffman, R. R., M. Detweiler, J. A. Conway, and K. Lipton. 1993. Some considerations in using color in meteorological displays. *Weather and Forecasting* 8:505–518.

Hoffman, R. R., and W. C. Elm. 2006, January/February. HCC implications for the procurement process. *IEEE Intelligent Systems* 21:74–81.

Hoffman, R. R., and S. M. Fiore. 2007, May/June. Perceptual (Re)learning: A leverage point for Human-Centered Computing. *IEEE Intelligent Systems* 22:79–83.

Hoffman, R. R., K. M. Ford, and J. W. Coffey. 2000a. *The Handbook of Human-Centered Computing. Deliverable on the Contract, "Human-Centered System Prototype."* National Technology Alliance.

Hoffman, R. R., K. M. Ford, and J. W. Coffey. 2000b. A case study in the research paradigm of Human-Centered Computing: Local expertise at weather forecasting. In *Report*. Pensacola, FL: Institute for Human and Machine Cognition, University of West Florida.

Hoffman, R. R., and L. F. Hanes. 2003, July–August. The boiled frog problem. *IEEE Intelligent Systems* 18:68–71.

Hoffman, R. R., J. K. Hawley, and J. M. Bradshaw. (March/April 2014). Myths of automation Part 2: Some very human consequences. *IEEE Intelligent Systems* 29:82–85.

Hoffman, R. R., S. Henderson, B. Moon, D. T. Moore, and J. A. Litman. 2011. Reasoning difficulty in analytical activity. *Theoretical Issues in Ergonomics Science* 12:225–240.

Hoffman, R. R., G. Klein, and J. E. Miller. 2011. Naturalistic investigations and models of reasoning about complex indeterminate causation. *Information, Knowledge, Systems Management* 10:397–425.

Hoffman, R. R., and G. Lintern. 2006. Eliciting and representing the knowledge of experts. In *Cambridge Handbook of Expertise and Expert Performance*, ed. K. A. Ericsson, N. Charness, P. Feltovich, and R. Hoffman (pp. 203–222). New York: Cambridge University Press.

Hoffman, R. R., G. Lintern, and S. Eitelman. 2004, March/April. The Janus Principle. *IEEE Intelligent Systems* 19:78–80.

Hoffman, R. R., and A. B. Markman, eds. 2001. *The Interpretation of Remote Sensing Imagery: Human factors*. New York: Lewis Publishers.

Hoffman, R. R., and M. J. McCloskey. 2013, July/August. Envisioning desirements. *IEEE Intelligent Systems* 28:82–89.

Hoffman, R. R., and L. Militello. 2009. *Perspectives on Cognitive Task Analysis*. Boca Raton, FL: Taylor and Francis/CRC Press.

Hoffman, R. R., K. N. Neville, and J. Fowlkes. 2009. Using cognitive task analysis to explore issues in the procurement of intelligent decision support systems. *Cognition Technology and Work* 11:57–70.

Hoffman, R. R., and R. J. Pike. 1995. On the specification of the information available for the perception and description of the natural terrain. In *Local applications of the ecological approach to human-machine systems*, ed. P. Hancock, J. Flach, J. Caird and K. Vicente (pp. 285–323). Mahwah, NJ: Erlbaum.

Hoffman, R. R., N. R. Shadbolt, A. M. Burton, and G. Klein. 1995. Eliciting knowledge from experts: A methodological analysis. *Organizational Behavior and Human Decision Processes* 62:129–158.

Hoffman, R. R., P. Ward, L. DiBello, P. J. Feltovich, S. M. Fiore, and D. Andrews. 2014. *Accelerated Expertise: Training for High Proficiency in a Complex World.* Boca Raton, FL: Taylor and Francis/CRC Press.

Hoffman, R. R., and D. D. Woods. 2005, January/February. Steps toward a theory of complex and cognitive systems. *IEEE Intelligent Systems* 20:76–79.

Hoffman, R. R., and J. F. Yates. 2005. Decision(?)Making. *IEEE Intelligent Systems* 20:22–29.

Hoffman, R. R., D. Ziebell, P. J. Feltovich, B. M. Moon, and S. F. Fiore. 2011, September/October. Franchise experts. *IEEE Intelligent Systems* 26:72–77.

Hoffman, R. R., D. Ziebell, S. M. Fiore, and I. Becerra-Fernandez. 2008, May/June. Knowledge management revisited. *IEEE Intelligent Systems* 23:84–88.

Hollnagel, E., and D. D. Woods. 1983. Cognitive systems engineering: New wine in new bottles. *International Journal of Man-Machine Studies* 18:583–600.

Hollnagel, E., D. D. Woods, and N. Leveson, eds. 2006. *Resilience Engineering: Concepts and Precepts*. Hampshire, England: Ashgate.

Hontarrede, M. 1998. Meteorology and the maritime world: 150 years of constructive cooperation. *WMO Bulletin* 47:15–26.

Hutchins, E. 1995. *Cognition in the Wild.* Cambridge, MA: The MIT Press.

Ibarra, H., R. Wolf, J. Meyer, and J. Miodzik. 1999. DateStreme: Weather education from AMS to the classroom. *Bulletin of the American Meteorological Society* 80:1901–1905.

Jackson, J. 2006, March. *Wal-Mart's Severe Weather Plan for Continuity of Operations.* National Severe Weather Workshop, Midwest City, OK.

Jang, J., S. B. Trickett, C. D. Schunn, and J. G. Trafton. 2012. Unpacking the temporal advantage of distributing complex visual displays. *International Journal of Human-Computer Studies* 70:812–827.

Jarboe, J. M., and R. M. Steadham. 1995. Preliminary results from the human factors evaluation of the Radar Analysis and Display System during the summer 1994 operational test in Phoenix. In *Proceedings of the 11th International Conference on Interactive Information and Processing Systems for Meteorology, Oceanography, and Hydrology*. Boston, MA: American Meteorological Society.

Jasperson, W. H., and D. E. Venne. 1990. A knowledge-based system architecture for short-range single-station weather forecasting. Report, Center for Atmospheric and Space Sciences, Augsburg College, Minneapolis, MN.

Jeffries, R., A. Turner, P. Polson, and M. Atwood. 1981. The processes involved in designing software. In *Cognitive Skills and Their Acquisition*, ed. R. J. Anderson (pp. 255–283). Hillsdale, NJ: Erlbaum.

Jenkins, J. J. 1953. Some measured characteristics of Air Force weather forecasters and success in forecasting. *Journal of Applied Psychology* 37:440–444.

Jenkins, J. J. 1985. Acoustic information for objects, places and events. In *Persistence and change: Proceedings of the First International Conference on Event Perception*, ed. W. H. Warren and R. E. Shaw (pp. 115–138). Hillsdale, NJ: Erlbaum.

Jesuroga, S., S. Drake, J. Cowie, and D. Himes. 1997. A software tool to display and manipulate meteorological data. In *Proceedings of the 13th Conference on Interactive Information and Processing Systems for Meteorology, Oceanography, and Hydrology* (pp. 359–361). Boston, MA: American Meteorological Society.

Johns, R. H. 1982. A synoptic climatology of Northwest-flow severe weather outbreaks. Part I: Nature and significance. *Monthly Weather Review* 110:1653–1663.

Johns, R. H. 1984. A synoptic climatology of Northwest-flow severe weather outbreaks. Part II: Meteorological parameters and synoptic patterns. *Monthly Weather Review* 112:449–464.

Johns, R. H. 1993. Meteorological conditions associated with bow echo development in convective storms. *Weather and Forecasting* 8:294–299.

Johns, R. H., and C. A. Doswell. 1992. Severe local storms forecasting. *Weather and Forecasting* 7:588–612.

Johnson, M., J. M. Bradshaw, R. R. Hoffman, P. J. Feltovich, and D. D. Woods. 2014, November/December. Seven cardinal virtues of human-machine teamwork. *IEEE Intelligent Systems* 29:74–79.

Johnson, V. C., and L. E. Spayd. 1996. The COMET outreach program: Cooperative research to improve operational forecasts. *Bulletin of the American Meteorological Society* 77:2317–2332.

Johnson-Laird, P. N. 1980. Mental models in cognitive science. *Cognitive Science* 4:71–115.

Johnson-Laird, P. N., and R. M. J. Byrne. 1991. *Deduction*. Hillsdale, NJ: Erlbaum.

Johnson-Laird, P. N. 1983. *Mental Models: Towards a Cognitive Science of Language, Inference, and Consciousness*. Cambridge, MA: Harvard University Press.

Johnson-Laird, P. N., S. S. Khemlani, and G. P. Goodwin. 2015. Logic, probability, and human reasoning. *Trends in Cognitive Sciences* 19:201–214.

Jolliffe, L. T., and D. B. Stephenson, eds. 2003. *Forecast Verification: A Practitioner's Guide in Atmospheric Science*. New York: John Wiley.

Jolliffe, I. T., and D. B. Stephenson. 2005. Comments on "Discussion for verification concepts in forecast verification: A practitioner's guide in atmospheric science." *Weather and Forecasting* 20:796–800.

Jones, D. W., and S. Joslyn. 2004. The MURI uncertainty monitor. In *Proceedings of the 84th American Meteorological Society Annual Conference*. Boston: American Meteorological Society. [Downloaded March 20, 2016, from https://ams.confex.com/ams/84Annual/webprogram/Paper72636.html]

Jones, J., and J. Gray. 2016. Why extreme weather is the new normal. CNN: Extreme Weather. [Downloaded 1 December 2016 from http://www.cnn.com/2016/11/28/us/extreme-weather-normal/]

Joslyn, S. L., and D. Jones. 2008. Strategies in naturalistic decision making: A cognitive task analysis of naval weather forecasting. In *Naturalistic Decision Making and Macrocognition*, ed. J. M. Schraagen, L. G. Militello, T. Ormerod and R. Lipshitz (pp. 183–201). Burlington, VT: Ashgate Publishing Co.

Joslyn, S. L., and J. E. LeClerc. 2012. Uncertainty forecasts improve weather related decisions and attenuate the effects of forecast error. *Journal of Experimental Psychology: Applied* 18:126–140.

Joslyn, S. L., L. Nadav-Greenberg, and R. M. Nichols. 2009, February. Probability of precipitation: Assessment and enhancement of end-user understanding. *Bulletin of the American Meteorological Society* 185–194.

Joslyn, S. L., K. Pak, D. Jones, J. Pyles, and E. Hunt. 2007. The effect of probabilistic information on threshold forecasts. *Weather and Forecasting* 22:804–812.

Jungerman, H. 1983. Two camps of rationality. In *Decision Making Under Uncertainty*, ed. R. W. Scholz (pp. 63–86). Amsterdam: North-Holland.

Kahneman, D., P. Slovic, and A. Tversky, eds. 1982. *Judgment Under Uncertainty: Heuristics and Biases*. New York: Cambridge University Press.

Kahneman, D., and A. Tversky, eds. 2000. *Values, Choices, and Frames*. Cambridge: Cambridge University Press.

Kain, J. S. (with 12 others). 2010. Assessing advances in the assimilation of radar data and other mesoscale observations within a collaborative forecasting-research environment. *Weather and Forecasting* 25:1510–1521.

Kaplan, E., and M. Kaplan. 2010. *Bozo Sapiens: Why to Err Is Human*. NewYork: Bloomsbury Press.

Karstens, C. D., T. M. Smith, and K. M. Calhoun. A. J. Clark, C. Ling, G. J. Stumpf, and L. P. Rothfusz. 2014. Prototype tool development for creating probabilistic hazard information for severe convective phenomena. In *Proceedings of the 94th Annual Meeting of the American Meteorological Society*, Boston, MA: American Meteorological Society. [https://ams.confex.com/ams/94Annual/webprogram/meeting.html]

Keenan, T., R. Potts, and T. Stevenson. 1992. An evaluation of the Darwin Area Forecast Experiment storm occurrence forecasts. *Weather and Forecasting* 7:515–523.

Kelly, J., and P. Gigliotti. 1997. The Australian Integrated Forecast System (AIFS): Overview and current status. In *Proceedings of the 13th Conference on Interactive Information and Processing Systems for Meteorology, Oceanography, and Hydrology* (pp. 141–144). Boston, MA: American Meteorological Society.

Kerr, R. A. 2012. Weather forecasts slowly clearing up. *Science* 38:734–737.

Kidd, A., and M. Welbank. 1984. Knowledge acquisition. In *Infotech State of the Art Report on Expert Systems,* ed. J. Fox. London: Pergamon.

Kirschenbaum, S. S. 2004. The role of comparison in weather forecasting: Evidence from two hemispheres. In *Proceedings of the Human Factors and Ergonomics Society 48th Annual Meeting* (pp. 306–310). Santa Monica, CA: Human Factors and Ergonomics Society.

Klein, G. 1989. Recognition-primed decisions. In *Advances in Man-Machine Systems,* vol. 5, ed. W. B. Rouse (pp. 47–92). Greenwich, CT: JAI Press.

Klein, G. 1992. Using knowledge engineering to preserve corporate memory. In *The Psychology of Expertise: Cognitive Research and Empirical AI,* ed. R. R. Hoffman, 170–190. Mahwah, NJ: Erlbaum.

Klein, G. 1993. A recognition-primed decision (RPD) model of rapid decision making. In *Decision Making in Action: Models and Methods,* ed. G. Klein, J. Orasanu, R. Calderwood, and C. E. Zsambok (pp. 138–147). Norwood, NJ: Ablex.

Klein, G. 1997a. The recognition-primed decision model: Looking back, looking forward. In *Naturalistic Decision Making,* ed. C. Zsambok and G. Klein (pp. 285–292). Mahwah, NJ: Erlbaum.

Klein, G. 1997b. Developing expertise in decision making. *Thinking & Reasoning* 3:337–352.

Klein, G. 1999. *Sources of Power.* Cambridge, MA: The MIT Press.

Klein, G. 2008. Naturalistic decision making. *Human Factors* 50:456–460.

Klein, G., and H. C. Baxter. 2009. Cognitive transformation theory: Contrasting cognitive and behavioral learning. In *The PSI handbook of virtual environments for training and education developments for the military and beyond: Vol. 1. Education: Learning, Requirements and metrics* (pp. 50–65). Santa Barbara, CA: Praeger Security International.

Klein, G., R. Calderwood, and A. Clinton-Cirocco. 1986. Rapid decision making on the fire ground. In *Proceedings of the 30th Annual Meeting of the Human Factors Society* (pp. 576–580). Santa Monica, CA: Human Factors Society.

Klein, G., R. Calderwood, and D. MacGregor. 1989. Critical decision method of eliciting knowledge. *IEEE Transactions on Systems, Man, and Cybernetics* 19:462–472.

Klein, G., and R. R. Hoffman. 1992. Seeing the invisible: Perceptual-cognitive aspects of expertise. In *Cognitive Science Foundations of Instruction,* ed. M. Rabinowitz (pp. 203–226). Mahwah, NJ: Erlbaum.

Klein, G., and R. R. Hoffman. 2008. Macrocognition, mental models, and cognitive task analysis methodology. In *Naturalistic decision making and macrocognition,* ed. J. M. Schraagen, L. G. Militello, T. Ormerod and R. Lipshitz (pp. 57–80). Aldershot, UK: Ashgate.

Klein, G., R. R. Hoffman, and J. M. Schraagen. 2008. The macrocognition framework of naturalistic decision making. In *Macrocognition*, ed. J. M. Schraagen (pp. 3–25). London: Ashgate.

Klein, G., B. Moon, and R. R. Hoffman. 2006a. Making sense of sensemaking 1: Alternative perspectives. *IEEE Intelligent Systems* 21:22–26.

Klein, G., B. Moon, and R. R. Hoffman. 2006b. Making sense of sensemaking 2: A macrocognitive model. *IEEE Intelligent Systems* 21:88–92.

Klein, G., J. Orasanu, R. Calderwood, and C. E. Zsambok, eds. 1993. *Decision Making in Action: Models and Methods*. Norwood, NJ: Ablex.

Klein, G., K. G. Ross, B. M. Moon, D. E. Klein, R. R. Hoffman, and E. Hollnagel. 2003. Macrocognition. *IEEE Intelligent Systems* 18:81–85.

Klein, G., D. D. Woods, J. D. Bradshaw, R. R. Hoffman, and P. J. Feltovich. 2004, November/December. Ten challenges for making automation a "team player" in joint human-agent activity. *IEEE Intelligent Systems* 19:91–95.

Klein, W. H., and H. R. Glahn. 1974. Forecasting local weather by means of model output statistics. *Bulletin of the American Meteorological Society* 55:1217–1227.

Klinger, D. W., E. B. Hahn, and E. Rall. 2007. Cognitive task analysis of the warning forecaster's task: An invitation for human-centered technologies. In *Expertise Out of Context*, ed. R. R. Hoffman (pp. 359–373). Boca Raton, FL: Taylor and Francis.

Knowles, M. 1975. *Self-Directed Learning: A Guide for Learners and Teachers*. Parsippany, NJ: Globe Fearon.

Knox, J. A., and S. A. Ackernan. 2005. What do introductory meteorology students want to learn? *Bulletin of the American Meteorological Society* 86:1431–1435.

Kocin, P. J., P. N. Schumacher, R. F. Morales, Jr., and P. J. Uccellini. 1995. Overview of the 12–14 March 1993 superstorm. *Bulletin of the American Meteorological Society* 76:165–182.

Kocin, P. J., and L. W. Uccellini. 2004. *The Cases: Vol. II . Northeast Snowstorms*. Boston: American Meteorological Society.

Koelsch, W. A. 1996. From Geo- to Physical Science: Meteorology and the American University, 1919–1945. In *Historical Essays on Meteorology, 1919–1995*, ed. J. R. Fleming (pp. 511–540). Boston: American Meteorological Society.

Koopman, P., and R. R. Hoffman. 2003. Work-arounds, make-work, and kludges. *IEEE Intelligent Systems* 18:70–75.

Kosslyn, S. M. 1994. Identifying objects seen from different viewpoints. *Brain* 117:1055–1071.

Kosslyn, S. 2006. *Graph Design for the Eye and Mind*. New York: Oxford University Press.

Kosslyn, S. M., K. E. Sukel, and B. M. Bly. 1999. Squinting with the mind's eye: Effects of stimulus resolution on imaginal and perceptual comparisons. *Memory & Cognition* 27:276–287.

Koval, J. P., and G. S. Young. 2001. Computer training for entrepreneurial meteorologists. *Bulletin of the American Meteorological Society* 82:875–888.

Krebs, M. J., and J. D. Wolf. 1979. Design principles for the use of color in displays. *Proceedings of the Society for Information Display* 20:10–15.

Krebs, W. K., and A. J. Ahumada, Jr. 2001. Air Traffic Control weather radar displays: Validation of a masking metric for prediction of text block identification. In *Proceedings of the Human Factors and Ergonomics Society 45th Annual Meeting* (pp. 1328–1332). Santa Monica, CA: Human Factors and Ergonomics Society.

Kriz, S., and M. Hegarty. 2007. Top-down and bottom-up influences on learning from animations. *International Journal of Human–Computer Studies* 65:911–930.

Kucera, P. C., and C. M. Lusk. 1996. Cool season product usage patterns from the DARE workstations at the Denver and Norman WSFOs. Technical Memorandum ERL-FSL-18, Forecast Systems Laboratory, Boulder, CO.

Kurz-Milcke. E. and Gigerenzer, G., eds. 2004. *Experts in Science and Society*. New York: Kluwer Academic Publishers.

Kusterer, K. C. 1978. *Know-How on the Job: The Important Working Knowledge of "Unskilled" Workers*. Boulder, CO: Westview.

Kyle, T. G. 1985. Expanding expertise by use of an expert system. In *Proceedings of the Conference on Intelligent Systems and Machines* (pp. 244–247). Center for Robotics and Advanced Automation, University of Michigan, Ann Arbor, MI.

LaChance, J. R. 2000, March 9. Statement before Subcommittee on Oversight of Government Management on Managing Human Capital in the 21st Century, U.S. Senate, Washington, DC. [Downloaded 25 June 2005 from http://www.opm.gov/testify/2000/Lachance3-09-00.htm].

LaDue, D. S. 2011. How meteorologists learn to forecast the weather: Understanding learning in complex domains. *Proceedings of the 20th Symposium on Education at the 91st Annual Meeting of the American Meteorological Society* (Article 2.4). Boston, MA: American Meteorological Society. [Downloaded from https://ams.confex.com/ams/91Annual/webprogram/Paper184220.html]

Lajoie, S. P. 2003. Transitions and trajectories for studies of expertise. *Educational Researcher* 32:21–25.

Lamos, J. P., and E. M. Page. 2012. Creating professional development series for operational forecasting. 21st Symposium on Education, American Meteorological Society, New Orleans, LA, J1.1. [Available at https://ams.confex.com/ams/92Annual/webprogram/Paper200964.html]

Larkin, J. H. 1983. The role of problem representation in physics. In *Mental Models*, ed. D. Gentner and A. Stevens (pp. 75–98). Hillsdale, NJ: Erlbaum.

Larkin, J. H., and H. A. Simon. 1987. Why a diagram is (sometimes) worth 10,000 words. *Cognitive Science* 4:317–345.

Lavin, S. J., and R. S. Cerveny. 1986. Identification and analysis of climactic fields through dot-density shading. *Journal of Atmospheric and Oceanic Technology* 3:552–558.

Layton, L., and A. Sipress. 2000a, January 28. Besieged metrorail sidelines 100 cars. *The Washington Post.*

Layton, L., and Sipress, A. 2000b, January 27. Snow removal runs gamut: Tough major roads are clear, some areas are still digging out. *The Washington Post.*

Lazo, J. K., R. E. Morss, and J. L. Demuth. 2009. 300 billion served: Sources, perceptions, uses, and values of weather forecasts. *Bulletin of the American Meteorological Society* 90:785–798.

Lazo, J. K., D. M. Waldman, B. H. Morrow, and J. A. Thacher. 2010. Assessment of household evacuation decision making and the benefits of improved hurricane forecasting. *Weather and Forecasting* 25:207–219.

Lazzara, M. A., J. M. Benson, R. J. Fox, D. J. Laitsch, J. P. Rueden, D. A. Santek, D. M. Wade, T. M. Whittaker, and J. T. Young. 1999. The Man-computer Interactive Data Access System: 25 years of interactive processing. *Bulletin of the American Meteorological Society* 80:271–284.

Lederberg, J., and E. A. Feigenbaum. 1968. Mechanization of inductive inference in organic chemistry. In *Formal Representation of Human Judgment*, ed. B. Kleinmuntz (pp. 187–267). New York: Wiley.

Lee, J. E. 1997. The importance of the Advanced Weather Interactive Processing System (AWIPS) to the modernized National Weather Service from a forecast office perspective. In *Proceedings of the 13th Conference on Interactive Information and Processing Systems for Meteorology, Oceanography, and Hydrology* (pp. 112–115). Boston, MA: American Meteorological Society.

Lee, R. R., and J. E. Passner. 1993. The development and verification of TIPS: An expert system to forecast thunderstorm occurrence. *Weather and Forecasting* 8:271–280.

Lenat, D. B., and E. A. Feigenbaum. 1987. On the threshold of knowledge. *Proceedings of the International Joint Conference on Artificial Intelligence* (pp. 1173–1182). Milano, Italy: International Joint Conference on Artificial Intelligence.

Lesgold, A. M. 1984. Acquiring expertise. In *Tutorials in Learning and Memory: Essays in Honor of Gordon Bower*, ed. J. R. Anderson and S. M. Kosslyn (pp. 31–60). San Francisco: W. H. Freeman.

Lesgold, A. M., H. Rubinson, P. Feltovich, R. Glaser, D. Klopfer, and Y. Wang. 1988. Expertise in a complex skill: Diagnosing X-ray pictures. In *The Nature of Expertise*, ed. M. T. H. Chi, R. Glaser, and M. J. Farr (pp. 311–342). Mahwah, NJ: Erlbaum.

Leutbecher, M., and T. Palmer. 2008. Ensemble forecasting. *Journal of Computational Physics* 227:3515–3539.

Lichtenstein, S., B. Fischhoff, and L. D. Phillips. 1982. Calibration of probabilities: The state of the art to 1980. In *Judgment Under Uncertainty: Heuristics and Biases*, ed. D. Kahneman, P. Slovic, and A. Tversky (pp. 306–334). New York: Cambridge University Press.

Liggins, M. E., D. L. Hall, and J. Llinas. 2008. *Multisensor Data Fusion: Theory and Practice*. Boca Raton, FL: Taylor & Francis CRC Press.

Lincoln, Y. S., and E. G. Guba. 1985. *Naturalistic Inquiry*. Thousand Oaks, CA: Sage.

Lindell, M. K., and H. Brooks. 2012. Workshop on Weather Ready Nation: Science Imperatives for Severe Thunderstorm Research. Report form the Hazard Reduction and Recovery Center, Texas A&M University, sponsored by the National Oceanic and Atmospheric Administration and the National Science Foundation.

Lipshitz, R., and O. Ben Saul. 1993. Schemata and mental models in recognition-primed decision making. In *Advances in Naturalistic Decision Making*, ed. C. E. Zsambok and G. Klein (pp. 293–303). Mahwah, NJ: Erlbaum.

Livezey, R. E., and M. M. Timofeyeva. 2008. The first decade of long-lead U.S. seasonal forecasts: Insights from a skill analysis. *Bulletin of the American Meteorological Society* 89:843–854.

Lohrenz, M. C., J. G. Trafton, M. R. Beck, and M. L. Gendron. 2009. A model of clutter for complex, multivariate geospatial displays. *Human Factors* 51:90–101.

Lorditch, E. 2009, January/February. Advances in weather analysis and forecasting. *Weatherwise* 62:22–27.

Love, G. G., and J. F. Mundy. 1997. Data visualization using the Navy's Weather Watch program. In *Preprints from the 13th Conference on Interactive Information and Processing Systems* (pp. 210–213). Boston, MA: American Meteorological Society.

Lowe, R. K. 1989. Search strategies and inference in the exploration of scientific diagrams. *Educational Psychology* 9:27–44.

Lowe, R. K. 1992. Dealing with graphic output from diagram processing tasks: Approaches to characterization and analysis. *Research in Science Education* 22:283–290.

Lowe, R. K. 1993a. Constructing a mental representation from an abstract technical diagram. *Learning and Instruction* 3:157–179.

Lowe, R. K. 1993b. *Successful Instructional Diagrams*. London: Kogan Page.

Lowe, R. K. 1994. Selectivity in diagrams: Reading beyond the lines. *Educational Psychology* 14:467–491.

Lowe, R. 2001. Components of expertise in the perception and interpretation of meteorological charts. In *Interpreting Remote Sensing Imagery: Human Factors*, ed. R. R. Hoffman and A. B. Markman (pp. 185–206). Boca Raton, FL: Lewis Publishers.

Lowe, R. K. 2015. Perceptual learning in the comprehension of animations and animated diagrams. In *Cambridge Handbook of Applied Perception Research*, ed. R. R. Hoffman, P. A. Hancock, M. Scerbo, R. Parasuraman, and J. Szalma (pp. 692–710). Cambridge: Cambridge University Press.

Lunch, P. 2008. The origins of computer weather prediction and climate modeling. *Journal of Computational Physics* 227:3431–3444.

Lusk, C. M. 1993. DARE-II workstation use at the Denver Weather Service Forecast Office. Technical Memorandum ERL-FLS-8, NOAA Forecast Systems Laboratory, Boulder, CO.

Lusk, C. M., and K. R. Hammond. 1991. Judgment in a dynamic task: Microburst forecasting. *Journal of Behavioral Decision Making* 4:55–73.

Lusk, C. M., P. Kucera, W. Roberts, and L. Johnson. 1999. The process and methods used to evaluate prototype operational hydrometeorological workstations. *Bulletin of the American Meteorological Society* 80:57–65.

Lusk, C. M., T. R. Stewart, K. R. Hammond, and R. J. Potts. 1990. Judgment and decision making in dynamic tasks: The case of forecasting the microburst. *Weather and Forecasting* 5:627–639.

Lynch, P. 2008. The origins of computer weather prediction and climate modeling. *Journal of Computational Physics* 227:3431–3444.

MacDonald, A. E. 1985. Design considerations of operational meteorological systems: A perspective based upon the PROFS experience. In *Preprints of the International Conference on Interactive Information and Processing Systems for Meteorology, Oceanography, and Hydrology* (pp. 16–23). Boston, MA: American Meteorological Society.

MacEachren, A. 1993. *Some truth with maps: A primer on cartographic symbolization and design*. New York: Association of American Geographers.

MacEachren, A. M. 1995. *How Maps Work: Representation, Visualization and Design*. New York, London: Guilford Press.

MacEachren, A., and J. H. Ganter. 1990. A pattern identification approach to cartographic visualization. *Cartographica* 27:64–81.

Macnamara, B. N., D. Z. Hambrick, and F. L. Oswald. 2014. Deliberate practice and performance in music, games, sports education, and professions: A metaanalysis. *Psychological Science* 25:1608–1618.

Maese, T., J. Hunziker, H. Owen, M, Harven, L. Wagner, R. Wilcox, K. Koehler, and G. Cavalieri. 2007. Hazardous weather detection and display capability for US Navy ships. *Proceedings of the 23rd Conference on Interactive Information Processing Systems for Meteorology, Oceanography and Hydrology*, Paper 3A.1. Boston, MA: American Meteorological Society. [Downloaded from https://ams.confex.com/ams/87ANNUAL/techprogram/session_19815.htm]

Mandl, H., and J. R. Levin, eds. 1989. *Knowledge acquisition from text and pictures*. New York: Elsevier.

Mandler, G. 1967. Organization and memory. In *The Psychology of Learning and Motivation*. vol. 1, ed. K. W. Spence and J. T. Spence (pp. 327–372). New York: Academic Press.

Manzato, A. 2005. An odds ratio parameterization for ROC diagram and skill score indices. *Weather and Forecasting* 20:918–930.

Market, P. S. 2006. The impact of writing area forecast discussions on student forecaster performance. *Weather and Forecasting* 21:104–108.

Martin, J. P., J. E. Swan, R. J. Moorhead, Z. Liu, and S. Cai. 2008. Results of a user study on 2D hurricane visualization. *Computer Graphics Forum* 27:991–998.

Mass, C. F. 2002. IFPS and the future of the National Weather Service. *Bulletin of the American Meteorological Society* 18:75–79.

Mass, C. F. 2003a. IFPS and the future of the National Weather Service. *Weather and Forecasting* 18:75–79.

Mass, C. F. 2003b. Reply. *Weather and Forecasting* 18:1305–1306.

Maunder, W. J. 1969. The consumer and the weather report. *Atmosphere* 7:15–22. [http://www.tandfonline.com/doi/abs/10.1080/00046973.1969.9676566]

Maximuk, L. P. 1997. The Kansas City EWFO perspective of the installation and evaluation of early AWIPS builds. In *Proceedings of the 13th Conference on Interactive Information and Processing Systems for Meteorology, Oceanography, and Hydrology* (pp. 131–135). Boston, MA: American Meteorological Society.

Mayer, R. E. 1989. Systematic thinking fostered by illustrations in scientific texts. *Journal of Educational Psychology* 81:240–246.

Mayer, R. E. 1993. Illustrations that instruct. In *Advances in Instructional Psychology,* Vol. 4, ed. R. Glaser (pp. 253–284). Mahwah, NJ: Erlbaum.

Mayer, R. E. 1995. *Using words and pictures to summarize scientific explanations*. Paper presented at the 36th Annual Meeting of the Psychonomic Society, Los Angeles, CA.

Mayer, R. E., and R. B. Anderson. 1991. Animations need narrations: An experimental test of a dual-coding hypothesis. *Journal of Educational Psychology* 83:484–490.

Mayer, R. E., and J. K. Gallini. 1990. When is an illustration worth ten thousand words? *Journal of Educational Psychology* 82:715–726.

Mayer, R. E., K. Steinhoff, G. Bower, and R. Mars. 1995. A generative theory of textbook design: Using annotated illustrations to foster meaningful learning of science text. *Educational Technology Research and Development* 43:31–43.

McCarthy, P. J., W. Purcell, and D. Ball. 2007. Project Phoenix: Optimizing the machine-person mix with high-impact weather forecasting. In *Proceedings of the 22nd Conference on Weather Analysis and Forecasting* (Paper 6A.5). Boston, MA: American Meteorological Society. [Downloaded March 20, 2016, from https://ams.confex.com/ams/22WAF18NWP/techprogram/paper_122657.htm]

McGraw, K. L., and K. Harbison-Briggs. 1989. *Knowledge Acquisition*. Englewood Cliffs, NJ: Prentice-Hall.

McIntyre, M. E. 1988. Numerical weather prediction: A vision of the future. *Weather* 43:294–298.

McIntyre, M. E. 1994. *Numerical Weather Prediction: An Updated Vision for the Future. The Life Cycles of Extratropical Cyclones*. Boston: American Meteorological Society.

McIntyre, M. E. 1999. Numerical weather prediction: A vision of the future, updated still further. In *The Life Cycles of Extratropical Cyclones* (pp. 337–355). Boston: American Meteorological Society.

McKeithen, K. B., J. S. Reitman, H. H. Reuter, and S. C. Hirtle. 1981. Knowledge organization and skill differences in computer programmers. *Cognitive Psychology* 13:307–325.

McNeese, M. D., B. S. Zaff, K. J. Peio, D. E. Snyder, J. C. Duncan, and M. R. McFaren. 1990. An advanced knowledge and design acquisition methodology: Application for the Pilot's Associate. Report AAMRL-TR-90–060, U.S. Air Force Armstrong Aerospace Medical Research Laboratory, Wright-Patterson Air Force Base, OH.

McPherson, R. D. 1991. An NMC odyssey. *Preprints of the American Meteorological Society 9th Conference on Numerical Weather Prediction* (pp. 1–4). Boston: American Meteorological Society.

Means, T. 2015. Vilhelm Bjerknes: Father of modern meteorology. [Downloaded January 2015 from http://weather.about.com/od/weatherhistory/a/Who-is-Vilhelm-Bjerknes.htm]

Meehl, P. 1954. *Clinical Versus Statistical Predictions: A Theoretical Analysis and Review of the Evidence*. Minneapolis: University of Minnesota Press.

Menzel, W. P., and J. F. W. Purdom. 1994. Introducing GOES-I: The first of a new generation of geostationary operational environmental satellites. *Bulletin of the American Meteorological Association* 75:757–781.

Merrem, F. H., and R. H. Brady. 1988. Evaluating an expert system for forecasting. In *Proceedings of the Fourth International conference on Interactive Information Processing Systems for Meteorology, Oceanography, and Hydrology* (pp. 259–261). Boston, MA: American Meteorological Society.

Meyer, T. J. 1986, May 21. High pressure and hot competition reign at National Weather-Forecasting Contest. *Chronicle of Higher Education*, pp. 29, 32.

Mieg, H. 2006. Social and psychological factors in the development of expertise. In *Cambridge Handbook of Expertise and Expert Performance*, ed. K. A. Ericsson, N. Charness, P. J. Feltovich, and R. R. Hoffman (pp. 743–760). New York: Cambridge University Press.

Mielke, P. W., K. J. Berry, C. W. Landsea, and W. M. Gray. 1997. A single-sample estimate of shrinkage in meteorological forecasting. *Weather and Forecasting* 12:847–858.

Militello, L. G., and R. J. B. Hutton. 2000. Applied cognitive task analysis (ACTA): A practitioner's toolkit for understanding cognitive task demands. In *Task Analysis*, ed. J. Annett and N. A. Stanton (pp. 90–113). New York: Taylor and Francis.

Miller, R. C. 1972. Notes on analysis and severe storm forecasting procedure of the Air Force Weather Service. Technical Report No. 200, Air Weather Service, Scott AFB, IL.

Moacdieh, N., and N. Sarter. 2015. Display clutter: A review of definitions and measurement techniques. *Human Factors* 57:61–100.

Mogil, H. M. 2001. The skilled interpretation of weather satellite images: Learning to see patterns and not just cues. In *Interpreting Remote Sensing Imagery: Human Factors*, ed. R. R. Hoffman and A. B. Markman (pp. 235–272). Boca Raton, FL: Lewis Publishers.

Mogil, H. M. 2015, 29 January. An almost perfect winter storm forecast. [Downloaded October 28, 2015, from http://www.weatherworks.com/lifelong-learning-blog/?p=491]

Moller, A. R., C. A. Dowsell, M. P. Foster, and G. R. Woodall. 1994. The operational recognition of supercell thunderstorm environments and storm structures. *Weather and Forecasting* 9:327–347.

Moninger, W. R. 1986. ARCHER: A prototype expert system for identifying some meteorological phenomena. Report, Environmental Sciences Group, Environmental Research Laboratory, NOAA, Boulder, CO.

Moninger, W. R. 1988. Summary report on the Second Workshop on Artificial Intelligence Research in the Environmental Sciences (AIRES). *AI Applications in Environmental Science* 2:65–72.

Moninger, W. R. 1990. The artificial intelligence shootout: A comparison of severe storm forecasting systems. In *Preprints of the 16th Conference on Severe Local Storms* (pp. 1–6). Boston, MA: American Meteorological Society.

Monmonier, M. 1999. *Air Apparent: How Meteorologists Learned to Map, Predict, and Dramatize Weather*. Chicago, IL: University of Chicago Press.

Moon, B., R. R. Hoffman, A. J. Cañas, and J. D. Novak, eds. 2011. *Applied Concept Mapping: Capturing, Analyzing and Organizing Knowledge*. Boca Raton, FL: Taylor and Francis.

Moon, B., R. R. Hoffman, and D. Ziebell. 2009, January/February. How did you do that? Utilities develop strategies for preserving and sharing expertise. *Electric Perspectives* 34:20–29.

Moore, P. 2015. *The Weather Experiment: The Pioneers Who Sought to See the Future*. New York: Farr, Strauss and Giroux.

Moore, W. L. 1922. *The New Air World: The Science of Meteorology Simplified*. Boston: Little, Brown.

Morris, V. R., H. M. Mogil, and Y. Tsann-Wang. 2012. A network of weather camps to engage students in science. *EOS: Earth and Space News* 93:151–154.

Morss, D. A. 2000. Introductory atmospheric sciences at a liberal arts university (1995–1999). *Bulletin of the American Meteorological Society* 81:2399–2415.

Morss, R. E., J. L. Demuth, A. Bostrom, J. K. Lazo, and H. Lazrus. 2015. Flash flood risks and warning decisions: A mental models study of forecasters, public officials, and media broadcasters in Boulder, Colorado. *Risk Analysis* 35:2009–2028.

Morss, R. E., J. K. Lazo, B. G. Brown, H. E. Brooks, P. T. Ganderton, and B. N. Mills. 2008, March. Societal and economic research and applications for weather forecasts: Priorities for the North American THORPEX Program. *Bulletin of the American Meteorological Society* 89:335–346.

Morss, R. E., and F. M. Ralph. 2007. Use of information by National Weather Service forecasters and emergency managers during CALJET and PACJET-2001. *Weather and Forecasting* 22:539–555.

Mostek, A. (with 10 others). 2004. VISIT: Bringing training to Weather Service forecasters using a new distance learning tool. *Bulletin of the American Meteorological Society* 85:823–829.

Mullendore, G. L., J. S. Tilley, and L. D. Carey. 2013. Leveraging field campaign resources to provide additional undergraduate forecasting and research experiences: A case study from Deep Convective Clouds and Chemistry (DC3). Poster presented at the American Meteorological Society 22nd Symposium on Education. [https://ams.confex.com/ams/93Annual/webprogram/22EDUCATION.html]

Murphy, A. H. 1985. Probabilistic weather forecasting. In *Probability, Statistics, and Decision Making in the Atmospheric Sciences*, ed. A. H. Murphy and R. W. Katz (pp. 337–377). Boulder, CO: Westview Press.

Murphy, A. H. 1988. Skill scores based on the mean square error and their relationships to the correlation coefficient. *Monthly Weather Review* 116:2417–2424.

Murphy, A. H. 1992. Climatology, persistence, and their linear combination as standards of reference in skill scores. *Weather and Forecasting* 7:692–698.

Murphy, A. H. 1993. What is a good forecast: An essay on the nature of goodness in weather forecasting. *Weather and Forecasting* 8:281–293.

Murphy, A. H., and B. G. Brown. 1983. Forecast terminology: Composition and interpretation of public weather forecasts. *Bulletin of the American Meteorological Society* 64:13–22.

Murphy, A. H., and B. G. Brown. 1984. A comparative evaluation of objective and subjective weather forecasts in the United States. *Journal of Forecasting* 3:369–393.

Murphy, A. H., and H. Daan. 1984. Impacts of feedback and experience on the quality of subjective probability forecasts: Comparison of results from the first and second years of the Zeikerzee experiment. *Monthly Weather Review* 112:413–423.

Murphy, A. H., and H. Daan. 1985. Forecast evaluation. In *Probability, Statistics, and Decision Making in the Atmospheric Sciences*, ed. A. H. Murphy and R. W. Katz (pp. 379–437). Boulder, CO: Westview Press.

Murphy, A. H., S. Lichtenstein, B. Fischhoff, and R. L. Winkler. 1980. Misinterpretations of precipitation probability forecasts. *Bulletin of the American Meteorological Society* 61:695–701.

Murphy, A. H., and R. L. Winkler. 1970. Scoring rules in probability assessment and evaluation. *Acta Psychologica* 34:273–286.

Murphy, A. H., and R. L. Winkler. 1971. Forecasters and probability forecasts: Some current problems. *Bulletin of the American Meteorological Society* 52:239–247.

Murphy, A. H., and R. L. Winkler. 1974a. Probability forecasts: A survey of National Weather Service forecasters. *Bulletin of the American Meteorological Society* 55:1449–1453.

Murphy, A. H., and R. L. Winkler. 1974b. Subjective probability forecasting experiments in meteorology: Some preliminary results. *Bulletin of the American Meteorological Society* 55:1206–1216.

Murphy, A. H., and R. L. Winkler. 1977. Reliability of subjective probability forecasts of precipitation and temperature. *Applied Statistics* 26:41–47.

Murphy, A. H., and R. L. Winkler. 1982. Subjective probabilistic tornado forecasts: Some experimental results. *Monthly Weather Review* 110:1288–1297.

Murphy, A. H., and R. L. Winkler. 1984. Probability forecasting in meteorology. *Journal of the American Statistical Association* 79:489–500.

Murphy, A. H., and R. L. Winkler. 1987. A general framework for forecast verification. *Monthly Weather Review* 115:1330–1338.

Murphy, G. L., and J. C. Wright. 1984. Changes in conceptual structure with expertise: Differences between real-world experts and novices. *Journal of Experimental Psychology: Learning, Memory, and Cognition* 10:144–155.

Myles-Worsley, M., W. A. Johnston, and M. A. Simons. 1988. The influence of expertise on X-ray image processing. *Journal of Experimental Psychology: Learning, Memory, and Cognition* 14:553–557.

Mylne, K. 2006. Predictability from a forecast providers perspective. In *Predictability of Weather and Climate,* ed. T. Palmer and R. Hagedorn (pp. 596–613). Cambridge: Cambridge University Press.

Nadav-Greenberg, L., S. L. Joslyn, and M. U. Taing. 2008. The effect of uncertainty visualizations on decision making in weather forecasting. *Journal of Cognitive Engineering and Decision Making* 2:24–47.

Namm, R. 1979. Study of the ability of meteorologists to communicate with the general public and the public's attitude toward various weather-related subjects. *National Weather Digest* 4:38–45.

National Public Radio. 2005, 22 March. Loss of line Workers. Report, "Morning Edition."

National Research Council. 2001. *Climate Change Science: An Analysis of Some Key Questions.* Washington, DC: National Academy Press.

National Research Council. 2006. *Completing the Forecast: Characterizing and Communicating Uncertainty for Better Decisions Using Weather and Climate Forecasts. Committee on Estimating and Communicating Uncertainty in Weather and Climate Forecasts.* Washington, DC: Board of Atmospheric Sciences and Climate, National Research Council.

National Research Council. 2010. *When Weather Matters: Science and Services to Meet Critical Societal Needs. Committee on Progress and Priorities of U.S. Weather Research and Research-to-Operations Activities.* Washington, DC: *Board on Atmospheric Sciences and Climate,* National Research Council.

National Science Board. 2004. *An Emerging and Critical Problem of the Science and Engineering Labor Force.* Arlington, VA: National Science Foundation (NSB 04–07). [Downloaded June 25, 2005, from http://www.nsf.gov/sbe/srs/nsb0407/nsb0407.pdf]

National Science Board. 2007. Hurricane Warning: The critical need for a National Hurricane Research Initiative. Report of the Task Force on Hurricane Science and Engineering, National Science Board, Washington, DC. [Downloaded 13 December 2016 from https://nsf.gov/nsb/publications/2007/hurricane/initiative.pdf]

National Weather Service. 2002. National Weather Service Instruction 20–103: Forecaster Development Program Training. [Downloaded October 23, 2006, from http://www.weather.gov/directives/020/pd02001003a.pdf]

National Weather Service. 2010. NOAA's National Weather Service Strategic Plan 2011–2020 (Final Copy for NEP Review). Silver Spring, MD: NOAA National Weather Service. [Downloaded 13 December 2016 from http://www.nws.noaa.gov/com/stratplan/files/plan_v01.pdf]

National Weather Service Training Center. 2006, August 31. Forecaster Development Program. [Downloaded October 23, 2006, from http://www.nwstc.noaa.gov/nwstrn/d.ntp/fdp/]

Navarra, J. G., J. Levin, and J. G. Navarra, Jr. 1993. An example of the use of meteorological concepts in the problem-based general education experiences of undergraduates. *Bulletin of the American Meteorological Society* 74:439–446.

Nebecker, F. 1995. *Calculating the Weather: Meteorology in the 20th Century*. New York: Academic Press.

Neumann, J. 1989. Forecasts of fine weather in the literature of classical antiquity. *Bulletin of the American Meteorological Society* 70:46–48.

Neville, K., R. R. Hoffman, C. Linde, W. C. Elm, and J. Fowlkes. 2008, January/February. The procurement woes revisited. *IEEE Intelligent Systems* 84:72–75.

Newell, A. 1985. Duncker on thinking: An inquiry into progress in cognition. In *A Century of Psychology as a Science*, ed. S. Koch and D. E. Leary (pp. 392–419). New York: McGraw-Hill.

Nickerson, R. S. 2014. Is global warming a challenge to human factors? *Ergonomics in Design* 22:4–7.

Nonaka, I., and H. Takeuchi. 1995. *The Knowledge Creating Company*. New York: Oxford University Press.

Norman, D. A. 1982. *Learning and Memory*. San Francisco: W. H. Freeman.

Norman, D. A. 1983. Some observations on mental models. In *Mental Models*, ed. D. Gentner and A. L. Stevens (pp. 7–14). Hillsdale, NJ: Lawrence Erlbaum.

Norman, D. A. 1988. *The Design of Everyday Things*. New York: Doubleday.

Norman, D. A. 1993. *Things That Make Us Smart*. New York: Perseus Books.

Norman, G. R., D. Rosenthal, L. R. Brooks, S. W. Allen, and L. J. Muzzin. 1989. The development of expertise in dermatology. *Archives of Dermatology* 125:1063–1068.

Northcraft, G. B., and M. A. Neale. 1987. Experts, amateurs, and real estate: An anchoring-and-adjustment perspective on property pricing decisions. *Organizational Behavior and Human Performance* 39:84–97.

Novak, D. R., C. Bailey, K. Brill, M. Eckert, D. Petersen, R. Rausch, and M. Schichtel. 2011. Human improvement to numerical weather prediction at the Hydrometeorological Prediction Center. *Proceedings of the 24th Conference on Weather Analysis and Forecasting/20th Conference on Numerical Weather Prediction* (Paper P330). Boston, MA: American Meteorological Society.

Novak, D. R., D. R. Bright, and M. J. Brennan. 2008. Operational forecaster uncertainty needs and future roles. *Weather and Forecasting* 23:1069–1084.

Novak, J. D. 1998. *Learning, Creating, and Using Knowledge*. Mahwah, NJ: Erlbaum.

Novak, J. D., and D. B. Gowin. 1984. *Learning How to Learn*. New York: Cambridge University Press.

Oakley, N. S., and B. Daudert. 2016. Establishing best practices to improve usefulness and usability of web interfaces providing atmospheric data. *Bulletin of the American Meteorological Society* 97:263–274.

O'Dell, C., and C. J. Grayson. 1998. *If We Only Knew What We Know: The Transfer of Internal Knowledge and Best Practice*. New York: The Free Press.

O'Hare, D., and N. Stenhouse. 2009. Under the weather: An evaluation of different modes of presenting meteorological information for pilots. *Applied Ergonomics* 490:688–693.

Olson, D. A., N. W. Junker, and B. Korty. 1995. Evaluation of 33 years of quantitative precipitation forecasting at the NMC. *Weather and Forecasting* 10:498–511.

Olson, J. M. 1987. Color and the computer in cartography. In *Color and the Computer*, ed. J. H. Durrett (pp. 205–219). New York: Academic Press.

Olson, R. 2014. Accuracy of three major weather forecasting services. [Downloaded November 19, 2014, from http://www.randalolson.com/2014/06/21/accuracy-of-three-major-weather-forecasting-services/]

Ooms, K., P. De Maeyer, and V. Fack. 2014. Study of the attentive behavior of novice and expert map users using eye tracking. *Cartography and Geographic Information Science* 41:37–54.

Orgill, M. M., J. D. Kincheloe, and R. A. Sutherland. 1992. The development of a prototype coupled analysis and nowcasting knowledge-based system for complex terrain interaction studies. *Weather and Forecasting* 7:353–372.

Oskamp, S. 1965. Overconfidence in case study judgments. *Journal of Consulting Psychology* 29:261–265.

Pachauri, R. K., and L. A. Meyer, eds. 2014. *Climate Change 2014: Synthesis Report*. Geneva, Switzerland: Intergovernmental Panel on Climate Change.

Palmer, T., and R. Hagedorn, eds. 2006. *Predictability of Weather and Climate*. Cambridge: Cambridge University Press.

Pane, J. F., A. T. Corbett, and B. E. John. 1996. Assessing dynamics on computer-based instruction. In *Proceedings of CHI 96: Computer-Human Interaction* (pp. 197–204). New York: Association for Computing Machinery.

Payne, J. W. 1982. Contingent decision behavior. *Psychological Bulletin* 92:382–401.

Peak, J. E. 1988. *An Expert System for Prediction of Maritime Visibility Obscuration. Report*. Monterey, CA: Naval Environmental Prediction Research Facility.

Peak, J. E., and P. M. Tag. 1989. An expert system approach for prediction of maritime visibility obscuration. *Monthly Weather Review* 117:2641–2653.

Pearce, M. L., and S. G. Hoffert. 1997. The SNAP weather information integrated forecast tool (SWIFT): Next generation software for research and operational meteorology. In *Preprints from the 13th Conference on Interactive Information and Processing Systems* (pp. 269–272). Boston, MA: American Meteorological Society.

Pearson, A. D. 1978. Meteorological Big Mac. Editorial, *The Kansas City Star*. Reprinted in L. Snellman, ed. Forum, *National Weather Digest, 3*, pp. 2–6.

Pearson, A. R., J. P. Schuldt, and R. Romero-Canyas. 2016. Social climate science: A new vista for psychological science. *Perspectives on Psychological Science* 11:632–650.

Perby, M.-L. 1989. Computerization and skill in local weather forecasting. In *Knowledge, Skill, and Artificial Intelligence*, ed. B. Göranzon and I. Josefson (pp. 39–52). Berlin: Springer-Verlag.

Peron, R. M., and G. L. Allen. 1988. Attempts to train novices for beer flavor discrimination: A matter of taste. *Journal of General Psychology* 115:402–418.

Petterssen, S. 1940. *Weather Analysis and Forecasting*. New York: McGraw-Hill.

Petterssen, S. 1956. *Weather Analysis and Forecasting*. 2nd ed. New York: McGraw-Hill.

Phelps, R. H., and J. Shanteau. 1978. Livestock judges: How much information can an expert use? *Organizational Behavior and Human Performance* 21:209–219.

Phoebus, P. A., D. R. Smith, P. J. Croft, H. A. Friedman, M. C. Hayes, K. A. Murphy, M. K. Ramamurthy, B. Watkins, and J. W. Zeitler. 2001. Ninth AMS Symposium on Education. *Bulletin of the American Meteorological Society* 82:295–303.

Pliske, R., B. Crandall, and G. Klein. 2004. Competence in weather forecasting. In *Psychological Investigations of Competent Decision Making*, ed. K. Smith, J. Shanteau, and P. Johnson (pp. 40–70). Cambridge: Cambridge University Press.

Pliske, R., D. W. Klinger, R. Hutton, B. Crandall, B. Knight, and G. Klein. 1997. *Understanding Skilled Weather Forecasting: Implications for Training and the Design of Forecasting Tools*, 122. Fairborn, OH: Klein Associates, Inc.

Postel, G. 2012, 24 April. Are weather forecasts beyond a few days any good? *The Washington Post.* [Downloaded November 13, 2012, from https://www.washingtonpost.com/blogs/capital-weather-gang]

Potter, S. S., D. D. Woods, E. M. Roth, and C. Elm. 2000. Bootstrapping multiple converging cognitive task analysis techniques in system design. In *Cognitive Task Analysis*, ed. J. M. Schraagen, S. F. Chipman, and V. L. Shalin (pp. 317–340). Mahwah, NJ: Erlbaum.

Pretor-Pinney, G. 2006. *The Cloudspotter's Guide: The Science, History and Culture of Clouds.* New York: Penguin Group.

Proctor, R. W., and K.-P. L. Vu. 2006. Laboratory studies of training, skill acquisition, and retention of performance. In *Cambridge Handbook of Expertise and Expert Performance*, ed. K. A. Ericsson, N. Charness, P. J. Feltovich, and R. R. Hoffman (pp. 265–286). New York: Cambridge University Press.

Pylyshyn, Z. 1973. What the mind's eye tells the mind's brain: A critique of mental imagery. *Psychological Bulletin* 80:1–24.

Quoetone, A. M., D. L. Andra, W. F. Bunting, and D. G. Jones. 2001. Impacts of technology and situation awareness on decision making: Operational observations from National Weather Service Warning Forecasters during the historic May 3, 1999, tornado outbreak. *Proceedings of the Human Factors and Ergonomics Society 45th Annual Meeting* (pp. 419–423). Santa Monica, CA: Human Factors and Ergonomics Society.

<bok>Rachlin, H. 1991. *Introduction to Modern Behaviorism* (3rd ed.). New York: Freeman.</bok>

Raffensberger, M. E., M. E. Cianciolo, E. O. Schmidt, and J. R. Stearns. 1997. The Cloud Scene Simulation Model—Recent enhancements and additions. In *Proceedings of the 13th Conference on Interactive Information and Processing Systems for Meteorology, Oceanography, and Hydrology* (pp. 398–410). Boston, MA: American Meteorological Society.

Ralph, F. M. (with 11 others). 2005, November. Improving short-term (0–48h) cool-season quantitative precipitation forecasting: Recommendations from a USWRP Workshop. *Bulletin of the American Meteorological Society* 86:1619–1632.

Ramage, C. S. 1978. Further outlook—Hazy. *Bulletin of the American Meteorological Society* 59:18–21.

Ramage, C. S. 1993. Forecasting in meteorology. *Bulletin of the American Meteorological Society* 74:1863–1871.

Rautenhaus, M., A. Schäfler, and R. Westerman. 2013/Winter. 3D visualization of ensemble forecasts. *ECMMF Newsletter* (European Centre for Medium-Range Weather Forecasts), pp. 34–38.

Rautenhaus, M., M. Kern, A. Schäfler, and R. Westerman. 2015a. Three-dimensional visualization of ensemble weather forecasts: Part 1. The visualization tool Met.3d (version 1.0). *Geoscientific Model Development* 8:2329–2353.

Rautenhaus, M., C. M. Grams, A. Schäfler, and R. Westerman. 2015b. Three-dimensional visualization of ensemble weather forecasts: Part 2. Forecasting warm conveyor belt situations for aircraft-based field campaigns. *Geoscientific Model Development* 8:2355–2377.

Reynolds, D. W. 1983. Prototype workstation for mesoscale forecasting. *Bulletin of the American Meteorological Society* 64:264–273.

Reynolds, D. 2003, July. Value-added quantitative precipitation forecasts: How valuable is the forecaster? *Bulletin of the American Meteorological Society* 876–878.

Rhyne, T., M. Bolstad, P. Rheingans, L. Petterson, W. Shackelford, M. E. Botts, E. Pepke, K. W. Johnson, W. Hibbard, C. R. Dyer, B. Paul, and L. A. Treinish 1992. Visualization requirements in the atmospheric and environmental sciences (five case study reports). In *Proceedings of the IEEE Conference on Visualization* (pp. 428–435). New York: IEEE.

Rieber, L., M. Boyce, and C. Assad. 1990. The effects of computer animation on adult learning and retrieval tasks. *Journal of Computer-Based Instruction* 17:46–52.

Ripenberger, J. T., C. L. Silva, H. C. Jenkins-Smith, and M. James. 2015, April. The influence of consequence-based messages on public responses to tornado warnings. *Bulletin of the American Meteorological Society* 96:577–590.

Risbey, J. S., and M. Kandlikar. 2002. Expert assessment of uncertainties in detection and attribution of climate change. *Bulletin of the American Meteorological Society* 83:1317–1326.

Risk Management Solutions. 2008. *Special Report: The 1993 Superstorm: 15-Year Perspective*. Newark, NJ: CAL Risk Management Solutions, Inc.

Roberts, W. F., P. C. Kucera, C. M. Lusk, L. E. Johnson, and D. C. Walker. 1997. WFO-Advanced. In Human Factors Applied to Graphical User Interface Design, instructors R. M. Steadham, M. Swartz, R. E. Schlegel, W. F. Roberts, and R. R. Hoffman. Short Course presented at the 77th meeting of the American Meteorological Society, Long Beach, CA. Boston, MA: American Meteorological Society.

Roberts, W. F., W. R. Moninger, B. de Lorenzis, E. Ellison, J. Flueck, J. S. McLeod, C. Lusk, et al. 1990. Shootout 89: A comparative evaluation of AI systems for convective storm forecasting. In *Preprints of the 6th International conference on Interactive Information Processing Systems for Meteorology, Oceanography, and Hydrology* (pp. 167–172). Boston, MA: American Meteorological Society.

Rockwood, A. A., J. F. Weaver, J. M. Brown, B. D. Jamison, and R. Holmes. 1992. An expert system for the prediction of downslope windstorms in northern Colorado. In *Preprints of the Symposium on Weather Forecasting* (pp. 210–211). Boston, MA: American Meteorological Society.

Roe, A. 1951. A study of imagery in research scientists. *Journal of Personality* 19:459–470.

Roebber, P. J. 1998. The regime dependence of degree day forecast technique, skill and value. *Weather and Forecasting* 13:783–794.

Roebber, P. J. April, 1999. Personal communication. Department of Geosciences, University of Wisconsin, Milwaukee, WI.

Roebber, P. J. 2009. Visualizing multiple measures of forecast quality. *Weather and Forecasting* 24:601–608.

Roebber, P. J. 2010. Seeking consensus: A new approach. *Monthly Weather Review* 138:4402–4415.

Roebber, P. J. 2015. Evolving ensembles. *Monthly Weather Review* 143:471–490.

Roebber, P. J., and L. F. Bosart. 1996a. The complex relationship between forecast skill and forecast value: A real world analysis. *Weather and Forecasting* 11:544–559.

Roebber, P. J., and L. F. Bosart. 1996b. The contributions of education and experience to forecasting skill. *Weather and Forecasting* 11:21–40.

Roebber, P. J., and L. F. Bosart. 1998. The sensitivity of precipitation to circulation details: Part I. An analysis of regional analogues. *Monthly Weather Review* 126:437–455.

Roebber, P. J., L. F. Bosart, and G. S. Forbes. 1996. Does distance from the forecast site affect skill? *Weather and Forecasting* 11:582–598.

Roebber, P. J., D. M. Schultz, B. A. Colle, and D. J. Stensrud. 2004. Towards improved prediction: High-resolution and ensemble modeling systems in operations. *Weather and Forecasting* 19:936–949.

Roebber, R. J., D. M. Schultz, and R. Romero. 2002. Synoptic regulation of the 3 May 1999 tornado outbreak. *Weather and Forecasting* 17:399–429.

Rogell, R. H. 1972. Weather terminology and the general public. *Weatherwise* 25:126–132.

Rogowitz, B. E., and L. A. Treinish. 1996. How not to lie with visualization. *Computers in Physics* 10:268–273.

Rosenholtz, R., Y. Li, J. Mansfield, and Z. Jin. 2005. Feature congestion: A measure of display clutter. In *Proceedings of the 2005 Conference for the Association for Computing Machinery Special Interest Group on Computer-Human Interaction* (pp. 761–770). New York: Association for Computing Machinery.

Roth, E. M., R. Scott, S. Deutsch, A. Kuper, V. Schmidt, A. Stilson, and J. Wampler. 2006. Evolvable work-centered support systems for command and control: Creating systems users can adapt to meet changing demands. *Ergonomics* 49:688–705.

Rothfusz, L. P., D. Devore, S. Amburn, S. Cooper, and J. Eise. 1992. Training issues in newly-created weather forecast offices. In *Proceedings of the Symposium on Weather Forecasting* (pp. 80–83). Boston, MA: American Meteorological Society.

Roulston, M. S., and L. A. Smith. 2004. The boy who cried wolf revisited: The impact of false alarm intolerance on cost-loss scenarios. *Weather and Forecasting* 19:391–397.

Rouse, W. B., and N. M. Morris. 1986. On looking into the black box: Prospects and limits on the search for mental models. *Psychological Bulletin* 100:349–363.

Rowe, G., and G. Wright. 2001. Expert opinions in forecasting: The role of the Delphi technique. In *Principles of Forecasting: A Handbook for Researchers and Practitioners*, ed. J. S. Armstrong (pp. 125–144). Norwell, MA: Kluwer.

Ruth, D. P. 2002. Interactive forecast preparation—the future has come. Preprints from the Interactive Symposium on the Advanced Weather Interactive Processing System (AWIPS) (pp. 20–22). Boston: American Meteorological Society.

Ryan, G. W., and H. R. Bernard. 2000. Data Management and Analysis Methods. In *Handbook of Qualitative Research*. 2nd ed., ed. N. K. Denzin and Y. S. Lincoln (pp. 769–802). Thousand Oaks, CA: Sage.

Sanders, F. 1958. The evaluation of subjective probability forecasts. Scientific Report No. 5., Department of Meteorology, Massachusetts Institute of Technology, Cambridge, MA.

Sanders, F. 1973. Skill in forecasting daily temperatures and precipitation: Some experimental results. *Bulletin of the American Meteorological Society* 54:1171–1178.

Sanders, F. 1986. Trends in skill of Boston forecasts made at MIT, 1966–1984. *Bulletin of the American Meteorological Society* 67:170–176.

Sanderson, P. M. 1989. Verbalizable knowledge and skilled task performance: Association, dissociation, and mental models. *Journal of Experimental Psychology: Learning, Memory, and Cognition* 15:729–747.

Sanger, S. S., R. M. Steadham, J. M. Jarboe, R. E. Schlegel, and A. Sellakannu. 1995. Human factors contributions to the evolution of an interactive Doppler radar and weather detection algorithm display system. In *Proceedings of the 11th International Conference on Interactive Information and Processing Systems for Meteorology, Oceanography, and Hydrology* (pp. 1–6). Boston, MA: American Meteorological Society.

Savelli, S., and S. Joslyn. 2013. The advantages of predictive interval forecasts for non-expert users and the impact of visualizations. *Applied Cognitive Psychology* 27:527–541.

Scaife, M., and Y. Rogers. 1996. External cognition: How do graphical representations work? *International Journal of Human–Computer Studies* 45:185–213.

Schlatter, T. W. 1985. A day in the life of a mesoscale forecaster. *ESA Journal* 9:235–256.

Schlatter, T. W. 1986. The use of computers for the display of meteorological information. In *Mesoscale Meteorology and Forecasting*, ed. P. S. Ray (pp. 752–775). Boston, MA: American Meteorological Society.

Schlatter, T. W., P. Schultz, and J. M. Brown. 1985. Forecasting convection with the PROFS system: Comments on the Summer 1983 Experiment. *Bulletin of the American Meteorological Society* 66:802–809.

Schmit, T. J., M. M. Gunshor, W. P. Menzel, J. J. Gurka, J. Li, and A. S. Bachmeier. 2005. Introducing the next-generation Advanced Baseline Imager on GOES-R. *Bulletin of the American Meteorological Society* 86:1079–1096.

Schmit, T. J., S. J. Goodman, D. T. Lindsey, R. M. Rabin, K. M. Bedka, M. M. Gunshor, J. L. Cintineo, C. S. Velden, A. S. Bachmeier, S. S. Lindstrom, and C. C. Schmidt. 2013. Geostationary Operational Environmental Satellite (GOES)-14 super rapid scan operations to prepare for GOES-R. *Journal of Applied Remote Sensing* 7. [http://spie.org/Publications/Journal/10.1117/1.JRS.7.073462]

Schmit, T. J., J. Steven, S. J. Goodman, M. M. Gunshor, J. Sieglaff, A. K. Heidinger, A. S. Bachmeier, S. S. Lindstrom, A. Terborg, J. Feltz, K. Bar, S. Rudlosky, D. T. Lindsey, R. M. Rabin, and C. C. Schmidt. 2015. Rapid refresh information of significant events: Preparing users for the next generation of geostationary operational satellites. *Bulletin of the American Meteorological Society* 96:561–576.

Schön, D. A. 1983. *The Reflective Practitioner*. New York: Basic Books.

Schön, D. A. 1987. *Educating the Reflective Practitioner*. San Francisco: Jossey-Bass.

Schraagen, J.-M., S. F. Chipman, and V. L. Shalin, eds. 2000. *Cognitive Task Analysis*. Mahwah, NJ: Erlbaum.

Schröder, F. 1993. Visualizing meteorological data for a lay audience. *IEEE Computer Graphics and Applications* 13:12–14.

Schumacher, R. M., and M. P. Czerwinski. 1992. Mental models and the acquisition of expert knowledge. In *The Psychology of Expertise: Cognitive Research and Empirical AI*, ed. R. Hoffman (pp. 61–79). Mahwah, NJ: Erlbaum.

Schwartz, S., and T. Griffin. 1986. *Medical Thinking: The Psychology of Medical Judgment and Decision Making*. New York: Springer Verlag.

Scott, R., E. M. Roth, S. E. Deutsch, E. Malchiodi, T. E. Kazmierczak, R. G. Eggleston, S. R. Kuper, R. D. Whitaker. 2002. Using software agents in a work centered support system for weather forecasting and monitoring. In *Proceedings of the Human Factors and Ergonomics Society 46th annual meeting* (pp. 433–438). Santa Monica, CA: Human Factors and Ergonomics Society.

Scott, R., E. M. Roth, S. E. Deutsch, E. Malchiodi, T. E. Kazmierczak, R. G. Eggleston, S. R. Kuper, and R. D. Whitaker. 2005, March/April. Work-centered support systems: A human-centered approach to intelligent systems design. *IEEE Intelligent Systems* 20:73–81.

Scribner, S. 1984. Studying working intelligence. In *Everyday Cognition: Its Development in Social Context*, ed. B. Rogoff and S. Lave (pp. 9–40). Cambridge, MA: Harvard University Press.

Scribner, S. 1986. Thinking in action: Some characteristics of practical thought. In *Practical Intelligence: Nature and Origins of Competence in the Everyday World*, ed. R. J. Sternberg and R. K. Wagner (pp. 14–30). Cambridge: Cambridge University Press.

Serafin, R. J., A. E. MacDonald, and R. L. Gall. 2002. Transition of weather research to operations: Opportunities and challenges. *Bulletin of the American Meteorological Society* 83:377–392.

Shadbolt, N., and A. M. Burton. 1990. Knowledge elicitation techniques: Some experimental results. In *Special Issue on Knowledge Acquisition, SIGART Newsletter*, No. 108, ed. C. R. Westphal and K. L. McGraw (pp. 21–33). New York: Special Interest Group on Artificial Intelligence, Association for Computing Machinery.

Shanteau, J. 1984. Some unasked questions about the psychology of expert decision makers. In M. E. El Hawary (Ed.), *Proceedings of the 1984 IEEE Conference on Systems, Man, and Cybernetics* (pp. 23–45). New York: Institute of Electrical and Electronics Engineers.

Shanteau, J. 1988. Psychological characteristics and strategies of expert decision makers. *Acta Psychologica* 68:203–215.

Shanteau, J. 1992a. How much information does an expert use? Is it relevant? *Acta Psychologica* 81:75–86.

Shanteau, J. 1992b. Competence in experts: The role of task characteristics. *Organizational Behavior and Human Decision Processes* 53:252–266.

Shanteau, J., P. Johnson, and K. Smith, eds. 2004. *Psychological Investigations of Competent Decision Making*. Cambridge: Cambridge University Press.

Shanteau, J., and R. H. Phelps. 1977. Judgment and swine: Approaches in applied judgment analysis. In *Human Judgment and Decision Processes in Applied Settings*, ed. M. F. Kaplan and S. Schwartz (pp. 255–272). New York: Academic Press.

Shanteau, J., and T. R. Stewart. 1992. Why study expert decision making? Some historical perspectives and comments. *Organizational Behavior and Human Decision Processes* 53:95–106.

Sheets, R. 1990. The National Hurricane Center—Past, present, and future. *Weather and Forecasting* 5:185–232.

Shields, M. D., I. Solomon, and W. S. Waller. 1987. Effects of alternative sample space representations on the accuracy of auditors' uncertainty judgments. *Accounting, Organizations and Society* 12:375–385.

Shortliffe, E. H. 1976. *Computer-Based Medical Consultations: MYCIN*. New York: Elsevier.

Shuman, F. G. 1989. History of numerical weather prediction at the National Meteorological Center. *Weather and Forecasting* 4:286–296.

Silva, A., B. S. Santos, and J. Madeira. 2011. Using color in visualization: A survey. *Computers & Graphics* 35:320–333.

Silver, N. 2012. *The Signal and the Noise: Why So Many Predictions Fail—But Some Don't*. New York: Penguin Press.

Simon, H. A. 1973. The structure of ill-structured problems. *Artificial Intelligence* 4:181–201.

Simon, H. A., and K. Gilmartin. 1973. A simulation of memory for chess positions. *Cognitive Psychology* 5:29–46.

Simonton, D. K. 1988. Age and outstanding achievement: What do we know after a century of research? *Psychological Bulletin* 104:251–267.

Simonton, D. K. 1999. Talent and its development: An emergenic and epigenetic model. *Psychological Bulletin* 106:435–457.

Simonton, D. K. 2000. Creative development as acquired expertise: Theoretical issues and an empirical test. *Developmental Review* 20:283–318.

Simonton, D. K. 2006. Historiometric methods. In *The Cambridge Handbook of Expertise and Expert Performance*, ed. K. A. Ericsson, N. Charness, P. J. Feltovich and R. R. Hoffman (pp. 319–335). New York: Cambridge University Press.

Sipress, A. 2000, January 26. Blindsided and snowed under: Area wakes up to white ... and many woes. *The Washington Post.*

Slovic, P. 1969. Analyzing the expert judge: A description of stockbrokers' decision processes. *Journal of Applied Psychology* 53:255–263.

Smallman, H. S., and M. Hegarty. 2007. Expertise, spatial ability and intuition in the use of complex visual displays. *Proceedings of the 51st annual meeting of the Human Factors and Ergonomics Society* (pp. 200–204). Santa Monica, CA: Human Factors and Ergonomics Society.

Smallman, H. S., and M. St. John. 2005. Naïve realism: Misplaced faith in the utility of realistic displays. *Ergonomics in Design* 13:6–13.

Smigielski, F. J., and A. M. Mogil. 1991a. A systematic satellite approach for estimating central pressures of mid-latitude oceanic cyclones. *Telus* 47A:876–891.

Smigielski, F. J., and A. M. Mogil. 1991b. Use of satellite information for improved ocean surface analysis. In *Proceedings of the First International Symposium on Winter Storms* (pp. 137–144). Boston, MA: American Meteorological Society.

Smith, D. L., F. L. Zuckerberg, J. T. Schafer, and G. E. Rasch. 1986. Forecast problems: The meteorological and operational factors. In *Mesoscale Meteorology and Forecasting*, ed. P. S. Ray (pp. 36–49). Boston, MA: American Meteorological Society.

Smith, D. R. 2000. Report on the 5th International Conference on School and Popular Meteorological and Oceanographic Education. *Bulletin of the American Meteorological Society* 81:1589–1598.

Smith, E. A. 1975. The McIDAS system. *IEEE Transactions on Geoscience Electronics* GE-13:123–136.

Smotroff, I. G. 1991. Meteorological classification of satellite imagery and ground sensor data using neural network data fusion. In *Preprints of the 7th International Conference on Interactive Information Processing Systems for Meteorology, Oceanography, and Hydrology*. Boston, MA: American Meteorological Society.

Snellman, L. W. 1977. Operational forecasting using automated guidance. *Bulletin of the American Meteorological Society* 58:1036–1044.

Snellman, L. W. 1978. Forum. *National Weather Digest* 4:3–4.

Snellman, L. W. 1982. Impact of AFOS on operational forecasting. In *Preprints of the 9th Conference on Weather Forecasting and Analysis* (pp. 13–16). Boston, MA: American Meteorological Society.

Snellman, L. W. 1991. An old forecaster looks at modernization—pros and cons. *National Weather Digest* 16:2–5.

Somerville, R. C. J. 2011. Computing the climate. *Science* 331:149–150.

Sonnentag, S., C. Niessen, and J. Volmer. 2006. Expertise in software design. In *Cambridge Handbook of Expertise and Expert Performance*, ed. K. A. Ericsson, N. Charness, P. J. Feltovich and R. R. Hoffman (pp. 373–387). New York: Cambridge University Press.

Spaid, E. L. 1994, 18 April. For these children, school is looking up. *The Christian Science Monitor*, p. 14.

Spangler, T. C., V. C. Johnson, R. L. Alberty, B. E. Heckman, L. Spayd, and E. Jacks. 1994. COMET: An education and training program in mesoscale meteorology. *Bulletin of the American Meteorological Society* 75:1249–1259.

Spiegler, D. B. 1996. A history of private sector meteorology. In *Historical Essays on Meteorology, 1919–1995*, ed. J. R. Fleming (pp. 417–441). Boston: American Meteorological Society.

Spiro, R., R. Coulson, P. Feltovich, and D. Anderson. 1988. Cognitive flexibility theory: Advanced knowledge acquisition in ill-structured domains. In *Proceedings of the 10th annual conference of the Cognitive Science Society*. Hillsdale, NJ: Erlbaum.

Spiro, R. J., P. J. Feltovich, M. J. Jacobson, and R. L. Coulson. 1992. Cognitive flexibility, constructivism, and hypertext: Random access instruction for advanced knowledge acquisition in ill-structured domains. In *Constructivism and the Technology of Instruction*, ed. T. Duffy and D. Jonassen (pp. 57–75). Hillsdale, NJ: Erlbaum.

Stanard, T. W., R. M. Pliske, A. A. Armstrong, S. L. Green, C. E. Zsambok, and D. P. McDonald. 2002. Collaborative development of expertise: Evaluation of an on-the-job (OJT) training program. In *Proceedings of the Human Factors and Ergonomics Society 46th annual meeting* (pp. 2007–2011). Santa Monica, CA: Human Factors and Ergonomics Society.

Stauffer, R., G. J. Mayr, M. Dabernig, and A. Zeileis. 2015. Somewhere over the rainbow: How to make effective use of colors in meteorological visualizations. *Bulletin of the American Meteorological Society* 96:203–216.

Steadham, R. 1998. *Operational needs for custom volume coverage patterns. Report, Applications Branch, Operational Support Facility*. Norman, OK: NOAA.

Steadham, R. M., M. Swartz, R. E. Schlegel, W. F. Roberts, and R. R. Hoffman. 1997. Human Factors Applied to Graphical User Interface Design. In *Short Course presented at the 77th meeting of the American Meteorological Society*, Long Beach, CA. Boston, MA: American Meteorological Society.

Stein, E. W. 1992. A method to identify candidates for knowledge acquisition. *Journal of Management Information Systems* 9:161–178.

Stein, E. W. 1997. A look at expertise from a social perspective. In *Expertise in Context*, ed. P. J. Feltovich, K. M. Ford, and R. R. Hoffman (pp. 181–194). Cambridge, MA: The MIT Press.

Stephenson, D. B. 2000. Use of the "odds ratio" for diagnosing forecast skill. *Weather and Forecasting* 15:221–232.

Sternberg, R. J., and P. J. Frensch. 1992. On being an expert: A cost-benefit analysis. In *The Psychology of Expertise: Cognitive Research and Empirical AI*, ed. R. R. Hoffman (pp. 191–203). Mahwah, NJ: Erlbaum.

Stevens, A. L., and A. Collins. 1978. Multiple conceptual models of a complex system. Report No. 3923, Bolt, Beranek, and Newman, Inc., Cambridge, MA.

Stevens, S. S. 1966. Matching functions between loudness and ten other continua. *Perception & Psychophysics* 1:5–8.

Stewart, A. E. 2009, December. Minding the weather: The measurement of weather salience. *Bulletin of the American Meteorological Society* 90:1833–1841.

Stewart, K. F., K. F. Heideman, W. R. Moninger, and P. Regan-Cirincione. 1992. Effects of improved information on the components of skill in weather forecasting. *Organizational Behavior and Human Decision Processes* 53:107–134.

Stewart, T. R. 1990. A decomposition of the correlation coefficient and its use in analyzing forecasting skill. *Weather and Forecasting* 5:661–666.

Stewart, T. R., and C. M. Lusk. 1994. Seven components of judgmental forecasting skill: Implications for research and the improvement of forecasts. *Journal of Forecasting* 13:579–599.

Stewart, T. R., W. R. Moninger, J. Grassia, R. H. Brady, and F. H. Merrem. 1989. Analysis of expert judgment in a hail forecasting experiment. *Weather and Forecasting* 4:24–34.

Stewart, T. R., R. Piekle, and R. Nath. 2004. Understanding user decision making and the value of improved precipitation forecasts. *Bulletin of the American Meteorological Society* 85:223–235.

Stewart, T. R., P. J. Roebber, and L. F. Bosart. 1997. The importance of the task in analyzing expert judgment. *Organizational Behavior and Human Decision Processes* 69:205–219.

Stokes, D. E. 1997. *Pasteur's Quadrant: Basic Science and Technological Innovation*. Washington, DC: Brookings Institution Press.

Stuart, N. A., P. S. Market, B. Telfeyan, G. M. Lackmann, K. Carey, H. E. Brooks, D. Nietfeld, B. C. Motta, and K. Reeves. 2006. The future of humans in an increasingly automated forecast process. *Bulletin of the American Meteorological Society* 87:1497–1502.

Stuart, N. A., D. M. Schultz, and G. Klein. 2007. Maintaining the role of humans in the forecast process: Analyzing the psyche of expert forecasters. *Bulletin of the American Meteorological Society* 88:1893–1898.

Stunder, M. J., R. M. Dyer, and R. Koch. 1987a. The use of an expert system in assisting forecasters in visibility decisions. In *Proceedings of the 3rd Conference on Interactive Information Processing*

Systems in Meteorology, Oceanography, and Hydrology (pp. 5206–5207). Boston, MA: American Meteorological Society.

Stunder, M. J., R. C. Koch, T. N. Sletten, and S. M. Lee. 1987b. Zeus: A knowledge-based expert system that assists in predicting visibility at air bases. Report No. AFGL-TR-87–0019, Air Force Geophysics Laboratory, Hanscom AFB, MA.

Sumner, T. 2015, May 2. Weather forecasting is getting a high-speed make over. *Science News*, pp. 20–23.

Suomi, V. E., R. Fox, S. S. Limaye, and L. W. Smith. 1983. McIDAS III: A modern interactive data access and analysis system. *Journal of Applied Meteorology* 22:766–788.

Sutcliffe, A. G. 1985. Use of conceptual maps as human–computer interfaces. In *People and Computers: Designing the Interface*, ed. P. Johnson and S. Cook (pp. 117–127). Cambridge: Cambridge University Press.

Sutton, O. G. 1954. The development of meteorology as an exact science. *Quarterly Journal of the Royal Meteorological Society* 80:328–338.

Swets, J. A., R. M. Dawes, and J. Monahan. 2000. Psychological science can improve diagnostic decisions. *Psychological Science in the Public Interest* 1:1–26.

Tak, S., A. Toet, and J. van Erp. 2015. Public understanding of visual representations of uncertainty in temperature forecasts. *Journal of Cognitive Engineering and Decision Making* 9:241–262.

Takle, E. S. 2000. University instruction in observational techniques: Survey responses. *Bulletin of the American Meteorological Society* 81:1319–1325.

Tanaka, J. W., T. Curran, and D. Sheinberg. 2005. The training and transfer of real-world, perceptual expertise. *Psychological Science* 16:145–151.

Targett, P. S. 1994. Predicting the future of the meteorologist—A forecaster's view. *Bulletin of the Australian Meteorological and Oceanographic Society* 7:46–52.

Tennekes, H. 1988. Numerical weather prediction: Illusions of security, tales of imperfection. *Weather* 43:165–170.

Tennekes, H. 1992. Karl Popper and the accountability of numerical weather forecasting. *Weather* 47:343–346.

Thompson, P. 1987. The maturing of the science. *Bulletin of the American Meteorological Society* 68:631–637.

Thorndike, E. L. 1912. *Education: A First Book*. New York: Macmillan.

Toth, Z., E. Kalnay, S. M. Tracton, R. Wobus, and J. Irwin. 1997. A synoptic evaluation of the NCEP ensemble. *Weather and Forecasting* 12:140–153.

Tracton, M. S. 1993. On the skill and utility of NMC's medium-range central guidance. *Weather and Forecasting* 8:147–153.

Tracton, M. S. 2015, January 26. Facebook posting.

Tracton, M. S., and E. Kalnay. 1993. Operational ensemble prediction at the National Meteorological Center: Practical aspects. *Weather and Forecasting* 8:379–398.

Trafton, J. G. 2004. Dynamic mental models in weather forecasting. In *Proceedings of the Human Factors and Ergonomics Society 48th annual meeting* (pp. 311–316). Santa Monica, CA: Human Factors and Ergonomics Society.

Trafton, J. G., and R. R. Hoffman. 2007. Computer-aided visualization in meteorology. In *Expertise Out of Context*, ed. R. R. Hoffman (pp. 337–358). Mahwah, NJ: Erlbaum.

Trafton, J. G., S. S. Kirschenbaum, T. L. Tsui, R. T. Miyamoto, J. A. Ballas, and P. D. Raymond. 2000. Turning pictures into numbers: Extracting and generating information from complex visualizations. *International Journal of Human–Computer Studies* 53:827–850.

Trafton, J. G., S. Marshall, F. Mintz, and S. B. Trickett. 2002. Extracting explicit and implicit information from complex visualizations. In *Diagramatic Representation and Inference*, ed. M. Hegarty, B. Meyer, and H. Narayanan (pp. 206–220). Berlin: Springer-Verlag.

Trafton, J. G., S. B. Trickett, and F. E. Mintz. 2005. Connecting internal and external representations: Spatial transformations of scientific visualizations. *Foundations of Science* 10:89–106.

Travis, D. 1991. *Effective Color Displays: Theory and Practice*. New York: Academic Press.

Treinish, L. A. 1994, November/December. Visualization of disparate data in the earth sciences. *Computers in Physics* 8:664–671.

Treinish, L. A. 1997. Three-dimensional visualization support of operational forecasting at the 1996 Centennial Olympic games. In *Proceedings of the 13th Conference on Interactive Information and Processing Systems for Meteorology, Oceanography, and Hydrology* (pp. 31–36). Boston, MA: American Meteorological Society.

Treinish, L. A. 2000. Multi-resolution visualization techniques for nested weather models. In *Proceedings of the conference on Visualization '00 (VIS '00)* (pp. 513–516). Los Alamitos, CA: IEEE Computer Society Press.

Treinish, L. A. 2002, January. Interactive, web-based three-dimensional visualizations of operational mesoscale weather models. In *Proceedings of the Eighteenth International Conference on Interactive Information and Processing Systems for Meteorology, Oceanography and Hydrology* (pp. 159–161). Boston, MA: American Meteorological Society.

Treinish, L. A., and B. E. Rogowitz. 1997. Perceptual guidance in atmospheric sciences visualization. In *Proceedings of the 13th Conference on Interactive Information and Processing Systems for Meteorology, Oceanography, and Hydrology* (pp. 425–428). Boston, MA: American Meteorological Society.

Treinish, L. A., and L. Rothfusz. 1997. Three-dimensional visualization for support of operational forecasting at the 1996 Centennial Olympic Games. In *Proceedings of the 13th International Conference on Interactive Information and Processing Systems for Meteorology, Oceanography and Hydrology* (pp. 31–34). Boston: American Meteorological Society.

Trickett, S. B., and J. G. Trafton. 2007. "What if…": The use of conceptual simulations in scientific reasoning. *Cognitive Science* 31:843–875.

Trickett, S. B., G. J. Trafton, and C. D. Schunn. 2009. How do scientists respond to anomalies? Different strategies used in basic and applied science. *Topics in Cognitive Science* 1:711–729.

Tufte, E. R. 1990. *Envisioning Information.* New York: Graphics Press.

Tufte, E. R. 2001. *The Visual Display of Quantitative Information.* New York: Graphics Press.

Turban, E., ed. 1992. *Expert Systems and Applied Artificial Intelligence.* New York: Macmillan.

Tversky, B., and J. B. Morrison. 2002. Animation: Can it facilitate? *International Journal of Human-Computer Studies* 57:247–262.

Tversky, A., and D. Kahneman. 1982. Judgment of and by representativeness. In *Judgment Under Uncertainty: Heuristics and Biases,* ed. D. Kahneman, P. Slovic, and A. Tversky (pp. 84–100). New York: Cambridge University Press.

Uccellini, L. W., S. F. Corfidi, N. W. Junker, P. J. Kocin, and D. A. Olson. 1992. Report on the Surface Analysis Workshop held at the National Meteorological Center 25–28 March 1991. *Bulletin of the American Meteorological Society* 73:459–472.

Uccellini, L. W., P. J. Kocin, R. S. Schneider, P. M. Stokols, and R. A. Dorr. 1995. Forecasting the 7–14 March 1993 superstorm. *Bulletin of the American Meteorological Society* 76:183–199.

Ulanski, S. L. 1993. An analysis of the liberal arts in introductory meteorology courses. *Bulletin of the American Meteorological Society* 74:2203–2209.

Vekirl, I. 2002. What is the value of graphical displays in learning? *Educational Psychology Review* 14:261–298.

Velden, C., B. Harper, F. Wells, J. L. Beven, R. Zehr, T. Olander, M. Mayfield, et al. 2006. The Dvorak tropical cyclone intensity estimation technique. *Bulletin of the American Meteorological Society* 87:1196–1210.

Vicente, K. J. 1999. *Cognitive Work Analysis: Toward Safe, Productive, and Healthy Computer-Based Work.* Mahwah, NJ: Erlbaum.

Vicente, K. J. 2000. Work domain analysis and task analysis: A difference that matters. In *Cognitive Task Analysis,* ed. J. M. Schraagen, S. F. Chipman, and V. L. Shalin (pp. 101–118). Mahwah, NJ: Erlbaum.

Vislocky, R. L., and J. M. Fritsch. 1995. Improved model output statistics forecasts through model consensus. *Bulletin of the American Meteorological Society* 76:1157–1164.

Vislocky, R. L., and J. M. Fritsch. 1997. Performance of an advanced MOS system in the 1996–97 National Collegiate Weather forecasting competition. *Bulletin of the American Meteorological Society* 78:2851–2857.

Vitart, F., F. Monteni, and R. Buiza. 2014. Have ECMWF monthly forecasts been improving? *ECMWF Newsletter* (European Centre for Medium-Range Weather Forecasts), No. 138, pp. 20–25.

Voss, J. M., S. Tyler, and L. Yengo. 1983. Individual differences in social science problem solving. In *Individual Differences in Cognitive Processes,* Vol. 1, ed. R. F. Dillon and R. R. Schmeck (pp. 205–232). New York: Academic Press.

Wagenaar, W. A., and J. G. Visser. 1979. The weather forecast under the weather. *Ergonomics* 22:909–917.

Wai, J. 2014. Experts are born then made: Combining prospective and retrospective longitudinal data shows that cognitive ability matters. *Intelligence* 45:74–80.

Walkup, L. E. 1965. Creativity in science through visualization. *Perceptual and Motor Skills* 21:35–41.

Wallsten, T. S. 1983. The theoretical status of judgmental heuristics. In *Decision Making Under Uncertainty*, ed. R. W. Scholz (pp. 21–37). Amsterdam: North-Holland.

Walsh, J. E., D. J. Charlevoix, and R. M. Rauber. 2014. *Severe and hazardous weather.* 4th ed. Dubuque, IA: KendallHunt.

Ware, C., and J. C. Beatty. 1988. Using color dimensions to display data dimensions. *Human Factors* 30:127–142.

Ware, R. H., D. W. Fulker, S. A. Stein, D. N. Anderson, C. K. Avery, R. D. Clark, K. K. Droegemeier, J. P. Kuettner, J. B. Minster, and S. Sorooshian. 2000. SuomiNet: A real-time national GPS network for atmospheric research and education. *Bulletin of the American Meteorological Society* 81:677–694.

Ware, C., G. W. Kelley, and D. Pilar. 2014. Improving the display of wind patterns and ocean currents. *Bulletin of the American Meteorological Society* 95:1573–1581.

Warner, T. T. 2011. *Numerical Weather and Climate Prediction.* Cambridge: Cambridge University Press.

Wash, C. H., R. L. DeSouza, M. Ramamurthy, A. Andersen, G. Byrd, J. Justus, H. Edmon, and P. Samson. 1992. Teaching with interactive computer systems: A report on the Unidata/COMET/STORM Workshop on Synoptic/Mesoscale Instruction. *Bulletin of the American Meteorological Society* 73:1440–1447.

Watkins, B. T. 1989. Many campuses now challenging minority students to excel in math and science. *Chronicle of Higher Education* 35:16–17.

Weaver, J. F., and R. S. Phillips. 1990. An expert system application for forecasting severe downslope winds at Fort Collins, Colorado, USA. In *Preprints from the 16th Conference on Severe Local Storms and the Conference on Atmospheric Electricity* (pp. 13–15). Boston, MA: American Meteorological Society.

Weitzenfeld, J. 1984. Valid reasoning by analogy: Technical reasoning. *Philosophy of Science* 51:137–149.

Wilks, D. S. 1995. *International Geophysics Series*. vol. 59. *Statistical Methods in the Atmospheric Sciences*. New York: Academic Press.

Wilkins, E. M., and R. E. Johnson. 1975. The AFOS experimental system. *IEEE Transactions on Geoscience Electronics* GE-13:99–110.

Williams, J. 2013. *Verifying weather forecasts*. Master's thesis, Northwest Missouri State University, Maryville, MO.

Williams, P. 1951. The use of confidence factors in forecasting. *Bulletin of the American Meteorological Society* 32:279–281.

Winkler, R. L., and A. H. Murphy. 1973a. Information aggregation in probabilistic prediction. *IEEE Transactions on Systems, Man, and Cybernetics* SMC-3:154–160.

Winkler, R. L., and A. H. Murphy. 1973b. Experiments in the laboratory and in the real world. *Organizational Behavior and Human Performance* 10:252–270.

Winkler, R. L., A. H. Murphy, and R. W. Katz. 1977. The consensus of subjective probability forecasts: Are two, three... heads better than one? In *Proceedings of the 5th Conference on Probability and Statistics in Atmospheric Sciences* (pp. 57–62). Boston, MA: American Meteorological Society.

Winston, P. H. 1984. *Artificial Intelligence*. 2nd ed. Reading, MA: Addison-Wesley.

Winters, H. A. 1998. *Battling the Elements: Weather and Terrain in the Conduct of War*. Baltimore, MD: Johns Hopkins University Press.

Wittgenstein, L. 1953. *Philosophical Investigations*. New York: Blackwell Publishing.

Woods, D. D., and S. Dekker. 2000. Anticipating the effects of technological change: A new era of dynamics for human factors. *Theoretical Issues in Ergonomics Science* 1:272–282.

World Meteorological Organization. 2016. Manual on the Global Data-Processing and Forecasting System. [Downloaded February 14, 2016, from https://www.wmo.int/pages/prog/www/DPFS/Manual/GDPFS-Manual.html]

Wynne, B. 1991. Knowledge in context. *Science, Technology & Human Values* 216:111–121.

Yarger, D. N., W. A. Gallus, Jr., M. Taber, J. P. Boysen, and P. Castleberry. 2000. A forecasting activity for a large introductory meteorology course. *Bulletin of the American Meteorological Society* 81:31–39.

Yoon, J.-H., S. Wang, R. R. Gilles, L. Hipps, B. Kravitz, and P. J. Rasch. 2015. Extreme fire season in California: A glimpse into the future? *Special Supplement to the Bulletin of the American Meteorological Society* 96:S5–S9.

Zakay, D., and S. Wooler. 1984. Time pressure, training, and decision effectiveness. *Ergonomics* 27:273–284.

Zhang, F., C. Snyder, and R. Rotunno. 2002. Mesoscale predictability of the "surprise" snowstorm of 24–25 January 2000. *Monthly Weather Review* 130:1617–1632.

Zrnić, D. S., J. F. Kimpel, D. E. Forsyth, A. Shapiro, G. Crain, R. Ferek, J. Heimmer, W. Benner, T. J. McNellis, R. J. Vogt. 2007. Agile-beam phased array radar for weather observations. *Bulletin of the American Meteorological Society* 88:1753–1766.

Zsambok, C. E., and G. Klein, eds. 1997. *Naturalistic Decision Making*. Mahwah, NJ: Erlbaum.

Zubrick, S. M. 1984. An expert system to aid in severe thunderstorm forecasting. Report ZU-RS-00025. Radian Corporation, Austin, TX.

Zubrick, S. M. 1988. Validation of a weather forecasting expert system. In *Machine Intelligence 11*, ed. J. E. Hayes, D. Michie, and J. Richards (pp. 391–422). Oxford, England: Clarendon Press.

Zubrick, S. M., and C. E. Riese. 1985. An expert system to aid in severe thunderstorm forecasting. In *Preprints of the 14th Conference on Severe Local Storms* (pp. 117–126). Boston, MA: American Meteorological Society.

Zuschlag, M. 2004. Quantification of visual clutter using a computational model of human perception: An application for head-up displays. In *Proceedings of The Conference on Human Performance, Situation Awareness and Automation (HPSAA II, vol. 2)* (pp. 143–148). Daytona Beach, FL: Embry-Riddle Aeronautical University.

Index

Printed in the United States
by Baker & Taylor Publisher Services